19.90

Thermal Analysis

Vol. 1   Theory
         Instrumentation
         Applied Sciences
         Industrial Applications

Proceedings of the Sixth International
Conference on Thermal Analysis

Bayreuth, Federal Republic of Germany
July 6–12, 1980

Vol. 1   Theory
         Instrumentation
         Applied Sciences
         Industrial Applications

Vol. 2   Inorganic Chemistry/Metallurgy
         Earth Sciences
         Organic Chemistry/Polymers
         Biological Sciences/Medicine/Pharmacy

 # Thermal Analysis

Vol. 1   Theory
Instrumentation
Applied Sciences
Industrial Applications

Editor:
H. G. Wiedemann
Mettler Instrumente AG
Greifensee/ZH
Switzerland

1980
Birkhäuser Verlag
Basel · Boston · Stuttgart

CIP-Kurztitelaufnahme der Deutschen Bibliothek

*Thermal analysis.* – Basel, Boston, Stuttgart: Birkhäuser.
Bd. 1971 mit d. Erscheinungsorten: Basel, Stuttgart. – Bd. 1977 im Verl. Heyden, London, Bellmawr (NJ), Rheine.
ISBN 3-7643-1202-5
1980.
Vol. 1. Theory, instrumentation, applied sciences, industrial applications: proceedings of the 6. Internat. Conference on Thermal Analysis, Bayreuth, Fed. Republic of Germany, July 6–12, 1980/ed. H. G. Wiedemann. –
1980.
ISBN 3-7643-1085-5
NE: Wiedemann, Hans G. [Hrsg.]; ICTA ‹06, 1980, Bayreuth›

All rights reserved. No part of this publication may be reproduced, stored in a retrieval system, or transmitted in any form or by any means, electronic, mechanical, photocopying, recording or otherwise, without the prior permission of the copyright owner.

© Birkhäuser Verlag Basel, 1980
ISBN 3-7643-1085-5
Printed in Switzerland

# Foreword

Although the use of heat to alter man's environment, habits and possessions began in the mists of prehistory, the deliberate and systematic study of the effects of heat upon materials and processes belongs largely to the 20th century. In the past 80 years thermal analysis has emerged as a broadly applicable and informative means for carrying out such studies. For a demonstration of this, the reader has only to explore the pages of this volume. From academic interest in reaction mechanisms to the day-to-day quality control of industrial products, thermal analysis provides a series of tools for the scientific and technical community which are unrestricted by field of endeavour. Chemists, biologists, geologists, industrialists and many others employ thermal analysis to examine the effects of heat upon physical and chemical processes occurring within systems of interest to them.

Growth in the value of thermal analysis is reflected in the rapidly expanding number of national societies whose interdisciplinary membership find common ground and purpose in meeting to discuss the development and application of these techniques. In its conferences, ICTA brings together a similar but broader membership for the same purpose, as well as providing an opportunity for personal friendships to form and grow based on common interests and mutual respect. Between conferences, ICTA teams of experts continue to provide leadership in the development of nomenclature, standards and publications of value to all thermal analysts. Thus, together ICTA and its Affiliated Societies are working, and will continue to work, to find still further approaches to, and application for, thermal analysis. The end is not yet in sight!

Ontario Research Foundation  H. G. McAdie
Mississauga, Ontario, Canada  President, ICTA

# Preface

The various methods of Thermal Analysis are well established in the many fields of natural science. An important number of efficient, high-performance instruments is available with sensitivities and temperature ranges which have been further improved during the past few years. The range of calorimetric applications is continuously increasing and a growing use of calorimetric methods is especially notable in the life sciences. Because of this, a special section on biological sciences, medicine and pharmacy has been introduced at the ICTA '80. We hope that this will lead to an exchange of ideas about new and interesting problems during the ICTA conference. The presentation of all such ideas will be made also possible thanks to the introduction of a poster session, which today offers the chance for improved communication between authors and scientists.

The conference has been organized by the Gesellschaft für Thermische Analyse (GEFTA) in collaboration with members of the Schweizerische Gesellschaft für Thermoanalytik und Kalorimetrie (STK). The editors express their thanks to all authors who have contributed papers to these Proceedings. Their co-operation has enabled these volumes to be ready for the conference. We further thank Dr. W.-D. Emmerich and the organizing committee for their collaboration.

Finally it is a pleasure to acknowledge the helpfulness and the great care of the Birkhäuser Verlag in production of these volumes.

        Greifensee/Zürich, März 1980        H. G. Wiedemann

# Contents

### Award and Plenary Lectures

H. R. Oswald:
Thermal Analysis in Inorganic Solid State Chemistry . . . . . . . . . . 1
P. K. Gallagher:
Some Applications of Thermal Analysis to the Communications
and Electronics Industry . . . . . . . . . . . . . . . . . . . . . . . . . 13
G. Lombardi:
Thermal Analysis in Earth Sciences . . . . . . . . . . . . . . . . . . . 583

### Theory

J. Šesták:
Kinetic Compensation Effect: Facts and Fiction of Linear Plots
Using Arrhenius Law . . . . . . . . . . . . . . . . . . . . . . . . . . . 29
J. J. Pysiak:
Regularities in the Kinetics of Thermal Dissociation of Solids . . . . 35
Z. S. Kolenda, J. Norwisz and N. Hajduk:
The Determination of the Straight Line Correlation Factors . . . . . 41
C. Comel, J. Veron, C. Bouster and P. Vermande:
Theoretical Thermogravimetric Analysis at Constant Heating Rates 51
J. Moll, D. Krug and D. Zepf:
A Comparison of Different Methods for the Estimation of
Kinetic Data . . . . . . . . . . . . . . . . . . . . . . . . . . . . . . . . 57
A. Reller and H. R. Oswald:
Experimental Requirements for the Determination of Kinetics and
Mechanism for Decomposition Reactions of Solids . . . . . . . . . . 63
M. Arnold, G. E. Veress, J. Paulik and F. Paulik:
Some Remarks on the Applicability of the Arrhenius Model in
Thermal Analysis . . . . . . . . . . . . . . . . . . . . . . . . . . . . . . 69
E. Koch and B. Stilkerieg:
Kinetic Characterization of the Oscillatory Belousov-Zhabotinsky
Reaction at Linearly Increased Temperature . . . . . . . . . . . . . . 75
A. Mokhlisse, G. Bertrand, M. Lallemant and
N. Roudergues:
Characteristics of Interface Endothermic Reactions Reconsidered
from a Few Simple Experiments . . . . . . . . . . . . . . . . . . . . . 81

G. Kiss, K. Seybold and T. Meisel:
Error Analysis in Thermal Purity Determinations . . . . . . . . . . . .  87
A. A. Van Dooren:
Influence of Experimental Variables on DSC Curves . . . . . . . . . .  93
Ž. D. Živković, B. Dobovišek and A. Rosina:
Influence of Mass and Grain Size on the Basic Geometry of the
DTA Curve . . . . . . . . . . . . . . . . . . . . . . . . . . . . . . . . . . . . . . . . . . .  99
R. B. Barendregt, J. Verhoeff and P. J. Van den Berg:
A Comparison of DTA Evaluating Methods of DTA Experiments
with Tertiary Butylperpivalate as a Testing Substance . . . . . . . . .  105
P. K. Gallagher and E. M. Gyorgy:
A Comparison of the Kinetics for the Thermal Decomposition of
$Ba(OH)_2 \cdot H_2O$ Obtained by TG and EGA Techniques . . . . . . . . .  113
T. P. Prasad and M. S. R. Swami:
Kinetic Analysis from Thermogravimetric Traces . . . . . . . . . . . . .  119
C. G. R. Nair and P. M. Madhusudanan:
A Spline Interpolation Method of Study Kinetics from Isothermal
and Non-Isothermal Thermogravimetry . . . . . . . . . . . . . . . . . . . .  127
P. H. Fong, S. P. Wong and D. T. Y. Chen:
Improvement of Differential Correction Method for Evaluation of
Kinetic Parameters from TG Curves . . . . . . . . . . . . . . . . . . . . . .  133
Z. Jerman:
Relation between the Kinetics and Thermodynamics in the
Endothermic Decomposition of Solids . . . . . . . . . . . . . . . . . . . . .  139
J. M. Criado:
Determination of the Mechanism of Thermal Decomposition
Reactions of Solids by Using the Cyclic and Constant
Decomposition Rate Thermal Analysis Method . . . . . . . . . . . . .  145
H. K. Cammenga and H.-J. Petrick:
A New Isothermal DTA Method for the Study of Surfactants at
Liquid Surfaces . . . . . . . . . . . . . . . . . . . . . . . . . . . . . . . . . . . . . . .  149
S. Suriñach, M. D. Baró and F. Tejerina:
The Ge-Sb-Bi Ternary Phase Diagram . . . . . . . . . . . . . . . . . . . . .  155

**Instrumentation**

M. Taniguchi, H. Moriguchi and S. Shimizu:
Dehydration of Crystallization Water in Salts Using a New
Designed Controlled-Water Vapor Micro DTA . . . . . . . . . . . . . .  163
P. Le Parlouër:
"Open System" DSC: A New Approach of Isothermal
Investigations . . . . . . . . . . . . . . . . . . . . . . . . . . . . . . . . . . . . . . . . .  169

W. W. Wendlandt:
Thermovoltaic Detection: A New Technique for the Study of
Thermal Decomposition Reactions . . . . . . . . . . . . . . . . . . . . . . 175

J. C. Tou, L. F. Whiting and D. I. Townsend:
Thermal Hazard Evaluation by an Accelerating Rate Calorimeter . 177

M. Nakanishi and S. Fujieda:
Estimation of Instantaneous Caloric Effect in the Heat Exchange
Type of Calorimetry . . . . . . . . . . . . . . . . . . . . . . . . . . . . . . . . . 183

S. Gál, J. Sztatisz and L. Fodor:
Thermoanalytical Investigations on Solid – Vapour Reactions . . . . 189

M. Ichihashi, A. Maesono, K. Takaoka and A. Kishi:
Isothermal Thermogravimetric Analyzer Using Infrared Image
Furnace and Microcomputer System . . . . . . . . . . . . . . . . . . . . . 195

P. D. Garn, P. Menis and H. G. Wiedemann:
Temperature Calibration in Thermogravimetry . . . . . . . . . . . . . . 201

B. O. Haglund and T. Luks:
Convection Effects in Thermogravimetry: Significance and
Corrections . . . . . . . . . . . . . . . . . . . . . . . . . . . . . . . . . . . . . . . . 207

E. Robens:
Errors in Thermogravimetric Experiments Resulting from
Absorption on the Counterweight . . . . . . . . . . . . . . . . . . . . . . . 213

A. Kishi, A. Maesono, M. Ichihashi, K. Takaoka,
Z. Hara and K. Akechi:
Microcomputer-Controlled Dilatometer with Infrared Rapid
Heating System and its Application . . . . . . . . . . . . . . . . . . . . . . 219

B. Andrejs, J. P. Schulz and E. Wappler:
Dynamic Elasticity Measurements on Plastics . . . . . . . . . . . . . . . 225

O. T. Sørensen:
Densification Studies of Ceramic Powder Compacts by Quasi
Isothermal Dilatometry . . . . . . . . . . . . . . . . . . . . . . . . . . . . . . . 231

E. L. Charsley, J. Joannou, A. C. F. Kamp,
M. R. Ottaway and J. P. Redfern:
A Simultaneous TG-DTA System for Operating from $-150\,°C$ to
$1500\,°C$ . . . . . . . . . . . . . . . . . . . . . . . . . . . . . . . . . . . . . . . . . . 237

J. Chiu and A. J. Beattie:
Techniques for Coupling Mass Spectrometry to Thermogravimetry 245

E. Kaisersberger:
Further Development and Application of a Combined System for
Simultaneous Thermal Analysis and Mass Spectrometry . . . . . . . . 251

M. Maruta, Y. Kunimatsu and K. Yamada:
High-Sensitivity Multi-Channel DTA Apparatus . . . . . . . . . . . . . 259

R. L. Fyans, J. S. Mayer and W. P. Brennan:
A Microcomputer Based DTA . . . . . . . . . . . . . . . . . . . . . . . . . . 265

E. Wappler and U. Kurpjuweit:
New Control Module for DSC, DTA, TMA and TG . . . . . . . . . . 273

W. Perron, G. Bayer and H. G. Wiedemann:
A New Instrument for Simultaneous Thermomicroscopy and
Differential Thermal Analysis ........................ 279
E. L. Charsley, A. C. F. Kamp and J. A. Rumsey:
Transmitted Light Hot Stage Microscopy in the Temperature
Range −180°C to 600°C ............................ 285
W. Ludwig:
Application of Area-Thermocouples in the Thermal Analysis of
Amorphous Thin Films .............................. 293
B. Krstic:
A New Apparatus for Study of Stabilities in Horizontal Air Layers
Heated from Below.................................. 299
E. Marti, A. Geoffroy, B. F. Rordorf and
M. Szelagiewicz:
Partial Pressure Measurement by the Flow Method ........... 305
B. F. Rordorf, A. Geoffroy, M. Szelagiewicz and
E. Marti:
Vapor Pressure Methods for Industrial Applications .......... 313
I. C. McNeill:
The Use of Subambient Thermal Volatilization Analysis to Study
Volataile Products of Polymer Degradation ................ 319

## Applied Sciences

S. B. Warrington and P. A. Barnes:
DTA/EGA Using a Specific Detector .................... 327
P. C. Jain and D. Chaubey:
Quantitative Differential Thermal Analysis at Elevated Pressures .. 335
J. S. Crighton and K. M. Li:
The Treatment of Thermoanalytical Data for Effective Presentation 341
H. J. Berndt and G. Heidemann:
Equilibrium Shrinkage-Force Measurement − A Method
Describing the State of Order of PET-Fibres ............... 345
L. Stäudel, G. Thiel and H. Wöhrmann:
Thermometric Titrimetry − A Suitable Way into Thermochemistry . 351
A. M. Abdel-Rehim:
Sintering of Corundum with Ammonium Fluoride ........... 357
A. Jarmontowicz and R. Krzywobłocka-Laurow:
Thermal Analysis of Concrete ......................... 363
I. Stebnicka-Kalicka:
Application of Thermal Analysis to the Investigations of Phase
Composition of Autoclaved Cement Pastes and Mortars ........ 369

V. Balek, J. Dohnálek and W. D. Emmerich:
Effect of Elevated Temperatures on the Hydration of Cement
Investigated by Emanation Thermal Analysis . . . . . . . . . . . . . . . 375
F. G. Buttler and S. R. Morgan:
A Thermogravimetric Method for Studying the Reaction between
Fly Ash and the $Ca(OH)_2$ Liberated on Hydration of Portland
Cement. . . . . . . . . . . . . . . . . . . . . . . . . . . . . . . . . . . . . . . . . . . . 381
E. Alsdorf, K. Habersberger, K. H. Schnabel and
W. Walkow:
Application of Temperature Programmed Sorption and Reaction
and of DTA Data to the Characterization of Bismuth Molybdate
Catalysts . . . . . . . . . . . . . . . . . . . . . . . . . . . . . . . . . . . . . . . . . . . 387
T. J. W. de Bruijn, W. A. de Jong and van den Berg:
Thermal Decomposition of Aqueous Manganese Nitrate Solutions. 393
A. Kołaczkowski and A. Biskupski:
Nitric Acid and Ammonia Concentrations as Factors Controlling
the Thermal Decomposition of Ammonium Nitrate . . . . . . . . . . . 399
S. Bordas, M. Geli, V. Balek and M. Vobořil:
Thermal Behaviour of $Ge_{0.25}Te_{0.60}Se_{0.15}$ Investigated by Emanation
Thermal Analysis and DTA . . . . . . . . . . . . . . . . . . . . . . . . . . . . 403
R. Halle and V. Carin:
Differential Thermal Analysis of Phosphate-Stabilized $Ca_2SiO_4$ . . . 409
Y. Saito, S. Sasaki and T. Maruyama:
Phase Study of the Praseodymium-Oxygen System by the Multiple
Thermal Analysis . . . . . . . . . . . . . . . . . . . . . . . . . . . . . . . . . . . . 415
C. G. Cordovilla and E. Louis:
Investigation of the Decomposition of the Solid Solution of a
Commercial Al-Zn-Mg Alloy (7015), by Means of DTA, Hardness
and Conductivity Measurements. . . . . . . . . . . . . . . . . . . . . . . . . 421
G. Liptay, L. Ligethy and E. Brandt-Petrik:
Thermal Investigation of Polyethylen Used in Power Cables . . . . . 427
V. Amicarelli, G. Baldassarre and L. Liberti:
Low Temperature Regeneration of Activated Carbon/Kinetic
Analysis of Thermodesorption of Phenol . . . . . . . . . . . . . . . . . . 433
R. R. Baker:
Development of Temperature Distribution Inside the Reaction
Zone of a Burning Cigarette . . . . . . . . . . . . . . . . . . . . . . . . . . . . 439
R. Halonbrenner:
Thermal Analysis as an Aid to the Criminalist . . . . . . . . . . . . . . . 445

## Industrial Applications

H. Möhler, K. D. Gauler, A. Henig, G. Janik and M. Jäth:
Thermoanalytical Characterization of Curing Behavior of Clear Varnishes .................................................. 453

H. Weber and U. Guggisberg:
Calorimetric Analysis of Polymer Fabrication ............... 461

P. K. Datta and T. R. Manley:
Thermo-Analytical Techniques in the Assessment of Mechanical Properties at Elevated Temperatures of Polyester-Glass Composites  467

G. Liptay, L. Ligethy and J. Nagy:
Thermal Analysis of RTV Silicone Rubbers used in High Voltage Cable Accessories .......................................... 477

Z. Kraus and V. Mentlík:
The New Method of the Power Current Elektrotechnology – The Method of Thermal Analysis ................................. 483

V. Mentlík and Z. Kraus:
Objective Determination of the State of Insulations by the Differential Thermal Analysis ............................. 489

V. Gottardi, G. Scarinci, G. Carturan, A. Marchetti and V. Frosini:
A DSC Study of Glasses Obtained from Organometallic Gels .... 493

L. Hälldahl and O. T. Sørensen:
Thermal Analysis Studies of the Decomposition of Ammonium Uranyl Carbonate (AUC) under Simulated Industrial Conditions .  499

C. Parton and D. Dollimore:
Thermal Studies of Contracting Gas Bubbles in Molten Glass .... 505

W. Hädrich and E. Kaisersberger:
Quantitative Phase Analysis in the System $CaSO_4 \cdot xH_2O$ by TG-DTA Methods ............................................ 511

D. R. Glasson and P. O'Neill:
Reactivity of Lime and Related Materials with Sulphur Dioxide .. 517

F. Oehme:
Determination of Nitric Acid in Nitrating Mixtures ............ 523

P. Marik-Korda:
Water Determination by "DIE" Method ........................ 529

I. Buzás, S. Gál and J. Simon:
Investigation of Vegetable Oils with Derivatograph ............ 535

R. Gygax, M. W. Meyer and F. Brogli:
Differential Scanning Calorimetry – Scope and Limitations of its Use as a Tool for Estimating the Reaction Dynamics of Potentially Hazardous Chemical Reactions ............................... 541

F. Brogli, R. Gygax and M. W. Meyer:
Differential Scanning Calorimetry – A powerful Screening Method
for the Estimation of the Hazards Inherent in Industrial Chemical
Reactions .................................... 549

H. S. Ray:
Non Isothermal Kinetics – A Generalized Approach .......... 555

W. Regenass:
Industrial Experience with Heat Flow Calorimetry ........... 561

M. L. Friman, M. Leskelä, L. Niinistö, E. Aalto,
A. Hörkkö and J. Poukari:
Thermoanalytical Study on Fertilizer Components and Their
Mixtures ..................................... 567

U. Kurpjuweit, E. Wappler and W. Keil:
DSC-Measurements on the Effect of Additives in Cutting Oils .... 571

E. García-Clavel and S. Goñi-Elizalde:
Thermoanalytical Study of $\beta$-FeOOH Obtained by Homogeneous
Precipitation with Urea ............................ 577

G. Lombardi:
Thermal Analysis in Earth Sciences ..................... 583

## Appendices

H. Kambe:
Memories of Kyoto ............................... 587

P. D. Garn:
Report of the Committee on Standardization ............... 593

J. P. Redfern:
Report of the Publications Committee ................... 595

Author Index .................................... 605
Subject Index ................................... 607

# Award and Plenary Lectures

# THERMAL ANALYSIS IN INORGANIC SOLID STATE CHEMISTRY

H.R.Oswald
Institute for Inorganic Chemistry
University of Zurich
Winterthurerstrasse 190
CH-8057 Zurich, Switzerland

## ABSTRACT

The conception presented in this paper aims at consistent investigations into problems which may be summarized under the term "Reactivity of Solids". In work of this kind, a careful consideration of relations between structure and morphology on the one hand and chemical reactivity (resp. behaviour in solid-solid phase transitions) on the other hand is indispensable. Structural and morphological results are mainly collected by X-rays and chemical electron microscopy, whereas thermoanalytical techniques are used to induce resp. follow reactions and phase transitions. Topochemical aspects and topotaxy are of central importance in our concept.

By a typical example is proved that it is often not possible to set up a realistic reaction mechanism from thermoanalytical curves only. Results on thermal decomposition of small single crystals of model compounds selected from coordination chemistry show how important a careful control of the experimental conditions is.

Furthermore, in situ thermogravimetry as an aid for preparative solid state chemical research in low pressure plasmas is described, and recent results on Jahn-Teller phase transitions are presented.

## INTRODUCTION

Quantitative thermal analysis is a powerful tool in contemporary solid state chemical research. Among the most important quantities which can be determined by thermoanalytical techniques are (refs. e.g. in [1]):

- Temperatures of phase transitions and chemical reactions
- Data about stoichiometry of reactions and purity of compounds
- Enthalpy differences
- Specific heat
- Details about kinetics and mechanism of chemical reactions and phase transitions.

In this lecture, these topics will not be given a systematic consideration. Instead, its aim lies rather in presenting a conception which has been established at our institute several years ago. By means of a few typical examples, it will be shown how thermoanalytical measurements, embedded in a broader framework of investigations, can contribute to solid state chemistry - in particular to solve questions concerning the problem complex "Reactivity of Solids", including not only chemical reactions, but also solid-solid phase transitions.

According to the scheme given in Fig. 1, a most central point in this

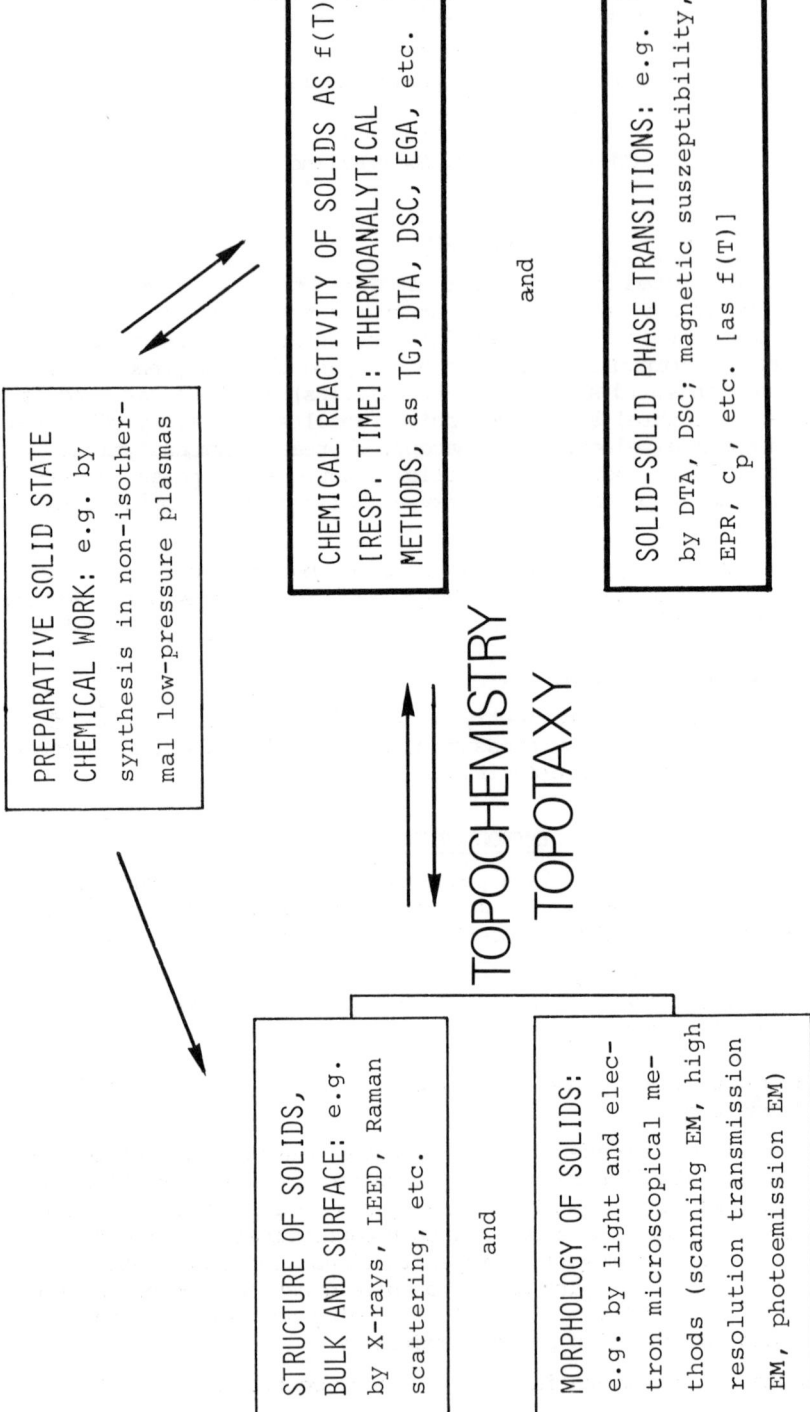

Fig. 1. Thermal Analysis and Investigations Into the Reactivity of Solids

field consists of the relationship between structure and morphology of solids on one hand and their behaviour in chemical reactions on the other hand. Particularly if reactions of the type

$$\text{solid 1} \longrightarrow \text{solid 2 + gas, resp.}$$
$$\text{solid 1 + gas} \longrightarrow \text{solid 2}$$

are considered under isothermal as well as dynamic conditions, thermal analysis plays an important, but by far not the only role. On the contrary it will be shown that additional results from structure-specific and morphological methods cannot only be valuable, but positively <u>must get used</u> in order to be able to interpret the thermoanalytical measurements at all.

Two further points of importance are enclosed in the cited scheme:

- In the context of solid state chemical research, preparative efforts are necessary in order to obtain new and interesting systems to work at. Often, such syntheses are following common paths - but in more original cases, e.g. chemical vapour deposition from plasmas, careful control by thermal analysis gets valuable as will be shown.

- Submitting solids to a controlled temperature program leads inevitably to the observation of solid-solid phase transitions. Whereas some of them are important for solid state physicists rather, other systems bear much chemical interest indeed.

## THE NECESSITY FOR ADDITIONAL DATA TO EXPLAIN TA-CURVES

Though there exist numerous more recent examples, this necessity gets drastically illustrated by the classical case of oxidizing high purity aluminum foil in streaming molecular oxygen at temperatures between 400 and 600°C [2,3]. Isothermal weight/time curves exhibit a typical sigmoid character - most pronounced in the region between 450 and 500° - which is not explicable by any of the usual growth laws, including the generally believed parabolic one. By transmission electron microscopy and selected area diffraction it is easily recognized that two kinds of products are formed: an amorphous $Al_2O_3$-layer growing at the phase boundary oxide/gas with cation diffusion as rate limiting step, and crystalline $\gamma$-$Al_2O_3$ nucleating and growing expitaxially in a thickness of about 200 Å at the phase boundary metal/amorphous oxide. The pronounced lateral growth of this second product can be expressed by a model for covering of a surface by films spreading out as expanding circles from predetermined nuclei with a random distribution:

$$W_{cryst.} = \rho \cdot d \, [1-\exp.(-\pi v^2 \omega t^2)]$$

($W_{cryst.}$: weight of crystals grown on unit area; $\rho$: density, d: thickness of crystals; t: time. The nucleation density $\omega$ is determined by counting, and the radial velocity of growth v by measuring the crystal size as a function of time, both on electron micrographs). The total weight gain is:

$$W_{total} = W_{cryst.} + W_{am.} \; ; \text{ with } W_{am(orphous)} = \sqrt{const. \cdot t}$$
$$\text{(parabolic law)}$$

At 450-500°C, both products are formed with about the same rate. It can be proven that there does not take place a mere partial recrystallisation of an amorphous oxide layer.

As a TG-curve cannot yield more than the weights' sum of the products formed, it is not possible to derive such a mechanism from thermogravimetry only. In view of the complexity of the occurring process, it is not advisable indeed to draw conclusions on the reaction mechanism solely from a mathematical fit of the experimental data within a selected (limited) interval of temperature and/or time.

## TOPOTACTIC DECOMPOSITION REACTIONS

The dominant influence of structural and morphological properties of a chemically reacting solid on the nature of the solid product as well as on the kinetics and mechanism of the reaction was early recognized by Kohlschütter [4] who first chose the expression "Topochemistry" to stress the localization of such processes. Later on Feitknecht and his school worked extensively in this field. As articles e.g. by Boldyrev [5] prove, topochemical reactions are nowadays as actual as ever.

The growing field of applications of X-rays and electron diffraction revealed a large number of topochemical reactions in which the property influenced most strikingly by the substrate is the product crystal structure orientation. Outgoing from Lotgering (supported by Gorter) [6] the more special term "Topotaxy" was introduced, the definition of which has been a subject of continuous discussion and refinement since. Taking the most relevant arguments into account, we have defined: "A process is called topotactic, if its solid product is formed in one or several crystallographically equivalent orientations relative to the parent crystal as a consequence of a chemical reaction or a solid-state transformation, and if it can proceed through the entire volume of the parent crystal". Considering a vast number of experimental data on topotactic reactions, we have undertaken a systematic classification of such processes [7], based on the overall importance of oriented nucleation (resp. the pre-existence of oriented nuclei) for topotaxy, the actual cause of which is the reduced energy required for oriented nucleation compared to random nucleation. Two principally different possibilities for the favouring of distinct nucleus orientation exist:

- Conservation of structural motives of the parent structure in the nucleus of the new phase, reducing the number of strong bonds which have to be broken, as well as the lengths of diffusion paths for most of the participating atoms. The structural motives or elements conserved may be three-dimensional (frameworks), two-dimensional (layers) or one-dimensional (chains).

- Epitactic nucleation, due to some metrical accord of otherwise unrelated parent and product structures in certain lattice planes. It may occur either on internal lattice planes or external surfaces.

Morphologically, topotaxy usually leads to the formation of more or less perfect pseudomorphs of product crystallites according to the shape of the initial crystals (cf. e.g. [8]).

Until quite recently, the main emphasis in our work on topotactic reactions has been laid on the elucidation of the structural-geometrical reaction mechanisms and on morphology, as a typical example shows [9]:

$$MoO_3 \cdot 2H_2O \xrightarrow{60-80°C} MoO_3 \cdot H_2O \xrightarrow{110-125°C} MoO_3$$
(yellow form)

(flowing air atmosphere; heating rate 0.2°/min)

The guiding element for the entire process is the conservation of planes of corner-sharing octahedra, as shown in Fig. 2. Between 60 and 80°C, only the interlayer water molecules are removed. In the second step at about 110-125°C, water molecules directly bound to molybdenum atoms are lost and the now empty coordination sites collapse to form double sheets with edge-sharing octahedra. This mechanism involves a minimum change in the structure of the individual layers, which even shows up in macroscopic observations - see Fig. 3 - the basal planes of the resulting pseudomorphs appearing unaffected.

The yellow form of $MoO_3 \cdot H_2O$ can only be obtained topotactically. If stoichiometrically the same compound grows from free solution, the needle shaped white modification is obtained, which, according to its crystal structure, is dehydrated in a different, chain controlled mechanism under formation of $MoO_3$ with another texture [10].

What can thermal analysis contribute to such work? Results from usual thermogravimetric measurements are of course necessary to survey the processes with respect to stoichiometry and reaction temperatures - but they cannot help to overcome the restriction to geometrical aspects.

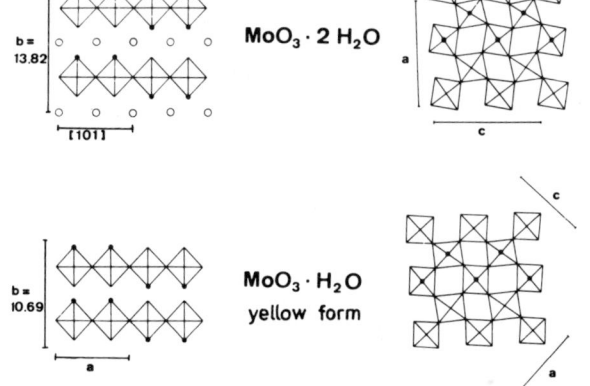

Fig. 2. Schematic crystal structures of phases involved in the thermal decomposition of $MoO_3 \cdot 2H_2O$, drawn in the relative orientation found experimentally [9]

left : projections parallel to the lattice layers

right: projections normal to the lattice layers

● : coordinated $H_2O$

O : interlayer $H_2O$

Fig. 3. a) Scanning electron micrograph of undecomposed prisms of $MoO_3 \cdot 2H_2O$
b) Similar micrograph of anhydrous $MoO_3$, pseudomorphous according to $MoO_3 \cdot 2H_2O$, lamellae ∥ (010), basal faces intact

Structural reaction mechanisms represent obviously part of a full insight only. We are therefore convinced that in addition, there must be made consequent use of very elaborate thermoanalytical investigations in order to collect the needed knowledge about reaction kinetics and energetics of solids.

Careful considerations, however, evoke serious reservations, as much theoretical thermoanalytical efforts in this respect have been made without taking seriously into account what <u>really happens</u> in the reacting system, and moreover a majority of experiments described in literature have been carried out under conditions far from being adequate. Indeed, too many results from kinetic and energetic thermoanalytical work appear not to be in accordance with the occurring physicochemical processes and thus fictive.

Being aware of this difficult situation, we have focused our recent interest on simple model systems selected from the broad field of inorganic and organometallic coordination chemistry. In order to avoid complications caused by the particular character of the water molecule, abstention from this ligand was decided for the present. Our conception has to meet requirements for the collection of realistic data and their consistent interpretation as well, and is based on the combination of independent methods:

- Detailed structural investigations of single crystals by X-rays

- Extensive morphological studies by electron microscopical techniques and optical thermomicroscopy with simultaneous micro-DTA, in order to reveal structural reaction mechanisms and topotaxy

- Quantitative thermal analysis with simultaneous mass spectrometry of evolved gases, using defined samples consisting of one single crystal of a few tenths of a mm in size and about 0.3 mg in weight. Appropriate results are e.g. strongly depending on the type of crystal face contacting the sample holder - explicable by the known structural reaction mechanism [11,12].

The investigated systems are selected according to their structural particularities (see also [13]), e.g.:

- $[Ni(SCN)_2(NH_3)_2]$, representing a case with one kind of <u>infinite chains</u>. Two ammonia molecules are lost in a single step decomposition reaction, and the linear velocity of the phase boundary moving into the crystal starting from the faces (001) resp. (00$\bar{1}$) along the c-axis (Fig. 4a) represents a realistic rate constant for the process. If the crystal contacts the sample pan or a second crystal by one of these faces, the reaction becomes much slower there, as the emission of ammonia gets hindered.

- $[Ni(SCN)_2(NH_3)_4]$, as a representative of a <u>spatial initial lattice with isolated molecular units</u>. Two ammonia molecules are lost near 110°C. The reaction starts on the entire surface of the crystal, and the phase boundary advances from all faces towards the centre (Fig. 4b). In such a case, the shape of the starting crystals - needles, plates or cubes - causes different time/weight curves, and hence, the over all weight loss of the sample registrated by TGA alone cannot lead to reliable kinetic and mechanistic data.

- $[Ni\{Pt(CN)_4\}(NH_3)_2]$, with a typically <u>layered structure</u>. The loss of two ammonia molecules has a highly topotactic character, determined by the conservation of the planes (001). Information about the true nature of the time-limiting step - diffusion of ammonia through the layered $[Ni\{Pt(CN)_4\}]$ product phase, or the break of $Ni-N_{(NH_3)}$ bonds - can again only be recognized by combined investigation.

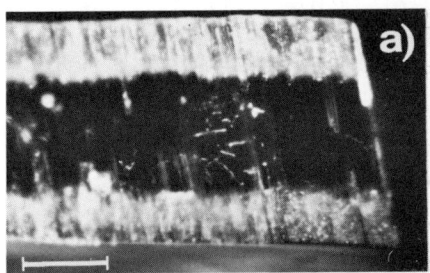 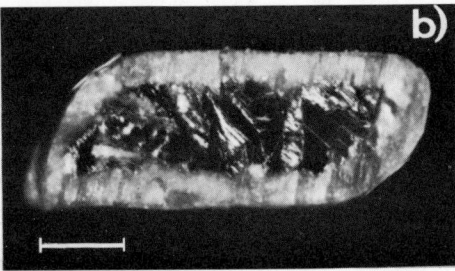

<u>Fig. 4.</u> Influence of structural properties on the course of thermal decomposition processes (single crystals, initial phases: dark; product: bright; ⊢⊣ = 0.1 mm)

a) $[Ni(SCN)_2(NH_3)_2] \rightarrow [Ni(SCN)_2]$, chain controlled mechanism; reaction front advances parallel (001) resp. (00$\bar{1}$) only

b) $[Ni(SCN)_2(NH_3)_4] \rightarrow [Ni(SCN)_2(NH_3)_2]$, initial lattice with isolated molecular units, reaction front moves towards the centre from all faces

We are convinced that the knowledge gained from work on model compounds will help to achieve meaningful quantitative interpretations of TA-curves from powder samples as well later on.

Within the field of layer lattices, similar detailed studies are effectuated on members of a clathrate complex family

$[M(diam)_n M'(CN)_4 \cdot 2G]$ 

M = Mn,Fe,Co,Ni,Cu,Zn; Cd
M' = Ni,Pd,Pt; Cd,Hg
diam = 2 x $NH_3$; en, etc.
G = "guest", i.e. benzene, aniline, etc.,

the crystal structures of which are either known or determined by ourselves.

For preliminary results on the stepwise thermal decomposition of $[Ni(NH_3)_2 Ni(CN)_4 \cdot 2 C_6H_6]$ and $[Cd(en)Ni(CN)_4 \cdot 2 C_6H_6]$ see [14].

As a morphological illustration, the nuclei of the new phase appearing when $[Cd(en)_3 Ni(CN)_4]$ looses its first ethylenediamine-ligand are shown in Fig. 5. It is evident that any model working e.g. with the conception of three dimensionally expanding spherical nuclei must be extremely rough compared to reality.

## SYNTHESIS AND EROSION OF SOLIDS IN LOW PRESSURE PLASMAS

Preparation of solid materials by chemical transport in nonisothermal low pressure plasmas and the interaction of plasmas with solid surfaces [15,16] represent an interesting new field of nonequilibrium chemistry. Exact monitoring the mass changes of the sample by TGA can, among other methods, contribute to achieve a better control over these complex systems.

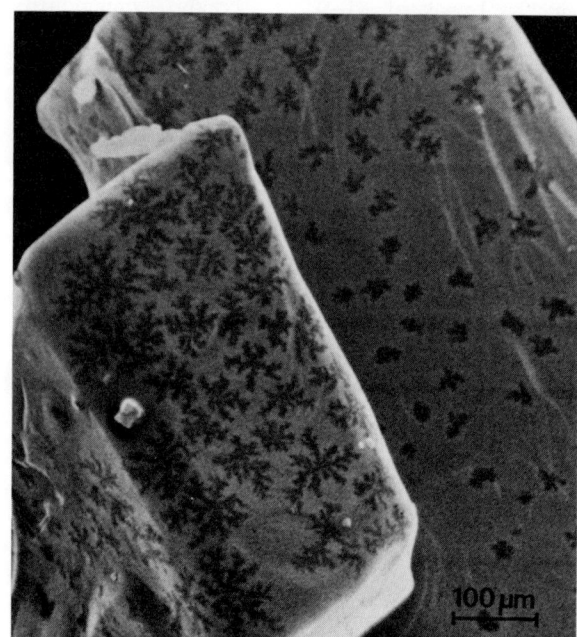

Fig. 5. Nucleation of $[Cd(en)_2 Ni(CN)_4]$ (dark contrast) by thermal decomposition of $[Cd(en)_3 Ni(CN)_4]$ single crystals

Scanning electron micrograph

In view of serious technical problems - e.g. to provide an adequate stability of the discharge and avoid any ignition of a discharge towards the microbalance itself - continuous in situ thermogravimetry in low pressure plasmas by means of electrically compensated microbalances became possible only recently at our institute [17,18]. Quite remarkable values for the long term stability were finally reached: noise and drift less than 0.1 $\mu$g in 10 hours without the glow discharge on, and $\simeq$ 0.1 $\mu$g under experimental conditions with a sample of pyrolytic graphite in burning hydrogen plasma at 0.04 torr. In similar examples, erosion rates of the order of $10^{-4}$ $\mu$g cm$^2$s$^{-1}$ which are equivalent to $\sim$0.01 monolayer per second could be reliably measured.

Besides nitridation of metals such as Mo, Nb [19], deposition of black phosphorus [20], our preparative interest is focused at present on the heterogeneous reaction

$$Si_{(s)} + xH(plasma) \rightleftarrows SiH_{x(g)}$$

where chemical transport can take place in either a temperature gradient, $T_1 \rightarrow T_2$, $T_1 < T_2$, or in a plasma gradient $\varepsilon_2 \rightarrow \varepsilon_1$, $\varepsilon_2 > \varepsilon_1$, or a combination of both [17,21]. By introducing a silicon charge into a zone of the plasma tube kept at $\sim$45°C, the forward reaction dominates and the gas phase becomes saturated with the gaseous reaction products. Depending on the temperature of the sample - a piece of a semiconductor grade silicon wafer hanging on a thin quartz fiber on the balance - either erosion (at lower temperature) or deposition of silicon (at higher temperature) results, as shown in Fig. 6. The importance of continuously weighing the sample in that kind of experiments is evident. Depositing thin layers of crystalline silicon at such low temperatures (down to 80°C by optimizing the experimental conditions) is of considerable interest for the development of silicon-based solar cells for terrestrial applications.

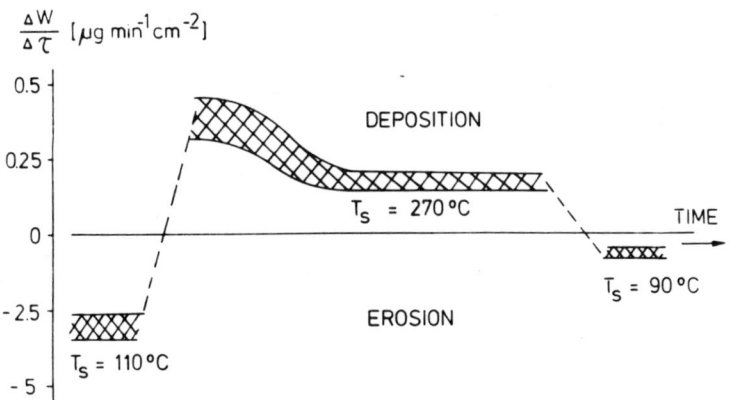

Fig. 6. Erosion and deposition of silicon in a hydrogen plasma, as measured by in situ thermogravimetry. $\Delta W/\Delta \tau$ = normalized weight changes of the sample; $T_s$ = sample temperature (see text)

Entirely based on the erosion of solids are current studies related to the material problems arising by plasma-wall interaction in devices for controlled thermonuclear fusion ("first wall" problem, see e.g. [22]). It has been shown [23] that the erosion rate of pyrolytic graphite (001) annealed at 2700°C by atomic hydrogen is strongly enhanced after irradiation with 2MeV $^4$He$^+$-ions. Due to induced lattice stress the enhancement of the reactivity is observed up to a depth of $\sim$ 30 $\mu$m, i.e. is much more than the projected range of the He$^+$-ions which amounts to $\sim$ 6 $\mu$m. Using the in situ gravimetry, the time requirements and accuracy are much improved compared to the previous periodic interruption of the experiment and conventional weighing of the sample. Other possible materials for the "first wall" are under investigation now.

## PHASE TRANSITIONS

Advantages of DSC and DTA in work on phase transitions are their sensitivity, requiring small sized samples only, and their rapidity in comparison e.g. to adiabatic calorimetry.

Our present activity concerns two aspects:

- Detailed kinetic and mechanistic studies of phase transitions in selected model compounds, by combining results of quantitative DSC with high resolution X-ray diffractograms as f(T) and hot stage optical microscopy. By proper choice of the models, phase transitions of variing degree of complexity - i.e. from nearly diffusionless to drastic reorganization of a crystal lattice - can be looked at, and the observation is facilitated if the two phases involved are differently coloured. An example is provided by the study of [NiBr$_2$(diazabutadiene)]-complexes, the yellow low temperature form with dimeric units of trigonal-bipyramidal coordination around Ni(II) of which is transformed at $\sim$ 380 K into a violet, tetrahedrally coordinated phase by breaking of two long nickel-bromine bonds only. Kinetic evaluation of the DSC curves yields different activation energies and mechanisms for powders and small single crystals, but only the second one can unequivocally be correlated with the hot stage microscopic observations [11,24].

- Investigations of a particular kind of phase changes taking place in various systems, i.e. work on so called "Jahn-Teller phase transitions":

$T_{low}$; static Jahn-Teller distortion of octahedra around Cu(II), Mn(III) $\quad \xrightarrow{endo} \atop \xleftarrow{exo} \quad$ $T_{high}$; dynamic Jahn-Teller distortion of octahedra

| Examples of systems | Temperature range | $\Delta$H(order of magnitude) |
|---|---|---|
| a) [Co(III)(NH$_3$)$_6$][Mn(III)F$_6$] (resp. Cr(III), Rh(III) instead of Co) | -170 - -80°C | 0.08 - 0.8 J/g |
| b) [K$_2$PbCu(NO$_2$)$_6$] (resp. Rb,Cs instead of K; Cu$_{1-x}$Zn$_x$ instead of Cu) | 1.5 - 120°C | 0.8 - 1.7 J/g |
| c) Mn$_3$O$_4$ Mn$_x$Cr$_{3-x}$O$_4$ | 1178°C $\sim$300-1178°C (for x=1.8-3.0) | 82.3 J/g $\sim$ 1.7 - 82.3 J/g (for x=2.1-3.0) |

In a), the interesting features are the smallness of $\Delta H$, and the behaviour of the anisotropic vibrational parameters within the $[MnF_6]^{3-}$ octahedron as determined by X-rays. For b), the DSC-data [25] can be correlated with numerous results which have been determined, in part by other scientists, with temperature dependent optical and EPR-spectra as well as X-ray and neutron diffraction. The manganese-chromium spinel system c) [26,27] bears chemical as well as physical interest. In the course of systematic work in order to achieve reproducible preparation of large and homogeneous samples, we have found the surprising effect, that at a particular composition the otherwise stable probe undergoes sudden weight losses during the phase transitions, see Fig. 7. They amount to $\sim 0.5$ % under atmospheric pressure, and up to 5 % at $10^{-4}$ torr, in uncovered crucibles. It is possible to empty a crucible by repeated cyclisation. The nature of the effect has to be sought in the sudden release of strong mechanical tensions, which can cause particles to leave their vessel, but at the present state of investigation, many questions remain open.

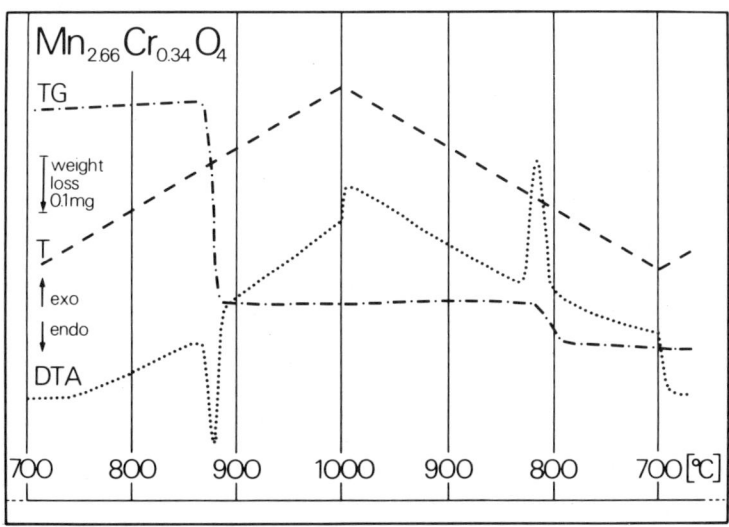

Fig. 7. Simultaneous TG and DTA curves of microcrystalline $Mn_{2.66}Cr_{0.34}O_4$ (sample 69.23 mg, 10°C/min, DTA sens. 50 $\mu$V full scale)

## ACKNOWLEDGEMENTS

The author wishes to thank numerous coworkers, colleagues and friends for their continuous support and encouragement. Financial help from the Swiss National Science Foundation is gratefully acknowledged.

## REFERENCES

[ 1] H.R.Oswald and E.Dubler, Reviews on Analytical Chemistry, EUROANALYSIS II, Budapest,1975, ed. W.Fresenius, Akadémiai Kiadó, Budapest, 1977, pp. 153-194.
[ 2] W.Meierhöfer, Ph.D. thesis, Univ. of Zurich, 1972.
[ 3] W.Meierhöfer and H.R.Oswald, Thermal Analysis, Proc.3rd ICTA, Davos, 1971, Vol. 2, ed. H.G.Wiedemann, Birkhäuser, Basel, 1972, pp. 409-421.
[ 4] V.Kohlschütter, Z.anorg.allg.Chem. $\underline{105}$ (1919) 1; ibid. 35.
[ 5] V.V.Boldyrev, J.Thermal Anal. $\underline{7}$ (1975) 685; $\underline{8}$ (1975) 175.
[ 6] F.K.Lotgering, J.Inorg.Nucl.Chem. $\underline{9}$ (1959) 113.
[ 7] J.R.Günter and H.R.Oswald, Bull.Inst.Chem.Res.Kyoto Univ. $\underline{53}$ (1975) 249.
[ 8] H.R.Oswald and J.R.Günter, 1976 Crystal Growth and Materials, ed. E. Kaldis and H.J.Scheel, North-Holland, Amsterdam, 1977, pp. 415-433.
[ 9] J.R.Günter, J.Solid State Chem. $\underline{5}$ (1972) 354.
[10] H.R.Oswald, J.R.Günter and E.Dubler, J.Solid State Chem. $\underline{13}$ (1975) 330.
[11] A.Reller, H.R.Beer and H.R.Oswald, Experientia Suppl. 37, eds. E.Marti, H.R.Oswald and H.G.Wiedemann, Birkhäuser, Basel, 1979, p. 61.
[12] H.R.Oswald and A.Reller, Proc. 9th Int.Sympos. on the React. of Solids, Cracow, 1980, Elsevier, Amsterdam, 1980, to be printed.
[13] A.Reller and H.R.Oswald, Thermal Analysis, Proc. 6th ICTA '80, Bayreuth, Vol. 1, ed. H.G.Wiedemann, Birkhäuser, Basel, 1980, p. 65.
[14] W.Bachmann, J.R.Günter and H.R.Oswald, Experientia Suppl. 37, eds. E.Marti, H.R.Oswald and H.G.Wiedemann, Birkhäuser, Basel, 1979, p. 36.
[15] S.Veprek, Current Topics in Material Science, Vol. 4, ed. E.Kaldis, North-Holland, Amsterdam, 1980.
[16] H.Winters, Plasma Chemistry, Vol. III, eds. S.Veprek and M.Venugopalan, Springer, Berlin, 1980.
[17] A.P.Webb and S.Veprek, Chem.Phys.Letters $\underline{68}$ (1980) 173.
[18] S.Veprek and A.P.Webb, Proc. 4th Int.Symp. on Plasma Chemistry, eds. S.Veprek and J.Hertz, Univ. of Zurich, 1979, p. 79.
[19] E.Wirz, H.R.Oswald and S.Veprek, Proc. 4th Int.Symp. on Plasma Chemistry, eds. S.Veprek and J.Hertz, Univ. of Zurich, 1979, p. 492.
[20] H.U.Beyeler and S.Veprek, Phil.Magazine, in press.
[21] Z.Iqbal, A.P.Webb and S.Veprek, Appl.Phys.Letters $\underline{36}$ (1980) 163.
[22] D.Gruen, S.Veprek and R.B.Wright, Plasma Chemistry, Vol. I, eds. S.Veprek and M.Venugopalan, Springer, Berlin, 1980.
[23] S.Veprek, A.P.Webb, H.R.Oswald and H.Stuessi, J.Nucl.Mat. $\underline{68}$ (1977) 32.
[24] H.R.Beer and H.R.Oswald, Thermal Analysis, Proc. 6th ICTA '80, Bayreuth, Vol. 2, ed. W.Hemminger, Birkhäuser, Basel, 1980, in press.
[25] E.Dubler, J.P.Matthieu and H.R.Oswald, Thermal Analysis, Proc. 4th ICTA, Budapest, 1974, Vol. 1, ed. I.Buzás, Akadémiai Kiadó, Budapest, 1975, p. 377.
[26] P.Holba, M.Nevřiva and E.Pollert, Mat.Res.Bull. $\underline{10}$ (1975) 853.
[27] H.R.Oswald, J.P.Matthieu and M.Wirz, Thermochim.Acta $\underline{20}$ (1977) 23, and so far unpublished work with A.Lüscher.

SOME APPLICATIONS OF THERMAL ANALYSIS
TO THE COMMUNICATIONS AND ELECTRONICS INDUSTRY

P. K. Gallagher
Bell Laboratories, Murray Hill, NJ 07974

I. INTRODUCTION

Thermal analysis encompasses a vast array of experimental techniques. Couple with this the broad spectrum of materials and processes of interest to the electronics and communications industries, one is left with an immense topic to summarize. The Bell System represents a reasonable cross section of this industry and I have chosen, therefore, to limit my survey to a broad selection of the practical uses of thermal analysis therein.

Examples are selected to include both the common thermal analytical techniques and a variety of materials, e.g., metals, ceramics, inorganic salts, semiconductors, polymers, and plastics. These specific materials are used for such things as magnets, batteries, capacitors, circuit boards and electrical insulation or encapsulants.

II. THERMOGRAVIMETRY (TG)

A. $Mn(NO_3)_2$ solution - Ta capacitors[1]

Capacitors prepared from an anodized Ta sponge are excellent high capacity, compact devices. Powdered Ta is partially sintered about a Ta wire which serves as one electrical contact. The metal sponge is then anodized to form a thin continuous oxide layer which serves as the dielectric. A counter electrode is applied over the oxide, by dipping the anodized metal into a solution of $Mn(NO_3)_2$ followed by thermally converting the adhering layer of solution to semiconducting $MnO_{2-x}$. This process is repeated until a continuous conducting contact layer is built and then contact is made to the external surface of the semiconductor and the device packaged for final use.

The nature and kinetics of the decomposition of $Mn(NO_3)_2$ solution were studied in considerable detail. Figure 1 shows the TG curve for a commercial solution. Gravimetric kinetic studies indicated that a moist atmosphere catalyzed the final stages of the decomposition by significantly reducing the activation energy associated with reaction (1).

$$MnONO_3 \rightarrow MnO_2 + NO_2 \qquad (1)$$

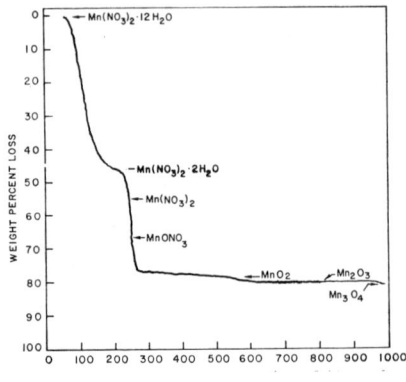

Fig. 1. TG of $Mn(NO_3)_2$ aqueous solution.

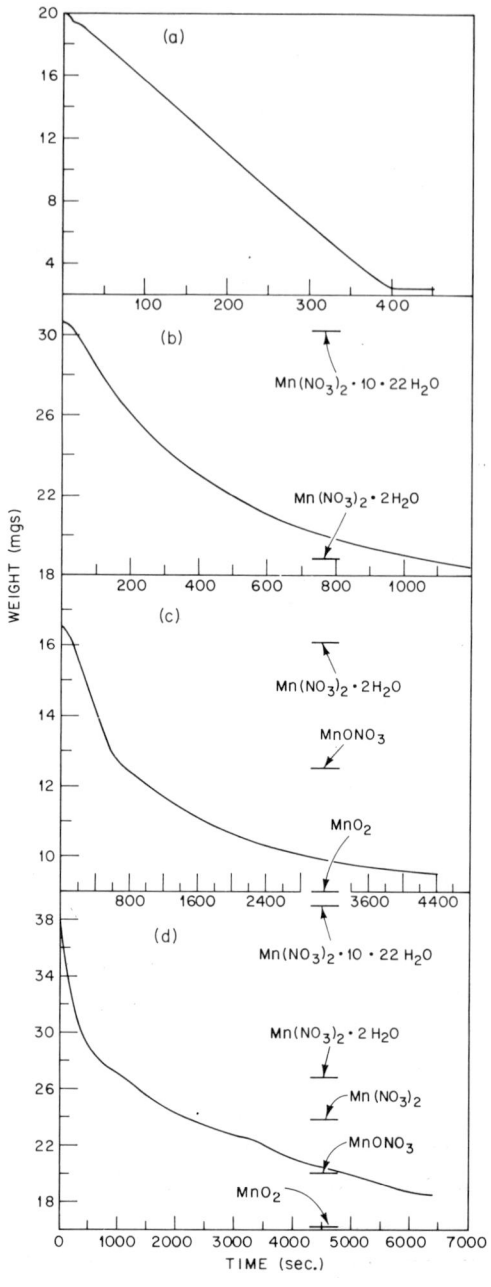

Fig. 2. Weight as a function of time for the thermal decomposition of aqueous manganese(II) nitrate ($d^{25°} = 1.528$ g/cm$^3$). (a) Loss of occluded water, 109°C; (b) dehydration of solution, 110°C, 1st dip; (c) decomposition of solution, 153°C, 1st dip; (d) combined dehydration and decomposition of solution, 109°C, 2nd dip.

Decomposition and kinetics were actually investigated on the tantalum devices similar to production. Typical isothermal weight loss curves for coating the Ta capacitors are shown in Fig. 2. The anodized Ta is stored under water which is driven off, Fig. 2a, prior to dipping in the $Mn(NO_3)_2$ solution. Overlap of the decomposition processes complicates interpretation. In Fig. 2b, the solution is dehydrated to a glass but does not decompose further. Higher temperatures, Fig. 2c, decompose the nitrate to the oxide. Subsequent dips decompose at much lower temperatures, Fig. 2d. Not only does the catalysis by moisture carry over to the device application leading to a reduction in the thermal treatment with a concommitant savings of time and energy, but it also yields a much cleaner, smoother, and more continuous electrode surface.

B. Chromindur alloys - Permanent magnets[2]

Thermomagnetometry (TM) has been applied to measuring the Curie temperature, $T_c$, of ferrites.[3] This type of measurement can also be utilized to help (a) describe the distribution of impurities,[4] (b) determine the formation of intermediates in a reaction path,[3,4] or (c) establish phase diagrams.[4]

High energy magnet alloys have been developed in the Fe-Cr-Co system (Chromindurs). Magnetic properties of these alloys are dependent upon their thermal history and achievement of the optimum properties requires close control of the spinodal decomposition of the alloy.

Because the useful magnetic properties of these materials depend upon the nature and degree of the phase separation, TM provides a rapid sensitive technique to study them. The method provides complimentary information to X-ray diffraction studies of changes in crystal structure and DSC studies which depend upon changes in enthalpy and heat capacity. The sensitivity of each technique is different. Several samples of Chromindur II (Fe-28Cr-10.5Co) having different thermal histories were studied under varying conditions.[2]

Type I material was annealed at 930°C and quenched while type II was given an aging heat treatment to optimize the isotropic magnetic properties. Type I material is essentially unstable single phase α material. Upon initial reheating it will pass through the region of immiscibility and will tend to disproportionate into a variety of $\alpha_1$ and $\alpha_2$ metastable spinodal phases depending upon the kinetics of the system and the heating rate. At longer times they will convert to more stable phases with greater compositional differences. As the temperature increases $\alpha_1$ and $\alpha_2$ phases will eventually redissolve. The rate of solution of the metastable spinodal phases should be much greater than that of the phases closer to equilibrium.

Figure 3 illustrates TM curves for the initial reheating of type I material. At high heating rates the material does not have time to underto spinodal decomposition and the $T_c$ of the pure α phase is obvious at ∿650°C. Slower heating rates allow for spinodal decomposition to occur and at the slowest rate even phases of greater stability with $T_c$ in excess of 700°C can form.

In contrast, type II has undergone conversion to spinodal products during preparation so that when both materials are heated, Fig. 4, the distinction is obvious. The curve for type II material reflects more the dissolution rate of spinodal products than the $T_c$ of the spinodal material which is probably higher still. The potential quality control aspects of such rapid TM scans are obvious.

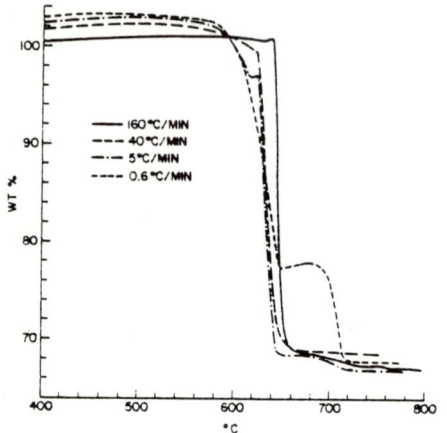

Fig. 3. TM curves of quenched Chromindur at various heating rates.

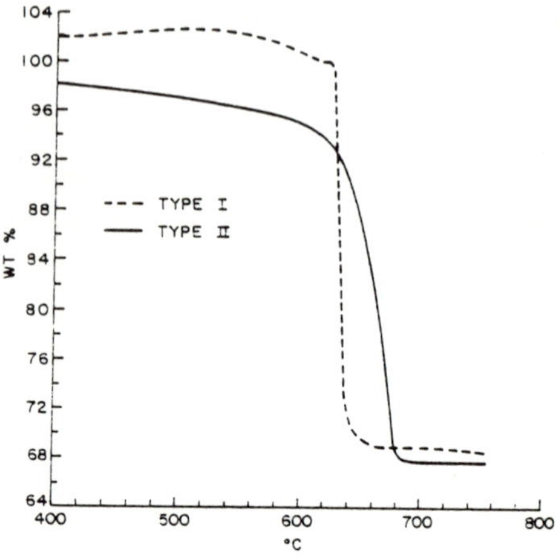

Fig. 4. TM curves of quenched (I) and annealed (II) Chromindurs (40°C min$^{-1}$).

## III. DIFFERENTIAL THERMAL ANALYSIS (DTA) AND DIFFERENTIAL SCANNING CALORIMETRY (DSC)

A. Water in polymers and polycarbonates - Electrical properties of insulation[5,6]

The bandwidth of communication systems is being expanded placing increased demands upon the dielectric materials used in cables. Changes in the processing led to increases in the dielectric loss. Initial studies traced these losses to a volatile impurity.[5] This led to extensive studies on the absorption of water by polymers and its effects upon their dielectric properties.[6]

The total quantity of water present in a particular polymer specimen was determined coulometricly. However, it was found necessary to distinguish between unassociated and "clustered" water within the polymers because their effects upon the dielectric loss spectrum were markedly different.[7] Clustered water can be determined by DSC.[6]

Figure 5 shows a heating and cooling trace for polycarbonate which was soaked in water at 97°C for 800h. The large undercooling was attributed to the small size of the water droplets ($\leq 1\mu m$) and lack of nucleation centers. The exact amount of clustered water is calculated from the area under the fusion peak.

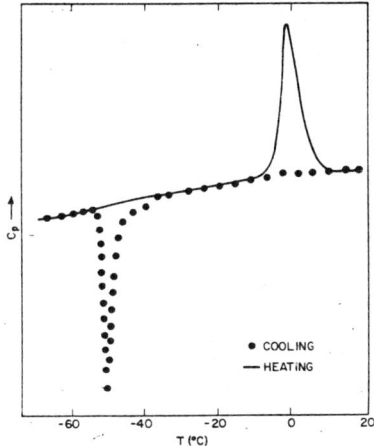

Fig. 5. DSC trace (20°C min$^{-1}$) of polycarbonate after soaking in $H_2O$ Q + 97°C for 800h.

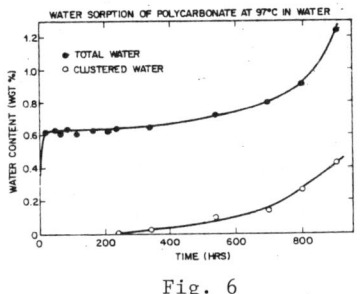

Fig. 6

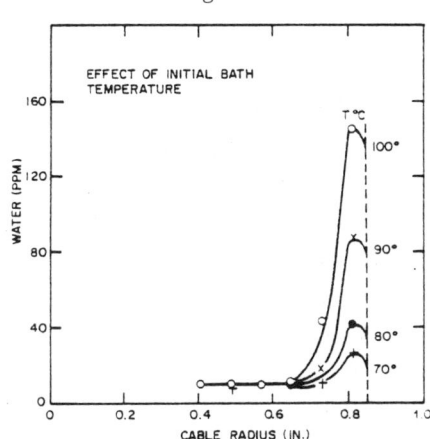

Fig. 7. Water in extruded polyethylene.

The total absorption process is shown in Fig. 6. There is a rapid absorption of unassociated water which levels off. A second water absorption process is evident later which leads to clustered water. Extruded polyethylene cable also shows the presence of both types of moisture with a pronounced surface gradient, see Fig. 7.[7] A successful compromise in the extrusion procedure was reached to achieve the necessary crystallinity for mechanical purposes and reduced water content for dielectric properties.

B. Polyolefins - Stabilization of insulation[8,9]

Polyolefins used as electrical insulation and cable jackets are subject to oxidative degradation and UV induced decomposition. Various light absorbers and antioxidants are added to improve the stability and lifetime of the polymers. Because of the exothermic nature of oxidative degradation, DTA and DSC are very useful techniques for evaluating various protective systems. The rapidity of such tests makes them ideal for quality control applications.[8]

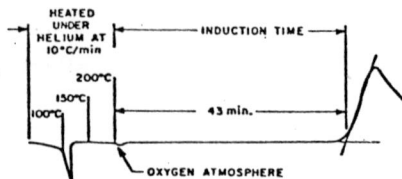

Fig. 8. Isothermal DTA trace for oxidative degradation.

Tests can be made in the conventional dynamic mode or isothermally. The dynamic test consists of a DTA scan of the polymer in flowing oxygen. The isothermal test involves heating samples to a preset temperature and establishing thermal equilibrium in an inert atmosphere prior to switching to the oxidizing atmosphere and measuring the induction time to oxidation. A typical experiment is shown in Fig. 8.[8] Table I compares results of dynamic and isothermal experiments.[8] The isothermal method is generally preferred. Figure 9 shows DSC results for three levels of protection in polyethylene.[9] Solubility and diffusion of antioxidants in polyethylene has been determined by TG.[10]

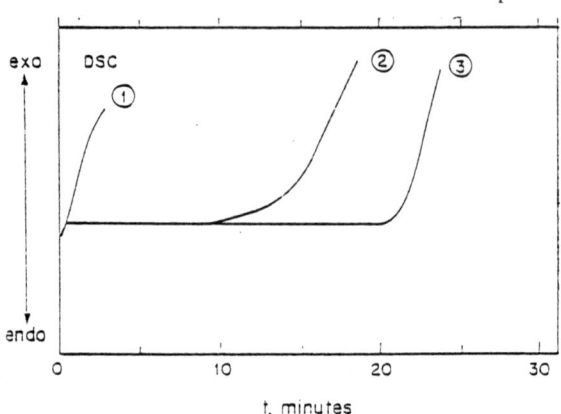

Fig. 9. DSC traces at 200°C for polyethylene with three levels of antioxidant.

## IV. EVOLVED GAS ANALYSIS (EGA)

A. GaAs/Au - Stability of contacts to semiconductor devices[11]

Low ohmic resistance, good adherence, and stability are required for electrical contacts. Frequently gold or gold based alloys are used for this purpose. Previous work using a variety of techniques had shown that for Au on GaAs, the GaAs decomposed at temperatures as low as 400°C with Ga diffusing into the Au forming a relatively low melting alloy. Mass spectroscopic EGA was used to follow the evolution of As from such a reaction and determine the effects of such variables as processing temperature and time relation-

TABLE I
COMPARISON OF THERMAL ANALYSIS METHODS FOR DETERMINING THE STABILITY OF POLYETHYLENE FORMULATIONS

| Stabilizer System | Dynamic ($10°C\ min^{-1}$) Oxidation Exotherm °C | Isothermal Induction Time Minutes at 200°C |
|---|---|---|
| None | 208 | 0 |
| 2.5% Carbon Black | 215 | 8 |
| 25% Carbon Black 0.05% Antioxidant A | 220 | 10 |
| 2.5% Carbon Black 0.1% Antioxidant A | 228 | 10 |
| 2.5% Carbon Black 0.1% Antioxidant B | 260 | 65 |

ships, cleaning and surface pretreatments, and Au thickness.[11] In order to observe such readily condensible and reactive vapors as As, it is necessary to minimize the diffusion path between sample and detector.[12]

Without Au electrodes GaAs begins to evolve As around 620°C in vacuum. With Au electrodes As evolution is detectable by 270°C and the rate peaks around 420°C depending on the surface pretreatment and thickness of the Au. This rate of maximum evolution agrees with the melting of Au-Ga eutectic formed by the decomposition. Figures 10 and 11 show the effects of Au thickness and surface pretreatment respectively for a $Ga_{0.7}Al_{0.3}As$ substrate. Clearly the amount of Au determines the amount of Ga dissolved and hence the amount of As evolved. It also influences the length of the diffusion path for As evolution. Surface treatment of $Ga_{0.7}Al_{0.3}As$ prior to the deposition of Au is also important. Cleaning with concentrated HCl and rinsing with water and methanol enhance interaction. A degree of interaction is desirable for adherence of the film, but, too much is detrimental to both adhesion and electrical properties.

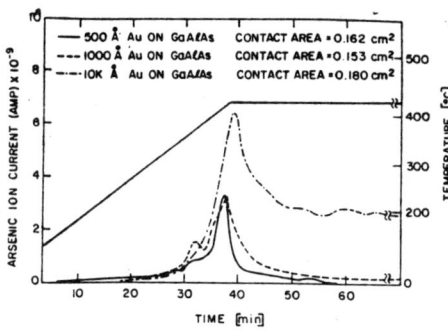

Fig. 10. EGA of Au on Ga(AlAs).

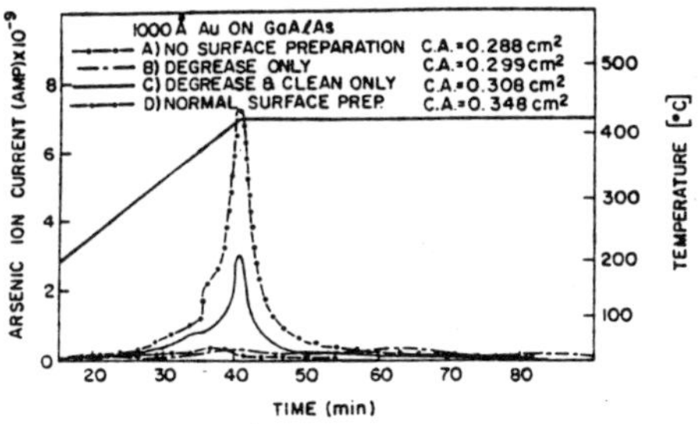

Fig. 11. Effects of surface preparation on the EGA of Ga(AlAs).

B. Epoxy resins and flame retardants - Encapsulating agents[13,14]

A more sophisticated mass spectrographic EGA apparatus is shown in Fig. 12.[13] This uses laser heating and differential pumping to provide a molecular beam of evolved products that is chopped prior to analysis by a quadrupole mass spectrometer. Chopping allows for elimination of the background signal and phase analysis of the products.

Molded plastic encapsulation is cheaper than a hermetically sealed metal or ceramic structure; however, the number of failures related to corrosion is greater. Besides greater permeation by the external environment, there are problems associated with outgassing and degradation of

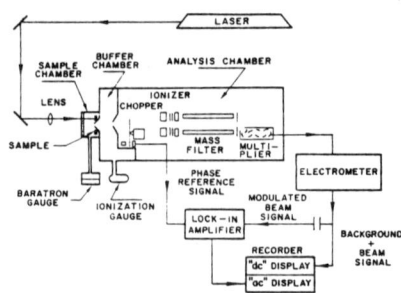

Fig. 12. Schematic of laser probe-molecular beam apparatus.

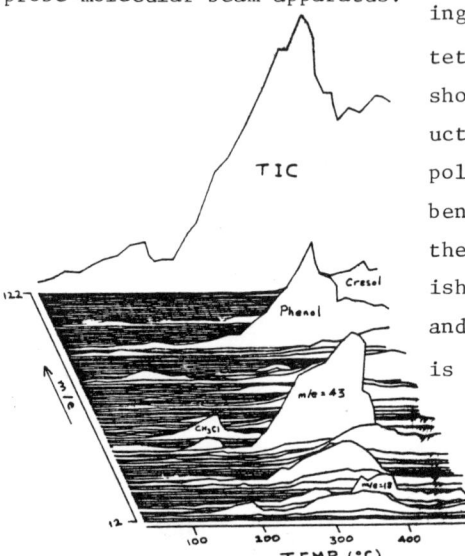

Fig. 13. Isometric plot of mass spectral ion profiles from novolac-epoxy with flame retardant.

the epoxy thermoset or additives, e.g., flame retardants. Comparison of the EGA of samples with and without flame retardants indicates very similar decomposition products <300°C; implying that the flame retardant is not responsible for failures during accelerated aging tests at 200°C. An isometric plot of the various mass profiles observed during pyrolysis of novolac-epoxy with a tetrabromobisphenol-A flame retardant is shown in Fig. 13.[14] At $\leq$150°C the products are related to low molecular-weight polymerization species, water, methanol, benzene, and toluene. By 200°C most of these species except water have diminished but methylchloride, carbon dioxide, and acetone are evident. Device failure is attributed to the joint presence of $Cl^-$ and $H_2O$ rather than anything derived from the flame retardant.

## V. THERMAL DILATOMETRY (TD) AND DYNAMIC MECHANICAL ANALYSIS (DMA)

A. Chromindur alloys - Powder metallurgical sintering[15]

Importance of Chromindur alloys was described earlier. The desirability of forming varied shapes makes powder metallurgical techniques of particular value in this system. Kinetics of sintering such alloys were studied by TD. Length and temperature outputs of the dilatometer were

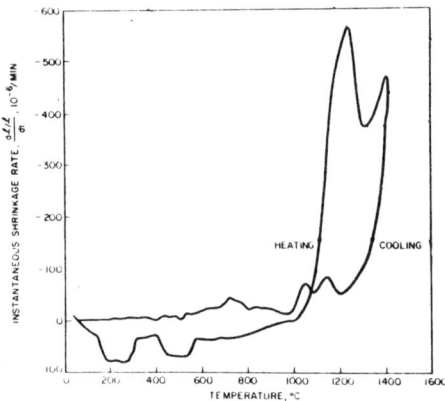

Fig. 14. Shrinkage rate vs. temperature for a Chromindur at 3.6°C $min^{-1}$.

punched on paper tape at preset, frequent intervals facilitating computer plotting and analysis of the data. A plot of instantaneous shrinkage rate $(dl/dt)l^{-1}$ versus temperature for a powder is given in Fig. 14 for a compact of 42.5Fe-29.5Cr-25Co-3Mo heated at 3.6°C $min^{-1}$. Phase relationships occurring during heating are very complex. At 1000° the compact consists of tetragonal, face centered cubic, and $Fe_2MoC$ phases. As the temperature is increased a body centered cubic phase grows and by 1400°C it is the only phase present.

Mathematical analyses were developed to determine the Arrhenius kinetic parameters for sintering under dynamic conditions[16] similar to treatments of thermal decomposition reactions. One such treatment utilizes a plot of log shrinkage $\Delta l/l_o$ versus $T^{-1}$ as shown in Fig. 15 for a sample heated at 4.95°C $min^{-1}$. There are two regions with markedly different activation energies. Shrinkage at T < 1250°C is small and less accurate, hence an alternative analysis is used for that data which also yields a value for the order for the shrinkage rate equation. An average of eight runs with heating rates from 1-11°C $min^{-1}$ gave activation energies of 63 Kcal/mole for the low temperature process and 49 Kcal/mole for the higher temperature one. These correlated well with diffusion energies in the face centered cubic phase and body centered cubic phase respectively in agreement with microscopic and X-ray phase analysis.

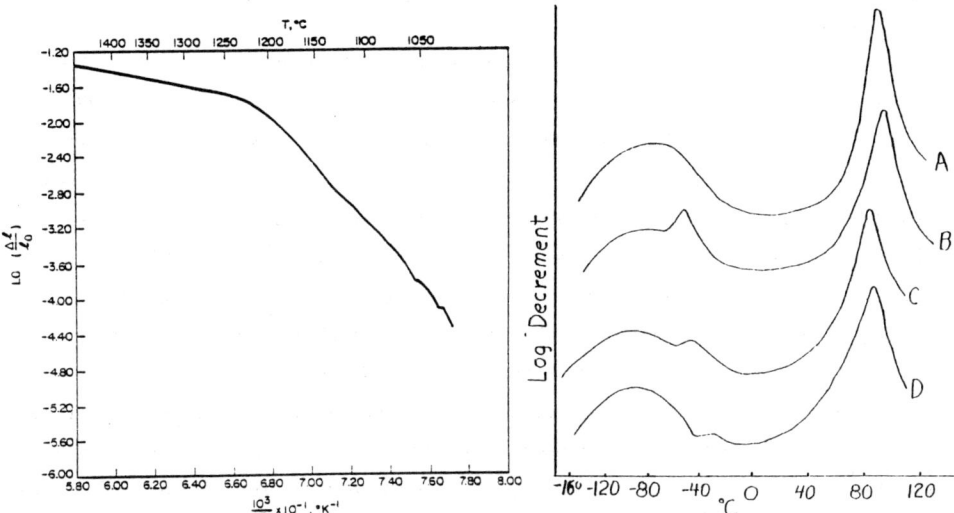

Fig. 15. Log shrinkage versus $T^{-1}$ for a Chromindur at 2.15°C $min^{-1}$.

Fig. 16. TBA of rubber-epoxy resins (see Table II).

B. Rubber-modified epoxy - Circuit boards[17]

Thermosetting epoxy resins are used to form circuit boards. Rubber additives improve the mechanical properties. These additives lead to phase separation and the resulting morphology is of considerable importance. Effects of such factors as the butadiene/acrylonitrile ratio of the additive, cure temperature, and gelation time upon the morphology and mechanical properties of bisphenol-A type epoxy resin were determined by DMA (torsion braid analysis, TBA), electron microscopy, and viscometry.

Results of TBA experiments performed on formulations which had been cured 14 h. at 120°C are shown in Fig. 16. Table II lists some properties of the rubber and morphology of the cured phase separated material. The glass transition temperature, $T_g$, of the rubber phase is evident and the magnitude of the effect is proportional to the volume fraction of rubber phase in the resin. Morphology and phase separation were described in terms of the equilibrium solubility of the rubber, gelation time, and the relative intensities of the glass transition of materials.

TABLE II
PROPERTIES AND MORPHOLOGY OF REACTIVE RUBBER-EPOXY RESINS CURED
14h AT 120°C (100 EPON 828 10 CTBN 5 PIPERDINE)

| | CTBN Rubber | Acrylonitrile | $T_g$[a] | VF[a] | S[b] |
|---|---|---|---|---|---|
| A | - | -% | -°C | 0% | 0μm |
| B | 1300x15 | 10 | -59 | 17 | 4.1 |
| C | 1300x8 | 17 | -45 | 12 | 1.4 |
| D | 1300x13 | 27 | -30 | 1 | 0.1 |

[a]Volume fraction of phase-separated rubber. [b]Average domain diameter.

## VI. COMBINED STUDY

β"-Al$_2$O$_3$ - Ionic conductor for batteries[18,19]

Relatively inert, refractory materials having high ionic conductivity are necessary as separator-electrolytes in high temperature battery systems, e.g., Na/S. The most commonly used conductor has been Li or Mg stabilized β"Al$_2$O$_3$. Conventional ceramic technology, milling-calcining-fabrication-sintering, has been used to prepare the thin walled tubes required; however, recently a more efficient, novel approach has been taken which eliminates the first two steps in processing.[18] Soluble Li and Na salts are added to an aqueous slurry of reactive Al$_2$O$_3$ prior to spray drying. Tubes are isostatically pressed from these powders which contain considerable amounts of acetates and/or hydroxides.

Thermal analytical techniques were used to follow the combined decomposition and reaction processes during the heating of green ceramic tubes.[19] Combined results of TG, DSC, DTA, and TD are shown in Fig. 17. Mass spectrographic EGA was also performed so that virtually all the techniques described herein were focussed on describing the overall pro-

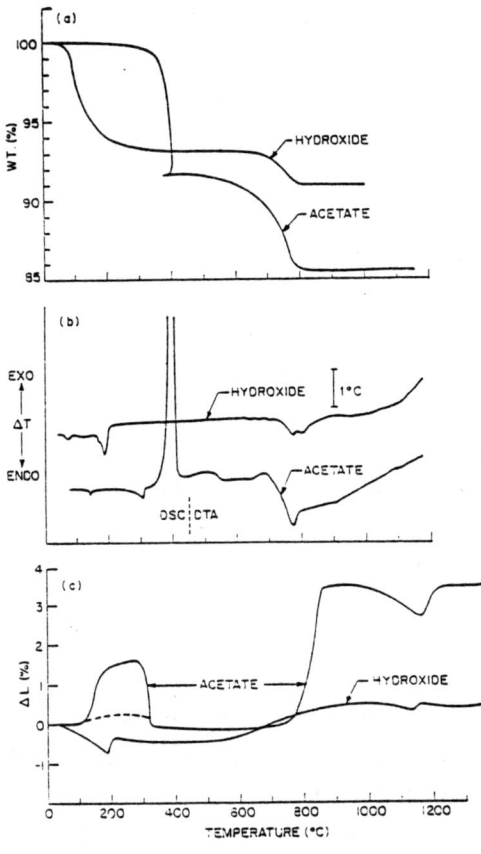

Fig. 17. Thermal analysis of spray dried powders a) TG, b) DSC/DTA, c) TD

cess. Table III summarizes the interpretation of the steps in the acetate firing. Large volume changes make the green ceramic fabricated from the acetate more subject to cracking during heating. Thermal analysis was of great value in understanding the processing and in pinpointing areas where the rate of temperature rise should be slowed to minimize cracking.

Hydroxide decompositions lead to smaller volume changes and the degree of conversion to $\beta''Al_2O_3$ is somewhat greater. General interpretation of the events is analogous to that in Table III. EGA indicated considerable $CO_2$ was evolved along with water during the decomposition ~700°C. Significant hydroxide had been converted to carbonate by atmospheric $CO_2$ during spray drying and other processing.

TABLE III
SINTERING OF $\alpha-Al_2O_3$ UNCALCINED ACETATES

| Temp (°C) | Dilatometer | TG | DTA | XRD | Mechanism |
|---|---|---|---|---|---|
| 100-200 | expansion | slight loss | endo | $\alpha-Al_2O_3$ + $NaC_2H_3O_2$(?) | acetate hydrate melting + $NaC_2H_3O_2 \cdot xH_2O \rightarrow NaC_2H_3O_2 + xH_2O$ |
| 275-325 | shrinkage | --- | endo | $\alpha-Al_2O_3$ + $NaC_2H_3O_2$(?) | acetate melting |
| 325-400 | --- | loss | exo | $\alpha-Al_2O_3$ | $2NaC_2H_3O_2 + 4O_2 \rightarrow Na_2CO_3 + 3H_2O + 3CO_2$ |
| 650-850 | expansion | loss | endo | $\alpha-Al_2O_3$ + $NaAlO_2$ | carbonate melting + $Al_2O_3 + Na_2CO_3 \rightarrow 2NaAlO_2 + CO_2$ |
| 1000-1150 | shrinkage | --- | --- | $\alpha-Al_2O_3$ + $NaAlO_2$ | sintering of $NaAlO_2$ |
| 1150-1200 | expansion | --- | --- | $\beta-\beta''-Al_2O_3$ | $NaAlO_2 + 5Al_2O_3 \rightarrow NaAl_{11}O_{17}$ |

## VII. CONCLUSIONS

These brief examples clearly indicate the usefulness of thermal analytical techniques to solve materials related problems associated with the electrical and communications industries. Usefulness for quality control stems largely from the savings of time due to scanning as opposed to stepwise isothermal studies. Forecasting the future is always a dangerous passtime. However, one thing is clear: minicomputers and microprocessors will have a major impact on the speed and simplicity of use as well as greatly expanding the potential for interaction with the process through feedback. This should lead to increased industrial use of these techniques.

## REFERENCES

1. P. K. Gallagher and D. W. Johnson, Jr., J. Electrochem. Soc. $\underline{118}$ (1971) 1530.
2. P. K. Gallagher, E. Coleman, S. Jin, and R. C. Sherwood, Thermochim. Acta, in press.
3. W. R. Ott and M. G. McLaren, "Thermal Analysis, Vol. 2," eds. R. F. Schwenker, Jr., and P. D. Garn, Academic Press, N.Y. 1969, pp. 1439-51.
4. P. K. Gallagher and S. St. J. Warne, Thermochim. Acta, in press.
5. H. E. Bair, G. E. Johnson, J. H. Daane, and E. W. Anderson, Proc. 25th Int'l Wire and Cable Symp., pp. 296-301 (1976).
6. H. E. Bair, G. E. Johnson and R. Merriweather, J. Appl. Phys. $\underline{49}$ (1978) 4976.
7. G. E. Johnson, H. E. Bair, E. W. Anderson, and J. H. Daane, 1976 Ann. Rept. Conf. on Electrical Insulation and Dielectric Phenomena (NAS-NRC).
8. M. G. Chan, H. M. Gilroy, I. P. Heyward, L. Johnson, and W. M. Martin, Proc. 36th ANTEC, Soc. Plast. Eng., pp. 381 (1978).
9. H. E. Bair in "Thermal Characterization of Polymeric Materials," E. A. Turi, ed., Academic Press, New York, 1980.
10. R. J. Roe, H. E. Bair, and G. Greniewski, J. Appl. Poly. Sci. $\underline{18}$ (1974) 843.
11. E. Kinsbron, P. K. Gallagher, and A. T. English, Solid-State Electronics $\underline{22}$ (1979) 517.
12. P. K. Gallagher, Thermochim. Acta $\underline{26}$ (1978) 175.
13. R. M. Lum, Thermochim. Acta $\underline{18}$ (1977) 73.
14. R. M. Lum and L. G. Feinstein, $\underline{In}$ "Proc. 30th Electronic Components

Conf." in press.
15. M. L. Green, to be published.
16. J. L. Woolfrey and M. J. Bannister, J. Am. Cer. Soc. $\underline{55}$ (1972) 390.
17. L. T. Manzione, J. K. Gillham, and C. A. McPherson, J. Appl. Poly. Sci., to be published.
18. D. W. Johnson, Jr., S. M. Granstaff, Jr., and W. W. Rhodes, Am. Cer. Soc. Bull. $\underline{58}$ (1979) 849.
19. E. M. Vogel, D. W. Johnson, Jr., and M. F. Yan, Am. Cer. Soc. Bull., in press.

# Theory

E. Koch
Max-Planck-Institute for Coal Research
Mühlheim/Ruhr, FRG

E. Grell
Max-Planck-Institute for Biophysics
Frankfurt/Main, FRG

# KINETIC COMPENSATION EFFECT: FACTS AND FICTION OF LINEAR PLOTS USING ARRHENIUS LAW

J. Šesták

Institute of Physics of the Czechoslovak Academy of Sciences, 180 40 Praha, ČSSR

## ABSTRACT

The deviation from the linear law $\ln k = A + B/T$ may be caused by either (i) insufficient choice of the particular form of basic kinetic equation, or, (j) virtual correlation of the terms $A$ and $B/T$. A quantitative estimation can be found through the angle at which these functions $A$ and $B/T$ meet each other in the Hilbert space.

## INTRODUCTION

The determination of the activation energy, $E$, and the preexponential factor, $Z$, is based on logarithmic treatment of Arrhenius law, $k = Z \exp(-E/RT)$, where $k$ and $T$ are the rate constant and temperature. In many cases the nonlinearity of the plot $\ln k$ vs $1/T$ is reported and often called <u>kinetic compensation effect</u> (1 - 5). The sources for such a nonlinearity are assumed to be the following:

Varying conditions of measurement due to inconstant pressure, external field, temperature, etc.

Undefined sample preparation yielding different fractions of solid compacts (shape and size distribution, concentration profiles, number of structural defects and other active sites, etc.).

Unappropriate choice of mathematical modelling for the process in question and the consequent treatment of associated kinetic equation.

Possible inadequacy of Arrhenius law to describe solid state reactions because of the invalid statistics of energy distribution (6).

Mathematical reasons hidden in the entire form of correlation employed (7).

## EFFECT OF EXPERIMENTAL CONDITIONS

This problem was first reported by Zawadski and Bretzsnajder (8) when studying thermal decomposition of $CaCO_3$ under various pressures of $CO_2$. They determined the dependence in the form $E = f(P_{CO_2})$. Later it was theoretically explained by Pavlyuchenko and Prodan (9), reinvestigated by Wist (10) and analyzed from the standpoint of chemical statistics by Roginskij and Chajt (11). This, in fact, is in agreement with a known dependence of the rate of the process ($\mathring{a}$) upon the pressure (P), which is accounted for in the form of a multiplying function f(P) to k (Barret (12)). The term f(P) may be expressed by the power-type ($\approx P^m$) and/or Langmuire-type ($\approx AP/(1 + BP)$) function, where P is frequently approached as $P_0(1 - a)T$ assuming an ideal case.

Another aspect of the E inconstancy is, e.g., the effect of varying defects concentration as shown by Olejnikov et al. (13) for the study of Mg-ferrite formation under the different partial pressures of oxygen. Similar findings were shown by Solymosi et al. (14) and Dollimore and Rogers (15). Guarrini et al. (16) pointed out that the nonlinearity increases with the mass of the sample and recommended the extrapolation to zero inweight. Lieser (17) investigated the change of E for a heterogeneous reaction proceeding from the surface to the center of solid NaJ. He found that the value of E increases three times reaching finally that of bulk diffusion.

## CHOICE OF KINETIC EQUATION

Hullet (18) made a search for the reasons of possible nonlinearities determining that any deviation from the straight line of the plot ln k vs 1/T is to be considered as an almost certain evidence that the observed process is a complex one. The alternative pathes of two or more simultaneous processes with the different E being concave upwards while the consecutive case concave downwards. Simple mathematical analysis of basic kinetic equation gives the following dependence (19): $E_{app} = E_{real} - RT \ln[F(a)/f(a)]$.

This indicates the <u>interdependence between apparent</u> ($E_{app}$) <u>and actual</u> ($E_{real}$) <u>activation energies</u>, if the false model relation F(a) is misused instead of the appropriate one, f(a). Similarly, neglecting the temperature dependence of $Z$ (= $Z' T^n$) yields $E_{app} = E_{real} - nRT \ln T$. In analogy, the absence of any additional multiplying term g(x) to k can be analyzed in the same way as $E_{app} = E_{real} - RT \ln g(x)$ (see, e.g., preceding paragraph, where x may equal to the pressure, concentration, mass, particle radii, etc.).

## PROBLEMS OF MATHEMATICAL CORRELATION

The result obtained by means of the regression analysis upon the general relation $\ln x = A - B/T$ (where x = k or = P) must be accepted critically because of the deformation of the normal Gauss distribution. Namely, if x increases ten times, the transformed term ln x rises only twice and in the same way are also changing the corresponding errors. The dispersion in ln x decreases with the increasing x so that the regression analysis is less reliable for the great k tending so to exhibit there an apparent nonlinearity.
Another problem is often interdependence between E and ln Z (1 - 5). Militký (7) and Pysiak (20) found by treating the correlation dependences of individual terms (ln k, ln Z and E/RT) that the <u>term ln Z appears to be a superfluous</u> and so <u>unnecessary parameter</u> in a narrow temperature interval. For dismissing such an effect the enormous extension of this working temperature interval is needed which, however, is not always convenient due to experimental reasons. Thus, a reference temperature should be introduced to be called <u>isokinetic temperature</u>, $\vartheta$, and established at $k = k_o$. The rate constant should then held the modified form of $k = k_o \exp -(E/R)(1/T - 1/\vartheta)$ (7). The practical value of $\vartheta$ can then be estimated from the intersection point of the straight lines read out from the original plot of ln k vs 1/T (20). Its value enters the correlation between E and ln Z as a multiplying constant ($\ln Z = E/(\vartheta R)$).

## DISCUSSION

It can be seen that the most serious and so far underestimated is the entire "mathematical" behaviour of the correlation employed (21). This may be proved quantitatively by means of the angle in which the functions in question (A and B/T or simply 1 and 1/T) meet each other in the Hilbert space (Vonka (22)). From the functional analysis follows that the cosinus of this angle $\theta$ equals to the ratio of the scalar multiplication of functions $|(1, 1/T)|$ with the product of their norms $\|1\|.\|1/T\|$. Hence (22)(see appendix (23), (24))

$$\cos \theta = T_0 T_F \ln(T_F/T_0)/(T_F - T_0),$$

which shows that the angle does not depend on the functions in question only, but also upon the working temperature interval (from $T_0$ to $T_F$). Furthermore, we can utilize the Taylor's expansion to obtain the simplified equation (22)

$$\theta_{(rad)} \approx 2(T_F - T_0)/(T_F + T_0)$$

if $(T_F - T_0) \ll (T_F + T_0)$. Now, at least, a rough calculus of $\theta$ is needed. According to Vonka (22), the angle below $5°$ is a clear indication of the mutual dependence of 1 and 1/T as demonstrated for two hypothetical cases:

| $(T_F - T_0)$ | $\theta_{(rad)}$ | $\theta\ (°)$ | $(T_F - T_0)$ | $\theta_{(rad)}$ | $\theta\ (°)$ |
|---|---|---|---|---|---|
| 310 - 300 | 0.033 | 1.9 | 920 - 900 | 0.022 | 1.26 |
| 320 - 300 | 0.065 | 3.7 | 940 - 900 | 0.044 | 2.5 |
| 325 - 300 | 0.08 | 4.6 | 960 - 900 | 0.065 | 3.7 |
| 330 - 300 | 0.095 | 5.45 | 980 - 900 | 0.085 | 4.9 |
| 340 - 300 | 0.125 | 7.2 | 990 - 900 | 0.095 | 5.5 |

It follows that for the temperatures around 300 K it is not recommendable to use the correlation (1, 1/T) within the temperature interval lower than 25 K and, similarly, at $T \approx 900$ K for $\Delta T < 80$ K.

## REFERENCES

(1) P.D. Garn, J. Thermal. Anal. **7** (1975) 475
(2) V.M. Gorbatchev, ibid **8** (1975) 588

(3) J. Pysiak, B. Sabalski and T. Zmijevski, in "Thermal Analysis" (4th ICTA), Vol. 1, p. 205, Budapest 1975; Rozc. Chem. 54 (1971) 263
(4) J. Zsakó, J. Thermal. Anal. 8 (1975) 593
(5) A.V. Nikolayev and V.A. Logvinenko, ibid 10 (1976), 363
(6) P.D. Garn, ibid 13 (1978) 581
(7) J. Militký, "Modelling of Nonisothermal Processes", Report VÚZ, Dvůr Králové n/L 1979; Proc. "TA and Calorimetry of Polymers", ČSVTS Silon, Planá n/L, 1979, p. 13
(8) J. Zawadski and S. Bretsznajder, Zeit. Electrochem. 41 (1935) 215
(9) M.M. Pavlyuchenko and E.A. Prodan, Dokl. AN USSR 136 (1961) 651
(10) A.O. Wist, in "Thermal Analysis" (2nd ICTA), Vol. 2, p. 1095, New York 1969
(11) S.Z. Roginskij and J.L. Chajt, Izvest. Akad. Nauk USSR, Khimia, 1961, p. 771
(12) P. Barret, "Cinetique heterogenne", Gauthier-Villars, Paris 1973
(13) N.N. Oleynikov, Ju.D. Tretyakov and A.V. Schumyancev, J. Solid State Chem. 11 (1974) 34
(14) F. Solymosi, K. Jáky and Z.G. Szabo, Zeit. anorg. allg. Chem. B 368 (1969), 211
(15) D. Dollimore and P.F. Rogers, Thermochim. Acta 30 (1979), 273
(16) G.G.T. Guarrini, R. Spinicci, F.M. Carlini and D. Donati, J. Thermal. Anal. 5 (1973), 307
(17) K.H..Lieser, in "Preprints of 8th Int. Symp. React. Solids", Götenborg, Sweden 1976, p. 174
(18) J.R. Hullet, Quarterly Reviews 18 (1964), 227
(19) J. Šesták, review "Philosophy of Nonisothermal Kinetics" in J. Thermal. Anal. 16 (1979) 503
(20) J. Pysiak, A. Glinka, in the Proc. "Termanal '79", SVŠT Bratislava, 1979, p. 323; in "Komunikaty" of 2nd Seminary of Bretsznajder, Plóck 1979, p. K4

(21) J. Šesták, "Thermophysical Properties of Solids; Their Measurement and Theoretical TA", Academia - Elsevier, in print
(22) P. Voňka, "Treatments of Measured Data in Physical Chemistry", Thesis at VŠCHT, Praha 1975
(23) B.Z. Vulich, "Uvedeniye v funkcionalnyi analyz", Nauka, Moskva 1965
(24) L. Collatz, "Funkcionální analýza a numerická matematika, SNTL, Praha 1970.

## APPENDIX

$$\cos \theta_{(1, 1/T)} = |(1, 1/T)|/( \|1\| \cdot \|1/T\| ) \langle T_o, T_F \rangle =$$

$$= \left[ \int_{T_o}^{T_F}(1.1/T)dT \right] \Big/ \sqrt{\int_{T_o}^{T_F}(1)^2 dT} \sqrt{\int_{T_o}^{T_F}(1/T)^2 dT}$$

# REGULARITIES IN THE KINETICS OF THERMAL DISSOCIATION OF SOLIDS

Janusz. J. Pysiak

Institute of Chemistry, Technical University Warsaw, Branch at Plock, Poland

## ABSTRACT

Theoretical considerations involving the theory of analytical geometry made it possible to represent the known Arrhenius equation as a projection correlation. It has been shown that the known regularities in the kinetics of thermal dissociation are simple consequencies of fundamental concepts of projection geometry.

## INTRODUCTION

The starting point in the derived equations is a line in plane $R^2$, that in any system of uniform projection coordinates may be represented by the equation:

$$a_1 x_1 + a_2 x_2 + a_3 x_3 = 0 \qquad (1)$$

The interpretation of the relation 1 referred to as the equation of a projection line, providing that $|a_1| + |a_2| + |a_3| \neq 0$, may be as follows:

1° - the equation is a necessary and sufficient condition for a point of coordinates $\{x_1, x_2, x_3\}$ to lie on a straight line of coordinates $\{a_1, a_2, a_3\}$;

2° - the equation represents a pencil of straight lines passthrough the point $\{a_1, a_2, a_3\}$ of coordinates $\{x_1, x_2, x_3\}$ and for this reason it may also be regarded as an equation of a point.

Hence the equation (1) may be regarded as an equation of a straight line or as an equation of a pencil of lines of the same coordinates 1/ depending on whether the coordinates $\{a_1, a_2, a_3\}$ are considered as the coordinates of a line or of a point.

The very shortly imaged results of considerations on projection geometry may be utilized for explanation of relationships observed in thermal dissociation of solids by using the rearranged form of the Arrhenius' equation:

$$\ln k = \ln A - E/RT \qquad /2/$$

where k - rate constant, A - pre-exponential coefficient, E - activation energy, R - gas constant, T - temperature.

## THE ARRHENIUS' EQUATION AS A PROJECTION CORRELATION

Multiplication of the equation 2 by the gas constant R gives, after rearrangement:

$$R \ln k + E \, 1/T - R \ln A = 0 \qquad /3/$$

It can easily be noticed that the rearranged Arrhenius equation becomes equivalent to the projection equation of a straight line (1). The fact enables the construction of a correlation table.

The Table 1 is constructed as follows: from each member of equation (3) one factor is taken to obtain the values $a_1$, $a_2$, $a_3$, the remaining being treated as $x_1$, $x_2$, $x_3$. One may obtain in this way 8 relationships. As these are dual in pairs, it is sufficient to take into consideration only 4 of them [2].

Table 1. Correlation relationships

|   | $a_1$ | $a_2$ | $a_3$ | $x_1$ | $x_2$ | $x_3$ |
|---|---|---|---|---|---|---|
| 1 | R | E | ln A | ln k | 1/T | R |
| 2 | ln k | E | ln A | R | 1/T | R |
| 3 | R | 1/T | ln A | ln k | E | R |
| 4 | R | E | R | ln k | 1/T | ln A |

Most of the presented relationships have not been known or investigated, as yet. Further consideration will therefore be limited to known and studied correlations.

## THE ZAWADZKI - BRETSZNAJDER RELATIONSHIP AND THE ISOKINETIC TEMPERATURE

In the study of thermal dissociation of calcium carbonate under various pressures of the gaseous reaction product Zawadzki and Bretsznajder [3] were first in 1935 to observe

that the activation energy of the process was a function of the gaseous reaction product:
$$E = f\left(p_{CO_2}\right) \qquad /4/$$
and increased with increasing $p_{CO_2}$. Theoretical interpretation of that relationship was given in 1961 Pavluchenko and Prodan [4].

Similar relationship were further observed by other workers for carbonates, hydroxides, oxides and peroxides, oxalates, ammoniates, and crystal hydrates. Relationships of this kind were recently found by A. Glinka [5] during dehydration of basic aluminium - potassium sulfate.

It has been shown [6], that the work by Zawadzki and Bretsznajder, the results given by E. Prodan and co-workers [7], and the results obtained recently by A. Glinka [5] represent pencil relationships of the type $k = g(T)$, where various lines may be obtained for different pressures of the gaseous reaction product /Figs. 1 and 2/.

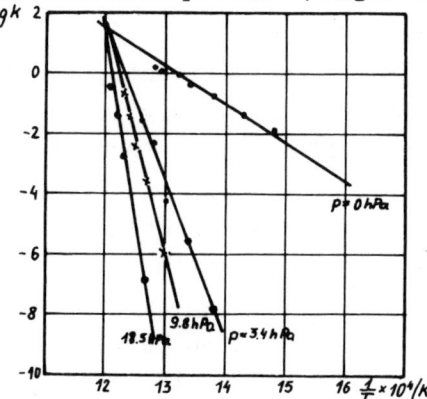 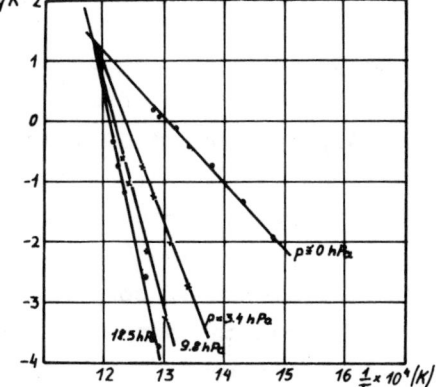

Fig.1. Dependence of log k vs. 1/T according to [5]

Fig.2. Dependence of log k vs. 1/T according to [5]

From first line in Table 1 it results that if the values $\{\ln k_o, 1/T_o, R\}$ are regarded as projection coordinates of a point, then the equation (3) becomes an equation of a pencil with one point common $(\ln k_o/R, 1/RT_o)$. It means, that the paires of magnitudes $(A,E)$ and $(k,T)$ are intercorrelated in appropriate scales and the equation (3) is valid for a certain pair of magnitudes $(k_o, T_o)$ - often referred to as an isokinetic pair - and any values of $(A,E)$. Having known $(k_o, T_o)$ it is possible to determine the temperature - the isokinetic temperature - the existence of which is a condition for occurrence of the compensation effect. The knowledge of the temperature enables the determination of the coefficient b in the compensation equation [6-10].

## COMPENSATION EFFECT AND THE DEVIATION FACTOR OF A CORRELA - TION

The observed change in activation energy due to increase of pressure of a gaseous reaction product is in many cases so large that the increase of E cannot be compensated by increase of T, and values of A in the Arrhenius equation $(2)$ must also be changed. The relationship between A and E is usually given by an empirical equation, usually referred to as a compensation equation [4]

$$\ln A = a + bE \qquad /5/$$

where a and b are constants; usually a = 0, and b $\in (0,1)$ [7]. The compensation equation is of general significance, since the compensation effect is observed in many catalytic reactions, in studies on viscosity and diffusion, in investigations of electrical conductivity and emission of electrons in biological processes, and during thermal dissociation of many solids see [4] and references cited therein.

The existence of compensation relationships may be proved on theoretical basis. Making use of correlation between the pairs $(A,E)$ and $(k,T)$ it is possible to regard the tercet $\{\ln k, 1/T_o, R\}$ as generalized coordinates of a straight line represented in Fig. 3; points on the line correspond to definite values of pressure $(p)$ of the gaseous reaction product under given experimental conditions [5].

If one assumes, according to experimental data [8,9], that $(A,E)$ are in pencil relationship (Fig.4), then for certain $(A_o, E_o)$ the values $(k,T)$ become, evidently, lineary related.

From the slope of the straight lines shown in Fig. 4 it is possible to find the values of b $(b_{exp})$, which is a coefficient in the compensation equation (5) and to compare with the values of b ($b_{calc}$) calculated /10/ from the equation:

$$b = 1/R \cdot \beta \qquad /6/$$

where $\beta$ is the isokinetic temperature found on the basic of pencil relationship between (k,T).

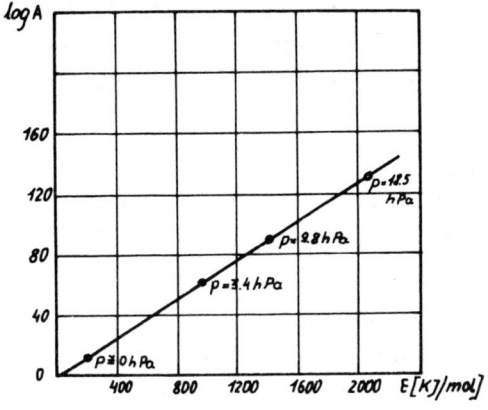
Fig.3. Compensation effect according to /5/

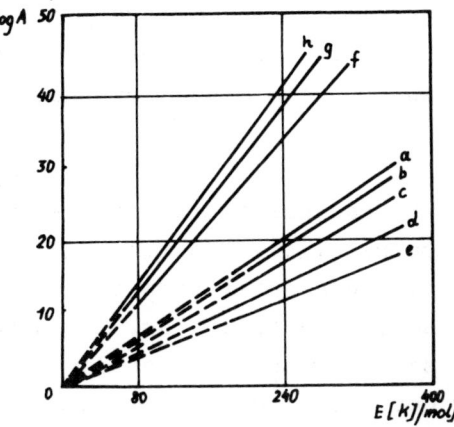
Fig.4. Compensation effect:
a-e-carbonates
t-h-tripoliphosfates

The above considerations suggest that in cases when pencil correlations exist between (k,T) and (A,E) there exist also deviation factors, which are decisive for the physical sense of those correlations /6/.

In one of such cases the deviation factor is pressure, as different slope angles of straight lines represented in Figs.1 and 2, as well as different points on straight line represented in Fig.3 correspond to different pressures. Besides, pressure as deviation factor is also decisive for the motive force of the process ($\Delta p = p_o - p$) and may regarded as a measure of a distance of a system from the state of equilibrium.

In another case the abstractively singled out deviation factor is an exceedingly complicated parameter, which is probably connected with the crystal structure of the starting material in particular with the nature and concentration of defects.

It seems therefore, that each correlation involves two dif-

ferent deviation factors - that give rise to the correlation although their physical sence is in many cases not clear. Despite of this, the use of such magitudes seems to be reasonable, as they are measurable quantities. Hence:
DEFINITION. If in any scales a pair of magitudes $(a_1, a_2)$ is in correlation with a pair $(x_1, x_2)$, then the slope of the lines in a pencil of one relation is referred to as a measure of the deviation factor of the given relationship.

## SUMMARY

The representation of the Arrhenius equation as a projection correlation is a first attempt to generalize some problems of fundamental significance in the kinetics of thermal dissociation of solids. The accepted approach: $1^o$ makes it possible to prove, that Zawadzki-Bretsznajder relationship, the compensation effect, and the isokinetic temperature $\beta$ are simple consequences of regarding the Arrhenius equation as a projection correlation; $2^o$ exhibits the existence of formerly unknown and therefore not investigated relationships and their mutual correlations; $3^o$ enables a theoretical isolation of eight deviation factors that impart these correlations a definite physical meaning.

## REFERENCES

1. K.Borsuk - Multidimensional analytica geometry, Warsaw,'66
2. B.Sabalski - Proceeding of the 2nd Session on Technology of Science, Plock, 1976. 3 - J.Zawadzki, S.Bretsznajder- - Z.Electrochem. 1935, 41, 215. 4 - M.M.Pavluchenko, E.A. Prodan - Dokl.AN SSSR 1961, 136, 651. 5 - A.Glinka - Doctoral Thesis, Tech.Univ.Warsaw, 1979. 6 - J.Pysiak, B.Sabalski - Proceedings of the 1st Seminar in the Honor of St. Bretsznajder, Plock, 1977; J.Thermal Analysis in print . 7 - - B.Sabalski, J.Pysiak - Proceeding of the 1st Seminar in the Honor of S.Bretsznajder, pp.25-29, Plock 1977; J.Thermal Analysis in print . 8 - J.Pysiak -A commentary to hybilitation thesis, Tech.Univ.Warsaw, 1975. 9 - J.Pysiak, M.Gawrońska -Proceed.2nd Polish Conf.on Calorimetry and T.A.Zakopane'76. 10-T.Żmijewski, J.Pysiak-Proceed.4th ICTA vol.I, Budapest'75.

# THE DETERMINATION OF THE STRAIGHT LINE CORRELATION FACTORS

Z.Sz. Kolenda
The St.Staszic University of Mining and Metallurgy,
Cracow, Poland

J.Norwisz, N.Hajduk
Institute of Non-ferrous Metals, Gliwice, Poland.

## ABSTRACT

When processing the experimental data it is essential to determine the factors of straight line equation $Y = AX+B$ which describes (in the best way) the given set of n measurement results. A certain defect of the used of least squares method is the necessity of choosing one from the quantity as a dependent variable with measurement error, while the other one is considered accurate. Because both variables are the results of measurements, thus there is no obstacle in principle to recognize the other variable as a dependent one and to determine the constants of the inverse equation

$$X = \frac{1}{A}Y - \frac{B}{A}$$

In the paper the possibilities of conversions of initial equation are presented and the effect of used conversions on the estimated kinetic parameters of copper oxidation in the conditions of linear temperature increase is evaluated.

## INTRODUCTION

The treatment of experimental results including measurements of the kinetics of chemical reactions comprises the two following stages:
- selection of a suitable reaction model (equation),
- determination of numerical values of the model parameters.
It is often the case that above two stages are combined

together and the essential features of the determined numerical values can be estimated by means of statistical methods. In so far as there exists general convictions as to the difficulty in finding an adequate model equation, there is no doubt, in principle, about the applied method of estimation of numerical values of coefficients of the model equation. The profound knowledge of linear correlation methods of two variables, based on the principle of least squares, offers the possibility to present a given equation in its linear form by means of known procedures (1). This may be accomplished, if the model equation is presented in the following form:

(1) $\qquad Y = AX + B$

where Y and X are known functions of the measured values $(\alpha_1, \alpha_2, \ldots, \alpha_N)$ and A and B are the known functions of the measured estimated values of the model equation. As the values X and Y are the results of measurements, there is no objection that instead of equation (1), the following reciprocal equation is considered:

(2) $\qquad X = \dfrac{1}{A} Y - \dfrac{B}{A}$

In this case, values A and B will be different from those determined by means of the method of least squares in equation (1). A more accurate analysis that equation (1) may be transformed on five different ways having the form:

(3) $\qquad y = ax + b$

where $y = y(Y,X)$, $x = x(Y,X)$, $a = a(A,B)$, $b = b(A,B)$. All transformation forms of equation (1) are shown in Table 1. Knowing the statistic values of equation (3), an estimation of the determined $T_\gamma$ values may be made using the standard deviation $s(T_\gamma)$.

(4) $\qquad s^2(T_\gamma) = s_o^2 \sum\limits_{i=1}^{n} \left[ \dfrac{\partial T_\gamma}{\partial A} \left( \dfrac{\partial A}{\partial a} \dfrac{\partial a}{\partial y_i} + \dfrac{\partial A}{\partial b} \dfrac{\partial b}{\partial y_i} \right) + \ldots \right.$

$$\ldots + \frac{\partial T_r}{\partial B}(\frac{\partial B}{\partial a}\frac{\partial a}{\partial y_i} + \frac{\partial B}{\partial b}\frac{\partial b}{\partial y_i})]^2$$

where n is the number of measuring data and $s_o^2$ is the average value of the variance of the dependent variable y. The form of the derivatives $\partial T_r/\partial A$, $\partial T_r/\partial B$ results from the equation defining the $T_r$ values, whereas that of derivatives $\partial A/\partial a$, $\partial A/\partial b$, $\partial B/\partial a$, $\partial B/\partial a$ is the result of the present character of the linear equation (Tab. 1). Derivatives $\partial a/\partial y$, $\partial b/\partial y$, on the other hand, result from the differentiation of the respective equations derived for their estimation by means of the method of least squares.

The effect of the change in the coordinate system of the linear equation on the result of estimation of value T is visualised below. This method has been used for estimating the activation energy and pre-exponential constant of the Arrhenius equation for the thermogravimetric measurements of the copper oxidation process under conditions of a linear temperature increase.

## DETERMINATION OF PARABOLIC EQUATION CONSTANTS UNDER THE CONDITION OF A LINEAR TEMPERATURE INCREASE

The differential form of the parabolic equation

(5) $$\frac{d \Delta m}{dt} = \frac{k_o}{2} \frac{\exp(-E/RT)}{\Delta m}$$

may be easily reduced to the linear form

(6) $$\ln(\frac{d \Delta m}{dt} \Delta m) = \ln \frac{k_o}{2} - \frac{E}{R}\frac{1}{T}$$

where: $\Delta m$ - change in weight, t - time, T - absolute temperature, $k_o$ - pre-exponential constant, E - activation energy, R - gas constant.

Comparing (6) with (1), one may obtain

(7) $$Y = \ln(\frac{d \Delta m}{dt} \Delta m), \quad X = 1/T$$
$$A = -E/R, \quad B = \ln k_o/2$$

As:

(8) $\quad E = -AR, \qquad k_o = 2\exp(B)$

hence

(9) $\quad \dfrac{\partial E}{\partial A} = -R, \qquad \dfrac{\partial E}{\partial B} = 0$

$\quad \dfrac{\partial k_o}{\partial A} = 0, \qquad \dfrac{\partial k_o}{\partial B} = 2\exp(B) = k_o$

Two integrating methods have been selected as a base of integration for equation (5). These are the Coats-Redfern's method (2) as well as the logarithmic are given by Mac Callum and Tanner (3).
The Coats-Redfern's method is based on equation

(10) $\quad 2\ln\left(\dfrac{\Delta m}{T}\right) = \ln\dfrac{k_o}{\beta}\dfrac{R}{E} - \dfrac{E}{R}\dfrac{1}{T}$

where $\beta$ is the heating rate equal to $dT/dt$.
The MacCallum-Tanner's method bases on the following equation:

(11) $\quad 2\ln \Delta m = \left(\ln\dfrac{k_o E}{R\beta} - 0.0297\,E^{0.435}\right) - (1034 + 0.1193\,E)\dfrac{1}{T}$

where E is expressed in J/mol.
Comparing equations (10),(11) with equation (1), the differences between the activation energy and the coefficient of the linear equation and of the respective derivatives may be easily determined.

## RESULTS OF CALCULATIONS

The equations presented above have served as a basis for numerical treatment of the measurement results in the oxidation process of a flat copper specimen with a surface of 5,06 cm$^2$ in the atmosphere of air at a heating rate of 9,7 K/min. In the calculations, the set of thirty eight experimental points have been used, resulting from the measurement of changes in weight.

Tab. 2 shows the calculated values of the activation energy and the pre-exponential constant. Standard deviations from these values and the correlation coefficient R, as well as the values of respective functional coefficients R1 and R2 are also presented in this table.

(12)
$$R1 = \sum_{i=1}^{n} (\Delta m_i - \Delta m_{ic})^2/n$$
$$R2 = \sum_{i=1}^{n} (\frac{d\Delta m_i}{dt} - \frac{d\Delta m_i}{dt})^2/n$$

where

$$\Delta m_{ic} = \sqrt{\frac{k_o}{\rho} \frac{E}{R} [L(z) - L(z_o)]}$$

Expression $\frac{d\Delta m_{ic}}{dt}$ is calculated from equation (5). $\Delta m_{ic}$, $\Delta m_i$ are the respective calculated and measured values of changes in weight. The value of integral $L(z) = \int_z^\infty \frac{\exp(-u)}{u^2} du$ for $z = E/RT$, $z_o = E/RT_o$, where $T_o$ is the initial temperature was calculated from Schlömilen's approximating equation (4).

Already a superficial analysis of data in Tab. 2 allows to conclude about the essential influence of the change of the coordinate system on the result of estimating the values of the activation energy and pre-exponential constant. If on the one hand changes in the activation energy are relatively small (~ 25 %), they are markedly larger with the pre-exponential constant, on the other. The respective values may differ by several orders one from another. Confining oneself merely to the results obtained by the differential method, it may be said that there is no exact criterion for the selection of one of the pairs of values E and $k_o$. The large coefficients of correlation suggest that all values are equally well. It could also be concluded that the values determined down to the last two forms of the linear equation are the best ones (5 and 6, tab.1). As the elaboration of experimental results is aimed at the determination of values describing a given set of results by the best possible method it is pertirent to compare the calculated values for the determined parameters E and $k_o$ with the experimental data. Factors R1 and R2 show that the last two pairs of values are very inaccurate. Thus the parameters of the statistic estimation are not always conclusive in considering the quality of obtained results. This is particularly emphosized in the case of results obtained by logarithmic methods. The achieved statistical values of the correlation coefficient or of the standard deviations are the resultant of an accidental arrangement of experimental data of the applied method and the assumed form of the linear expression.

Instead of equation (6), every other equation may be considered, which results from transformations presented in Tab. 1.

Employing different forms of the linear equation, the accidental and all-to-good sets of data, contributing to a faulty estimation of the values of the determined activation energy E and pre-exponential constant, may be brought into the foreground. Through the application of the method of least squares for the linear correlation of two variables, the sum of squares of corrections may be minimized to the value of the current dependent variable y

$$S = \sum_{i=1}^{n} (y_i - y_{ic})^2$$

and not the sum of squares of corrections to the measured N value

$$S = \sum_{u=1}^{N} \sum_{i=1}^{n} (\alpha_{ui} - \alpha_{uic})^2$$

Thus for equation (6) e.g. the sum

$$S = \sum_{i=1}^{n} (\ln \frac{d\Delta m_i}{dt} \Delta m_i - \ln \frac{d\Delta m_{ic}}{dt} \Delta m_{ic})^2$$

is minimized.

Even if values $\alpha_{ui}$ feature a standard distribution, value y is characterized by a distribution that is visibly the standard one. As in general, no study has been made so far of the type of distribution of the dependent variable y, an error may be committed resulting from the use of methods referring to the standard or Student's distribution when determining the accuracy of the estimation.

The ease of programming the automatic transformattion of the output linear equation makes it possible to use methods of quick inspection of all three variants in the determination of the most probable values, in order to reduce the possibility of committing errors in the estimation procedure to a minimum.

## DISCUSSION

Summarizing the problems presented above, it must be stated that the use of different forms of the transformed output linear equation may be considered as an additional measure against the occurence of faulty results. In some cases, however, the statistic estimation parameters, the high correlation coefficient and low standard deviation may suggest the best values of errors from the point of view of a quantitative description of given phenomena. It must also be mentioned, that the method of estimation of kinetic constants, basing on the results of thermogravimetric measurements under conditions of a linear temperature increase, is particularly vulnerable to the this type of error. This results from the form of the differential equation. Even in the case of the differential method, the effect of the exponential function is obvious. In the case of the integral method function $L(z)$ increases at a particularly high rate and although the linear function derived from different mathematical operations, is relatively stable, the values of unknown parameters related with one another by exponential functions, may be very sensitive to small changes in output values. Thus, the apparently high statistical estimating accuracy of the straigh-line equation does not necesserily carry with it a good estimating capacity of parameters.

## REFERENCES

(1) R.Deutsch, Estimation Theory, Prentice Hall, Inc., Englewood Cliffs, New York, 1965.
(2) A.W.Coats, J.P.Redfern, Nature, 201 (1964), 68.
(3) J.P.MacCallum, J.P.Tanner, Eur. Polym. J., 6 (1970) 1033.
(4) J.Norwisz, N.Hajduk, J.Therm.Anal., 13 (1978) 223.

Table 1. Transformation of the linear output equation
$Y = AX + B$

| Number of equation | $y = ax+b$ | | | | Partial derivatives | | | |
|---|---|---|---|---|---|---|---|---|
| | y | x | a | b | $\partial A/\partial a$ | $\partial A/\partial b$ | $\partial B/\partial a$ | $\partial B/\partial b$ |
| 1 | Y | X | A | B | 1 | 0 | 0 | 0 |
| 2 | X | Y | 1/A | -B/A | $-1/a^2$ | 0 | $b/a^2$ | -1/m |
| 3 | Y/X | 1/X | B | A | 0 | 1 | 1 | 0 |
| 4 | 1/X | Y/X | 1/B | -A/B | $b/a^2$ | -1/a | $-1/a^2$ | 0 |
| 5 | X/Y | 1/Y | -B/A | 1/A | 0 | $-1/b^2$ | -1/b | $a/b^2$ |
| 6 | 1/Y | X/Y | -A/B | 1/B | -1/b | $a/b^2$ | 0 | $-1/b^2$ |

Table 2. Activation energy and pre-exponential constants determined by means of different methods

| Method | Number of equations | Kinetic constants | | Standard deviations | | Correlation coefficient | Functions | |
|---|---|---|---|---|---|---|---|---|
| | | E kJ/mol | $k_o \cdot 10^{-3}$ mg$^2$/min cm$^2$ | S(E) kJ/mol | S($k_o$) mg$^2$/min cm$^2$ | $|R|$ | R1 mg$^2$ | $R2 \cdot 10^2$ mg$^2$/min |
| Differential | 1 | 86.6 | 44.7 | 1.81 | 11.5 | 0.9924 | 1.10 | 1.93 |
| | 2 | 88.0 | 53.6 | 1.83 | 14.0 | 0.9924 | 0.90 | 1.79 |
| | 3 | 87.3 | 48.5 | 2.15 | 13.5 | 0.9902 | 1.02 | 1.86 |
| | 4 | 89.0 | 61.1 | 2.19 | 17.3 | 0.9902 | 0.84 | 1.71 |
| | 5 | 74.0 | 59.4 | 3.93 | 3.0 | 0.9999 | 9.54 | 5.73 |
| | 6 | 74.0 | 59.6 | 3.93 | 3.0 | 0.9999 | 9.51 | 5.72 |
| Integral logarythmic | 1 | 88.7 | 89.4 | 1.15 | 15.6 | 0.9975 | 4.32 | 2.50 |
| | 2 | 89.2 | 96.1 | 1.16 | 16.9 | 0.9975 | 4.67 | 2.61 |
| | 3 | 88.2 | 83.7 | 1.31 | 15.2 | 0.9979 | 4.03 | 2.42 |
| | 4 | 88.7 | 89.6 | 1.31 | 16.3 | 0.9979 | 4.29 | 2.50 |
| | 5 | 88.5 | 108.3 | 1.60 | 30.1 | 0.9999 | 11.37 | 4.13 |
| | 6 | 88.5 | 108.2 | 1.60 | 30.1 | 0.9999 | 11.36 | 4.13 |
| Coats-Redfern's | 1 | 82.8 | 26.1 | 1.25 | 5.0 | 0.9960 | 1.91 | 2.49 |
| | 2 | 83.4 | 28.8 | 1.26 | 5.6 | 0.9960 | 1.68 | 2.36 |
| | 3 | 81.8 | 22.4 | 1.32 | 4.2 | 0.1693 | 2.28 | 2.71 |
| | 4 | 126.2 | 12 480.0 | 45.13 | 78 940.0 | 0.1693 | 49.24 | 34.28 |
| | 5 | 82.4 | 24.8 | 1.34 | 4.7 | 0.2586 | 2.02 | 2.56 |
| | 6 | 127.2 | 14 340.0 | 46.91 | 94 050.0 | 0.2586 | 5.28 | 36.92 |

# THEORETICAL THERMOGRAVIMETRIC ANALYSIS
# AT CONSTANT HEATING RATES

C. COMEL, J. VERON, C. BOUSTER and P. VERMANDE
Laboratoire de Chimie Appliquée
Département de Génie Energétique - Bât. 404
Institut National des Sciences Appliquées
69621 Villeurbanne Cedex   FRANCE

## ABSTRACT

Let the general reaction rate equation be a function $h(\alpha,T)$ of the conversion $\alpha$ and the temperature $T$. Then the evolution, with the heating rate $\phi$, of $\alpha$, $T$ and the experimental rate $\upsilon$ can be forecasted at the maximum of the rate under peculiar conditions. For instance, if $h(\alpha,T) = k(T)(1-\alpha)^n$, it can be shown that an increase of $\phi$ compels related increase of $T_m$ and decrase of $\upsilon_m$. Regarding $\alpha_m$, it increases first with $\phi$ up to a value $\alpha_{mo}$ and decreases thereafter.

## INTRODUCTION

The kinetic study of a chemical reaction by means of thermogravimetric methods is very often founded on the variation of the temperature $T_m$ and the conversion $\alpha_m$ with the heating rate $\phi$ at the maximum of the reaction rate (1), (2), (3), (4).

Till now, it seems to us that no theory accounts for the general features of the variation of those parameters with the heating rate $\phi$. At the very most, we have noticed here and there authors pointing out on the one hand that the conversion $\alpha_m$ is fairly constant and independent of the heating rate (5), not beeing so for the temperature $T_m$ and the reaction rate $\upsilon_m$, and on the other hand that the variations of $\alpha_m$ depend on experimental errors or diffusion phenomena.

In spite of all these disturbances, which are very difficult to get rid of, we assert that those variations are mainly governed by rules which can be mathematically derived from the fundamental reaction rate equation at constant heating rate.

From a rather general point of view, the reaction rate can ever be written whatever the experimental conditions are

$$\frac{d\alpha}{dt} = h(\alpha, T).$$

Hence it is reasonable to write at a constant heating rate $\phi$

$$\left(\frac{\partial \alpha}{\partial T}\right)_\phi = \frac{h(\alpha, T)}{\phi} \qquad (1)$$

and to define the "experimental reaction rate" as follows

$$\mathcal{V} = \left(\frac{\partial \alpha}{\partial T}\right)_\phi \qquad (2)$$

Studying equation (1) when the following common relation holds : $h(\alpha, T) = k(T) f(\alpha, T)$ (3), we are allowed to get some general conclusions and express in the peculiar case where $f(\alpha, T) = (1-\alpha)^n$ the variations of $\alpha m$, $Tm$ and $\mathcal{V}m$ functions of the heating rate $\phi$.

## MATHEMATICAL STUDY OF THE RATE EQUATION

To the partial derivative equation (1), the following differential system can be associated :

$$\frac{d\phi}{dt} = 0 \; ; \quad \frac{dT}{dt} = \phi \; ; \quad \frac{d\alpha}{dt} = k(T) f(\alpha, T)$$

where the parameter t is readily identified to the time. The general solution of equation (1) is any implicit function of two prime integrals $I_1$ and $I_2$ of the associated differential system.

First we obtain directly $I_1 = \phi$ from $d\phi/dt = 0$. Then the other two equations of the system yield the new one $\phi \frac{d\alpha}{f(\alpha,T)} = k(T) dT$ which can be integrated by using the fact that the temperature is a function $T(\alpha, \phi)$ and

defining the two functions
$$F(\alpha,\phi) = \int_{\alpha_0}^{\alpha} \frac{d\alpha}{f(\alpha,T)} \quad \text{and} \quad K(T) = \int_{T_0}^{T} k(T) \, dT$$

Therefore we obtain readily the expression
$$I_2 = \phi F(\alpha,\phi) - K(T).$$

Hence the general solution is written $I_2 = G(I_1)$.

The physical solution of equation (1) depends on the initial conditions which imply $I_2 = 0$.

Therefore we have $\phi F(\alpha,\phi) = K(T)$ \quad (4)

which is the equation obtained for the surface solution of (1) by putting $F(\alpha_0, \phi) = 0$ when $\alpha(T_0, \phi) = \alpha_0$ for every $\phi$.

At the maximum of the rate, if $\phi$ is constant, the following relation holds $\left(\frac{\partial v}{\partial T}\right)_\phi = \left(\frac{\partial^2 \alpha}{\partial T^2}\right)_\phi = 0$ which can be expressed taking the expression of $v$ into account by the relation

$$-\left[\frac{k(\partial f/\partial\alpha)_T}{d\ln k/dT + (\partial \ln f/\partial T)_\alpha}\right]_m = \phi \quad (5).$$

Thus we obtain the equation of a surface which could be told "the surface of the maxima". Eliminating $\phi$ between the equation of the surface solution of (1) and the equation of the surface of the maxima, we obtain the implicit equation relating $\alpha m$ and $Tm$ for every $\phi$:

$$\left[K + \frac{k F(\partial f/\partial\alpha)_T}{d\ln k/dT + (\partial \ln f/\partial T)_\alpha}\right]_m = 0 \quad (6)$$

Furthermore transferring the equation of the surface of the maxima in the gereral equation of the rate we obtain the expression of the maximum

$$v_m = -\left[\frac{d\ln k/dT + (\partial \ln f/\partial T)_\alpha}{(\partial \ln f/\partial\alpha)_T}\right]_m \quad (7).$$

## VARIATION LAWS OF $\alpha_m$, $T_m$ AND $\mathcal{V}_m$.

All the preceding relations can be simplified when f and hence F are only functions of $\alpha$. So $\alpha_m$ and $T_m$ are related by the equation

$$\left[ K + \frac{k\, F\, (df/d\alpha)}{d\ln k/dT} \right]_m = 0 \qquad (8).$$

The maximum of the rate can be written in a more simple way

$$\mathcal{V}_m = \left( \frac{k\, f}{\phi} \right)_m = - \left[ \frac{d\ln k/dT}{d\ln f/d\alpha} \right]_m \qquad (9)$$

when the relation $\phi(d\ln k/dT) + k\,(df/d\alpha) = 0$ holds.

Now, if $f(\alpha) = (1 - \alpha)^n$, it is very easy to compute successively $df/d\alpha = -n\,(1 - \alpha)^{n-1}$, $d\ln f/d\alpha = -n/(1-\alpha)$ and finally obtain by integration for $\alpha_o = 0$

$$F(\alpha) = \frac{1}{n-1} \left[ (1 - \alpha)^{1-n} - 1 \right] \quad \text{if } n \neq 1$$

$$F(\alpha) = -\ln(1 - \alpha) \qquad \text{if } n = 1.$$

Assume that $k(T) = k_o \exp(-E/RT)$ and put it in (8). Then we have the relations

for $n \neq 1$ 
$$\alpha_m = 1 - \left[ 1 - \frac{n-1}{n}\, \frac{E}{RT^2}\, \frac{K}{k} \right]_m^{\frac{1}{n-1}} \qquad (10)$$

for $n = 1$ 
$$\alpha_m = 1 - \exp\left( -\frac{E}{RT^2}\, \frac{K}{k} \right)_m \qquad (11).$$

More over, with (10) and (11) put in (9), we get

for $n \neq 1$ 
$$\mathcal{V}_m = \left[ \frac{E(1 - \frac{n-1}{n}\, \frac{E}{RT^2}\, \frac{K}{k})^{\frac{1}{n-1}}}{n\, R\, T^2} \right]_m \qquad (12)$$

for $n = 1$ 
$$\mathcal{V}_m = \left[ \frac{E \exp(-\frac{E}{RT^2}\, \frac{K}{k})}{RT^2} \right]_m \qquad (13)$$

and yet for every $n > 0$

$$\frac{E\, \phi}{RT_m^2} = \left[ nk\left(1 - \frac{n-1}{n}\, \frac{E}{RT^2}\, \frac{K}{k}\right) \right]_m \qquad (14)$$

This last equation gives us by differenciation for every n > 0

$$\frac{dTm}{d\phi} = \frac{E}{k(E+2nRT_m)} > 0 \qquad (15).$$

Thus, we conclude that $Tm$ is an ever increasing function of $\phi$.

Hence forth, $\alpha m$ and $\vartheta_m$ vary similarly as well in function of $\phi$ as in function of $Tm$.

Furthermore, it can be shown that the shape of the curves which represent $\alpha_m(Tm)$ and $\vartheta_m(Tm)$ is similar for any value of the ratio $E/RT^2$ in equations (10), (11), (12), (13). To fix ideas, we have computed and solved the equations (10) and (12) taking $n = 1.2$ and $E = 40$ kcal mol$^{-1}$ for different values of $T_0$. The corresponding curves are represented at the figures 1 and 2.

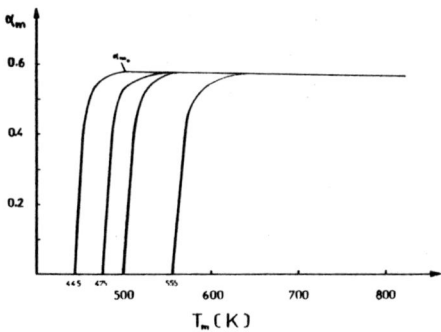 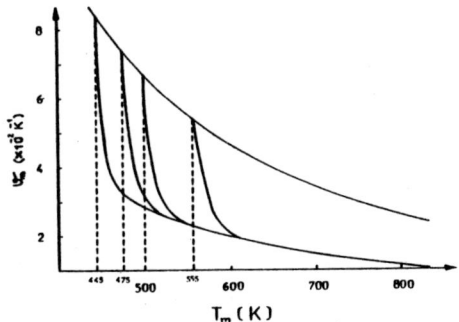

Fig. 1 : $\alpha_m$ versus $T_m$ for different values of $T_0$

Fig. 2 : $\vartheta_m$ versus $T_m$ for different values of $T_0$

It can be seen straight off that the conversion $\alpha m$ increases first up to a value $\alpha m_0$ and decreases afterwards rather slowly provided that the increasing values of $Tm$ are sufficiently different from $T_0$. That is, for sufficiently high heating rates. In the other hand, the maximum rate $\vartheta_m$

always decreases as Tm or φ therefore increases.

## CONCLUSION

We have established the variations laws of the characteristic kinetic parameters $\alpha_m$, $T_m$, $\vartheta_m$ corresponding to the maximum of the rate for a reaction carried out at different constant heating rates. The conversion $\alpha_m$ increases first up to a value $\alpha_{m_0}$ function of E, n and $T_o$ and decreases afterwards slowly for increasing heating rates. Simultaneously $T_m$ increases and $\vartheta_m$ decreases.

All these theoretical results will be corroborated by an other method in a next paper (6). To date, they have been checked to a fairly great extent in polystyrene pyrolysis kinetics. They will be applied soon to other polymer kinetics and then all the experimental results will be published in the just upper mentioned paper.

## REFERENCES

(1) D.W. Van Krevelen, C. Van Heerden and F.J. Huntjens, Fuel 30 (1951) 256

(2) H.E. Kissinger, Analyt. Chem. 29 (1957)) 1702

(3) H.H. Horowitz and G. Metzger, Analyt. Chem. 35 (1963) 1464

(4) R.M. Fuoss, I.O. Salyer and H.S. Wilson, J. Polym. Sci. 2A (1964) 3147

(5) T. Ozawa, J.Thermal Anal. 2 (1970) 301

(6) J. Veron et al. (in preparation).

# A COMPARISON OF DIFFERENT METHODS FOR THE ESTIMATION OF KINETIC DATA

Jürgen Moll, Detlef Krug* and Dorothee Zepf

Institut für Anorganische Chemie der Eberhard-Karls-Universität Tübingen,
Auf der Morgenstelle 18, D-7400 Tübingen, FRG.

## ABSTRACT

As could be shown in previous publications, the course of less complicated heterogeneous solid reactions can be described by the classical Arrhenius equation, modified for Thermal Analysis by introducing the linearly increasing temperature.

If n = 1, the interesting kinetic data, preexponential factor A and energy of activation E, can be found very easily from the position of the rate maximum.

Also if n ≠ 1, there is an easily practicable approximation method with the help of a plotter.

Looking for other easier procedures we tested the possibility to estimate E from the ratio of the rates $v_1$ and $v_2$ at the temperatures $T_1$ and $T_2$, respectively. This method is limited by the inaccuracy in evaluating the mentioned data. As the accuracy increased in the meantime /4/, this method also produces very satisfactory results. The very less expenditure is to be emphasized.

The different, purposeful simple, methods are proved by practical measurements.

The evident discrepancy of the results of one of these methods indicates the existence of various mechanisms in the different sections of the course of reaction.

*Correspondence author

# INTRODUCTION

Applicability of the Arrhenius equation

The most simple way to describe the course of solid reactions is to use the Arrhenius equation. This formula reveals the correlation between the rate v, the preexponential factor A, the energy of activation E, order parameter n and the remaining fraction of substance P as

$$v = \frac{d[P]}{dt} = A \cdot e^{-E/RT} \cdot [P]^n \quad . \tag{1}$$

This way of applying the Arrhenius equation for heterogeneous reactions causes one to make a series of idealized preconditiones:
- there is only one mechanism during the whole course of the reaction,
- the energy of activation and the preexponential factor remain constant,
- the transfer of energy is not affected during the course of the reaction and
- the transport of material of all compounds does not change while the reaction proceeds.

It is to be emphasized, that these prerequisites shall also be valid, if, as a rule in Thermal Analysis, the temperature is rising during the course of the reaction. In fact this may cause difficulties.

However, the Arrhenius equation, modified for linearly increasing temperature $T^*$

$$v = \frac{d[P]}{dT} = \frac{A}{m} \cdot e^{-E/RT^*} \cdot [P]^n \tag{2}$$

(linear rate of heating: m = dT/dt)
is such an effortless mathematical modell, that tentative use of the equation is justified in all cases. Problems, that are brought about by the contrary courses of the parameters E, A and n, can be solved even if using only small modells of computers.

Utilization of the Arrhenius equation

It is a precondition for the afore mentioned simple application of the Arrhenius equation, that easy and reliable methods for its interpretation should be available. The following three methods for the evaluation of kinetic data have been suggested in previous publications:
1) By analyzing the reaction-index $R_i$, which is specified by the position of the maximum and the corresponding rate /2/.
2) If n = 1 by analysis of the position of the temperature of the rate maximum, the single relevant parameter in this case.
3) In case of n ≠ 1 according the so-called "Gitty-Method" /3/, /5/.


The method of the ratio of rates

In this paper we submit an additional possibility to utilize the Arrhenius equation. This method is based on the old chemists rule of thumb, that the rate of reaction is doubled, if the reaction temperature is increased by 10 K. It has been pointed out earlier, that with a specific energy of activation this rule has only one point of temperature of exact validity /1/. The principle of an evaluation method to be used at any optional temperature is as follows:
At the temperature $T_1$ the rate $v_1$ is

$$v_1 = \frac{d[P]}{dT} = \frac{A}{m} \cdot e^{-E/RT_1} \cdot [P_1]^n \qquad (3)$$

and at the higher temperature $T_2$ the rate $v_2$ is

$$v_2 = \frac{d[P]}{dT} = \frac{A}{m} \cdot e^{-E/RT_2} \cdot [P_2]^n \qquad (4)$$

The difference between $P_1$ and $P_2$ is very small, if there is no vast difference between $T_1$ and $T_2$. And the difference will diminish even further in $[P]^n$, if n < 1.

Thus, with $v_2/v_1$

$$\log v_2/v_1 = \frac{E}{2.303\ R} (1/T_1 - 1/T_2) \qquad (5)$$

E is equal to

$$E = \frac{2.303\ R\ \log v_2/v_1}{1/T_1 - 1/T_2} \qquad (6)$$

Knowing nothing about A and n we can thus calculate the energy of activation E solely from the ratio of the rates $v_2/v_1$ and the corresponding temperatures $T_1$ and $T_2$.
Tests with calculated rates, formula (2), demonstrate, that the accuracy is better than 1 %, even if the temperature difference is about 50 K. In practical operation the accuracy is limited by the inaccuracy of the determination of the rates and the temperatures. The precondition $P_1 = P_2$ is less well met at higher temperatures, i.e. higher rates. In addition, while the temperature decreases the ratio $v_2/v_1$ increases, because the term $(1/T_1 - 1/T_2)$ increases for a constant difference of $T_1$ and $T_2$. This course is plotted in figure 1 at E = 70 kJ. For both reasons formula (6) should be applied at low temperatures. On the other hand the relative inaccuracy in determining the rates -very small absolute values at lower temperatures- increases as a result of the limited precision of measurement. Therefore the optimal range of application will be near the point of inflection of the curve.

Experimental Procedure

For testing purposes the decomposition of CO and $CO_2$ of the oxalate of calcium and strontium has been used. The EGA experiments were carried out in a helium atmosphere using an apparatus of the firm Netzsch/Selb. The sample mass was between 10 and 40 mg; the grain size between 80 and 90 µm. The heating rate was always 5 K/min. Method 1) was evaluated according to /2/ with the master plot. For methods 2) to 4) the data T and EGA were also digitally recorded on a tape and subsequently plotted according to /4/ as EGA = f(T) on

a larger scale.
The results are listed in the following chapter.

## RESULTS AND DISCUSSION

The results for the energy of activation E (in kJ) as obtained by the different methods, are listed in the following table:

| step | n = | method 1 | method 2 | method 3 | method 4 |
|---|---|---|---|---|---|
| $CaC_2O_4 \rightarrow CaCO_3$ | 1.0 | 261.7 /2/ | 120.6 /5/ | 122.3 /5/ | 123.7* |
| $CaCO_3 \rightarrow CaO$ | 0.5 | 238.2 /2/ | ** | 107.8 /5/ | 105.4* |
| $SrC_2O_4 \rightarrow SrCO_3$ | 1.0 | 257.1* | 106.9 /5/ | 106.0 /5/ | - |
| $SrCO_3 \rightarrow SrO$ | 0.4 | 335.1* | ** | 191.4 /5/ | - |

\* results of this work
\*\* method not applicable, because $n \neq 1$.

It is to be pointed out, that there is a good correspondence between the methods 2) - 4) and an obvious discrepancy of the results of method 1).
The methods 2) - 4) differ from method 1) by using data only from a small range of the ascending part of the curve. This is especially marked in method 4). In method 1) the whole course of the reaction is included in the calculation as the whole plane underneath the peak is evaluated. These results are in agreement with those, obtained from mechanism-based equations /6/. The differing results suggest the presence of different mechanism of reaction. This kind of dependence on the curve segments has also been discovered by other authors /7/.

The application of the methods 1) to 4) on a theoretical curve, calculated by using equation (2) produce completely the same results for all four methods.
Now, knowing E and v by using method 4), the preexponential factor A can be easily calculated. The evaluation method 4), ratio of rates method, which requires very little effort, seems to be exceptionally suitable, to yield the wanted kinetic data of small segments of reaction.
The methods 2) and 3), which require the use of a plotter, make it possible to controll all recieved kinetic data. Their use is made more easy, as method 4) yields a first approximation value with little effort.

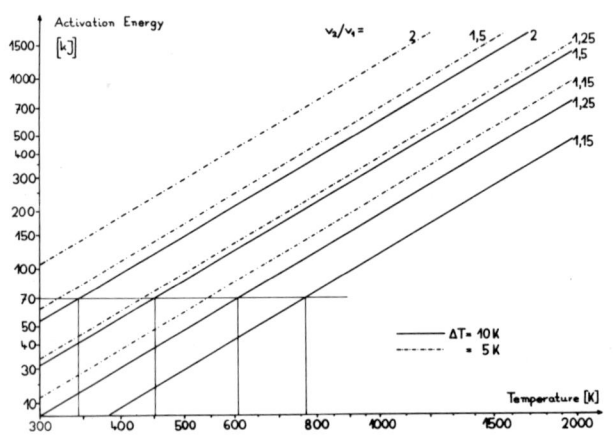

Fig.: 1

E vs. T for different ratios of rates

REFERENCES

/1/ D. Krug, Habilitationsschrift, Tübingen 1971.
/2/ D. Krug, Thermochimica Acta 10 (1974), 217.
/3/ D. Krug, ibid. 20 (1977), 53.
/4/ J. Moll, Diplomarbeit, Tübingen 1978.
/5/ D. Zepf, Zulassungsarbeit, Tübingen 1979.
/6/ S. Boy a. K. Böhme, Thermochimica Acta 28 (1979), 249.
/7/ K.N. Ninan a. C. G. R. Nair, ibid. 30 (1979), 25.

# EXPERIMENTAL REQUIREMENTS FOR THE DETERMINATION OF KINETICS AND MECHANISM FOR DECOMPOSITION REACTIONS OF SOLIDS

A. Reller and H.R. Oswald
Institute for Inorganic Chemistry
University of Zurich
Winterthurerstrasse 190
CH-8057 Zurich

## ABSTRACT

Kinetics and mechanism of thermal decomposition reactions of solids cannot be interpreted by sole evaluation of thermo-analytical measurements. It is rather necessary to correlate structure, morphology and macroscopic reaction mechanism with quantities obtained from time dependent thermoanalytical experiments.
In this paper we present three decomposition processes of model compounds selected from Ni(II)-complexes with ammine ligands. The results prove that structural and morphological properties as well as experimental conditions strongly determine the course of thermal decompositions.

## STATEMENTS

- In order to obtain reproducible and consistent results, all experiments have been carried out using crystalline initial substances with known structure.
- Extensive morphological studies by electron and light microscopy have been taken in account to get insight into the structural reaction mechanism and topotaxy [1].
- To avoid influences of experimental inadequacies on quantitative thermogravimetric analyses - e.g. hindrance of the emission of volatile products on faces contacting the sample holder or neighbouring crystals [2], as well as effects of thermal conductivity - all experiments have been

performed isothermally by using isolated single crystals of a few tenth of a mm in size and about 0.2 - 0.5 mg in weight. For each measurement, one individual was selected and, in order to minimize the contact of the sample holder with the reacting material, set on a net of fine platinum wires.
- The qualitative composition of volatiles evolved during the reactions has been identified by simultaneous mass spectrometry.

### EXPERIMENTAL PART

From a number of investigated systems, three examples of decomposition reactions of initial crystals belonging to principally different structural types are presented: a phase with one kind of infinite chains (I); one with isolated molecular units (II); and another with a layered structure (III).

I. The thermal decomposition of [Ni(SCN)$_2$(NH$_3$)$_2$]

The complex compound [Ni(SCN)$_2$(NH$_3$)$_2$] crystallizes in dark blue monoclinic plates (001). The unit cell parameters have been determined as follows: a = 8.6986 Å b = 7.9911 Å c = 5.6917 Å β = 105.914° space group C 2/m [3]. Structural details are given in Fig. 1a, b.

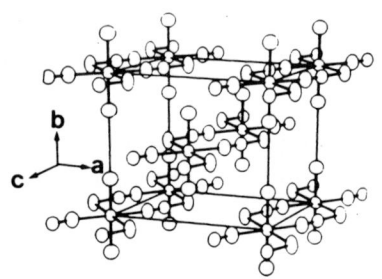
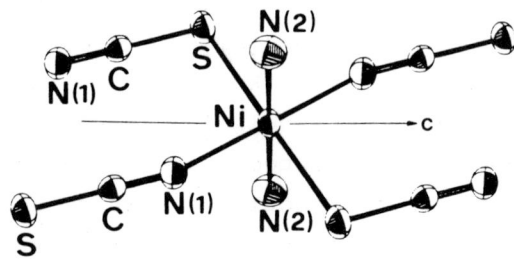

Fig. 1a. Unit cell of [Ni(SCN)$_2$(NH$_3$)$_2$], chains [001]

Fig. 1b. Coordination of Ni(II) in [Ni(SCN)$_2$(NH$_3$)$_2$]; N(2) is part of the NH$_3$ ligand

The compound undergoes the following single step decomposition reaction:

$$[Ni(SCN)_2(NH_3)_2]_{(s)} \longrightarrow [Ni(SCN)_2]_{(s)} + 2\ NH_3{(g)}$$

Electron and light microscopy prove that the reaction begins on the crystallographic faces (001) resp. (00$\bar{1}$). The nickel-thiocyanate product layers grow along the crystallographic c - axis (which corresponds to the direction of the infinite molecular chains) into the crystal. At a temperature of 190°C the velocity of the advancing phase boundary is linear in the range of $0.15 < \alpha < 0.80$. Accordingly the time limiting step is not controlled by the diffusion of ammonia molecules through the product layer but by the break of Ni - $N_{(NH_3)}$ bonds. The number of Ni - $N_{(NH_3)}$ bonds broken per time unit can be determined by evaluating the velocity of the phase boundary moving along the c - axis into the crystal and is therefore the realistic rate constant for the process. The only evaluation of thermogravimetric measurements by means of the methods published [4,5] would obviously yield rate constants related to an over all reaction, but not to the time limiting reaction step.

## II. The thermal decomposition of $[Ni(SCN)_2(NH_3)_4]$

$[Ni(SCN)_2(NH_3)_4]$ crystallizes in blue monoclinic prisms. The molecular units and their spatial arrangement are presented in Fig. 2a, b [6].

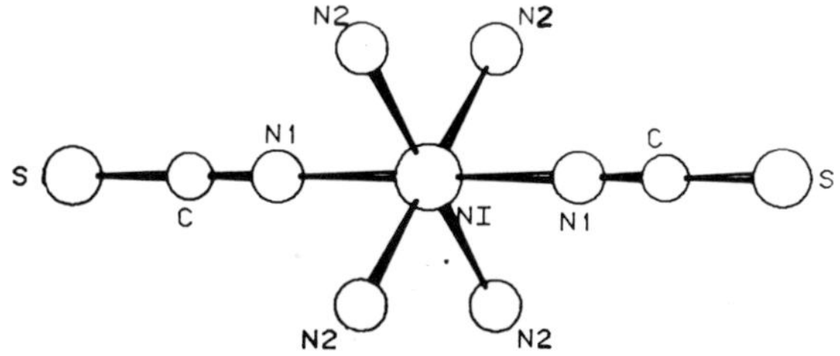

Fig. 2a. Coordination of Ni(II) in the molecular unit of $[Ni(SCN)_2(NH_3)_4]$; N(2) is part of the $NH_3$ ligand

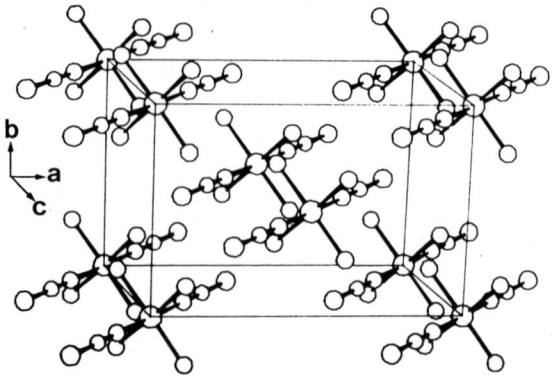

Fig. 2b. Unit cell of $[Ni(SCN)_2(NH_3)_4]$

In the temperature range of 110°C the following decomposition reaction takes place:

$$[Ni(SCN)_2(NH_3)_4]_{(s)} \longrightarrow [Ni(SCN)_2(NH_3)_2]_{(s)} + 2\, NH_{3(g)}$$

The reaction starts on the entire surface of the crystal and the phase boundary advances subsequently towards the centre. As mentioned above, thermogravimetry only registrates the over all weight loss of the sample, and the time dependent course of the measurement is therefore basically influenced by the shape of the starting material: needles, plates or cubes will produce different time/weight curves and hence any evaluation of kinetic parameters and reaction mechanisms must lead to fictive results, caused by morphological properties of the initial compound.

Pursuing the time dependent movement of the phase boundary along all of the crystallographic axes will yield realistic rate constants.

III. The thermal decomposition of [Ni{Pt(CN)$_4$}(NH$_3$)$_2$]

[Ni{Pt(CN)$_4$}(NH$_3$)$_2$] is tetragonal and cristallizes in truncated quadratic bipyramids (001). The unit cell parameters have been determined as follows: a = 7.24 Å c = 17.45 Å space group I 4$_1$ / amd (without proton positions) [7]. This corresponds to a layered structure (Fig. 3).

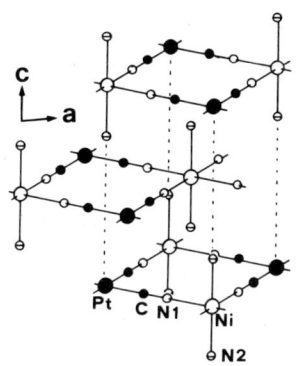

Fig. 3. Structural principle of [Ni{Pt(CN)$_4$}(NH$_3$)$_2$]; N(2) is part of the NH$_3$ ligand

Thermogravimetry and simultaneous mass spectrometry reveal in the temperature range of 260°C the single step decomposition reaction

[Ni{Pt(CN)$_4$}(NH$_3$)$_2$]$_{(s)}$ ⟶ [Ni{Pt(CN)$_4$}]$_{(s)}$ + 2 NH$_3$$_{(g)}$

During the reaction, the product phase grows parallel to the (001) planes into the crystal. Comparative X-ray and microscopical investigations reveal a high degree of topotaxy: the ammonia molecules diffuse parallel to the conserved planes (oo1). By measuring the time dependence of the movement of the phase boundary, realistic kinetic parameters as well as informations concerning the time limiting step - diffusion of ammonia molecules through the [Ni{Pt(CN)$_4$}] product layer, or the break of Ni - N$_{(NH_3)}$ bonds - will be obtainable.

## CONCLUSION

The interpretation of the course of thermal decomposition reactions requires the knowledge of the topochemical mechanism, influenced by the structural and morphological properties of both the initial compound and product. Together with quantitative thermoanalytical measurements, accomplished under careful control of the experimental conditions, the kinetics of the processes occurring - e.g. phase boundary reactions, growth of nuclei or diffusion of volatiles through product layers - can be evaluated.

Investigations on isolated single crystals of model compounds allow to work out the principles of solid state decomposition processes. This will contribute to achieve meaningful quantitative interpretations of the thermal behaviour of chemically reacting powders as well.

## LITERATURE

[1] J.R. Günter and H.R. Oswald, Bull. Inst. Chem. Res. Kyoto Univ. $\underline{53}$ (1975) 249.
[2] A. Reller, H.R. Beer and H.R. Oswald, in Experientia Supplementum 37; E. Marti, H.R. Oswald and H.G. Wiedemann eds.; Birkhäuser, Basel, 1979, p. 61.
[3] E. Dubler and A. Reller, to be published.
[4] J. Šesták and G. Berggren, Thermochim. Acta $\underline{3}$ (1971) 1.
[5] V. Šatava, Thermochim. Acta $\underline{2}$ (1971) 423.
[6] E.K. Yukhno and M.A. Poraj-Košic, Kristallografija, SSSR, $\underline{2}$ (1957) 239.
[7] A. Reller, Ph.D. thesis, Univ. of Zurich, in progress.

# SOME REMARKS ON THE APPLICABILITY OF THE ARRHENIUS MODEL IN THERMAL ANALYSIS

M. Arnold, G.E. Veress, J. Paulik and F. Paulik
Institute for General and Analytical Chemistry, Technical University Budapest, Hungary

## ABSTRACT

On the basis of mathematical considerations authors found that the parameters of the Arrhenius-model cannot be calculated from the course of thermoanalytical curves by curve-fitting because the parameters cannot well be estimated and between the estimated parameters and the measured curves there is no one-to-one correspondence.

## INTRODUCTION

In thermal analysis it is a general view that the parameters of the Arrhenius equation

$$\frac{d\alpha}{dt} = A \cdot \exp(-E/RT) \cdot (1-\alpha)^n$$

valid for homogeneous reactions under isothermal conditions, can reliably be calculated from the course of thermoanalytical curves by curve-fitting in the case of heterogeneous systems and under non-isothermal conditions too.
It is presumed that
1./ the parameters of the model have a physico-chemical meaning,
2./ the parameters of the model can be well estimated from the curve measured,
3./ between the measured curves and the estimated parameters there is a one-to-one correspondence.

## ON THE SUPPOSITIONS

However, it can be proved that neither of the listed assumptions are fulfilled.
ad 1./ It is known that thermal reactions are composed of

many chemical and physical processes /chemical reactions, heat- and gastransport, motion of phase boundaries, nucleus formation and nucleus growth, etc./. These elementary processes significantly influence the shape of non-isothermal thermoanalytical curves (1) . However, the Arrhenius model does not take into account these circumstances. On the basis of these considerations we doubted already earlier (1,2) that to kinetic parameters any physico-chemical meaning could be attributed.

ad 2./ On mathematical basis it can also be proved (3) that the second supposition is not fulfilled either, since the independent estimation of the parameters from the measured curve is very poor.

It is well known that the measure of the goodness of the estimation is the so-called "condition number". In the case of estimating the Arrhenius parameters the calculated value of this number is $10^5$ order of magnitude, what means that the measured data should be known with an error not greater than 0,0001% in order to ensure that the uncertainty of the parameters should not be greater than 10 %.

ad 3./ It can be proved that there exists a strong correlation between the parameters of the model. From this follows however, that between the parameters and the measured curve the connection is not unequivocal. With other words, while a parameter triplet defines a single curve, a single curve may correspond to many dissimilar parameter triplets. Accordingly, the third supposition is not fulfilled either. We wish to demonstrate this in Fig.1 where the confidence domain

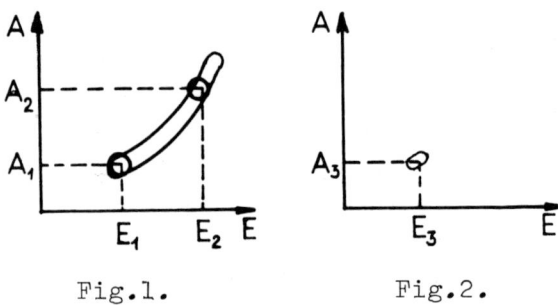

Fig.1.        Fig.2.

of the related parameter pairs /A,E/, belonging to probability level 0,97 is shown; n is taken as known and constant. From Fig. 1 it should follow that any pairs of parameters lying in the mentioned confidence domain is "good" since each corresponding curve differs from the measured curve by less than 3 %. However, it is self-evident that an unequivocal relation between the parameters and the measured curve can only exist if the ideal confidence domain is similar to the one which can be seen in Fig. 2.

## ON THE CALCULATION METHODS

Accordingly, the question may arise how is it possible that with the help of various calculation methods well reproducible parameter triplets can be obtained after all.
a./ The principle of one of the mathematical methods for the estimation of the parameters of the Arrhenius model is schematically illustrated in Fig. 3. According to this procedure first the "n" value is defined by estimation and furtheron taken as constant. Thereafter, again by estimation, a probable numerical value $/E_4/$ is stated, upon which

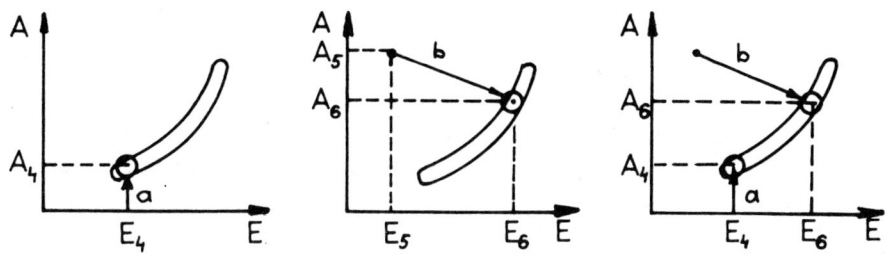

Fig.3. Fig.4. Fig.5.

with the help of the "trial and error" method the pre-exponential factor $/A_4/$ is determined at which the calculated and measured values fit best together. In Fig. 3 the direction and path of estimation is marked with line "a" while the confidence domain by a circle.
b./ The principle of the second, very often applied calcul-

ation procedure is illustrated in Fig. 4. In this case the
n value is also supposed to be known and constant - at
least in first approximation - . For the pair of parameters
E and A arbitrary guessed values /$A_5$, $E_5$/ are given. Thereafter both of them are systematically changed according to
the given equation until the calculated curve well approaches the measured one. Then this parameter pair /$E_6$, $A_6$/
is accepted as the suitable one. In Fig. 4 the path /b/ and
direction resp. of the approach is defined by the applied
equation. Since other calculation methods apply other
equations, the direction of the approach changes too.
By comparing the parameter values /$E_4$, $A_4$ and $E_6$, $A_6$/
obtained by the two different kind of estimation methods
/Fig. 5/ we generally find a significant difference between them. Should there be no difference between these
values after all, this could be attributed solely to chance.
Accordingly, it is evident that the calculated parameter
values are characteristic for the applied mathematical
procedure and not for the thermal reaction itself.
c./ The principle of the third type kinetic calculation
is illustrated in Fig. 6 schematically. According to this,

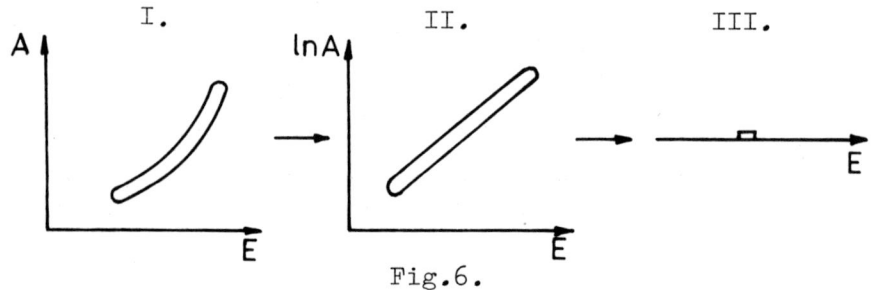

Fig.6.

linearization is performed with logarithmic transformation
/I-II/ and from the straight obtained in this way the kinetic parameters are estimated by linear regression /II-III/.
However, the independence of parameter estimation would
become lost also here in the course of the mathematical
operations, since the number of parameters to be calculated

is reduced. In the case of Fig. 6 e.g. the "A" is omitted and therefore diagram III became a one-dimensional one.
d./ Similarly to the former types of calculation, with the same consequence one parameter is transitionally omitted also in the case of calculation methods operating with difference-differential, or second derivative curves or with various heating rates resp.

Many researchers suppose that the above discussed phenomenon is actually a compensation effect. However, we do not know whether nature produces such a compensation effect. But what we actually observe is certainly a mathematical compensation effect which is a trivial consequence of an ill-conditioned parameter estimation and as such renders in an indirect way the applicability of the model questionable.

### REFERENCES
(1) F.Paulik, J.Paulik, J. Thermal Anal. 5 /1973/ 253
(2) F.Paulik, J.Paulik, Anal.Chim.Acta 56 /1971/ 328
(3) M.Arnold, G.E.Veress, J.Paulik, F.Paulik, J.Thermal Anal. 17 /1979/ 479

# KINETIC CHARACTERIZATION OF THE OSCILLATORY BELOUSOV-ZHABOTINSKY REACTION AT LINEARLY INCREASED TEMPERATURE

E. Koch and B. Stilkerieg
Institut für Strahlenchemie im Max-Planck-Institut für
Kohlenforschung, Mülheim a.d. Ruhr, FRG

## INTRODUCTION

In the recent years, the phenomenon of chemical oscillations has become a subject of general interest because such reactions act as internal clock in the biochemistry of living organisms. Chemical oscillations often reveal a direct analogy to electric switching (flip-flop oscillation), and their occurence has initiated a vigorous advancement of the theories of thermodynamics far from equilibrium (1,4).

Although very interesting also from the kinetic viewpoint, all chemical oscillations found up to now are of very complicated nature. Hence, they represent new types of complex reactions and are appropriate candidates for a test of such non-isothermal techniques as are summarized in thermal analysis, if these are applied with the required criticism. Therefore, uniformity of temperature is necessary as is guaranteed by the use of our all-liquid DTA apparatus (2).

The Ce-ion-catalyzed oscillatory oxidation of malonic acid by bromate/bromide in 1M aequeous sulfuric acid (=BZ-reaction) is the mostly studied example of a periodical chemical process. For the description of essential features of the BZ-reaction, the so-called "Oregonator"-mechanism was proposed by Field and Noyes (3)(restricted to species of essentially varying concentration):

$A + Y = X$ (M1) Formation of $HBrO_2$ (=X) from $BrO_3^-$ and $Br^-$
$X + Y = P$ (M2) Termination reaction $Br^-$ (= Y) + $HBrO_2$
$A + X = 2X + Z$ (M3) Autocatalytic formation of $HBrO_2$ and Ce(4+) (=Z) from $BrO_3^-$ and $HBrO_2$
$2 X = A + P$ (M4) Disproportionation of $HBrO_2$
$Z = f\ Y$ (M5) Regeneration of $Br^-$; here: $f = 1$

In this scheme, the bromide ion acts as a control intermediate by which either reaction 1 or the autocatalytic reaction 3 is preferred. Undamped oscillations can only occur in an open system, i. e. if the bromate consumed is currently supplemented. This was realized in flow reactors (4).

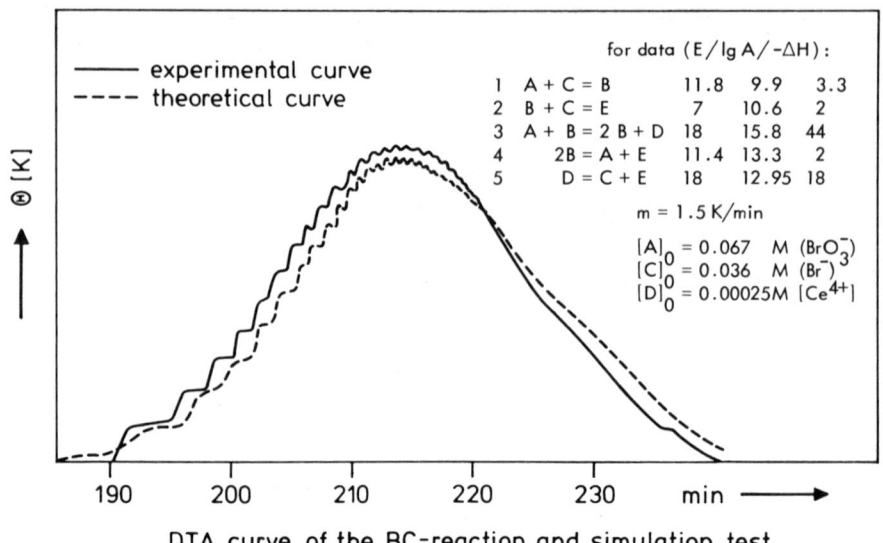

DTA curve of the BC-reaction and simulation test

### RESULTS AND DISCUSSION

If an oscillating system is continuously heated as a sample, we are dealing with a partially closed system. The bromate will be consumed at the highest temperature at the end of the experiment. Hence, the DTA curve obtained will include doublefold information: The envelopping curve must characterize the kinetics and the energetics of the continuous bromate consumption process, whereas the periodical "fine structure" which shows a monotonous decrease and vanishes towards the end of the reaction should reflect the limit cycle behaviour, which is typical for such reactions.

When performing our calorimetric experiments using the hundredfold of the usual amount of catalyst $(Ce(SO_4)_2 \cdot 6H_2O)$, we observed, indeed, DTA curves of this type. Using reduced amounts of the catalyst, we stated that the temperature coefficient of the oscillation frequency as well as initial temperature $(T_o)$ and oscillation-disappearance temperature $(=T_D)$ remained constant; however, the amplitudes of the oscillatory structure became smaller, until (beyond .0002 M catalyst) the periodicity disappeared within the noise level of the equipment. Then, conditions are reached which were typical of our first experiments (5); the corresponding DTA curves could be described by a simple bimolecular or a three-step model only involving reactions 1,3 and 5 of the Oregonator (= neglection of the termination reactions).

Basing on the Oregonator mechanism and on the very efficient integration subroutines of Gear which were specially developed for such mathematically "stiff" systems of differential equations (6), we succeeded in simulating the total oscillatory DTA signals (dotted line in Fig) by adaption of the activation energies, frequency factors and enthalpies of the particular 5 processes. However, this model is not universal enough to describe the special influence of the catalyst as an amplitude-regulating factor. The frequent similarity of the envelopping curve with that of an bimolecular process enables us to calculate approximate overall activation data and other parameters (cf. (7)). The dependence of such data on the catalyst concentration again reveals that the Oregonator scheme is incorrect, since Ce-ion is not a bimolecular reactant in it; but the tendencies of the different type-dependent parameters, especially

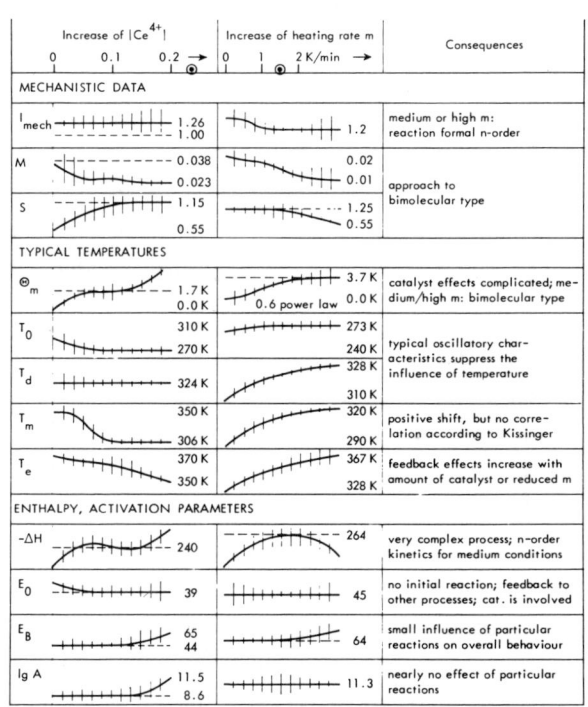

| Various parameters | | |
|---|---|---|
| What type of reaction? | n-order indicator | Mechanistic coordinates |
| Signal height as indicator? | Onset, end of oscillatory phase, maximum, end of curve | |
| Total enthalpy | Initial and overall activation parameters | |

PROPERTIES OF THE ENVELOPPING CURVES

Standard conditions (=⊙): $|Ce^{4+}|_0 = 0.025$ M; $m = 1.5$ K/Min; $c_{298} = 0.1$ $Min^{-1}$; vertical lines: degree of scatter. $I_{mech}$ = reaction coefficient (=1 for formal reaction of n-order); M = reaction type index; $\Theta_m$ = maximum signal height; $E_0$ = initial activation energy; $E_B$ = overall activation energy; A = Arrhenius factor; $T_{0,d,m,e}$ cf. text. Units: $Mol/dm^3$; K; kJ/Mol; Min.

the mechanistic coordinates may exclude a further increase of the
complexity of the mechanism when the catalyst concentration is raised.

Interesting aspects arise from increasing the heating rate. Whereas
the peak temperature and the final temperature are lifted, the initial
temperature and the initial activation energy remain remarkably constant.

The overall enthalpy is maximum for a medium heating rate of $\sim$ 1.5
K/min, namely 264 kJ/Mol $BrO_3^-$. Many other results can be summarized by
the statement that for lower heating rates, the features of the curves
reflect increasing variety, whilst for high heating rates the results
are less reproducible so that no reliable conclusions can be drawn for
the moment.

From the oscillatory structures of the curves in Table 1 the temperature
coefficient of the frequency of the oscillation was calculated. The
corresponding activation energies are 49.4 kJ/Mol (1.5 K/Min) and
32.22 kJ/Mol (0.75 K/Min), respectively. These values are related to
the activation energies of reactions 2,5 and (less effective), to 1 and
3 in a very complex manner, but the real frequencies can be simulated by
the computer when mainly the activation parameters of the bromide-
generating reaction 5 are adapted. If the experiments are performed
under Argon instead of air, higher signals, but a lower temperature
coefficient for the increase of this frequency was observed.

## STUDY OF SUBSYSTEMS

In order to gain additional results on special steps of the mechanism of
the BZ reaction, the particular components and combinations of two or
three components of the list ($Br^-$, $BrO_3^-$, malonic acid, mono- or dibromo-
malonic acid and Ce(3+) or Ce(4+) salts) were studied in 1M aequ. $H_2SO_4$
(last figure). Surprisingly, the simpler systems lead to more complicated
and less reproducible results than the complete system. With respect to
the bromide regenerating reaction 5, the studies of solutions of Ce(4+)-
sulphate with excess monobrommalonic acid in dil. $H_2SO_4$ revealed that
the initial activation energy ($\sim$ 68 kJ/Mol) is strongly increased for
higher amounts of catalyst than 0.025 M. Bad reproducibility of the
results for such high concentrations, especially in the enthalpy values,
and an escaping of the mechanistic coordinates from the AB zone suggest
a complicated mechanism. Similarly, the behaviour of the reaction between bromide and bromate ion disagrees with the concept of a normal bimolecular process for a 1:1 or even the postulated 3:1 ratio of the

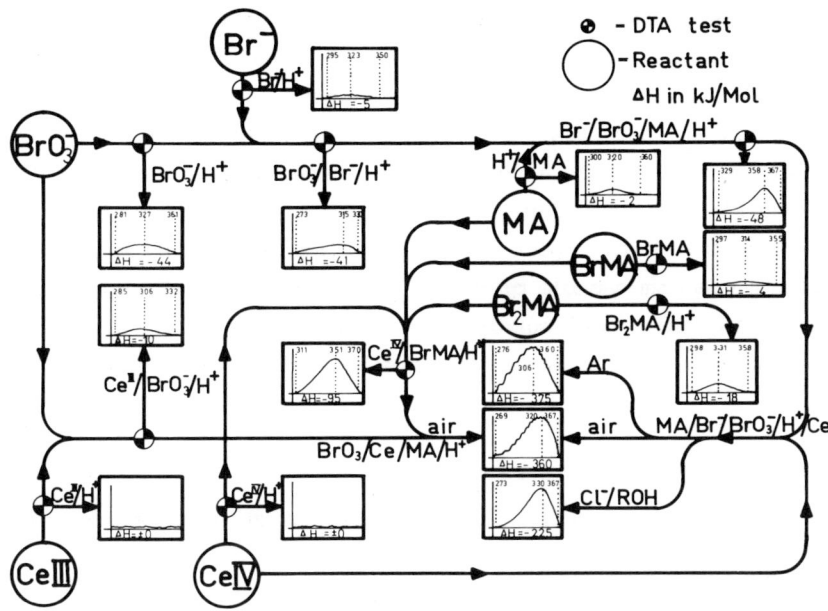

*Kinetic circuit and positions for thermoanalytical probes*

reactants because shape indices rather due to a first-order than to a second-order process are obtained. The overall enthalpy shows a strong increase with the bromate, but a decrease with the bromide concentration.

Rather conclusive results were obtained when bromide was absent in the complete BZ system. In this case, the oscillations which require $Br^-$ at least as an intermediate do not die out which strengthens the assumption of a consecutive radicalic reaction involving $Br\overset{\bullet}{O}_2$ radicals in process 3. The only probable way for the appearance of $Br^-$ under such conditions is passing through the intermediates $HBrO_2$, $HBrO$ (by disproportionation) and monobrommalonic acid; the latter may react with Ce(4+) in reaction 5 to regenerate $Br^-$ ions. The concept of an involved radicalic process is supported by the sensitivity of the system to oxygen, $Cl^-$ ions or methanol since these chemicals reduce or prevent the oscillations.

## CONCLUSIONS

The non-isothermal study of the BZ-reaction reveals that the Oregonator scheme is a good starting basis for understanding this reaction also over a wide temperature range. Of special interest, because in contrast with elementary and "normal" complex reactions, is the independency of the initial temperature on the heating rate which represents a new type of complex behaviour, whereas other parameters show the expected trends.

Although many results agree with the Oregonator model, the influence of the catalyst concentration on DTA signals cannot be described by it. This statement which was already known before and had initiated Edelson, Noyes and Field to consider up to 17 additional reactions (8), should not reduce the considerable pedagogical significance of this simplest chemical reaction mechanism capable of continuous oscillations. Despite of its incompleteness, this model surely offers a key for understanding such peculiar events even considered as responsible for self-organization processes.

## REFERENCES

(1) G. Nicolis and I. Prigogine, "Self-Organization of Nonequilibrium Systems", Wiley, New York (1977)

(2) E. Koch, Chem.-Ing.-Tech. 37, 1004 (1965); "Non-isothermal Reaction Analysis", Monograph, Academic Press, London (1977), espec. Ch. 5/V; E. Koch, L. Carlsen and B. Stilkerieg, Thermochim. Acta 33, 387 (1979); E. Koch, L. Carlsen and B. Stilkerieg, Ber. Bunsenges. Phys. Chem. 83, 1238 (1979)

(3) R.J. Field and R.M. Noyes, J. Chem. Phys. 60, 1877 (1974)

(4) Lectures on "Kinetics of Physicochemical Oscillations", held by the Deutsche Bunsengesellschaft für Physikalische Chemie, 19.-22. Sept. 1979 in Aachen, Germany

(5) E. Koch and B. Stilkerieg, Thermochim. Acta 29, 205 (1979)

(6) W. Gear, "Numerical Initial Value Problems in Ordinary Differential Equations", Prentice Hall, Englewood Cliffs (1971). We thank Prof. Dr. R.J. Field for submitting a modern version of a Gear program

(7) E. Körös, Nature 251, 703 (1974); E. Körös, M. Orbán and Zs. Nagy, Acta Chim. Acad. Sci. Hungaricae, Tomus, 100, 449 (1979); I. Lamprecht and B. Schaarschmidt, Thermochim. Acta 22, 257 (1978)

(8) D. Edelson, R.M. Noyes and R.J. Field, Int. J. Chem. Kinetics 11, 155 (1979)

CHARACTERISTICS OF INTERFACE ENDOTHERMIC REACTIONS RECONSIDERED
FROM A FEW SIMPLE EXPERIMENTS

A. Mokhlisse, G. Bertrand, M. Lallemant and N. Roudergues
Laboratoire de Recherches sur la Réactivité des Solides, (LA. 23),
Faculté des Sciences, Dijon, FRANCE.

## ABSTRACT

Some experiments chosen from given decomposition of hydrates, evaporation of liquids, sublimation of ice suggest that the problem of interpreting the kinetics of heterogeneous, endothermic reactions should be reconsidered taking into account the part played by transfers at the interface so as to give an expression of rate which in addition to the affinity extent, contains a term accounting for interface couplings.

## INTRODUCTION

The profile of the progress curve of a solid-gas decomposition reaction is often interpreted from morphological considerations and it is assumed that the actual conditions of evolution of the system are those of an ideal isothermic reactor. This comes to ignore the possible effect of the reaction enthalpy on the transformation kinetics itself, by the thermal flows involved. This approach is described below from a few simple experiments.

## EXPERIMENTAL

The results below relate to **reactions condensed phase → gas** (with or without solid residue). The temperature and pressure fixed ($T_c$ and $P_c$) were constant during an experiment ; the thermal exchange between reactor and sample mainly occurs through gas conduction ; the temperatures were measured by Thermocoax ($\emptyset$ = 0.25 mm) and the reaction was initiated by rapid lowering of pressure (t = 0) from the equilibrium value $P_{s(T_c)}$ to $P_c$.

## RESULTS

Three main types of change appear depending on the morphology and conditioning of the sample. The first type corresponds to instantaneous reactions, with constant interface area (Fig. 1) (e.g. evaporation of liquid). From initial time, the liquid cools and its temperature becomes constant, while the interface reaction rate falls from the maximum value $v_0$ to value $v_1$ in stationary conditions. The relation between reaction progress and temperature evolution of the condensed phase appears clearly. The same holds for the decomposition of monocrystalline cubes of hydrates coated with resin on five faces the remaining being sown by friction. Even in this case rate and temperature keep constant despite the increase in thickness of the reaction product thus excluding the occurrence of diffusion process.

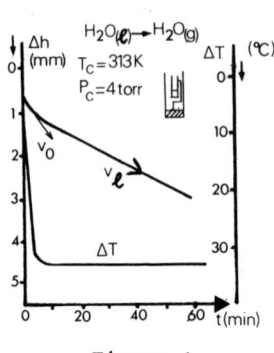
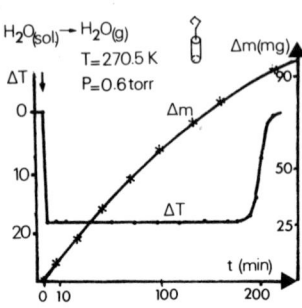

Figure 1    Figure 2

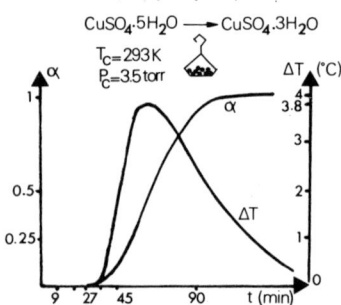
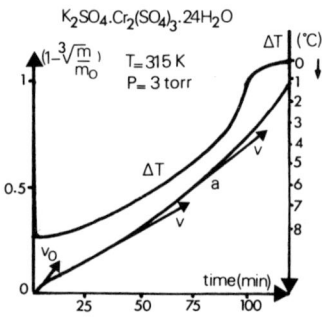

Figure 3    Fig. 4. Cubic sample sown on all its faces.

The second type (figure 2) differs from above in that the interface surface varies : sublimation of cylindrical samples of ice or decomposition of a monocrystalline cube of hydrate sown on all its faces. A transform of the weight loss curve, associated with the sample morphology, is here necessary to obtain v.s. time the change in the interface progress rate (figure 4), it still varies in the same way as the sample temperature : maximum ($v_0$) at initial and final times it takes on a mini-

mum value at maximum cooling.
The third type occurs during the decomposition of polycrystalline powders or monocrystalline samples that have not been subjected to any prior treatment (figure 3) ; the reaction progress corresponding to a sigmoïdal curve is still accompanied by a maximum thermal effect at the inflexion point of the α = f(t) curve ; the influence of the endothermicity of the reaction upon the various parts of the progress curve is easy to demonstrate :

- reduction of the latency period (τ) by introduction of a good conducting gas (figure 5).
- linear development of the dehydrated domains ; the decomposition is frequently shown by the development of isolated polycrystalline domains (picture 6a). After a short acceleration period their development becomes constant along each direction (figure 6b) recalling the above results.
- progression of the continuous interface : after the inflexion point it shows the same features as those in figure 2, except for the initial times.

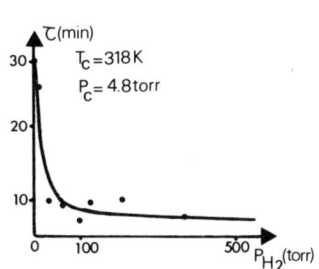

Fig. 5. Latency time of the dehydration : $CuSO_4, 5H_2O \rightarrow CuSO_4, 3H_2O$.

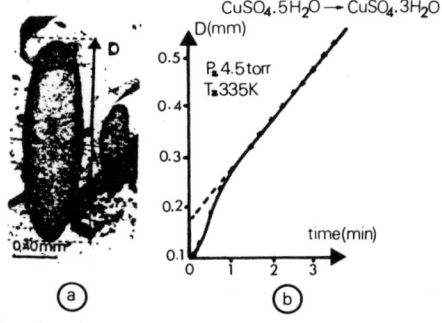

Figure 6

In all the reactions considered the system evolution causes cooling which in return acts upon the transformation progress. This influence is more or less explicit depending on the characteristics of the condensed phase and the reaction area. The regrouping of the reactions into a single class, the interface endothermic transformations is confirmed by the specific profile of the $v(P)_T$ curves (1). The reaction progress being governed by possible energy input at the interface a minute examination of temperature is required.

## INTERFACE TEMPERATURE ANALYSIS

Thermal profiles. Let us take for example the evaporation of liquids. We notice (figure 7) that the temperature near the interface, the lowest in all the reactor, is of an other order of magnitude than $T_c$ and that thermal gradients occur on either part of the interface ; the farther $P_c$ moves from $P_{s(T_c)}$, the sharper the profiles.

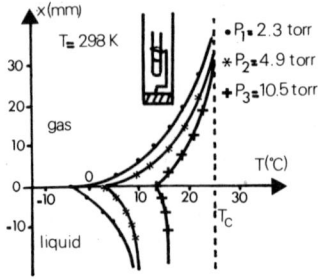

Fig. 7. Evaporation of water.

Value of the interface temperature $(\hat{T})$. Experiments on the sublimation of ice yield a remarkable result : the ice sample is uniformly cooled and, in stationary conditions (table I) its temperature is

$$\hat{T} = T_{s(P_c)} \qquad [1]$$

Can this result be extended to the class of endothermic interface transformations ?

| $P_c$ (torr) | 0.03 | 0.6 | 2 |
|---|---|---|---|
| $T_{s(P_c)}$ (K) | 223.1 | 250.4 | 263.3 |
| $\hat{T}$ (K) | 222.5 | 250.5 | 263.0 |

| $P_c$ (torr) | 2.3 | 15.5 | 29.2 |
|---|---|---|---|
| $T_{s(P_c)}$ (K) | 264.9 | 291 | 301.5 |
| $\hat{T}$ (K) | 269.8 | 291.4 | 301.5 |

Table I. Sublimation of Ice $T_c$ = 270.5 K.

Table II. Evaporation of water $T_c$ = 313K.

During evaporation, the above relation is only observed in the pressure ranges that are very close to equilibrium (evaporation of water ; 313 K, table II). In the decomposition of the salt hydrates already mentioned, the accurate measurement of the local interface temperature is faced with experimental difficulties. Overall coolings performed on solid or powder samples (figures 3 and 4) from 5 to 20°C according to pressure $P_c$ do not allow to think, at present, that relation [1] is verified in the latter case.

All these results confirm that it is necessary to take into account the part played by thermal transfers at the interface in order to interpret realistically the kinetics of endothermic interface transformations. The model below takes this consideration into account.

## DISCUSSION : AN INTERFACE THERMODYNAMIC MODEL

Developed for the evaporation of liquids it rests on the coupling (T, P, I) at the interface of heat and matter flows.
Let us recall, here, the expression of rate (2)

$$v = \underbrace{- Kr \ln \frac{P_c}{P_s(\hat{T})}}_{A} - \underbrace{\left[\frac{C_1 C_2}{\Lambda}(-r \ln \frac{P_c}{P_s(\hat{T})}) + \frac{C_1}{\Lambda} \lambda_o . G.(1-\exp-\frac{P_c}{\pi_o})\right]}_{B} \quad [2]$$

where K, $C_1$, $C_2$, $\Lambda$ are the phenomenological coefficients, $\lambda_o$, $\pi_o$, G characterizing the thermal behaviour of the gas.
It consists of a first term (A) resulting from the chemical disequilibrium of the interface, the single component of rate in isothermic system ($\hat{T} = T_c$) and of a second term (B) taking into account the thermal gradients and the coupling of flows ; (B) represents globally the "braking" of the reaction due to the endothermic effect of the transformation.
This expression describes satisfactorily the kinetics of evaporation of a liquid (2). For ice sublimation, equality [1] allows to simplify [2] and to write

$$v = \beta(1 - \exp - \frac{P_c}{\pi_o}) \cdot \ln \frac{P_c}{P_s(T_c)} \quad \text{with} \quad \beta = - \frac{C_1 \lambda_o G_o}{\Lambda} \quad [3]$$

The comparison of the experimental rate curve and curve calculated from equation [3] (figure 8) shows that the model easily extends to a state change such as sublimation.

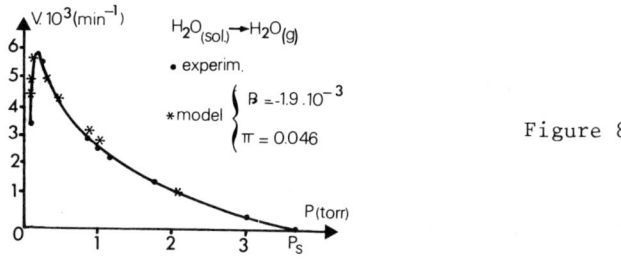

Figure 8

This model cannot be applied to the decomposition of mineral solids for it does not consider the processes and specific features of the solid phase (adsorption, structure ...). It may be regarded as a particular case of a general relation inherent to endothermic interface transformations $v = v_o - \Delta v$ where rate $v_o$ only depends on experimental constraints ($P_c$, $T_c$, crystal ...) and may be expressed in different forms depending on the nature of the chemical processes involved and where $\Delta v$ represents the "braking" due to the endothermicity of the transformation. The kinetic behaviour of the class including endothermic interface transformations will then be described clearly.

## REFERENCES

(1) G. Bertrand, M. Lallemant, G. Watelle, J. Inorg. Nucl. Chem., **36** (1974) 1303.
ibid. J. Inorg. Nucl. Chem., **40** (1978) 819.
(2) G. Bertrand, R. Prud'homme, J. Non-Equilib. Thermodyn., **4** (1979) 1.

# ERROR ANALYSIS IN THERMAL PURITY DETERMINATIONS

Gy.Kiss, K.Seybold and T.Meisel
Institute of General and Analytical Chemistry
Technical University of Budapest

## ABSTRACT

We studied by simulation and experimentaly the deviations occuring at purity determinations. We propose a method for the proper choice of range for evaluation. The continuous and stepwise measuring techniques were compared.

## INTRODUCTION

Calorimetric purity test is a general method by which the degree of purity of a substance can be determined without knowing the nature of the contaminants if the molecular weight of the main component is known. In practice the measurements are carried out by DSC or QDTA instrument. In evaluating the experimental results two problems may arise:
a., The enthalpy loss which means that the curves fitted to the enthalpies measured deviate from the theoretical curves in a way as if an enthalpy loss occurred in the initial part of the process.
b., The temperature lag which is due to the operation principle of DSC and QDTA devices and it means that the temperature of the sample is lower than recorded value.
In the following section the possible reasons for the enthalpy loss will be discussed and an equation deduced by which the problems outlined can readily be treated by mathematical means. In addition to this the to error sensitivity of the equation used to estimate the degree of purity will be studied by simulation for the continuous and stepwise method of measurement and the results will be compared.
The conclusions were tested by measurements on the phenacetine-benzamide system.

## THEORY

Let us start with the well known equation valid for equilibria in eutectics.

$$T_o - T = xR_o \frac{T_o^2}{H_o} \frac{1}{F/T/} \qquad /1/$$

The melted fraction $F/T/$ can be calculated as the ratio of the enthalpy change up to temperature T $(h/T/)$ and the total enthalpy change $/h_o/$ as follows:

$$F/T/ = \frac{h/T/}{h_o} \qquad /2/$$

considering that $h_o = nH_o$ where n = moles of the sample

$$F/T/ = \frac{h/T/}{nH_o} \qquad /3/$$

Inserting /3/ into /1/ and rearrangeing, we get

$$T_o - T = xnR_o T_o^2 \frac{1}{h/T/} \qquad /4/$$

The sensitivity to the accuracy of enthalpy measurement is clearly shown by Eq.4.

A generally accepted explanation for the enthalpy loss is that in the initial section of the process the signal merges into the base line due to the small signal intensity. The change of the enthalpy of melting with the degree of contamination may also contribute to the enthalpy loss. The enthalpy of melting is between that of the eutectic and of the pure substance. Accordingly, Eq.4., being based on Eq.1. i.e. on the assumption of the constancy of the heat of melting, can only be considered as an approximation. The greater the degree of contamination of the melted fraction, /at the beginning of the reaction/ the less valid Eq.1.

Now, the equestion arises whether instead of Eq.1. or Eq.4. an equation yielding a better approximation can be derived? Unfortunately, the answer is no, as by calorimetry a signal is proportional to the derivative of the product the melted fraction and the melting heat function of impurity, whereas $F/T/$ and $H/T/$ would be needed separately to obtain a more reliable equation.

Let us find the section of the process where Eq.4. yields a sufficiently good approximation. In Fig.1. the approximated and phase-diagram based melting curves are shown.

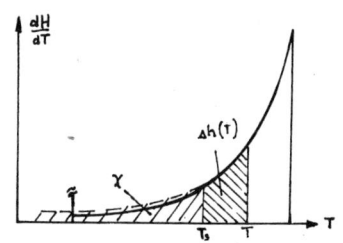

Fig.1. Approximated /--/ and real /-/ melting curves.

Let $T_s$ be the temperature where the approximated and phase-diagram based melting curves practically coincide. $T_s$ depends on the nature of the system studied, but in any case it is within the range of smaller degrees of contamination, and corresponds to the section of the phase diagram where the equlibrium curve is practically rectilinear.

Let $\chi$ denote the enthalpy change before $T_s$ according to the approximating equation, and $\Delta h/T/$ the real enthalpy change between $T_s$ and T. Then

$$T_o - T = xnR_oT_o^2 \frac{1}{\chi + \Delta h/T/} \quad ; \quad T > T_s \quad /5/$$

It should be pointed out that the deduction refers to eutectics, but similar results can be obtained using the equation deduced by Mastrangelo and Dornte /1/ for solid solutions. Accordingly, the following procedure is proposed for processing the experimental data:

$$\min_{x,T_o,\chi} \sum_{i=1}^{n} \left( T_o - T_i - xnR_oT_o^2 \frac{1}{\chi + \Delta h/T_i/} \right)^2$$

The minimum was found by iteration with respect to $\chi$.

### Investigation of the error sensitivity

The procedure for the evaluation of the results obtained by continuous technique are shown Fig.2. /2/,/3/. In the simulation the effect of the random error of the DSC measurement and of the heat resistance determined by calibration was investigated. Experiments were conducted to determine the error of calibration and it was found to change between 20 and 26 mJ/sK. The system of differential equations describing the processes taking place during melting was

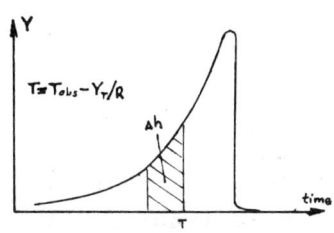

Fig.2. Mode of evaluation at continuous technique.

numerically solved and a random error was imposed on the data. The confidence intervals of the x values estimated from the data thus obtained are shown in Fig.3/a.

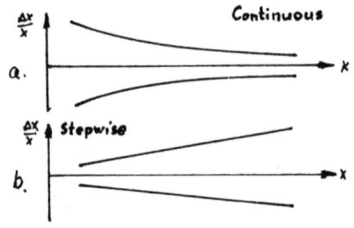 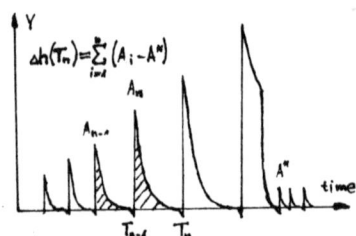

Fig.3. Results of error simulation.

Fig.4. Mode of evaluation at stepwise technique.

The evaluation of the data of stepwise measurements is schematically shown in Fig.4. /3/,/4/. In this case the effect of random error imposed on the peaks was studied. The theoretical peak areas were calculated and random error was imposed. The confidence intervals of the x data obtained by estimation are given in Fig.3/b. The confidence intervals of the two methods were plotted in a system of coordinates in which the ordinate was not graduated, the reason being that the magnitude of the limits is a function of the standard deviations chosen arbitrarily. Accordingly, only the general shape of the curves can be compared.

The results indicate that the stepwise technique provides better results at a lower degree of contamination, the tendency of the change of accuracy is the opposite for the continuous technique. For the stepwise technique this means a coincidence with the range was Eq.5. is valid, but not for the continuous method.

## EXPERIMENTAL PROCEDURE

The measurements were made on the phenacetine-benzamide system. Two mixtures were prepared: the mole fraction of benzamide was 0,0208 in the one and 0,0019 in the other. The series of experiments were carried out on repetitive samples.

An important problem connected with the evaluation of the

recording yielded by the continuous technique is the proper
choice of the range.The views in this respect are diverse /4/.In the generally accepted procedure the range is
chosen between certain values of the melted fraction,al-
though no special reason is given for the choice in the
literature.The authors of the present paper propose the
following procedure.
It is known that any derivative of the $\frac{dH}{dT}$ vs.T curve is
strictly monotonouos.However,this is not true for the experimental curves as,due to the random noise in the initial section and to appreciable deviations from equilibrium in the final part of the curve,the first derivative
is not monotonous.
Obviously,if these sections is also evaluated,the error is
greater than when evaluating on the basis of the central
section.
The range evaluated was chosen as follows:curve smoothing
and numerical derivation was carried out and the portion
determined where the first derivative is strictly monotonous.

## RESULTS

The results are summarized in Figs.5.and 6.

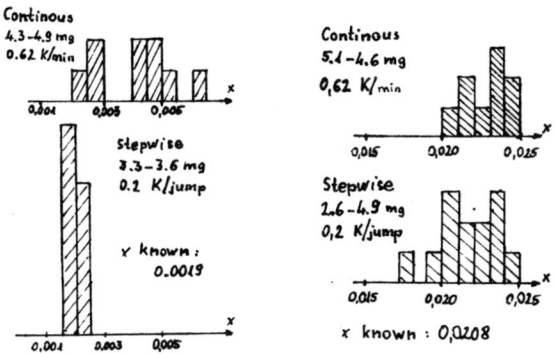

Fig.5 and 6. Histograms of measured impurity by both
methods in two series.

Based on the results,the following conclusions can be drawn:
1.,The tendency of the standard deviations corresponds to that yielded by simulation;standard deviations are smaller for the stepwise technique.
2.,The accuracy of the measurements is higher in the case of the stepwise technique.
Summing up it can be concluded that the stepwise technique is more suited to calorimetric purity tests,the evaluation is simpler,which compensates the 2-3 times of measurement.

## Acknowledgement

The authors wish to express their thanks to prof.J.Varga for his valuable help with the evaluation of the experimental results.

## REFERENCES

/1/ S.W.R.Mastrangelo,R.W.Dornte,J.A.C.S. 77 /1955/ 6200
/2/ E.E.Marti,Th.Acta 5 /1972/ 173
/3/ E.F.Palermo,J.Chiu,Th.Acta 14 /1976/ 1
/4/ H.Straub,W.Perron,Anal.Chem. 46 /1974/ 128

# INFLUENCE OF EXPERIMENTAL VARIABLES ON DSC CURVES
A.A. van Dooren
PHILIPS DUPHAR B.V., WEESP, THE NETHERLANDS

## 1. INTRODUCTION

Differential Scanning Calorimetry (DSC) is widely regarded as an important analytical tool in pharmaceutical research. It is well-known, however, that experimental factors can influence the results considerably. These factors relate to:
1. The adjustment of the apparatus, such as heating rate and calorimetric sensitivity.
2. The sample's mass, packing and porosity, particle size (distributions), pretreatment and dilution.
3. The reference material's mass, nature and pretreatment.
4. The atmosphere: thermal conductivity, oxidising or inert nature, static or flowing conditions.

In our experiments we investigated the main effects and possible interactions of these factors in a statistically justified manner.

## 2. EXPERIMENTAL

### 2.1. Materials

As models for substances with interesting thermodynamic behaviour we used the following compounds:

a. <u>Adipic acid</u>, $HOOC(CH_2)_4COOH$, which should melt without noticeable decomposition with an extrapolated onset temperature of $150.0°C$.
b. <u>Naphazoline nitrate</u>, (nitrate of 4,5 dihydro-2-(1-naphtalenyl-methyl)-1H-imidazole) which melts at $167-170°C$ with decomposition.
c. <u>Potassium nitrate</u>, $KNO_3$, which has a solid-solid transformation at $128°C$.
d. <u>Sodium citrate dihydrate</u>, $C_6H_5Na_3O_7 \cdot 2H_2O$, which loses water of crystallisation at $150 - 160°C$.

As the reference compounds/ diluents a) purified carborundum and b) neutral aluminum oxide were used.
For calorimetric and temperature calibration we used ultrapure indium.

### 2.2. Equipment

The apparatus was a Mettler TA 2000A heat-flux DSC system with a constant supply of gas and a constant check of the percentage of oxygen in the atmosphere. In preliminary experiments, the performance of the equipment was found to be acceptable.
Weighings were done on an electronic microbalance, they were reproducible to within 0.040 mg of the required values.

### 2.3. Study design

Table 2.3.1. gives a summary of the study design.

Table 2.3.1. Study design

| Experiment number | Variables to be investigated |
|---|---|
| 1. | Sample mass (0.1 - 15.0 mg) |
| 2.1. | Heating rate (0.01 - 0.495 K/s) |
| 2.2. | Heating rate (0.01 - 0.495 K/s), triturated samples |
| 3. (adipic acid) | Sample mass (2 and 5 mg) x reference mass (0 mg, 2 and 5 mg). x sieve fraction (small, non-fractionated and large particles) |
| 3. (other compounds) | Sample mass (2 and 5 mg) x sieve fraction (small, non-fractionated and large particles) x atmosphere (air, $O_2$, $N_2$) |
| 4. | Concentration in dilution (10, 30 and 50 %) x preparation of dilution (whirlmixer, mortar and pestle, dissolution/recrystallisation) x type of diluent (carborundum, aluminum oxide, air) x sieve fraction of active ingredient (small, non-fractionated and large particles). |
| 5. | Atmospheres (static air, flowing air, $N_2$, $O_2$, He, vacuum) x sample holders (completely closed, with pierced lids, without lids). |

## 2.4. Characterization of the curves

The evaluation of the DSC curves obtained was done on the following aspects. See fig. 2.4.1.

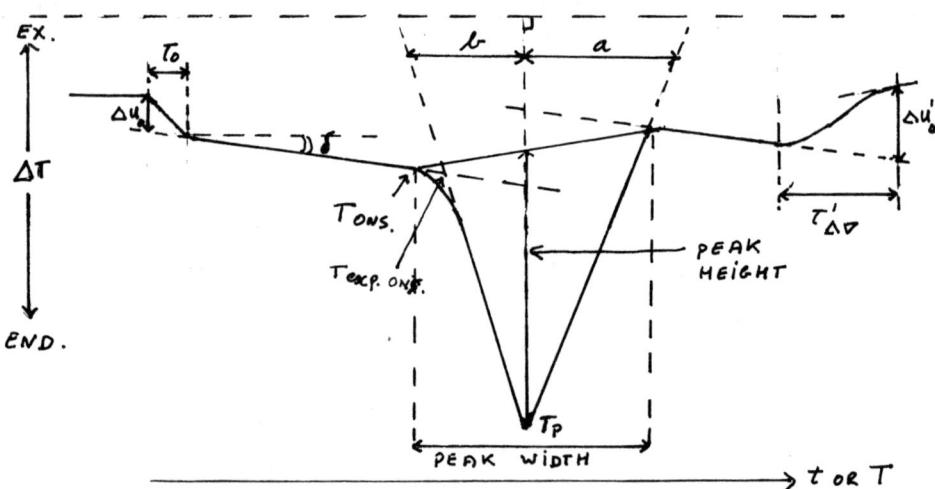

Fig. 2.4.1. Formalized DSC curve

- initial baseline deflection (isothermal → heating) $\Delta U_o$ (in $\mu V$)
- lag time (isothermal → heating) $\tau_o$ (in s)
- base line drift angle $\delta$ (in $\mu V/K$), after Gäumann and Oswald (1)
- onset temperature Tons (in $^oC$)
- extrapolated onset temperature Texp.ons (in $^oC$), as defined by ICTA (2)
- peak temperature Tp (in $^oC$)(temperature at peak top)
- peak height (in $\mu V$), as defined by ICTA (2)
- peak width (in K), as defined by ICTA (2)
- shape index s= a/b, as defined by Kissinger (3)
- specific enthalpy $\Delta H_s$ (in J/g), computed from the peak area (n.b. The interpolated baseline has always been drawn as a straight line. The baseline displacement after the peak was always very small)
- baseline displacement (in $\mu V$), being the difference between the extrapolated baselines before and after the peak, measured at the line through the peak top vertical to the time axis
- baseline deflection $\Delta u'_{\Delta \triangledown}$ (heating → cooling)(in $\mu V$)
- lag time $\tau'_{\Delta \triangledown}$ (heating → cooling(in s)

## 3. RESULTS AND DISCUSSION

### 3.1. Calibration checks

Every day the calorimetric sensitivity ($E_{in}$) was determined with a sample of indium. Each indium sample was used for only one week. Table 3.1.1. gives some results of the calibration tests.

Table 3.1.1. Calorimetric sensitivity $E_{in}$ mean values

| area determination | $E_{in}$ ($\mu V/mW$) | n | s($\mu V/mW$) | v.c. (%) |
|---|---|---|---|---|
| cut/weigh | 14.49 | 97 | 0.35 | 2.40 |
| planimeter | 14.50 | 97 | 0.23 | 1.60 |

The table shows that a planimeter for the area determination gave a significantly lower coefficiënt of variation, but had no influence on the mean value.

### 3.2. Effects of experimental variables on base-line related aspects

The theory of DSC states that the initial baseline deflection $\Delta U_o$ depends essentially on the masses and specific heats of sample and reference (including holders), the heating rate and the assembly of the apparatus (calorimetric sensitivity). This was partly confirmed in our experiments. The results, however, were not always significant as the coefficiënts of variation could be very large (up till 50 %).
The type or particle size distribution of the compound did not show a significant effect on $\Delta U_o$, contrary to dilution. The effect of dilution is nonsystematical and may come from the concentration of active ingredient, the type of diluent or the method of dilution or a combination of these.
There does not appear to be a simple relationship between $\Delta U_o$ and $\tau_o$. The value of $\tau_o$ was not influenced by sample mass and only

slightly by the heating rate.
For the $\Delta U'_{\Delta\tau}$ and $\tau'_{\Delta\tau}$ parameters, the same conclusions could be drawn, mutatis mutandis, as for $\Delta U_o$ and $\tau_o$, respectively. No simple relationship between $\Delta U'_{\Delta\tau}$ and $\Delta U_o$ was found.

Our results demonstrated a small drift angle which could only be determined with a fairly large standard deviation. A small increase could be observed with increasing sample mass and heating rate and, moreover, a large dependency on thermopile adjustment could be seen. By using a reference compound like carborundum (instead of air), the drift angle was decreased nonlinearly.
Dilution also influences the drift angle. With adipic acid diluted with aluminum oxide, the 100 % increase in drift angle is highly significant. Moreover, aluminum oxide sometimes gave rise to a nonlinear base-line with varying drifts. Carborundum, on the other hand, gave a significant decrease in drift angle.

### 3.3. Effects of experimental variables on temperatures

The coefficients of variation for the temperature determinations were always less than 1 %. Especially for the onset temperature, however, large differences in response per different experiment could be seen. The value of Tons is apparently more dependent on the experimental setting than those of Texp.ons. and Tp.
In general, no effects of sample and reference masses on temperatures could be seen. The heating rate, however, influences the temperature values significantly, in a manner depending on the type of compound: they increase for potassium nitrate and sodium citrate but they generally decrease for the melting peaks of adipic acid and naphazoline nitrate.
The atmosphere may also play a role. In the case of naphazoline nitrate the peak starts at a higher temperature in nitrogen atmosphere. This may be explained by the oxidation which accompanies the melting of this compound. In the presence of oxygen or air, this oxidation already starts before the beginning of fusion.
An interesting effect appeared to be the influence of particle size. For the two solid-liquid transitions, the Texp.ons. and Tp decreased with decreasing particle size. In the case of sodium citrate both Texp.ons. and Tp are smaller for the unsieved sample than for the fractionated samples. For potassium nitrate, the effects of particle sizes on temperatures are varying.
The effect of dilution is enormous for all compounds. This can partly be explained by the concurrent influence of the preparation of dilutions on particle sizes, but the concentration level also has an effect on the temperature responses. The lower the concentration the higher the temperature needed for the transition to start.

### 3.4. Effects of experimental variables on peak characteristics

Peak height, peak width and shape index could only be determined with relatively large standard deviations. They were subject to many influences, expecially sample mass and heating rate, but simple relations could not be found. Therefore, it is more meaningful to describe the peak in terms of specific enthalpy $\Delta H_s$ (4).
$\Delta H_s$ was found to be independent of sample mass above a critical value of 0.2 mg. For masses below, $\Delta H_s$ tends to become too low. An independency of heating rate, however, could not always be

found. For naphazoline nitrate a significant decrease with 4 % (at the 99 % level) in response was shown if the heating rate was low (0.02 K/s). This decrease could only be seen in air or oxygen, but not in nitrogen. For $KNO_3$ and sodium citrate it was found that $\Delta H_s$ was independent of heating rate only if rates $< 0.32$ K/s were used.
Helium atmosphere always gave a sharp decrease in $\Delta H_s$ (to about 40%) It is known that its thermal conductivity is approximately 5.5 times greater than that of air. Another effect of atmosphere was found with naphazoline nitrate. In this case the presence of an inert atmosphere increased the $\Delta H_s$ with approximately 5 %.
This may be explained by the fact that oxidation evolves extra energy which must be subtracted from the melting endotherm. For the same compound, a lower response was found for smaller particles, as smaller particles are more easily oxidised. The specific enthalpies of the polymorphic transition and dehydration appeared to be independent of particle size distribution, but for undiluted samples of adipic acid, a significantly higher response was found if the unsieved material was used. If comminuted samples of adipic acid were used instead of untreated samples, no difference in response was observed.
For adipic acid there are also effects of atmosphere on the results. Especially in vacuum, almost all of the sample had disappeared after the DSC treatment. In oxidising atmospheres we saw that small particles of adipic acid were subject to degradation.

Dilution also had an enormous effect on the specific enthalpy. The calculated mean values for all 4 compounds were lower than the values in undiluted samples and the standard deviations were much higher. Except for the dehydration reaction the responses in aluminum oxide were significantly smaller. Furthermore, it was concluded that the procedures that we used to prepare the dilutions were not adequate to ensure homogeneous and reproducible mixing.

## 4. SUMMARY AND CONCLUSIONS

In a factorial design we investigated the influences of a number of experimental variables on the results in quantitative DSC. The variables related to the equipment, sample, reference and atmosphere.
It was found that if DSC curves have to be described with base-line or peak characteristics other than temperatures and specific enthalpies, one should allow for relatively large standard deviations.
Furthermore, it was shown that calculated mean values in quantitative DSC depend heavily on the adjustment of the experimental variables. Sample mass, heating rate, particle size distribution, comminution, the atmosphere applied, mass of reference compound and dilution (type of diluent, concentration in dilution, preparation of dilution), they may all influence the results in a systematic or nonsystematic manner which also depends on the type of transition.
It was not always easy and sometimes even impossible to explain all the results obtained. More work is necessary in this field.
However, it can already be concluded that the influence of experimental variables on the results of DSC data of pharmaceutically interesting compounds is greater than to date was presumed.

## 5. REFERENCES

1. A. Gäumann, J. Oswald, Chimia 21 (1967), 421-426
2. ICTA recommendations on nomenclature and reporting of thermal analysis data (1977)
3. H.E. Kissinger, Anal.Chem. 29 (1957), 1702 - 1706
4. E.M. Barrall, L.B. Rogers, Anal.Chem. 34 (1962), 1101 - 1106

# INFLUENCE OF MASS AND GRAIN SIZE ON THE BASIC GEOMETRY OF THE DTA CURVE

Živan D. ŽIVKOVIĆ

Technical Faculty in Bor, University of Beograd, Yugoslavia

Bogomir DOBOVIŠEK and Andrej ROSINA

Faculty for Natural Sci.& Technology, Ljubljana, Yugoslavia

## ABSTRACT

The paper presents the investigation results on the influence of the sample mass and grain size on the basic geometry of the DTA curve used in qualitative phase analysis. Powder magnesite and calcite were taken for samples.

## INTRODUCTION

In the qualitative phase analysis of powdered materials the DTA curve obtained is compared to standard etalons[1,2]. This method requires careful selection of experimental conditions: heating rate, initial furnace temperature, the type and position of termocouple, reference body, furnece atmosphere and the mass and grain size of the sample. The effect of the mentioned parameters on the character of DTA curve has been thoroughly investigated except for the influence of sample mass and grain size. A number of investigators[3-7] have observed and underlined the influence of mass and grain size. however the influence of these parameters have not yet been accurately determined. Naturally, this problem is very important for DTA of powdered samples, therefore this investigation was aimed to clarify it as much as possible.

## EXPERIMENTAL

All tests were carried out on samples of natural calcite and magnesite of 99.5% purity. Calcite and magnesite were selected because of thier different behaviour during thermal decomposition-calcite disintegrates while magnesite remains compact even after completed decomposition. Furthermore, both can be found in natural state in comparatively pure state and can easily be prepared in powdered state. Calcined $\alpha Al_2O_3$ was used as a reference. All tests where carried out with heating rate of $10°$/min in air and the DTA apparatus used has already been described[8].

## RESULTS AND DISCUSSION

Important geometrical elements of a DTA curve representing the case of an endothermic reaction are seen in fig.1.

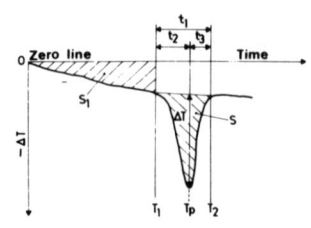

Fig.1.Some geometrical elements of the DTA curve for the endothermic reaction

Where: $S;S_1;t_1;t_2;t_3;T_1;T_2;T_p$ and $\Delta T$, peak area($mm^2$); area betwen DTA curve and zero line ($mm^2$); duration of transformation (min); time from the beginning of transformation to the maximum declination of DTA curve (min); time from the maximum to the and of transformation(min); temperature in the begining, at the maximum and in the end of transformation (K) and peak heigh (mV), respectively.

### Influence on peak symmetry

Geometrical elements $t_2$ and $t_3$ define the peak symmetry i.e. the position of the maximum in respect to the beginning and the end of a peak. The influence of sample mass and granulation on $t_1, t_2$ and $t_3$ in the case of magnesite and calcite can be seen in fig.2 and 3, respectively.

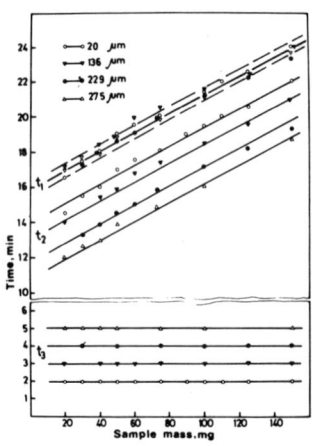
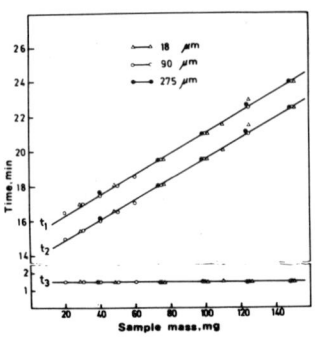

Fig.2.Influence of magnesite mass and grain size on $t_1$ $t_2$ and $t_3$

Fig.3.Influence of calcite mass and grain size on the $t_1$, $t_2$ and $t_3$

From figs.2 and 3 it can be seen that for magnesite $t_3$ increases with grain size. The relationship betwen $t_3$ and

the average diameter of magnesite $d_{av.}$ ($\mu$ m) is:
$$t_3 = 1.682 + 0.01012 \cdot d_{av.}$$
For calcite $t_3$ does not depend on average grain size, which can be explained by the fact that magnesite during thermal decomposition is not subjected to disintegration. Therefore, at higher grain size diffusion resistance to the liberating $CO_2$ at definite reaction stage has stronger influence on reaction rate. A similar asymmetry i.e. an increase in $t_3$ with grain size can also be expected at other minerals which do not disintegrate during thermal decomposition.

<u>Influence on $T_1$, $T_p$ aand $T_2$</u>

The influence of sample mass and grain size on temperature of the beginning, maximum and end of a peak for magnesite and calcite is seen in figs. 4 and 5, respectively.

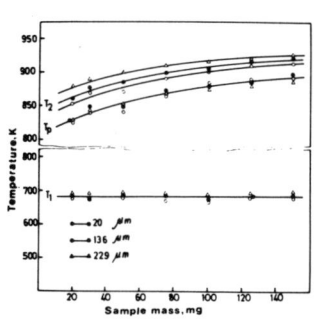 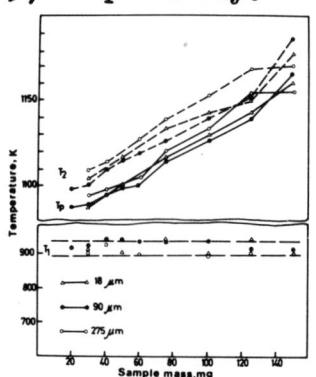

Fig.4. Influence of sample mas and grain size of magnesite on $T_1$, $T_p$ and $T_2$

Fig.5. Influence of sample mass and grain size of calcite on $T_1$, $T_p$ and $T_2$

Naturally, a change in the peak symmetry is associated with changes in temperature of the beginning, maximum and end of the peak. The results obtained show that $T_p$ increases with sample mass according to the hyperbolic law for magnesite and almost linearily for calcite. In both cases $T_2$ increases equaly with sample mass while $T_1$ remains unaffected. The analysis of the influence of sample mass and grain size of magnesite and calcite on $T_1$, $T_p$ and $T_2$ has revealed that temperature of the peak maximum $T_p$, which is most frequently used in qualitative DTA for comparasion with the standardized etalon peak, is dependent from grain size. On the contrary, a strong influence of sample mass on $T_p$ has been observed since increase in sample mass from 20 to 150mg, increase $T_p$ in both cases by 60 to 70 K which is quite a high increase for qualitative phase analysis.

### Influence on $\Delta T$

Investigation of reaction kinetics has been made possible only after thoroughful examinations of the influence of heating rate on the peak height $\Delta T$. However, there is no literature data on the influence of sample mass and grain size on the peak height at constant heating rate. The influence of sample mass and grain size of magnesite and calcite on peak height is given in figs. 6 and 7, respectively.

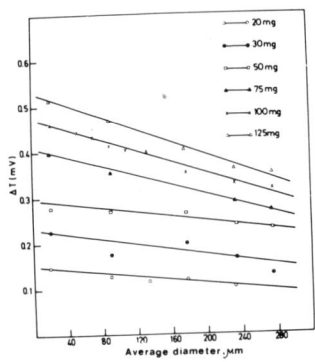 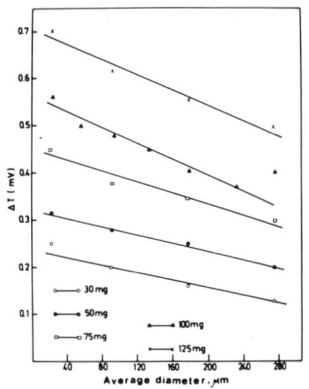

Fig.6. Relationship between $\Delta T$ and average grain size of magnesite at different sample weights

Fig.7. Relationship between $\Delta T$ and average grain size of calcite at different sample weights

As can be seen from figs. 6 and 7, $\Delta T$ linearily decreases with increasing average grain size. The decrease is higher at higher sample weight. Consequently, in the investigation of reaction rates based on DTA peak height special attention must be paid to sample weight and grain size, otherwise wrong informations on kinetics of the observed reaction may be obtained.

### Influence on deviation from zero line

A DTA curve usually continuously deviates from zero line during the heating up to the beginning of transformation. In many cases the deviation continues also after the end of

a peak. Any improvement in the deviation from zero line is
os special interest both for qualitative and quantitative
DTA since the peak is much more accurately defined in the
case of small deviation of DTA curve from zero line. Basic
geometrical elements of a peak are also more easily determined in this case. The area $S_1$ can be taken as a measure of
the deviation from zero line (see fig.1). The deviation from
zero line dependence on average grain size and sample mass
for magnesite and calcite is seen in figs.8 and 9, respectively.

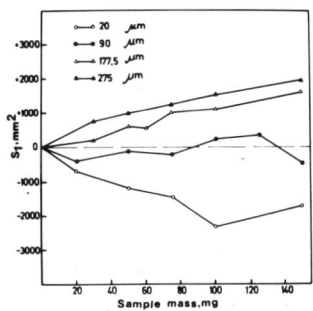
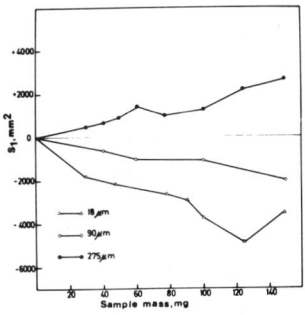

Fig.8. Relationship betwen the area $S_1$ and sample mass for different grain size of magn.  Fig.9. Relationship betwen the area $S_1$ and sample mass for different grain size of calc.

Based on these results it can be concluded that the deviation from zaro line can be reduced by proper selection of the
grain size of sample and reference. The deviation is reduced
to zero in the moment when the coefficient of heat transfer
of the sample equals that the reference[9].
Consequently, for given grain size of a sample there is a
definite grain size of reference at which the both coefficient of heat transfer have the same value, i.e. by changing
grain size of reference it is possible to make $S_1$ to tend to
zero value.
The influence of the grain size of reference on S and $S_1$ at
a given grain size of magnesite is given in fig.10.
As seen from fig.10, $S_1 \longrightarrow 0$ when average grain size of
the reference tends to 90 - 100 $\mu$m at the average grain
size of magnesite 112.5 $\mu$m. Optimum grain size of the reference has to be experimentally determined for each particular sample. The area S is independet from the grain size of

reference.

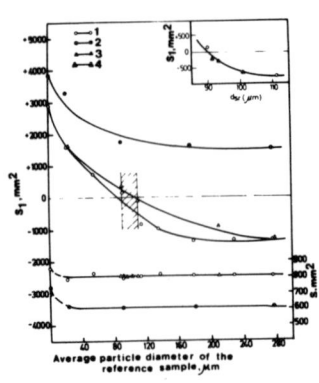

1-average grain size of magnesite 112.5 μm
2-average grain size of magnesite 275 μm
3-average grain size of magnesite 112.5 μm, reference being a mixture of classes 90 and 112.5 μm
4-average grain size of magnesitel 112.5 μm, reference being a mixture of classes 22 and 275 μm

Fig.10. The influence of the grain size of reference on S and $S_1$ at (reference $\alpha Al_2O_3$)

## CONCLUSIONS

Based on the results obtained it can be concluded that sample mass and grain size have a definite influence on the geometrical elements of DTA curve which can be seen in fig.1. Therefore, the authors propose that the existing etalons[1,2] for use in qualitative DTA should also include the data of sample mass and average grain size of both the sample and reference in order to improve the usability of experimental DTA results for qualitative investigations of powdered samples.

## REFERENCES

(1) R.C.Mackenzie, SCIFAX Differential Thermal Analysis Data Index, London, 1965.
(2) G.Liptay, Atlas of Thermoanalytical Curves, Academiai Kiado, Budapest, Vol.1(1971); Vol.5(1976).
(3) P.F.Kerr and J.L.Kulp, Amer. Mineral., 33 (1948) 387.
(4) R.A.Rowland and E.C.Jonas, Amer. Mineral., 34 (1949) 550.
(5) G.T.Faust, Amer. Mineral., 35 (1950) 207.
(6) J.L.Kulp et al., Amer. Mineral., 36 (1951) 643.
(7) M.I.Pope and D.I.Suton, Thermochim. Acta, 23 (1978) 188.
(8) Ž.D.Živković, Thermochim. Acta, 34 (1979) 91.
(9) Ž.D.Živkovic and B.Dobovisek, Min.& Met.Quat., 26 (1979) 4329

# A COMPARISON OF DTA EVALUATING METHODS OF DTA EXPERIMENTS WITH TERTIARY BUTYLPERPIVALATE AS A TESTING SUBSTANCE

R.B.Barendregt, J.Verhoeff
Prins Maurits Laboratory TNO, P.O.Box 45,
2280 AA Rijswijk (Z.H.), The Netherlands

P.J.Van den Berg
Laboratory of Chemical Technology, Delft
University of Technology, Julianalaan 136,
2628 BL Delft, The Netherlands

## ABSTRACT

To compare several methods, used to evaluate DTA experiments, the peroxide tertiary butylperpivalate was degraded. Non-isothermal conditions were applied and kinetic constants were evaluated. The results show that the methods of Borchardt-Daniels and Coats-Redfern turned out to be most appropriate. These were the most accurate and the least time-consuming. These are also comparable to isothermal results. The activation energy of the decomposition of tertiary butylperpivalate is 123 kJ mole$^{-1}$ and the frequency factor is $3,8.10^{14}$ s$^{-1}$.

## INTRODUCTION

With DTA the temperature difference between a sample holder and a reference holder is measured as a function of time and/or temperature. With non-isothermal experiments the sample as well as the reference holder are subject to a temperature program. With isothermal experiments the temperature is kept constant.

Several attempts have been made to evaluate kinetic constants (activation energy and frequency factor) from DTA experiments. This requires a number of experiments at various temperatures. With non-isothermal experiments, however, only one experiment may be sufficient.

The objective of this article is to compare well-known non-isothermal methods mutually and with the isothermal method. Therefore a testing substance (liquid) was degraded non-isothermally at different heating rates. Kinetic parameters were obtained, using different methods, and compared with isothermal results, obtained with former experiments (1).

The non-isothermal evaluation methods, which were used were those of Freeman/Carroll (2), Borchardt/Daniels (3), Kissinger (4), Rogers/Smit (5), Coats/Redfern (6), Guylai/Greenhow (7) and Horowitz/Metzger (8). Also a method based on heat effects at equal conversion was included.

## EXPERIMENTAL PART

All DTA experiments were performed in a Mettler TA 2000 thermoanalyzer. As a testing substance tertiary butylperpivalate (an organic peroxide) was used with a concentration of 25 vol % in shellsol T (a mixture of higher alkanes). Of this solution 20 to 40 mg were weighed in a NiCr high pressure vessel with an inner vessel of glass. The vessel was tightly closed before measuring. Non-isothermal curves were recorded with the following heating rates:
A. 5 K min$^{-1}$; B. 1 K min$^{-1}$; C. 0,8 K min$^{-1}$; D. 0,6 K min$^{-1}$; E. 0,4 K min$^{-1}$; F. 0,2 K min$^{-1}$; G. 0,1 K min$^{-1}$.

## THEORETICAL ASPECTS

With DTA several heat exchange phenomena play a role. These result in a heat belance equation for the reference as well as the sample holder:

Reference: $C_p^R \dfrac{dT}{dt} = k_1^R (T_F - T_R) + k_2^R (T_F^4 - T_R^4) + k_3^R (T_T - T_R)$

Sample: $C_p^S \dfrac{d(T+\Delta T)}{dt} + \dfrac{dQ_t}{dt} = k_1^S \{T_F - (T_R + \Delta T)\} + k_2^S \{T_F^4 - (T_R + \Delta T)^4\}$

$$\left[ + k_3^S \{T_T - (T_R + \Delta T)\} \right.$$

In these two equations only heat exchange phenomena between the containers and the oven ($k_1$, $k_2$) and between the containers and the thermocouples ($k_3$) were taken into account. $T_F$, $T_R$ and $T_T$ are the temperature of the oven, reference holders and thermocouple, $\Delta T$ the temperature difference between sample and reference holder, $C_p$ the heat capacity, k the heat exchange coefficients and $\dfrac{dQ_t}{dt}$ the heat generation term. Combination leads to the DTA equation (9):

$$\Delta T = \dfrac{(C_p^S - C_p^R)}{k(T)} \cdot \dfrac{dT}{dt} + \dfrac{C_p^S}{k(T)} \cdot \dfrac{d(\Delta T)}{dt} + \dfrac{dQ_t}{dt} \cdot \dfrac{1}{k(T)} - \dfrac{\Delta k(T)}{k(T)}$$

where $\Delta k(T) = (k_1^S - k_1^R)(T_F - T_R) + (k_2^S - k_2^R)(T_F^4 - T_R^4) + (k_3^S - k_3^R)(T_T - T_R)$

and $k(T) = k_1^S + 4 k_2^S T_R^3 - k_3^S$

A few simplifications may be made.

1. Reference and sample holder are placed symmetrically in the oven and there are no temperature gradients in the sample and reference mass, hence: $\frac{\Delta k(T)}{k(T)} = 0$
2. The difference in heat capacity of sample and reference mass may be neglected with respect to other terms: $c_p^S = c_p^R$.
3. The change in heat capacity of the sample during the process is directly proportional to x (degree of conversion).

With a linear heating rate ($\phi = \frac{dT}{dt}$) the DTA equation becomes:

$$\Delta T = \frac{\phi}{k(T)} \{x \cdot \Delta c_p^S + (c_p^S + x\Delta c_p^S) \cdot \frac{d\Delta T}{dt} + \frac{dQ_t}{dT}\}$$

If it is further assumed that the heat capacity and heating terms are small compared with the heat generation then the following equation remains: $\Delta T = \frac{\phi}{k(T)} \cdot \frac{dQ_t}{dt}$ which is achieved by the use of small samples and low heating rates. If only one process is involved than it follows that $\Delta T = \frac{Q_o}{k(T)} \cdot \frac{dx}{dt}$, which means that the DTA signal is directly proportional to the process rate. In the equation $Q_o$ may be regarded as the heat of reaction and k(T) as the calibration constant of the cell. The latter equation was used with the experiments.

With the DTA experiments the peak was approximated by a triangle and the baseline was assumed to be a straight line through the inflection points (Figure 1).

The melting heat of indium was used for calibration.

## RESULTS AND DISCUSSION

The curves of the non-isothermal experiments with tertiary butyl perpivalate are drawn in Figure 2. The results of the different evaluation methods as well as the comparison with isothermal results are listed in Figure 3. The method of Borchardt/Daniels and of Coats/Redfern are the best and lead to nearly the same result. Other methods lead to results that are dependent on the heating rate or have a rather great standard deviation. Non-isothermal methods which need more than one curve do not have advantages above the isothermal method; besides they lead to a great diversity in results.

To compare the methods of Borchardt/Daniels and Coats/Redfern with the isothermal results rate constants were calculated at three temperature levels (see Table 1). The three methods lead to nearly the same results. As was to be expected the experiments with the lowest heating rate lead to the best results, but it appeared that all heating rates below 1 K min$^{-1}$ are usable.

It does not appear from the results that integral methods would be better than differential methods. The Borchardt-Daniels method as a differential method seems to be suitable for working by hand, because of the easy applicability. The Coats-Redfern method as an integral method will involve more work but could easily be programmed for use in a computer. It may be seen as a disadvantage that the analytical rate expression and the order have to be known, but the isothermal methods have this same draw back. Besides, it is possible to try some orders and to select the order leading to the best results. Methods that calculate the order (e.g. Freeman and Carroll) often lead to erroneous results. It will be more useful to use Kissinger's method of shape index or the method of Guylai and Greenhow for determination of the order.

## CONCLUSIONS

1. With certain assumptions the DTA signal may be regarded as directly proportional to the rate of the process, especially if small samples and low heating rates are involved.
2. With non-isothermal experiments the methods of Borchardt/Daniels and Coats/Redfern lead to the best results. With heating rates smaller than 1 K min$^{-1}$ the calculated rate constants are nearly the same as those, formerly obtained with isothermal experiments.

## ACKNOWLEDGEMENT

The authors would like to thank Mr.A.C.Hordijk and Mr.J.H.M.Van Liempt of PML-TNO for the performance and evaluation of the isothermal DTA experiments.

## REFERENCES

(1) Internal Report PML M9203, M8009, To be published in Thermochim.Acta
(2) E.S.Freeman and B.Carroll, J.Phys.Chem., 62 (1958) 394
(3) H.J.Borchardt and F.Daniels, J.Am.Chem.Soc., 79 (1957) 41
(4) H.E.Kissinger, Anal.Chem., 29 (1957) 1702
(5) R.N.Rogers and L.C.Smit, Thermochim.Acta, 1 (1970) 1
(6) A.W.Coats and J.P.Redfern, Nature (London), 210 (1964) 68
(7) G.Guylai and E.J.Greenhow, J.Thermal Anal., 6 (1974) 279
(8) M.H.Horowitz and G.Metzger, Anal.Chem., 35 (1963) 1464
(9) M.Neriva, P.Holba and J.Sesták, Thermal Anal.Proc.Int.Conf. 4th 1974, 3 (1975) 981

## TABLE 1

Comparison of kinetic constants of tertiary butyl perpivalate, as obtained by isothermal methods, the method of Borchardt-Daniels and the method of Coats-Redfern. B. 1 K min$^{-1}$; G. 0,1 K min$^{-1}$.

| | | Isothermal method | Method Borchardt/Daniels | | Method Coats/Redfern | |
|---|---|---|---|---|---|---|
| | | | experiment B | experiment G | experiment B | experiment G |
| Activation energy [kJ mole$^{-1}$] | | 123 | 130 | 121 | 130 | 126 |
| Frequency factor [s$^{-1}$] | | $3,8.10^{14}$ | $3,1.10^{15}$ | $1,5.10^{14}$ | $3,7.10^{15}$ | $8,1.10^{14}$ |
| Logarithm of the rate constant (log k(T)) at: | 323 K | -5,3 | -5,5 | -5,4 | -5,5 | -5,5 |
| | 348 K | -3,9 | -4,0 | -4,0 | -3,9 | -4,0 |
| | 373 K | -2,6 | -2,7 | -2,8 | -2,6 | -2,7 |

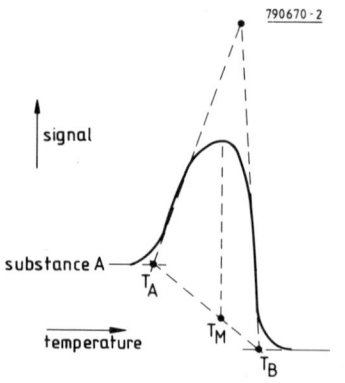

Fig.1. Estimation of the DTA peak by a triangle $T_M$=temperature of maximum $T_A$, $T_B$=temperatures of the inflection points.

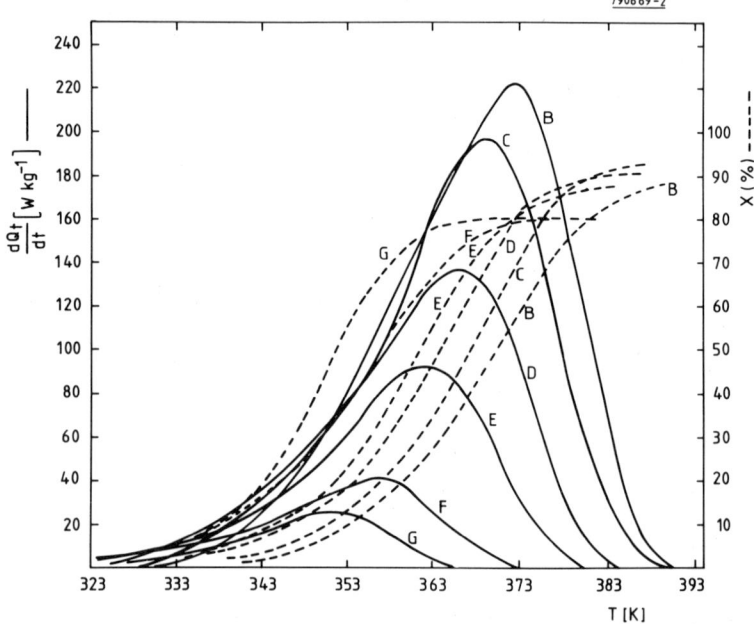

Fig.2. Thermal degradation of tertiary butyl perpivalate. The heat generation ($\frac{dQ_t}{dt}$) and degree of conversion (x) as a function of the temperature. B: 1 K min$^{-1}$; C: 0,8 K min$^{-1}$; D: 0,6 K min$^{-1}$; E: 0,4 K min$^{-1}$; F: 0,2 K min$^{-1}$; G: 0,1 K min$^{-1}$.

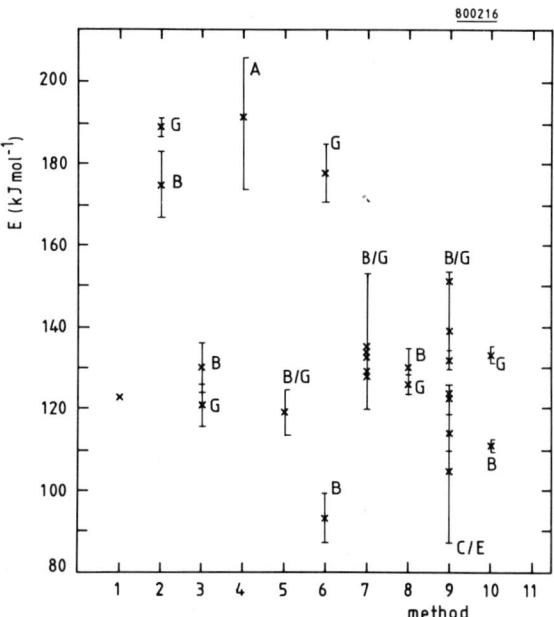

Fig.3. The activation energy (E), calculated with evaluating methods to DTA. A: 5 K min$^{-1}$; B: 1 K min$^{-1}$; C: 0,8 K min$^{-1}$; D: 0,6 K min$^{-1}$; E: 0,4 K min$^{-1}$; F: 0,2 K min$^{-1}$; G: 0,1 K min$^{-1}$.

1. Isothermal[x]; 2. Freeman/Carroll; 3. Borchardt/Daniels; 4. Piloyan; 5. Kissinger; 6. Rogers/Smit; 7. Equal conversion; 8. Coats/Redfern; 9. Guylai/Greenhow; 10. Horowitz/Metzger.

[x] Obtained formerly (lit. 1)

A COMPARISON OF THE KINETICS FOR THE THERMAL DECOMPOSITION
OF $Ba(OH)_2 \cdot H_2O$ OBTAINED BY TG AND EGA TECHNIQUES

P. K. Gallagher and E. M. Gyorgy
Bell Laboratories, Murray Hill, N.J. 07974

## ABSTRACT

Simultaneous measurements were made of the weight loss of the sample and the dew point of the $N_2$ stream as $Ba(OH)_2 \cdot H_2O$ was heated at 1°C/min. A computer program converted the dew point to weight of water/min and integrated this over the time of the experiment. The amount of water due to the decomposition was small compared to the natural content of the predried gas stream. Consequently, accurate subtraction of this background was required and achieved by using a B spline polynomial fit to the EGA data in the regions where no weight loss occurred.

The corrected EGA curve compared very well with the TG results. Kinetic analysis of the simple loss of the water of hydration (60-100°C) was performed. Agreement was excellent for the simple dehydration.

## INTRODUCTION

Magnetic factors have been observed to influence the reactivity of materials. An external magnetic field has markedly changed the rate of reaction when either a product or reactant is below its Curie temperature, $T_c$.[1-3] In the absence of an external magnetic field anomalies have been observed in oxidation rates at $T_c$.[4-6]

There are obvious difficulties following the reaction rate of magnetic materials in the presence of an external magnetic field by weight changes. This work is intended to determine the feasibility of using moisture analysis of the gas stream to determine the rate of reduction of oxides by hydrogen. Kinetics of the dehydration and decomposition of $Ba(OH)_2 \cdot H_2O$ was investigated simultaneously by TG and EGA techniques to determine if the same activation energy was obtained by both methods. The choice of $Ba(OH)_2 \cdot H_2O$ was based on its two distinct evolutions of water which are not only widely separated in temperature but are also different from a mechanistic point since they arise from different chemical processes.

## EXPERIMENTAL PROCEDURE AND RESULTS

The sample was Fisher Scientific Co. reagent grade $Ba(OH)_2 \cdot H_2O$. The TG experiment was performed using a Fisher Scientific Co. Thermobalance Model 100. Water content of the exit gas stream was measured by a Panametrics Model 700 Moisture Analyzer. The path between the sample and detector was made as short as feasible based upon the furnace size. Output from the Cahn RG balance, from a Chromel Alumel thermocouple in close proximity to the sample, and from the analog output of the Panametrics Moisture Analyzer were inputs to a digital data acquisition system described elsewhere.[7] Measurements were averaged over one minute intervals and the temperature was programmed to rise at $1°C\ min^{-1}$. Sample size was 20 mg and $N_2$ was flowed through the system at a rate of $365\ cm^3 min^{-1}$. This relatively rapid flow was necessary to minimize the time lag and diffusional broadening of the evolution peak.

Figure 1 shows the TG and DTG curves for the experiment. The dew point of the gas stream was converted to ppm of $H_2O$ and coupled with the flow rate to yield mgs of $H_2O$ per minute analogous to the DTG curve. The values were progressively summed to give an integrated value of mgs evolved. These two curves are presented in Fig. 2.

It is obvious from inspection of Fig. 2 that the background moisture in the gas stream is large relative to that evolved during the decomposition of the sample. Peaks associated with the temporary evolution are clearly evident but they are overwhelmed in the integrated curve by the background. To subtract the background, a B spline routine was used to fit an eighth order polynomial to the evolution curve using the areas where there was no weight loss, i.e., 40-50, 110-375 and 540-760°C. This background was subtracted correcting the EGA curve as shown in Fig. 3. The base line is now flat and at the appropriate level. The DTG curve is reproduced along with the corrected EGA curve for comparison. The corrected EGA curve is integrated to give the total water evolved for comparison with the TG curve in Fig. 4. An alternative to this mathematical correction would be to use a second analyzer prior to the sample and then use the differential output. This however does not take into account degassing of the apparatus with increasing temperature or the change in detector sensitivity with temperature.

## DISCUSSION

Inspection of both Figs. 3 and 4 indicates that the amount of water determined from the corrected EGA is less than that measured by TG. However, to the extent that this is due to constant errors in the flow rate, calibration of the moisture sensor, or background correction; the kinetic analysis will be unaffected. If the shapes of the DTG and EGA peaks are identical then values of fraction reacted, $\alpha$, will be the same regardless of absolute values of the weight loss.

The TG and integrated EGA curves were subjected to the kinetic analysis developed in earlier work.[8] This dynamic kinetic analysis ($0.1 \leq \alpha \leq 0.9$) utilizes three common methods: difference-differential (FC), integral (CR), and differential (ABS). Results for the dehydration are summarized in Table I. The contracting area best fit the numerous models used by both the ABS and CR methods. The FC approach does not allow for various models but the optimum order of the reaction, n, is very close to 0.50 predicted by the contracting area model. Agreement between the two experimental methods is excellent and suggests that the EGA approach can be used successfully.

The decomposition step is more complex because the melting of $Ba(OH)_2$ occurs and the product BaO is insoluble in the melt.[9] The complexity is obvious from the shapes of DTG and EGA curves in Fig. 3. A multistep decomposition was also observed by Judd and Pope.[10] Consequently a kinetic analysis of this data was not made.

## REFERENCES

(1) G. S. Krinchik, R. M. Shavartsman, and A. Y. Kipnis, JETP Lett., 19, (1974) 231

(2) M. W. Rowe, S. M. Lake, and R. Fanick, Nature (London), 266 (1977) 612

(3) R. S. Mehta, M. S. Dresselhaus, G. Dresselhaus, and H. J. Zeiger, Phys. Rev. Lett. 43 (1979) 970

(4) P. K. Gallagher, E. M. Gyorgy, and H. E. Bair, J. Chem. Phys. 71 (1979) 830

(5) B. C. Scales and M. B. Maple, Phys. Rev. Lett. 39 (1977) 1636

(6) J. C. Measor and K. K. Afzulpurkar, Philos. Mag., 10 (1964) 817

(7) P. K. Gallagher and D. W. Johnson, Jr., Thermochim. Acta 4 (1972) 283

(8) D. W. Johnson, Jr., and P. K. Gallagher, J. Phys. Chem. 76 (1972) 1474
(9) J. M. Criado and J. Morales, J. Therm. Anal. 10 (1976) 103
(10) M. D. Judd and M. I. Pope, "Thermal Analysis, Vol. 2" (R. F. Schwenker, Jr. and P. D. Garn, eds.), Academic Press, N. Y. 1969, p. 1423

TABLE I

Kinetic Arrhenius Parameters for the Reaction[a]

$Ba(OH)_2 \cdot H_2O \rightarrow Ba(OH)_2 + H_2O$

|  | ABS[b] | | CR[b] | | FC | | |
|---|---|---|---|---|---|---|---|
|  | E | A | E | A | E | A | n |
| TG | 21.2 | $4.5 \times 10^9$ | 24.6 | $5.1 \times 10^{11}$ | 21.4 | $5.0 \times 10^9$ | 0.52 |
| EGA | 21.7 | $8.0 \times 10^9$ | 25.6 | $2.0 \times 10^{12}$ | 22.5 | $2.3 \times 10^{10}$ | 0.56 |

a. Units of E are Kcal Mole$^{-1}$ and A are sec$^{-1}$.
b. These results are for the contracting area model.[8]

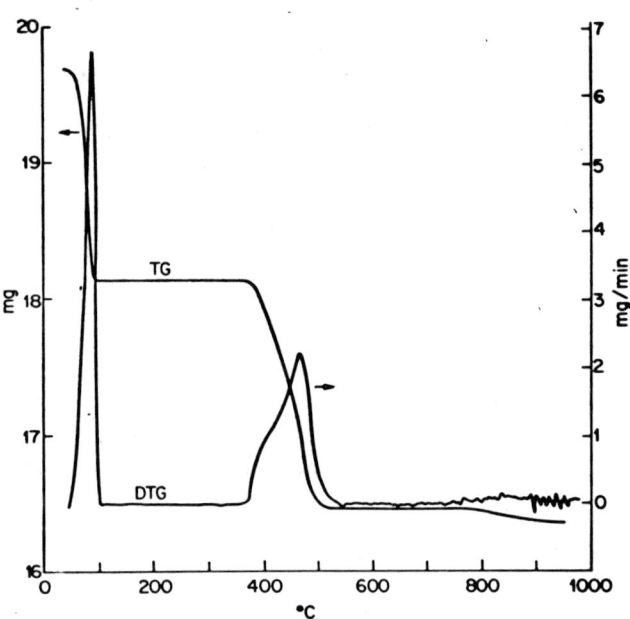

Fig. 1  TG and DTG curves for $Ba(OH)_2 \cdot H_2O$ heated at 1°C/min in $N_2$.

117

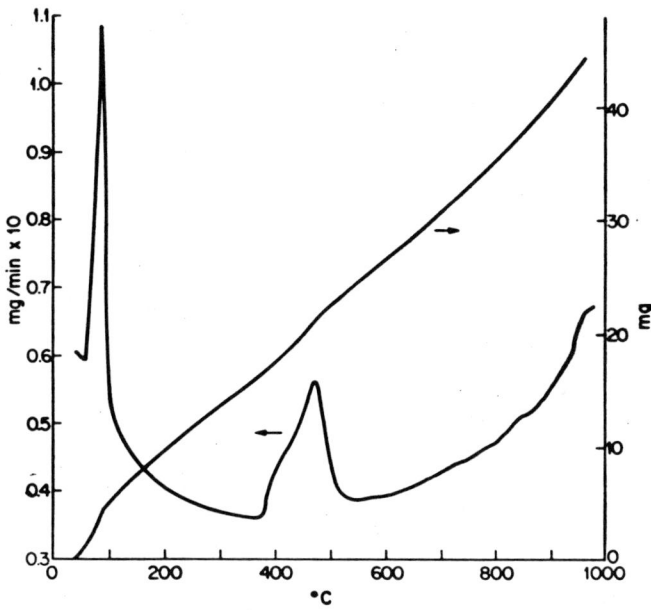

Fig. 2
EGA of $Ba(OH)_2 \cdot H_2O$ heated at 1°C/min in $N_2$.

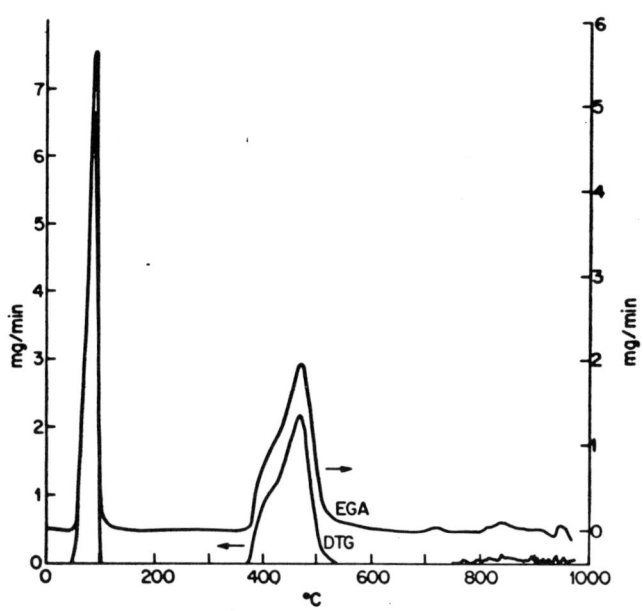

Fig. 3
Comparison of the corrected EGA with the DTG.

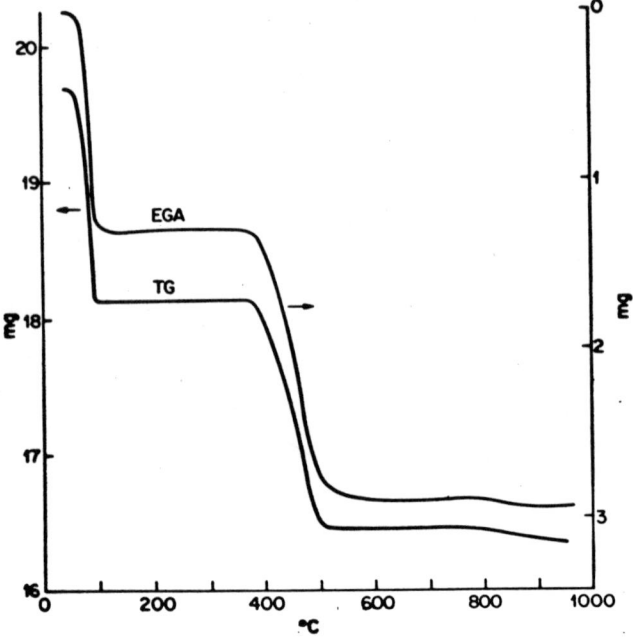

Fig. 4  Comparison of the corrected integrated EGA with the TG.

# KINETIC ANALYSIS FROM THERMOGRAVIMETRIC TRACES

T.P.Prasad and M.S.R.SWAMI
REGIONAL RESEARCH LABORATORY, BHUBANESWAR-751013
INDIA

## INTRODUCTION

KINETIC ANALYSIS USING THE THERMAL ANALYSIS TECHNIQUES IS WELL KNOWN AND IS AN ESTABLISHED PRACTICE. SEVERAL TREATMENTS HAVE BEEN PROPOSED FROM TIME TO TIME. IN THE RECENT PAST, ZSAKO[1] PROPOSED A METHOD WHICH HAS BEEN FOUND TO BE WIDELY APPLICABLE. IN THE PRESENT APPROACH, ZSAKO'S TRIAL AND ERROR METHOD HAS BEEN FURTHER SIMPLIFIED SO AS TO ENABLE TO ASSIGN MECHANISM AT EACH TEMPERATURE OF THE PROCESS UNDER INVESTIGATION.

## METHOD

THE STARTING EQUATION FOR THIS PURPOSE HAS BEEN:

$$\log \frac{z E_a}{R q} = \log g(\alpha) - \log p(x) - B \quad\quad (1)$$

THE FORM OF $g(\alpha)$ WAS TAKEN FROM TABLE 1 AND ITS VALUE WAS CALCULATED FROM THE THERMOGRAVIMETRIC CURVE. THE VALUE OF $-\log p(x)$ AT EACH TEMPERATURE OF THE PROCESS UNDER INVESTIGATION WAS CALCULATED AS SUGGESTED BY ZSAKO. THE VALUE OF B WAS CALCULATED AT EACH TEMPERATURE USING EQUATION (1) FOR EACH OF THE $g(\alpha)$ FUNCTIONS PRESENTED IN TABLE 1. WE GET AS MANY B VALUES AS THERE ARE $g(\alpha)$ FUNCTIONS AT EACH TEMPERATURE. THE STANDARD DEVIATION, $\delta$, WAS CALCULATED USING THE EQUATION:

$$\delta = \sqrt{\frac{(\bar{B} - B_i)^2}{r}} \quad\quad (2)$$

WHERE, $B_i$ IS THE INDIVIDUAL B VALUE FOR A PARTICULAR $g(\alpha)$ FUNCTION, $\bar{B}$ IS THE ARITHMATIC MEAN OF ALL THE $B_i$ VALUES AND r REPRESENTS THE NUMBER OF $g(\alpha)$ FUNCTIONS TESTED. THUS WE GET AS MANY $\delta$ VALUES AS THERE ARE $g(\alpha)$ FUNCTIONS AND THE $g(\alpha)$ FUNCTION CORRESPONDING TO THE MINIMUM VALUE OF $\delta$, $\delta_{min}$, IS TAKEN TO REPRESENT THE MECHANISTIC EQUATION AT THAT TEMPERA-

TURE. THE PROCEDURE IS REPEATED AT EACH TEMPERATURE IN THE ENTIRE TEMPERATURE RANGE OF THE PROCESS UNDER INVESTIGATION. THUS AT EACH TEMPERATURE, A $\delta_{min}$ VALUE AND THE CORRESPONDING g ($\alpha$) FUNCTION WERE OBTAINED. THE PRESENT APPROACH HAS BEEN FOUND TO BE SIMPLER WITH THE ADDITIONAL ADVANTAGE OF TESTING THE MECHANISM AT EACH TEMPERATURE OF THE PROCESS.

THE EXPERIMENTAL ACTIVATION ENERGY MAY BE CALCULATED USING ANY OF THE METHODS PROPOSED BY COATS AND REDFERN [2], FREEMAN AND CAROLL [3], HOROWITZ AND METZGER [4], ETC.

TO CALCULATE THE FREQUENCY FACTOR FOR THE ENTIRE PROCESS UNDER INVESTIGATION, THE VALUE OF $\overline{B}$ CORRESPONDING TO THE MINIMUM $\delta_{min}$ WAS TAKEN AS REQUIRED BY THE EQUATION :

$$\log Z = \overline{B} - \log Rq - \log Ea \quad\quad (3)$$

WHERE, R IS THE UNIVERSAL GAS CONSTANT, q IS THE HEATING RATE AND $E_a$ IS THE ACTIVATION ENERGY. THE APPARENT ACTIVATION ENTROPY IS CALCULATED BY THE EQUATION :

$$S^{\ddagger} = 2.303 \log \frac{Zh}{kT} \quad\quad (4)$$

WHERE h IS THE PLANCK'S CONSTANT, T IS $T_{1/2}$ AND k IS THE RATE CONSTANT. $T_{1/2}$ IS THE TEMPERATURE AT WHICH $\alpha = 0.5$. THE VALUE OF k IS CALCULATED FROM THE EQUATION :

$$k = Z \exp(-Ea/RT_{1/2}) \quad\quad (5)$$

<u>RESULTS</u>

THE PROCEDURE IS ILLUSTRATED BY TAKING THE DEHYDRATION OF CALCIUM OXALATE MONOHYDRATE INTO CONSIDERATION.

USING THE EQUATION OF DHARWADKAR AND KHARKHANAWALA [5], AN EXPERIMENTAL ACTIVATION ENERGY OF 19 KCAL/MOLE WAS OBTAINED. VALUES OF B FOR VARIOUS MECHANISTIC EQUATIONS ALONGWITH $\delta$- VALUES ARE GIVEN IN TABLE 2. FOR THE TEMPERATURE OF 473° K. FROM THE TABLE IT IS EVIDENT THAT $\delta$ HAS A MINIMUM VALUE OF 0.039 AT 473°K. THE CORRESPONDING MECHANISTIC EQUATION IS, THEREFORE, $R_3$ AND THE RATE-CONTROLLING PROCESS IS

"PHASE BOUNDARY REACTION-ASSUMES SPHERICAL SYMMETRY"
(FOR EXPLANATION OF MECHANISTIK EQUATIONS PLEASE REFER TABLE 1 ). THE MECHANISTIC EQUATIONS AT OTHER TEMPERATURES ARE GIVEN IN TABLE 3. FROM TABLE 3, IT IS EVIDENT THAT THE MECHANISM CHANGES AFTER $513^{\circ}K$ AND THE MINIMUM $\delta_{min}$ VALUE FOR THE WHOLE PROCESS IS 0.002. USING THE VALUE OF $\bar{B}$ CORRESPONDING TO THIS VALUE, A VALUE OF $1.4 \times 10^{6}$ $SEC^{-1}$ WAS OBTAINED FOR Z AND A VALUE OF - 65.5 e.u. FOR $S^{\ddagger}$.

TABLE 4 PRESENTS MECHANISTIC EQUATIONS AT EACH TEMPERATURE BY WIDELY VARYING THE VALUE OF $E_{\alpha}$. IT CAN BE SEEN THAT THE PRESENT APPROACH GIVES THE SAME MECHANISTIC EQUATION INDICATING THAT THE PRESENT METHOD IS INDEPENDENT OF $E_{\alpha}$.
THUS, E MAY BE CALCULATED FROM ANY ONE OF THE METHODS AVAILABLE WITHOUT RECOURSE TO TRAIL AND ERROR METHODS TO ARRIVE AT "CORRECT" $E_{\alpha}$ VALUE.

## REFERENCES

(1) J. ZSAKO, JOURNAL OF PHYSICAL CHEMISTRY, 72 (1968) 2406
(2) A.W. COATS AND J.P. REDFERN, NATURE, 101 (1964) 68
(3) E.S. FREEMAN AND B. CARROLL, JOURNAL OF PHYSICAL CHEMISTRY, 62 (1958) 394
(4) H.H. HOROWITZ AND G. METZGER, ANALYTICAL CHEMISTRY 35 (1963) 1464
(5) S.R. DHARWADKAR AND M.D. KHARKHANAWALA, 'THERMAL ANALYSIS', EDITED BY G.F. SCHWENKER AND P.D. GARN, ACADEMIC PRESS, NEWYORK, 2 (1969) 1049

TABLE 1

MECHANISTIC EQUATIONS

| TYPE OF EQUATION | SYMBOL | FORM OF $f(\alpha)$ | FORM OF $g(\alpha)$ | RATE-CONTROLLING PROCESS |
|---|---|---|---|---|
| PARABOLIC LAW | $D_1$ | $\frac{1}{2}\alpha$ | $\alpha^2$ | * |
|  | $D_2$ | $[-\ln(1-\alpha)]^{-1}$ | $\alpha + (1-\alpha)\ln(1-\alpha)$ |  |
| JANDER | $D_3$ | $\frac{3}{2}(1-\alpha)^{2/3}\left[1-(1-\alpha)^{1/3}\right]^{-1}$ | $\left[1-(1-\alpha)^{1/3}\right]^2$ |  |
| GINSTLING-BROUNSTEIN | $D_4$ | $\frac{3}{2}(1-\alpha)^{1/3}\left[1-(1-\alpha)^{1/3}\right]^{-1}$ | $\left(1-\frac{2}{3}\alpha\right)-(1-\alpha)^{2/3}$ |  |
| MAMPEL | $F_1$ | $(1-\alpha)$ | $-\ln(1-\alpha)$ |  |
| AVRAMI | $A_2$ | $2(1-\alpha)[-\ln(1-\alpha)]^{1/2}$ | $[-\ln(1-\alpha)]^{1/2}$ |  |
| AVRAMI | $A_3$ | $3(1-\alpha)[-\ln(1-\alpha)]^{2/3}$ | $[-\ln(1-\alpha)]^{1/3}$ |  |
|  | $R_2$ | $2(1-\alpha)^{1/2}$ | $1-(1-\alpha)^{1/2}$ |  |
|  | $R_3$ | $3(1-\alpha)^{2/3}$ | $1-(1-\alpha)^{1/3}$ |  |

* PLEASE SEE C.G.R.NAIR AND P.M.MADHUSUDANAN, THERMOCHIMICA ACTA, 14 (1976) 373

## TABLE 2

### TYPICAL $\delta$ VALUE FOR DIFFERENT MECHANISTIC EQUATIONS AT SPECIFIED TEMPERATURE

$$CaC_2O_4 \cdot H_2O \longrightarrow CaC_2O_4 + H_2O$$

$T$ = $473°K$
$\alpha$ = 0.147
$-\log p(x)$ = 11.429
$\overline{B}$ = 10.026

| MECHANISTIC EQUATION | B | $\delta$ |
|---|---|---|
| $D_1$ | 9.766 | 0.087 |
| $D_2$ | 9.487 | 0.180 |
| $D_3$ | 9.857 | 0.390 |
| $D_4$ | 9.842 | 0.395 |
| $F_1$ | 10.632 | 0.202 |
| $A_2$ | 11.030 | 0.303 |
| $A_3$ | 11.163 | 0.379 |
| $R_2$ | 10.313 | 0.096 |
| $R_3$ | 10.143 | 0.039 * |

\* INDICATES THE VALUE OF $\delta_{min}$ AT THE SPECIFIED TEMPERATURE

## TABELE 3

### MECHANISTIC EQUATIONS CORRESPONDING TO $\delta_{min}$ VALUES AT DIFFERENT TEMPERATURES

| T, °K | $\alpha$ | $\delta_{min}$ | MECHANISTIC EQUATION CORRESPONDING TO $\delta_{min}$ |
|---|---|---|---|
| 453 | 0.074 | 0.071 | $R_3$ |
| 473 | 0.147 | 0.039 | $R_3$ |
| 493 | 0.263 | 0.013 | $R_3$ |
| 503 | 0.368 | 0.002 * | $R_3$ |
| 513 | 0.474 | 0.014 | $R_3$ |
| 523 | 0.611 | 0.023 | $R_2$ |
| 533 | 0.716 | 0.013 | $R_2$ |
| 543 | 0.810 | 0.003 | $R_2$ |
| 553 | 0.947 | 0.007 | $R_2$ |

$$E_a = 19 \text{ Kcal/mole}$$
$$Z = 1.4 \times 10^8 \text{ sec}^{-1}$$
$$S^{\neq} = -65.475 \text{ e.u.}$$

\* INDICATES THE VALUE OF MINIMUM $\delta_{min}$

## TABLE 4

## NON-DEPENDENCE OF MECHANISM ON ACTIVATION ENERGY

SYSTEM : $CaC_2O_4 \cdot H_2O \longrightarrow CaC_2O_4 + H_2O$

| Temperature, °K | Activation energy Kcal/mole | Mechanism Symbol |
|---|---|---|
| 453 | 10 | $R_3$ |
|  | 26 | $R_3$ |
| 473 | 10 | $R_3$ |
|  | 26 | $R_3$ |
| 493 | 10 | $R_3$ |
|  | 26 | $R_3$ |
| --- | -- | -- |
| 533 | 10 | $R_2$ |
|  | 26 | $R_2$ |
| 543 | 10 | $R_2$ |
|  | 26 | $R_2$ |
| 553 | 10 | $R_2$ |
|  | 26 | $R_2$ |

# A SPLINE INTERPOLATION METHOD TO STUDY KINETICS FROM ISOTHERMAL AND NON-ISOTHERMAL THERMOGRAVIMETRY

C.G.R. Nair and P.M. Madhusudanan
Department of Chemistry, Kerala University
Trivandrum-695001, India

## ABSTRACT

A spline interpolation method, free from subjective errors, is suggested for the determination of slopes $d\alpha/dT$ used for the evaluation of kinetic parameters from thermogravimetry.

## INTRODUCTION

The mathematical methods in vogue for the evaluation of kinetic parameters from thermogravimetric data (1),(2) can be classified into three categories: a) the approximation methods, b) the integral methods and c) the differential methods. In two earlier communications (3),(4) we had proposed mathematical refinements for an approximation method (3) and for an integral method (4). In the present communication, we propose a mathematically elegant procedure which may be adapted for differential methods.

The differential methods involve the measurement of the slope of the TG curve at chosen points. Their main drawback is the difficulty in obtaining reliable values of slopes from TG traces. The common procedure employed for this purpose is the method of tangents or numerical differentiation (5),(6),(7), eg., with the aid of central difference formulae. Both these depend on the personal judgement of the experimentor and are thus prone to subjective errors. The procedure suggested in this paper is free from such subjective errors.

## DERIVATION AND DISCUSSION

The differential equation for the thermogravimetric trace can be written in the form

$$\frac{d\alpha}{dT} = \frac{kT}{h\phi} e^{\Delta S/R} f(\alpha) e^{-E/RT} \qquad \cdots (1)$$

where $\alpha$, $f(\alpha)$, $\phi$, $\Delta S$ and E represent the fractional decomposition at temperature T, the conversion function, the heating rate, the entropy of activation and the energy of activation, respectively. Eq. (1) can be written in the logarithmic form, after some convenient transformation, as follows:

$$\log \frac{1}{Tf(\alpha)} \frac{d\alpha}{dT} = \log \frac{k}{h\phi} e^{\Delta S/R} - \frac{E}{2.303RT} \qquad \cdots (2)$$

A plot of the LHS versus 1/T will give a straight line from the slope and intercept of which E and $\Delta S$ can be determined. What form of $f(\alpha)$ shall we use? The usual form of $f(\alpha) = (1-\alpha)^n$, well-known in homogeneous kinetics, will be not quite relevant in solid state decomposition kinetics, for various reasons. (For one thing, the order parameter "n" has only a formal use in solid state kinetics; its physical significance is questionable and undefined as yet (6-10)). Therefore, we shall have to use the form of $f(\alpha)$ from mechanism-invoking equations. This is done by curve-fitting techniques employing each of the nine-mechanism-invoking equations (11) and choosing the best fit value.

Now, the only difficulty associated with the use of eq. (2) is the evaluation of the slope, $d\alpha/dT$, from the TG trace at various points $(T_i, \alpha_i)$. Earlier workers obtained the slopes either by drawing tangents or by using central difference formulae. We propose the following spline interpolation method (12) for the slope determination.

Let $(T_0, \alpha_0)$, $(T_1, \alpha_1)$ ..... $(T_n, \alpha_n)$ be the set of experimental points through which the TG curve passes. (The T values may be equally spaced or not). We note that the function $\alpha$ is such that $d\alpha/dT$ and $d^2\alpha/dT^2$ are continuous between $T_0$ and $T_n$. Now we assume that $\alpha$ coincides with a cubic polynomial in each closed subinterval between $T_i$ and $T_{i+1}$ where i = 0,1,2, .... (n-1) and define $\alpha = \alpha_i$ at temperature $T_i$. The cubic polynomial for the interval between $T_i$ and $T_{i+1}$ may be written in the form

$$\alpha = a_i (T-T_i)^3 + b_i (T-T_i)^2 + c_i (T-T_i) + d_i \quad .. (3)$$

where $a_i$, $b_i$, $c_i$ are coefficients and $d_i$ is a constant. Now employing the usual spline interpolation technique, the values of the first derivative $(\frac{d\alpha}{dT})_j$ and $\alpha_j$ at any point $(T_j, \alpha_j)$ intermediate between the selected points $(T_i, \alpha_i)$, $(T_{i+1}, \alpha_{i+1})$ can be evaluated by a simple computer program.

The method described above pertains to non-isothermal thermogravimetry. But it can be equally well applied to isothermal TG traces also. For isothermal TG, equation (1) can be modified, using the time differential dt (instead of the temperature differential dT) and omiting the heating rate $\phi$ from RHS. The procedure thereafter is similar.

## TESTING THE VALIDITY OF THE METHOD

The validity of the method has been tested by numerical evaluation carried out on (a) a theoretical TG trace (13) and (b) an experimental TG trace of the dehydration of calcium oxalate monohydrate. The data are presented in Tables I to V.

The salient advantage of the method suggested here is that the slope is determined very accurately by using the spline interpolation formula, avoiding subjective errors altogether. Further, it may be noted that the use of eq. (2) has the following additional incidental features: (1) the mechanism-based form of $f(\alpha)$ is used, (2) the temperature dependence of the pre-exponential factor is taken into account by invoking the logic of the absolute rate theory.

## REFERENCES

(1) W.W. Wendlandt, "Thermal Methods of Analysis", second edition (Interscience, New York, 1974).
(2) R.C. Mackenzie (Ed.), "Differential Thermal Analysis", Vol. I and II, (Academic Press, New York, 1970).
(3) P.M. Madhusudanan and C.G.R. Nair, Thermochim. Acta, 12 (1975) 97
(4) C.G.R. Nair and P.M. Madhusudanan, Current Science (India), 44 (1975) 212

(5) J. Sestak, A. Brown, V. Rihak and C. Berggren, in "Thermal Analysis", Vol.II, (Eds.) R.F. Schwenker, Jr., and P.D. Garn (Academic Press, New York, 1969), p. 1036.
(6) W.B. Hillig, in "Kinetics of High Temperature Processes", (Ed.), W.D. Kingery (Wiley, New York, 1959), p. 311.
(7) W. Gomez, Nature, 192 (1961) 865
(8) B.N.N. Achar, G.W. Brindley and J.H. Sharp, in "Proceedings of the International Clay Conference, 1966", (Eds.), L. Heller and A. Weiss (Israel,Program for Scientific Translations, Jerusalem, 1966), p. 128.
(9) T.A. Clarke and J.M. Thomas, Nature, 219 (1968) 149
(10) T.A. Clarke, R.L. Evans, K.G. Robins, and J.M. Thomas, Chem. Comm.,(1969) 266
(11) F. Skvara and V. Satava, J. Thermal Anal., 2 (1970) 330
(12) E.V. Krishnamurthy and S.K. Sen, "Computer-based Numerical Algorithms", (Affiliated East-West Press, New Delhi, 1976), p. 117
(13) G. Gyulai and E.J. Greenhow, J. Therm. Anal., 6 (1974) 279
(14) E.S. Freeman and B. Carroll, J. Phy. Chem., 62 (1958)394
(15) H.H. Horowitz and G. Metzger, Anal. Chem., 35 (1963)1464
(16) A.W. Coats and J.P. Redfern, Nature, 201 (1964) 68

Table I: THEORETICAL TG DATA[13]

(Values of log A, E and $\phi$ and form of $f(\alpha)$ are assumed. From a computerised program, values of T corresponding to chosen values are calculated)
(Assumed values: log A = 14.00; E = 60 k cal; $f(\alpha) = (1-\alpha)$; $\phi = 2°/min$)

| Temperature, K | Chosen value of $\alpha$ | Temperature, K | chosen value of $\alpha$ |
|---|---|---|---|
| 819.5 | 0.1 | 867.8 | 0.6 |
| 835.6 | 0.2 | 874.3 | 0.7 |
| 846.1 | 0.3 | 881.3 | 0.8 |
| 854.2 | 0.4 | 890.1 | 0.9 |
| 863.3 | 0.5 | | |

## Table II: THEORETICAL TG DATA

(The coefficients $a_i$, $b_i$, $c_i$ and $d_i$ in eq.(3) are calculated by feeding data from Table I. Now, using spline fit, $\alpha$ and $d\alpha/dT$ are calculated at chosen values of T). Calculated values of slope and $\alpha$ at specified temperatures:

| Chosen temperature, K | $\alpha$ calculated from spline fit | slope $(\frac{d\alpha}{dT})$ x $10^3$ calculated from spline fit |
|---|---|---|
| 820 | 0.1024 | 4.794 |
| 830 | 0.1591 | 6.656 |
| 840 | 0.2379 | 9.268 |
| 850 | 0.3461 | 12.354 |
| 860 | 0.4808 | 14.780 |
| 870 | 0.6343 | 15.498 |
| 880 | 0.7823 | 13.725 |
| 890 | 0.8997 | 9.121 |

values calculated by spline fit method
log A = 13.84     E = 59.32 k cal.

## Table III: EXPERIMENTAL DATA

(Values of T and $\alpha$ from actual experimental TG trace are tabulated for the dehydration of Calcium oxalate monohydrate, $CaC_2O_4 \cdot H_2O$). $\phi$ = 5°/min, $f(\alpha) = (1-\alpha)^{2/3}$.

| Temperature, K | $\alpha$ | Temperature, K | $\alpha$ |
|---|---|---|---|
| 393 | 0.041 | 419 | 0.403 |
| 398 | 0.078 | 425 | 0.560 |
| 403 | 0.119 | 431 | 0.754 |
| 408 | 0.187 | 437 | 0.944 |
| 413 | 0.269 | | |

## Table IV

Values of slope and $d\alpha/dT$ calculated from experimental data using spline fit for $CaC_2O_4 \cdot H_2O$. (Coefficient $a_i$, $b_i$, $c_i$ and $d_i$ in eq. (3) are calculated by feeding data from Table III. Now, using spline fit, $\alpha$ and $d\alpha/dT$ are calculated at chosen values of T).

| Chosen temperature, K | $\alpha$ | $(\frac{d\alpha}{dT}) \times 10^3$ |
|---|---|---|
| 390 | 0.01512 | 6.610 |
| 400 | 0.09235 | 7.500 |
| 405 | 0.14350 | 13.470 |
| 410 | 0.21758 | 15.730 |
| 415 | 0.30979 | 21.650 |
| 420 | 0.42754 | 24.720 |
| 425 | 0.56007 | 28.970 |
| 430 | 0.71917 | 34.390 |

## Table V

Comparison of activation energies of dehydration of $CaC_2O_4 \cdot H_2O$ reported in literature with the value obtained from the present spline fit method.

| Method | E. kJ mole$^{-1}$ |
|---|---|
| Freeman-Carroll[14] | 92.05 |
| Hovowitz-Metzger[15] | 84.52 |
| Coats-Redfern[16] | 89.54 |
| The present spline fit method | 85.86 |

IMPROVEMENT OF DIFFERENTIAL CORRECTION METHOD FOR EVALUATION
OF KINETIC PARAMETERS FROM TG CURVES

P.H. Fong, S.P. Wong and D.T.Y. Chen
Department of Chemistry, The Chinese University of Hong Kong, Hong Kong

ABSTRACT

A new differential correction method for evaluating kinetic parameters from thermogravimetric data, based on the minimization of the sum of squares difference between the observed and calculated temperatures, is proposed. The new method was tested by the experimental data of acid catalyzed iodination of acetone.

INTRODUCTION

In a recent paper[1], a combined numerical method was suggested to evaluate kinetic parameters from TG curves. The secondary $\frac{d\alpha}{dT}$ data are first treated by the Linear Least Squares Method of Gay[2] to find out the correct type of reaction mechanism and approximate values of its kinetic parameters. The latters are, then, improved by differential correction method using the more exact experimental data of $\alpha$ and T. The method has been applied successfully to analyse a set of synthetic data and thermal dehydration data of gypsum by Sestak et al. and also the TG curve of a $Mg(OH)_2$ sample[3]. However, the method is not always successful, divergence may occure in some cases especially when it is applied to analyse solution data[4]. It is therefore, necessary to improve the method.

THEORETICAL PART

The basic equation for a dynamic kinetic method is

$$\int_o^\alpha \frac{d\alpha}{f(\alpha)} = \frac{A}{C}\int_o^T e^{-E/RT} dT \qquad (1)$$

where $\alpha$ is the fraction reacted at time t at which the temperature of the sample is T, A is the pre-exponential factor for the Arrhenius equation, C is the heating rate and E, the activation energy. The left hand side of the above equation may be symbolised by F and the right hand side, by G. In the original DC method, the principle is to find out a set of kinetic parameters m, n, p, A and E so as to make the sum of squares difference

$$S = \Sigma \left[ F(\alpha_i, m, n, p) - G(T_i, A, E) \right]^2 \qquad (2)$$

a minimum.

The reason for divergence may be explained as follows. Suppose a set of parameters m, n and p give a value of F smaller than the desired accuracy, then, any set of A and E which give G value smaller than, not necessarily close to F, would yield S satisfy our condition mathematically but might not give rational parameters and would, thus, cause overflow or underflow in computer programme.

Now let us think of the problem in a slight different way. For a certain set of parameters, a graph of $\alpha$ vs T curve may be calculated from equation (1). An $\alpha$ vs T curve may also be constructed from thermal analysis data. So, for a given value of $\alpha_i$, there are two values of T; $T_e^i$ read from the experimental curve and $T_c^i$, from the calculated curve. The best set of parameters must be those which would make the following sum of least squares difference

$$S_T = \Sigma \left[ T_e^i - T_c(\alpha_i, m, n, p, A, E) \right]^2 \qquad (3)$$

a minimum. In this treatment $T_e^i$ are experimental values which are definite and real. Therefore, there could only be one set of kinetic parameters that would give $S_T$ a minimum value; we might expect no divergence to occur. Beside this merit, kinetic parameters obtained by minimizing $S_T$ are believed to be more accurate than those obtained by minimizing S. As we have noticed that, kinetic parameters evaluated by minimizing S give the differences of $(T_e - T_c)$ larger at lower values of $\alpha$ than at higher values. Minimization of $S_T$ should be a better criterion than the other.

In order for equation (3) to be hold, the necessary conditions are:

$$\frac{\partial S_T}{\partial m} = 0, \quad \frac{\partial S_T}{\partial n} = 0, \quad \frac{\partial S_T}{\partial p} = 0, \quad \frac{\partial S_T}{\partial A} = 0 \quad \text{and} \quad \frac{\partial S_T}{\partial E} = 0 \;.$$

Let $m^o$, $n^o$, $p^o$, $A^o$ and $E^o$ be the approximate values of the corresponding parameters derived from the LLS method as described in the previous paper[1], and $\Delta m$, $\Delta n$, $\Delta p$, $\Delta A$ and $\Delta E$ be the differential corrections that must be added to the corresponding approximate parameters in order to achieve the condition of "best fit", namely,

$$m = m^o + \Delta m \qquad (4)$$

$$n = n^o + \Delta n \qquad (5)$$

$$p = p^o + \Delta p \tag{6}$$
$$A = A^o + \Delta A \tag{7}$$
$$E = E^o + \Delta E \tag{8}$$

Substitute these equations into eq. (3), apply the Taylor's expansion, and neglect the terms which have higher powers in $\Delta m$, $\Delta n$, $\Delta p$, $\Delta A$ and $\Delta E$; also write T instead of $T_c$, for the sake of simplicity, we have

$$S_T = \Sigma \left[ T_e^i - \left( T(\alpha_i, m^o, n^o, p^o, A^o, E^o) + \left(\frac{\partial T}{\partial m}\right)_i \Delta m + \left(\frac{\partial T}{\partial n}\right)_i \Delta n + \left(\frac{\partial T}{\partial p}\right)_i \Delta p + \left(\frac{\partial T}{\partial A}\right)_i \Delta A \right.\right.$$
$$\left.\left. + \left(\frac{\partial T}{\partial E}\right)_i \Delta E \right) \right]^2 \quad \ldots \quad (9)$$

The necessary conditions to make $S_T$ a minimum are

$$\frac{\partial S_T}{\partial(\Delta m)} = 2\Sigma \left[ T_e^i - (T^o + \left(\frac{\partial T}{\partial m}\right)_i \Delta m + \left(\frac{\partial T}{\partial n}\right)_i \Delta n + \left(\frac{\partial T}{\partial p}\right)_i \Delta p + \left(\frac{\partial T}{\partial A}\right)_i \Delta A + \left(\frac{\partial T}{\partial E}\right)_i \Delta E) \right] \left(\frac{\partial T}{\partial m}\right)_i = 0 \tag{10}$$

$$\frac{\partial S_T}{\partial(\Delta n)} = 2\Sigma \left[ T_e^i - (T^o + \left(\frac{\partial T}{\partial m}\right)_i \Delta m + \left(\frac{\partial T}{\partial n}\right)_i \Delta n + \left(\frac{\partial T}{\partial p}\right)_i \Delta p + \left(\frac{\partial T}{\partial A}\right)_i \Delta A + \left(\frac{\partial T}{\partial E}\right)_i \Delta E) \right] \left(\frac{\partial T}{\partial n}\right)_i = 0 \tag{11}$$

$$\frac{\partial S_T}{\partial(\Delta p)} = 2\Sigma \left[ T_e^i - (T^o + \left(\frac{\partial T}{\partial m}\right)_i \Delta m + \left(\frac{\partial T}{\partial n}\right)_i \Delta n + \left(\frac{\partial T}{\partial p}\right)_i \Delta p + \left(\frac{\partial T}{\partial A}\right)_i \Delta A + \left(\frac{\partial T}{\partial E}\right)_i \Delta E) \right] \left(\frac{\partial T}{\partial p}\right)_i = 0 \tag{12}$$

$$\frac{\partial S_T}{\partial(\Delta A)} = 2\Sigma \left[ T_e^i - (T^o + \left(\frac{\partial T}{\partial m}\right)_i \Delta m + \left(\frac{\partial T}{\partial n}\right)_i \Delta n + \left(\frac{\partial T}{\partial p}\right)_i \Delta p + \left(\frac{\partial T}{\partial A}\right)_i \Delta A + \left(\frac{\partial T}{\partial E}\right)_i \Delta E) \right] \left(\frac{\partial T}{\partial A}\right)_i = 0 \tag{13}$$

$$\frac{\partial S_T}{\partial(\Delta E)} = 2\Sigma \left[ T_e^i - (T^o + \left(\frac{\partial T}{\partial m}\right)_i \Delta m + \left(\frac{\partial T}{\partial n}\right)_i \Delta n + \left(\frac{\partial T}{\partial p}\right)_i \Delta p + \left(\frac{\partial T}{\partial A}\right)_i \Delta A + \left(\frac{\partial T}{\partial E}\right)_i \Delta E) \right] \left(\frac{\partial T}{\partial E}\right)_i = 0 \tag{14}$$

where $T^o$ denotes $T(\alpha_i, m^o, n^o, p^o, A^o, E^o)$. These expressions lead to five equations which are linear to these differential corrections. These five equations can be expressed in the following form

$$[A][\Delta] = [B] \tag{15}$$

where $[A]$, $[\Delta]$ and $[B]$ are all matrices; they are defined as follows:

$$[A] = \begin{bmatrix} \Sigma(\frac{\partial T}{\partial m})_i^2 & \Sigma(\frac{\partial T}{\partial n})_i(\frac{\partial T}{\partial m})_i & \Sigma(\frac{\partial T}{\partial p})_i(\frac{\partial T}{\partial m})_i & \Sigma(\frac{\partial T}{\partial A})_i(\frac{\partial T}{\partial m})_i & \Sigma(\frac{\partial T}{\partial E})_i(\frac{\partial T}{\partial m})_i \\ & \Sigma(\frac{\partial T}{\partial n})_i^2 & \Sigma(\frac{\partial T}{\partial p})_i(\frac{\partial T}{\partial n})_i & \Sigma(\frac{\partial T}{\partial A})_i(\frac{\partial T}{\partial n})_i & \Sigma(\frac{\partial T}{\partial E})_i(\frac{\partial T}{\partial n})_i \\ & & \Sigma(\frac{\partial T}{\partial p})_i^2 & \Sigma(\frac{\partial T}{\partial A})_i(\frac{\partial T}{\partial p})_i & \Sigma(\frac{\partial T}{\partial E})_i(\frac{\partial T}{\partial p})_i \\ & & & \Sigma(\frac{\partial T}{\partial A})_i^2 & \Sigma(\frac{\partial T}{\partial E})_i(\frac{\partial T}{\partial A})_i \\ & & & & \Sigma(\frac{\partial T}{\partial E})_i^2 \end{bmatrix}$$

$$\ldots\ldots\ldots(16)$$

$$[\Delta] = \begin{bmatrix} \Delta m \\ \Delta n \\ \Delta p \\ \Delta A \\ \Delta E \end{bmatrix} \quad \ldots\ldots\ldots(17)$$

$$[B] = \begin{bmatrix} \Sigma(T_e^i - T^o)(\frac{\partial T}{\partial m})_i \\ \Sigma(T_e^i - T^o)(\frac{\partial T}{\partial n})_i \\ \Sigma(T_e^i - T^o)(\frac{\partial T}{\partial p})_i \\ \Sigma(T_e^i - T^o)(\frac{\partial T}{\partial A})_i \\ \Sigma(T_e^i - T^o)(\frac{\partial T}{\partial E})_i \end{bmatrix} \quad \ldots\ldots\ldots(18)$$

$\Delta m$, $\Delta n$, $\Delta p$, $\Delta A$ and $\Delta E$ values can be obtained by solving equations (15). The coefficients in the above equations contain T which must be calculated from equation (1). This involves the evaluation of two integrals F and G by an numerical method e.g. Gauss quadrature. Partial derivatives of T with respect to m, n, p, A and E are related to F and G by the following equations:

$$\frac{\partial T}{\partial m} = \frac{\partial F}{\partial m} / \frac{\partial G}{\partial T} \qquad (19)$$

$$\frac{\partial T}{\partial n} = \frac{\partial F}{\partial n} / \frac{\partial G}{\partial T} \qquad (20)$$

$$\frac{\partial T}{\partial p} = \frac{\partial F}{\partial p} / \frac{\partial G}{\partial T} \qquad (21)$$

$$\frac{\partial T}{\partial A} = \frac{\partial G}{\partial A} / \frac{\partial G}{\partial T} \qquad (22)$$

$$\frac{\partial T}{\partial E} = \frac{\partial G}{\partial E} / \frac{\partial G}{\partial T} \qquad (23)$$

The obtained $\Delta m$, $\Delta n$, $\Delta p$, $\Delta A$ and $\Delta E$ values are combined with the first set of $m^o$, $n^o$, $p^o$, $A^o$ and $E^o$ to give a second set of $m^o$, $n^o$, $p^o$, $A^o$ and $E^o$. The process is repeated until the desired accuracy is acquired.

## TESTING OF THE SUGGESTED METHOD

The suggested method was tested by the data of a zero order reaction in solution---acid catalyzed iodination of acetone[4]. The original data are expressed in x and dx/dt where x is concentration in moles per liter. They are converted to $\alpha$ and $d\alpha/dT$ by the following transformations:

$$\alpha = \frac{x}{a} \qquad (24)$$

where a is the initial concentration.

$$\frac{d\alpha}{dT} = \frac{1}{C} \frac{d\alpha}{dt} = \frac{1}{aC} \frac{dx}{dt} \qquad (25)$$

The data used for testing are shown in table 1.

Table 1. Experimental data of iodination of acetone in solution

| t (min) | T(°K) | $\alpha$ | $\frac{d\alpha}{dT}$ | T cal. |
|---|---|---|---|---|
| 30 | 286.2 | | | |
| 35 | 287.8 | 0.133 | 1.34 | 287.7 |
| 40 | 289.5 | 0.172 | 1.60 | 289.5 |
| 45 | 291.2 | 0.219 | 1.92 | 291.2 |
| 50 | 292.8 | 0.277 | 2.27 | 293.0 |
| 55 | 294.5 | 0.342 | 2.68 | 294.5 |
| 60 | 296.2 | 0.413 | 3.35 | 296.0 |
| 65 | 297.8 | 0.522 | 3.89 | 297.8 |
| 70 | 299.5 | 0.642 | 4.75 | 299.5 |

The basic kinetic equation should also be transformed as follows:

$$\frac{dx}{dt} = Ae^{-E/RT}(a-x)^n \tag{26}$$

$$\frac{d\alpha}{dt} = \frac{1}{a}\frac{dx}{dt} = Aa^{n-1}e^{-E/RT}(1-\alpha)^n \tag{27}$$

$$\frac{d\alpha}{dT} = \frac{1}{C}\frac{dx}{dt} = \frac{Aa^{n-1}}{C}e^{-E/RT}(1-\alpha)^n \equiv Be^{-E/RT}(1-\alpha)^n \tag{28}$$

$$\ln\frac{d\alpha}{dT} = \ln B - \frac{E}{R}(\frac{1}{T}) + n\ln(1-\alpha) \tag{29}$$

By linear least squares method, we obtain $B=1.1433 \times 10^{13}$, $E=71.21$ kJ, $n=-0.1172$. Results obtained from differential correction method are shown in Table 2.

Table 2. Results from DC method.

| Count | n | B ($10^{-13}$) | E (kJ) |
|---|---|---|---|
| 0 | -0.11720 | 1.1433 | 71.21 |
| 1 | 0.37482 | 2.7960 | 96.30 |
| 2 | 0.25948 | 3.3650 | 96.28 |
| 3 | 0.25842 | 3.4799 | 96.31 |
| 4 | 0.25860 | 3.9640 | 96.32 |
| 5 | 0.25858 | 3.4941 | 96.32 |
| 6 | 0.25839 | 3.4875 | 96.32 |
| 7 | 0.25844 | 3.4886 | 96.32 |

From the above table, the value of n and E are 0.26 and 96.3 kJ respectively. The theoretical value of n is 0 and accepted value of E is 81.6 kJ. This discripency must be attributed to experimental error and not to the method. As we can see from Fig. 3 of the original paper, the straight line could be drawn a little steeper which will lead to higher values of n and E. Calculated values of temperature at the corresponding values of α are listed in the last column of Table 1, for comparison. It is seen that they are essentially the same as the experimental values except for two points with a difference of 0.2 °K which is less than 0.1% in error.

## REFERENCES

(1) David T.Y. Chen and Pong H. Fong, Thermochim. Acta 18(1977)161.
(2) I.D. Gay, J. Phys. Chem., 75(1971)1610.
(3) Pong H. Fong and David T.Y. Chen, Thermochim. Acta 18(1977)273.
(4) P. H Fong and D.T.Y. Chen, J. Thermal Anal. 8(1975)305.

# RELATION BETWEEN THE KINETICS AND THERMODYNAMICS IN THE ENDOTHERMIC DECOMPOSITION OF SOLIDS

Z. Jerman

Chemopetrol, Výzkumný ústav anorganické chemie
400 60 Ústí nad Labem, Czechoslovakia

## ABSTRACT

Decomposition measurements with $Mg(OH)_2$ in water vapour, performed by a technique based on recording the temperature inside the specimen, on its surface and in the furnace metal block, were evaluated by calculating the heat-flow into the specimen, maximum reaction rate and first-order rate constants. The temperature dependence of the latter was correlated using the Planck radiation law and respecting the influence of the equilibrium proximity. The results were: $A=6.6 \times 10^7 s^{-1}$, $E=135.7$ kJ/mole. Since $E = \Delta H^{\neq} + RT$, the activation enthalpy $\Delta H^{\neq} = 130.3$ kJ/mole at 650 K is greater than the enthalpy change from tabulated data ($\Delta H^o_{650} = 118.3$ kJ/mole) by 12 kJ/mole. It corresponds to the enthalpy difference between a low-temperature (active) and high-temperature (stabile) MgO from $Mg(OH)_2$. Therefore, the activation enthalpy here is equivalent to the enthalpy change of the reaction $Mg(OH)_2(s) = MgO(active) + H_2O(g)$. This type of mechanism may be common for the endothermic decomposition of solids.

## INTRODUCTION

The extent of the influence of the geometric mechanism (nucleation, growth), physical processes (mass and heat transfer), thermodynamics (equilibrium proximity) as well as of the chemical mechanism proper on the kinetics of endothermic decomposition of solids is rather difficult to analyse. So far one has rarely succeeded in arranging kinetics measurements in such a way that the chemical reaction alone would be the rate controlling process, which would allow us to draw conclusions as to the type of the elementary chemical process.

The linearity of log k vs. $T^{-1}$ dependence within narrow temperature intervals cannot be taken as a proof of a single rate-controlling process. The inconsistency of results often leads to efforts to correlate at any price (the compensation effect between the Arrhenius parameters) and even to scepticism as concerns the interpretation of the kinetic results at all. In many cases the reason for the inconsistency of the results is the proximity of the chemical equilibrium affecting considerably the apparent Arrhenius parameters. A further misleading factor is the ignorance of the exact mean temperature of the sample; this problem cannot be solved even by decreasing the weight of the sample to milligram quantities.

In the case of a reaction of the type $A(s) = B(s) + C(g)$, where the reverse reaction has a zero activation enthalpy, the activation enthalpy $\Delta H^{\neq}$ and the enthalpy change of the direct reaction will be equal. The temperature dependence of the activation enthalpy at the rate constant $k_1 = 10^{-3}$ $s^{-1}$ for monomolecular decomposition in gaseous state is shown in Fig.1. This dependence is virtually identical with the dependence according to Polanyi-Wigner equation (1). Also shown in Fig.1 is the dependence derived on the basis of the Planck radiation law applied to all the crystal planes (2); this is very close to the points indicating the decomposition reaction enthalpies $\Delta H^{o}_{298}$ and decomposition temperatures of some solids. The predictions of high values of Arrhenius parameters A and E for the thermal decomposition of solids, as obtained from the Polanyi-Wigner equation and from the more recent studies (3,4), are probably due to the fact that the great difference between the entropy of the solid and gaseous states are not considered. Since this difference is 80-120 $J.mole^{-1}.K^{-1}$, the preexponential term A for solids should be lower by $\exp(\Delta S/R) = 10^4 - 10^6$ than for gases and thus its value should be $10^9 - 10^7$ $s^{-1}$, which conforms with the application of the Planck radiation law:

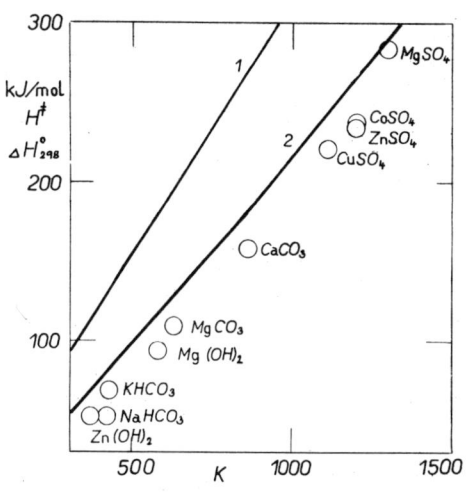

Fig.1. Temperature dependence of activation enthalpy at $k_1 = 10^{-3} s^{-1}$

1 Monomolecular reaction in gas and Polanyi-Wigner eq.
2 Application of the Planck radiation law
O Reaction enthalpy change $\Delta H^0_{298}$ and approximate decomposition temperature of some solids

$$k_1 = (4\pi k/M\ N_A^2\ h^3\ c^2)\Delta H^{\neq 2}\ T\ \exp(-\Delta H^{\neq}/RT) \qquad [s^{-1}]$$

where $\Delta H^{\neq} = E - RT$, $k_1$ is the rate constant, M is the number of the structural units per 1 $m^2$ of the crystal plane; the other symbols have usual physical meaning.

The verification of a theory concerning the decomposition reactions of solids cannot be done without the knowledge of the geometric mechanism because the rate constant should always be related to the reaction zone only. There is an evidence that the decomposition of magnesium hydroxide proceeds within the whole mass of the crystal(5) and, therefore, this solid appears to be the proper substance for proving the theory.

## EXPERIMENTAL PROCEDURE

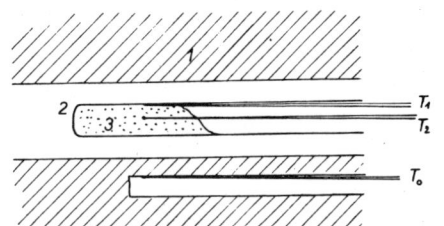

Fig.2 Diagram of the measuring set
1 Al block, 2 glass cell, 3 sample, $T_0, T_1, T_2$ thermocouples

Magnesium hydroxide precipitated from magnesium nitrate solution by ammoniak was used. The thermal decomposition was conducted by two methods. The method based on the measuring of the temperatures inside and on the surface of the sample as well as of the enveloping furnace metal block was applied at higher temperatures and higher reaction rates (6). The diagram of the measuring arrangement is in Fig.2. The cell containing 1 g $Mg(OH)_2$ and two thermocouples was placed into the furnace maintained at constant temperature and the time-dependence of the temperature was recorded. Finally the weight-loss was determined.

## RESULTS AND DISCUSSION

The evaluation of the reaction rate was derived from the relation for the heat flow

$$q = \alpha (T_o - T_1) + \beta (T_o^4 - T_1^4) \qquad [Wm^{-2}]$$

where $\alpha$ is the coefficient for heat transfer by convection and conduction and $\beta$ is the thermal radiation coefficient. Corrections were made for the heat flow consumed for heating up the cell with the sample and for the heat losses and the values of the constants $\alpha$ and $\beta$ for the experimental assembly were determined by measuring the decomposition of $NaHCO_3$ and $Mg(OH)_2$. The first order rate constant was calculated from the maximum rate and the corresponding mean temperature of the sample $(T_1 + T_2)/2$ was assessed. Considering the high reaction rate and a limited contact of the sample with the surrounding atmosphere, the water vapour pressure $p_{H_2O}=1$ was assumed. At lower temperatures and lower reaction rates only the weight loss was determined after isothermal heating of the sample for a certain period of time using a ceramic boat pan and the same furnace and both the water vapour atmosphere and air.

The results are demonstrated in Fig.3. The values of $k_1$ and T obtained by means of the former method were correlated using the following relations:

$$k_1 = 2.2 \times 10^{-6} \Delta H^{\neq 2} \times T \times \exp(-\Delta H^{\neq}/RT)(1-\exp(\Delta G^o/RT))$$

$$\exp(\Delta G^o/RT) = \exp(\Delta H/RT) \times \exp(-\Delta H^o/606 R)$$

$$\Delta H^{\neq} = \Delta H^o \qquad \text{(presumption)}$$

where $\Delta G^o$ is the standard free enthalpy change, $\Delta \bar{H}^o$ is the mean standard enthalpy change in the examined temperature interval, 606K is the equilibrium temperature for $p_{H_2O} = 1$. The curve 1 in Fig.3 corresponds to $\Delta H^{\neq} = 130.3$ kJ/mole. The calculation based on tabulated data gave $\Delta H^o_{650} = 118.3$ kJ/mole. The difference 130.3 - 118.3 = 12 kJ/mole is in agreement with the difference between the enthalpy of the low-temperature (active) and high-temperature magnesium oxide, which is 11.7 kJ/mole (7). The curve 2 was constructed for $p_{H_2O}=0$ and it corresponds to Arrhenius parameters E=135.7 kJ/mole, A=6.6 x $10^7$ s$^{-1}$.

The curve 1 complies also with the results obtained with the ceramic vessel in water atmosphere. The results obtained in air lie between the curve 1 and the straight line calculated for $\Delta H^{\neq} = 130.3$ kJ/mole and $p_{H_2O} = 0$. This is obviously due to certain partial pressure of $H_2O$ within the reacting material and, under these conditions, the system at a low temperature would not be far from the equilibrium.

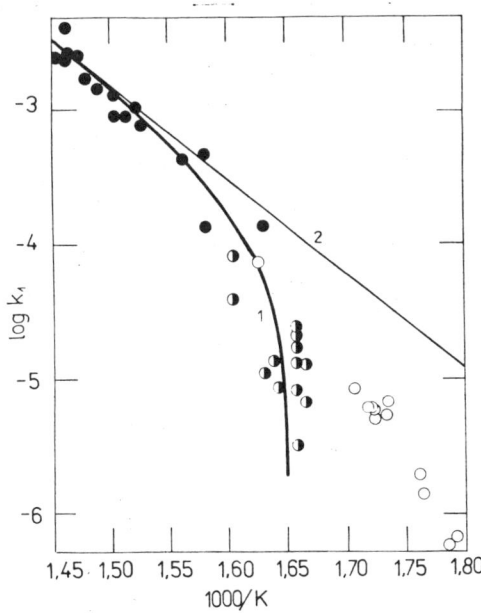

Fig.3. Kinetics of $Mg(OH)_2$ decomposition
The results obtained by:
● new method, ◐ weighing (decomposition in $H_2O$ vapour)
○ weighing (decomp. in air),
curve 1 experimental, $p_{H_2O}=1$,
curve 2 calculated, $p_{H_2O}= 0$

The results indicate that energy fluctuations in the lattice, governed by the Planck radiation law, are controlling the thermal decomposition of solids in the cases when the effect of heat and mass transfer can be neglected. A difference between the determined activation enthalpy and the enthalpy change of an endothermic reaction calculated from tabulated data (this difference should always be a positive one) may indicate that an active product is formed during the thermal decomposition. A too great difference, however, is suspicious as it might originate in errors.

When the temperature region of the kinetic measurements extends to temperatures at which the equilibrium of the decomposition reaction is established at a low partial pressure of the gaseous product, the simple linear correlation $\log k$ vs. $T^{-1}$ gives too high apparent Arrhenius parameters.

### REFERENCES

(1) M. Polanyi and E. Wigner, Z.Phys.Chem. 139A (1928) 439
(2) Z. Jerman, Collection Czechoslov.Chem.Commun. 38 (1973) 3210
(3) R.D. Shannon, Trans.Faraday Soc. 60 (1964) 1902
(4) H.F. Cordes, J.Phys.Chem. 72 (1968) 2185
(5) J.C. Niepce and G. Watelle, J.Chem.Soc. 74 (1978) 1530
(6) Z. Jerman, Proceedings: Zborník VIII. celoštátnej konferencie o termickej analýze, Vysoké Tatry 1979, p.79
(7) R. Fricke and J. Lücke, Ztschr.Elektrochem. 41 (1935) 174

# DETERMINATION OF THE MECHANISM OF THERMAL DECOMPOSITION REACTIONS OF SOLIDS BY USING THE CYCLIC AND CONSTANT DECOMPOSITION RATE THERMAL ANALYSIS METHOD

J. M. Criado

Departamento de Química Inorgánica and Departamento de Investigaciones Físicas y Químicas, Centro Coordinado del C. S. I. C., Facultad de Química, Sevilla (Spain)

## ABSTRACT

The cyclic and Constant Rate Thermal Analysis technique (CRTA) has being used. The two rates automatically selected in the cyclic curve are small enough to allow the two states of the sample to be compared have nearly the same reacted fraction, $\alpha$. Thus, the activation energy, E, can be calculated without prior knowledge of the actual reaction mechanism.

A procedure is developed for determining the kinetic law followed by the reaction by means of master curves that represent the values of $\alpha$ as a function of ln $A/C$ - $E/RT$, once E has been obtained from the cyclic CRTA diagram. The plots of $\alpha$ against $-E/RT$ must be superpossed upon the curve corresponding to the actual kinetic of the reaction by a lateral shift. The length of the lateral shift must be equal to ln $A/C$, C being the constant rate and A the preexponential factor of Arrhenius. This procedure has been checked by studying the thermal decomposition of $BaCO_3$.

## INTRODUCTION

It has been proved in a previous paper (1) that the kinetic analysis of a single thermoanalytical curve obtained using the Constant Decomposition Rate Thermal Analysis technique (CRTA), developed by Rouquerol et al. (2), allows to discriminate without ambiguities among Avrami-Erofeev, "n order" and diffusion controlled reactions. This does not occur if the conventional TG technique is employed, as was demonstrated before (3)(4). Therefore, CRTA holds a number of advantages with regards to TG in order to perform the kinetic analysis of thermal decomposition reactions of solids. However, it has been shown in the literature (1) that, when "n order" reactions are involved, one standard CRTA curve does not supply enough information for calculating $\underline{n}$, what is a limitation of this method of kinetic analysis.

In order to overcome the above limitation a method is developed in this paper that allows the accurate discernment of the kinetic law followed by the reaction from the analysis of a cyclic CRTA curve.

## EXPERIMENTAL PROCEDURE

$BaCO_3$ from d'Hemio (a. r. grade) was used.

It was employed the CRTA system, developed by Rouquerol et al., that was described elsewhere (5-7). This apparatus permits to maintain cons tant at values previously selected both the rate of decomposition and the residual pressure in the close vicinity of the sample. On the other hand, an automatic device (including a timer and an air-operated butterfly valve) joined to the equipment allows an alternative control of the reaction at two different rates in order to carry out cyclic CRTA experiments.

Cyclic CRTA curves of $BaCO_3$ were recorded by using a weight of sample of about 150 mg and a residual pressure of $5.10^{-5}$ torr. The two states of the sample to be compared have almost the same reacted fraction. Thus, the activation energy can be calculated without prior knowledge of the actual reaction mechanism (2). These conditions allowed to perform about 15 cycles, i.e. 15 succesive measurements of E, during the decomposition of $BaCO_3$. By way of example, Fig. 1 shows one of these cycles.

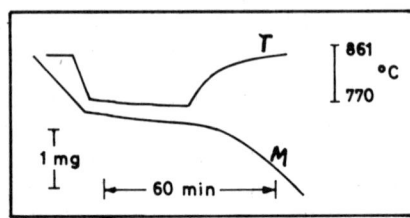

Fig. 1. A cycle in a CRTA curve of $BaCO_3$

## RESULTS AND DISCUSSION

Bearing in mind that CRTA diagrams are obtained at a constant decomposition rate, C, and taking into account the general expression of the reaction rate of thermal decomposition of solids, we can write:

$$d\alpha/dt = C = A..\exp(-E/RT).f(\alpha) \qquad (1),$$

that can be rearranged in the form:

$$\ln(1/f(\alpha)) = \ln(A/C) - (E/RT) \qquad (2),$$

where $\alpha$ is the reacted fraction of solid at the time $t$; $f(\alpha)$ a function depending on the reaction mechanism and the other letters have their usual meaning.

In the case of a cyclic CRTA experiment carried out under the conditions above described we can easily derive from eqn. (2):

$$\ln(C_1/C_2) = -(E/R)[(1/T_1) - (1/T_2)] \qquad (3),$$

where $C_1$ and $C_2$ are the two rates in the cycle to which correspond the temperatures $T_1$ and $T_2$, respectively.
Therefore, E can be calculated without knowledge of $f(\alpha)$.
On the other hand, as $\ln 1/f(\alpha)$ is a function of $\alpha$, which is equal to –

$\ln(A/C) - (E/RT)$, $\alpha$ can be calculated as a function of $\ln(A/C)-(E/RT)$. Fig. 2 shows the theoretical master curves calculated by this way from the $f(\alpha)$ functions more commonly used in the literature (8) for describing thermal decomposition reactions of solids. Furthermore, the plot of the experimental values of $\alpha$ against $-E/RT$, once E has been obtained - from the cyclic CRTA experiment, must be superpossed upon the master curve corresponding to the actual kinetic of the reaction by a lateral - shift. The length of this lateral shift must be equal to $\ln(A/C)$.

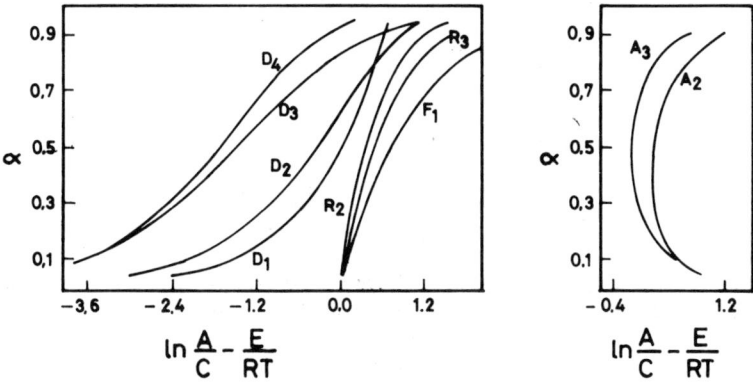

Fig. 2. Master curves for the kinetic analysis of CRTA curves. The symbols of Sharp et al. (8) have been used for naming the $f(\alpha)$ functions.

The above method has been applied to study the thermal decomposition - of $BaCO_3$ and an activation energy E = 64 Kcal/mol has been calculated from the cyclic CRTA curve. By using this activation energy, the $\alpha$ values taken from a CRTA curve recorded at $C = 1.1 \cdot 10^{-3}$ min$^{-1}$ have been drawn against $-E/RT$ and compared with the master curves included in Fig. 2. From these comparisons, it may be concluded that the decomposition of $BaCO_3$ follows a first order kinetic law ($F_1$) as shown in Fig. 3. On the other hand, the length of the lateral shift corresponds to $\ln(A/C) = 31.2$ and, therefore, $A = 4 \cdot 10^{10}$ min$^{-1}$.

The calculated kinetic parameters of the thermal decomposition of $BaCO_3$ are in very good agreement with those previously determined (9) from a comparison of TG and standard CRTA data.

In summary, we can conclude that the method of kinetic analysis of thermal decomposition of solids developed in the present paper supplies an easy and quick way for discriminating the actual reaction mechanism and determining the values of both the activation energy and the preexponential factor of Arrhenius.

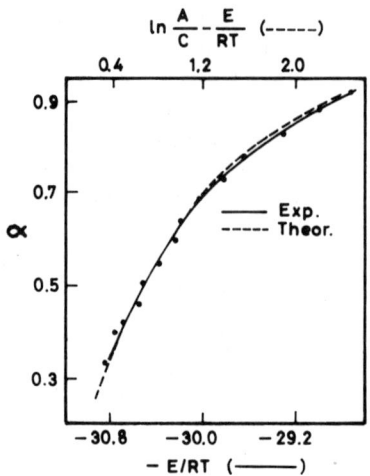

Fig. 3. Kinetic analysis of a CRTA curve of $BaCO_3$. A comparison with the master curve corresponding to a first order law ($F_1$)

## REFERENCES

(1) J. M. Criado, Thermochim. Acta, **28** (1979) 307
(2) F. Rouquerol and J. Rouquerol, in "Thermal Analysis" (G. H. Wiedeman Ed.), Verlag, Basel and Stüggart, 1972, Vol. 1, p. 373
(3) J. M. Criado and J. Morales, Thermochim. Acta **16** (1976) 382
(4) J. M. Criado and J. Morales, Thermochim. Acta **19**, (1977) 305
(5) J. Rouquerol, Proceedings 2nd International Conference on Thermal Analysis, Worcester, U. S. A., 1968, Academic Press, New York 1969, Vol. 1, p. 281
(6) J. Rouquerol, J. Thermal Anal. **5** (1973) 203
(7) F. Rouquerol, S. Regnier and J. Rouquerol, in "Thermal Analysis" (E. Buzag Ed.), Heyden, London 1975, Vol. 1, p. 313
(8) J. H. Sharp, G. W. Brindley and N. N. Achar, J. Am. Ceram. Soc., **47** (1966) 379
(9) J. M. Criado, F. Rouquerol and J. Rouquerol, Thermochim. Acta, in the press

## ACKNOWLEDGEMENT

The author is indebted to Drs. F. and J. Rouquerol for the facilities given to him for carrying out the experiments in the Centre de Recherches de Microcalorimetrie et de Thermichimie du C. N. R. S., Marseille (France) and for many helpful discussions.

A NEW ISOTHERMAL DTA METHOD FOR THE
STUDY OF SURFACTANTS AT LIQUID SURFACES

H.K. Cammenga and H.-J. Petrick
Institut für Physikalische Chemie, TU Braunschweig,
D-3300 Braunschweig, FRG

## ABSTRACT

A new isothermal DTA method for the detection and study of surfactants at the liquid/vapour interface is presented. In this method a certain mass of liquid is slightly superheated above its boiling point in two identical vessels by a constant heat input. If now a surfactant is added to one of the vessels, it more or less strongly retards the evaporative mass (and thus heat) flux from the corresponding surface and this liquid warms up with respect to the reference liquid. The time-dependent temperature difference can be used to obtain information about amount and nature of the surfactant, its adsorption isotherm, the area per molecule in the monolayer etc.

## INTRODUCTION

To the knowledge of the authors no use of the well established DTA methods has so far been made to study phenomena at liquid surfaces, presumably because it was felt that surface effects would be too small to be accessible to these methods. However, this is not generally true.

It is well known that amphiphilic molecules may strongly adsorb at liquid/liquid or liquid/vapour interfaces. If the adsorbing substance (called tenside, detergent or surfactant) is only very slightly soluble in the underlying liquid it may form a monolayer, in which the adsorbed layer is not more than one molecule "thick"[1]. In the case of the water/vapour (or/air) interface the adsorbed substances must possess a hydrophilic end solvated by water molecules and a hydrophobic end which is expelled from the liquid. At low number density $N_s$ of adsorbed molecules these are far from each other and one speaks of gaseous monolayers. At increasing $N_s$ molecules may interact to form clusters and this is called the liquid (expanded) state. On further compression many surfactants form condensed monolayers, in which the monolayer resembles a two-dimensional crystal. The molecular cross sectional area in the case of a n-alkane(1) derivative is then $a_m = 0.185$ nm$^2$.

## BASIC PROCEDURE

The adsorption of monolayers at interfaces may often considerably retard the rate of transport processes, e. g. evaporation, across these interfaces and this effect and its consequences can be followed by DTA. The basic arrangement is schematically illustrated in Fig. 1. Two identical ves-

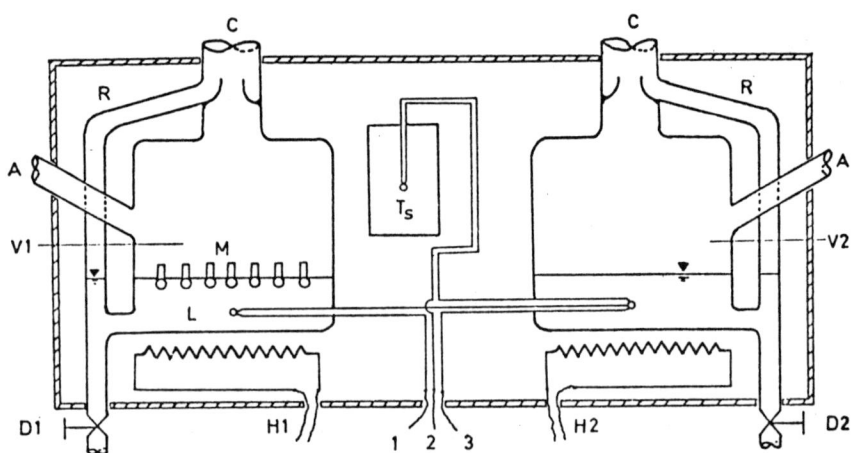

Fig. 1. A sample addition port, C condenser (not shown on the figure), D1, D2 drains, H1, H2 adjustable heaters, L liquid, M monolayer on surface, R return tube from condenser, $T_s$ reference boiler for equilibrium boiling temperature (for details see Fig. 2), V1, V2 vessels containing sample and reference, 1, 2, 3 thermocouple leads.

sels V1, V2 contain approximately the same mass of highly purified liquid L exposing a surface area S. Both vessels are contained in an "oven" kept just above the boiling point of the liquid studied. By two heaters H1, H2 the liquid can be quiescently distilled at a constant desired rate $G_{vo}$. The vapour is condensed in the condensers C (not shown) and the condensate through R returned into the vessels. Since the latent heat of evaporation must be transported to the surfaces by conduction and convection, the liquids of mean inner temperature $\overline{T}_F$ are superheated with respect to their equilibrium boiling point $T_s$ by $\overline{\Delta T}_F = \overline{T}_F - T_s$. $\overline{\Delta T}_F$ can be measured with a thermocouple (2,3 in Fig. 1, or another sensor), $T_s$ being established in the reference boiler shown in Fig. 2.

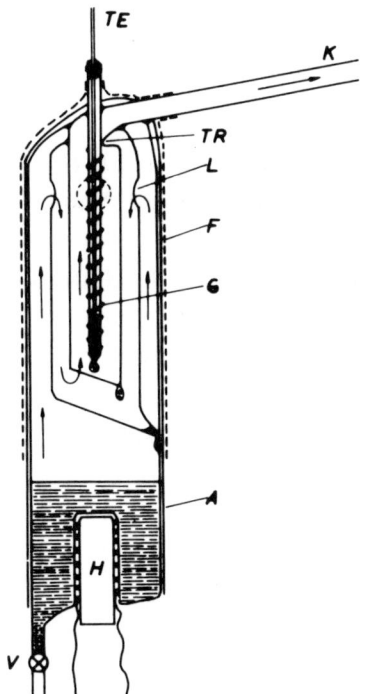

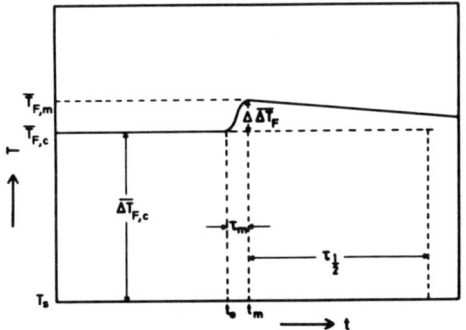

Fig. 3 (above). Schematic illustration of the signals obtained in the present method. $T_{F,c}$, $T_{F,m}$ inner liquid temperature without ("clean") and with monolayer, t time, $t_o$ time of sample addition, $t_m$ time at which maximum response is reached.

Fig. 2 (left). Reference boiler. TE thermocouple, D condenser, TR drop applicator, L holes for vapour passage, F aluminium radiation shield, G glass spiral, A asbestos isolation, H heater ($\simeq$2W), V valve, → path of the vapour.

If now a surfactant (or a sample containing a surfactant) is added through A in vessel V1 onto the surface by suitable means [2], it usually forms a monolayer M with an evaporation resistance $R_s$, and at first the evaporation rate drops from $G_{vo}$ to $G_v$. However, the constant heat supply $\dot{q}$ is still coming from H1 and this can only be carried away by evaporation. Thus $G_v$ must rise to the same value as $G_{vo}$ was before, which owing to $R_s$ can only be achieved by an appropriate increase in the driving force, i.e. $\overline{\Delta T_F}$ rises by $\Delta \overline{\Delta T_F}$, see Fig. 3. After a maximum increase at time interval $\tau_m$ the signal drops due to slow volatilization, solution etc. of the surfactant. $\Delta \Delta T_F$ can be measured with junctions 1,3 (Fig.1). Both $\Delta T_F$ and $\Delta \Delta T_F$ show oscillations due to convective liquid motion and the mean signals $\overline{\Delta T_F}$ and $\Delta \overline{\Delta T_F}$ are obtained with an electronic averager, see block diagram in Fig. 4. Since the oscillating signal may contain valuable information (e.g. oscillation damping by the monolayer) both signals are recorded.

The method presented here is based on observations of HICKMAN and WHITE [3,4], a phenomenon which they called "modified superheating", and qualitatively explained by a reduction of the surface tension driven convection (MARANGONI convection) of mass and heat transfer in the vicinity of the liquid surface. A phenomenological theory has recently been given by CAMMENGA et al. [2,7], only the results of which can be reported here because of limited space.

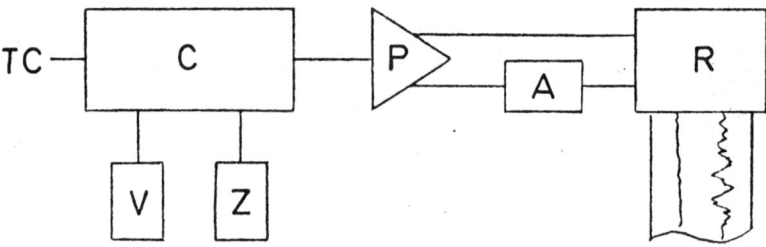

Fig. 4. Block diagram of electronics. TC thermocouples, C compensating unit, V constant voltage supply, Z zero detector, P preamplifier, A averager, R two-channel recorder.

## THEORY

The evaporation resistance of a monolayer is given by

$$R_s = \left(\frac{1}{G_v} - \frac{1}{G_{vo}}\right)(\overset{+}{\varrho} - \overset{+}{\varrho}{}') \; ; \qquad (1)$$

here the driving force is expressed as the partial vapour density difference between surface ($\overset{+}{\varrho}$) and vapour ($\overset{+}{\varrho}{}'$). If the total heat flux to the surface is

$$\dot{q} = \frac{\Delta H_v}{M}\frac{dm}{dt} = \frac{\Delta H_v}{M} S \, G_{vo} \qquad (2)$$

($\Delta H_v$ enthalpy of evaporation, M molar mass, dm/dt evaporative mass flux) it can be shown that $R_s$ is given by [2, 7]

$$R_s = \frac{\Delta H_v \, \varrho_v}{G_{vo} \, RT_s^2} \, T_F = \frac{\Delta H_v^2 \, S \, \varrho_v}{M \, RT_s^2 \, \dot{q}} \, T_F \, . \qquad (3)$$

$\varrho_v$ is the vapour density at temperature $T_s$, R the molar gas constant. If $\Delta$ is the per cent reduction in evaporation rate and $\delta$ the per cent increase in liquid temperature caused by the monolayer, these are related by [2, 7]

$$\Delta \equiv \frac{G_v - G_{vo}}{G_{vo}} 100 = -\frac{\Delta \overline{\Delta T_F}}{\overline{\Delta T_F} + \Delta \overline{\Delta T_F}} 100 \approx -\delta \equiv \frac{\Delta \overline{\Delta T_F}}{\overline{\Delta T_F}} 100 \, . \qquad (4)$$

As WHITE has shown, the following equation holds for $\delta$, if $A_s > a_m$ ($A_s$ is the area per molecule in the surface) [6]

$$\delta = A \ln \frac{B}{A_s} = A \ln (B \cdot N_s) \, , \qquad (5)$$

in which the constants show little chemical specifity and are obtained by calibration. For $A_s \lesssim a_m$ higher additions of surfactant cause no further increase in $\delta$, which then remains constant.

The beginning of the plateau region in $\delta$ presumably corresponds to the completion of the monolayer [6]. In this case the weight concentration of surfactant in the sample is given by [7]

$$c = \frac{S \, M}{N_A \, a_m \, V} \qquad (6)$$

($N_A$ AVOGADRO constant, V sample volume).

## POSSIBLE APPLICATIONS

The following informations may inter alia be obtained in using the present method: Qualitatively, it can be detected, whether a sample contains any surfactants; the sensitivity is of the order of $10^{-10}$ mol or in the nanogramme range[8]. For a known surfactant the amount can be determined[5]. If a known amount of a known surfactant is applied, its per cent evaporation reduction can be obtained and its evaporation resistance be calculated. Furthermore, either the cross sectional area of the adsorbed molecule or its molecular weight can be determined. By eq. (5) the molecular concentration $N_s$ and thus the surface pressure and surface tension (and eventually the adsorption isotherm) may be evaluated. For all these purposes only trace amounts of substance are needed the sample remaining uneffected by the procedure and after the DTA measurements may be used for further (e.g. analytical, chemical) examination.

## REFERENCES

[1] G.T. Barnes, Specialist Periodical Reports: Colloid Science 2 (1975) 173
[2] H.K. Cammenga, "Evaporation Mechanisms of Liquids", in Curr. Top. Mat. Sci., Ed. E. Kaldis, Vol. V, 325-446
[3] K. Hickman and I. White, Science 172 (1971) 718
[4] K. Hickman, I. White, W.V. Kayser, H.K. Cammenga, Office of Saline Water Report 808, Springfield, VA, 1973
[5] I. White, J. Colloid Interf. Sci. 56 (1976) 613
[6] I. White, Ind. Eng. Chem. Fundam. 15 (1976) 53
[7] E. Wolf, F.-W. Schulze, H.-J. Petrick, H.K. Cammenga and G.T. Barnes, Tenside Detergents 16 (1979) 57
[8] H.K. Cammenga, Ber. Bunsenges. Phys. Chem. 78 (1974) 1264

This work has in part been financially supported by funds granted by the Bundesminister für Forschung und Technologie, Bonn, FRG, which is gratefully acknowledged.

# THE Ge-Sb-Bi TERNARY PHASE DIAGRAM

S. Suriñach, M.D. Baró and F. Tejerina
Departamento de Termología.Facultad de Ciencias.Universidad
Autónoma de Barcelona. Bellaterra (Barcelona) SPAIN

## ABSTRACT

The ternary phase diagram Ge-Sb-Bi has been calculated based on the model of regular associated solutions. In the calculation on the ternary T-x diagram five parameters are needed: the three interaction parameters $\alpha_{12}$, $\alpha_{23}$ and $\alpha_{13}$ in the liquid, and the interaction parameter, $\Omega$, in the solid, derived from the binary phase diagrams, and the dimer dissociation constant, K, for the Bi. Good agreement with experimental data is obtained for the binary and ternary systems.

## INTRODUCTION

Chalcogenide systems are increasingly used, mainly in electronics, both in the crystalline and non-crystalline states. They show the interesting property of the dependence of the energy gap on alloy composition (1,2). The control of the composition of the alloy permits gaining control of its electronic properties.

Some different techniques for growing alloy crystals have already been developped (3,4). The bigger difficulty in the growth of homogeneous alloy crystals from dilute or stoichiometric melts results from the differences in the equilibrium compositions of the liquid and solid phases, which leads to segregation during solidification and compositional changes in the grown crystals. A knowledge of the phase relationships that govern the growth of such alloys from solution is therefore essential. The experimental determination of the phase diagrams including liquidus curves and tie lines can be prohibitively time-consuming and lead to significant experimental errors.

It is the principal objetive of this work to construct the Ge-Sb-Bi liquidus surface consistent with the experimental

data.
Usually, in the calculations of solid-liquid equilibria in ternary systems involving semiconductors, the needed component activities in the liquid phase were obtained from the ternary regular solution model (5,6). Generally, the reasonableness of the regular solution approach is demonstrated by the good agreement between theoretical and experimental liquidus isotherms. Nevertheless, the thermodynamic properties of some systems involving Bi do not obey regular solution behaviour.

Although the structure of liquid Bi has not yet been entirely resolved, to consider Bi as a partly diatomic liquid is reasonable. Some papers indicate that the structure of Bi (l) is anomalous in comparison with typical monatomic liquid metals (7,8). The existence of straight chains with covalent character in Bi(l) was deduced from studies of electron diffraction patterns (9). In view of this, we can postulate the existence of stable complexes ($Bi_2$) in the liquid state, and the ternary phase diagram of liquid-solid equilibria in the Ge-Sb-Bi system can be well described by the model of regular associated solution (R.A.S.) derived from Jordan's model (10).

## EXPERIMENTAL PROCEDURE

Appropiate weighed amounts of the elements (Balzers 5N purity), introduced into a precleaned quartz tube, sealed in vacuum, were held at 1000°C for 12h, constantly agitated to make the melt homogeneous, and then quenched in air. Differential thermal analysis were performed with a STA 429 Netzsch thermal analyser, on about 200 mg powdered samples heated at 5°C/min in a dynamic argon atmosphere. The reference material was carborundum.

## CALCULATION OF THE Ge-Sb-Bi PHASE DIAGRAM

A schematic phase diagram for the system Ge-Sb-Bi is shown in fig. 1. In view of it, the ternary phase diagram was calculated by using the equilibrium condition between the liquid solution and the Ge solid solution in the Ge rich region, which obeys the following thermodynamic equation

$$- \ln a_{Ge}^l = (\Delta H_{Ge}^F / R)(1/T - 1/T_{Ge}^F)$$

and the condition for equilibrium between the liquid solution and the solid solution Sb,Bi in the Sb-Bi rich region which obeys subsequent thermodynamic expressions

$$\ln a_{Sb}^s / a_{Sb}^l = (\Delta H_{Sb}^F / R)(1/T - 1/T_{Sb}^F)$$

$$\ln a_{Bi}^s / a_{Bi}^l = (\Delta H_{Bi}^F / R)(1/T - 1/T_{Bi}^F)$$

Where $a_i^l$, $a_i^s$, $T_i^F$ and $\Delta H_i^F$ are the activity in liquid phase, activity in solid phase, temperature and enthalpy of fusion respectively. For the calculation of the activities in the solid phases we supposed a regular solution (11) between the Sb and the Bi and an ideal behaviour for the others. The activities in the liquid phases were calculated by using the R.A.S. theory (12) provided that the monomers Ge,Sb,Bi and the dimer $Bi_2$ exist in liquid equilibrium. The two liquidus surfaces were superimposed, this allowed the determi

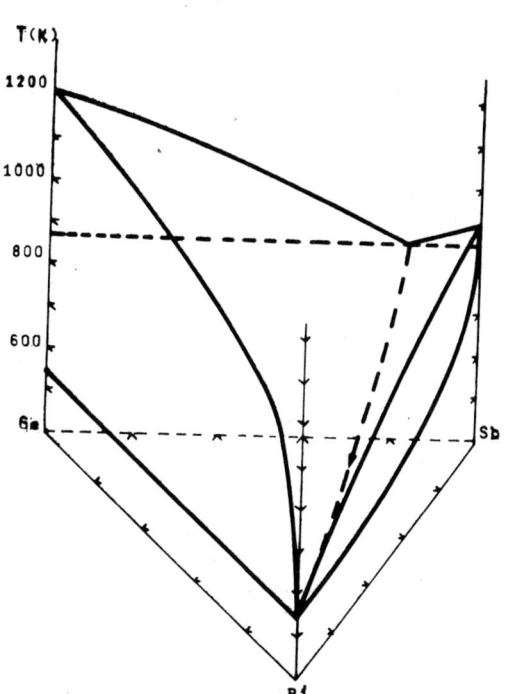

Fig. 1. Schematic Ge-Sb-Bi ternary phase diagram.

nation of the eutectic line in the Ge-Sb-Bi system.
The thermodynamic data needed were selected by (13) and they are listed in table I.

Table I - Values of the enthalpy and melting temperature for Ge,Sb,Bi elements used in the calculation of the ternary phase diagram (the data come from ref.13)

| Element | $\Delta H_i^F$ (cal/mol) | $T_i^F$ (K) |
|---|---|---|
| Ge | 8100 | 1210 |
| Sb | 4770 | 903,5 |
| Bi | 2600 | 544,5 |

The interaction parameters were determined by making a best fit to the experimental data on the binary systems and they are reported in table II.

Table II - Interaction parameters used for the R.A.S. calculation of Ge-Sb-Bi phase diagram

| System | $\alpha$ (cal/mol) | $\Omega$ (cal/mol) |
|---|---|---|
| Ge-Sb | 2640-1,98 T | -- |
| Ge-Bi | 5872-3,17 T | -- |
| Sb-Bi | 3358-6,12 T | 2116-1,07 T |

The $\alpha_{Ge-Sb}$ and $\alpha_{Ge-Bi}$ were calculated from the experimental data given in (14), the $\alpha_{Sb-Bi}$ and $\Omega$ had been determined by us (15). In all cases the calculated liquidus curves seem to fit the experimental data well.

The constant K for the dissociation reaction $Bi_2(l) \rightleftarrows 2Bi(l)$ were obtained from (16).

## RESULTS AND DISCUSSION

The phase diagram for the Ge-Sb-Bi system calculated is recorded in fig.2. The experimental results obtained for this system are also shown in the figure. The good agreement be-

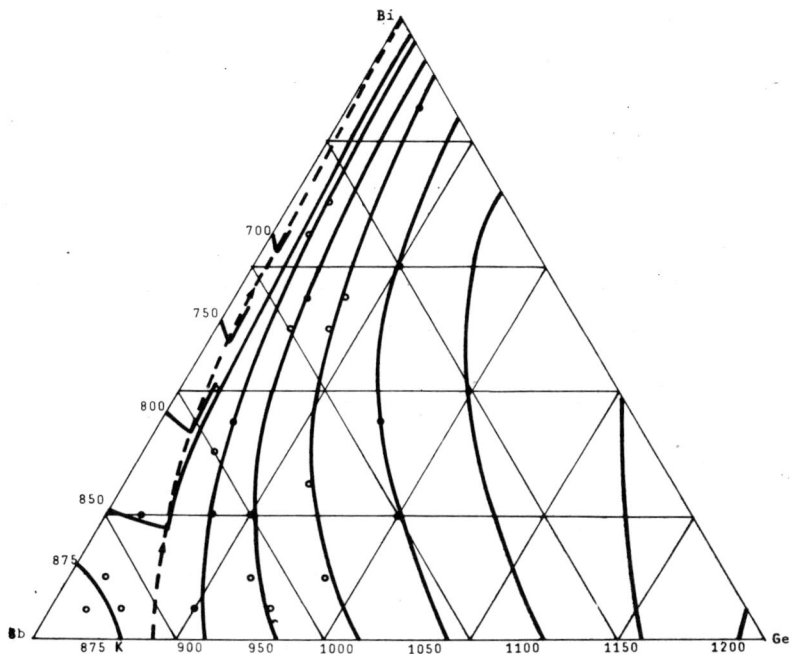

Fig.2. Ge-Sb-Bi ternary phase diagram: (—) liquidus isothermes; (---) eutectic line (Extrapolated; (.) experimental data.

tween the calculations and the experimental data is proved. The standard error of estimate between the calculated and experimental liquidus temperature is 6K. We can affirm that the R.A.S. model give reasonably good fit to the experimental data and this model is attractive for ternary systems because of its relative simplicity and of the small number of parameters involved.

As a conclusion, it seems that the association is a very useful and fruitful concept to take into account, when dealing with the properties of liquid solutions of chalcogens.

## REFERENCES

1) C. Hilsum, Electronic Materials, Eds. N. B. Hannay & U. Colombo, Plenum Press, New York 1973.
2) N.Ch.Abrikosov, V.F. Bankina, L.V. Porestskaya, L.E. She-

limova and E.V. Hendnova Semiconducting II-VI, IV-VI, and V-VI compounds Ed. Soviet Physics-Semiconductors.Plenum press, New York 1969.

3) A. Laugier, J. Cadoz, M. Faure and M. Moulin. J. Crystal Growth 21 (1974) 235.

4) G.A. Antypas and L.W. James. J. Appl. Phys. 41 (1970) p. 2165.

5) M.B. Panish and M. Ilegems. Progress in Solid State Chemistry, Ed.H. Reiss and J.O. McCaldin, Pergamon Press, New York, 1972.

6) A.S. Jordan, Met. Trans. 2 (1971) p. 1965.

7) V.G. Busch and Y. Tieche, Helv. Phys. Acta 35 (1962) p. 273.

8) S.P. Isherwood and B.R.Orton, Phil, Mag. 17 (1968) p.561

9) H. Richter, The properties of liquid metals, ed. S. Takeuchi. Halstead Press, New York 1973.

10) A.S. Jordan, Met. Trans, 1, (1970) p. 239.

11) A.S. Jordan, J. Electrodum. Soc., 119, (1) (1972), 123

12) A.S. Jordan, Met. Trans. 7B, (1976), p. 191.

13) Metals Reference Book, Ed. C.J. Smithells, Butterworths, London-Boston, 1976.

14) C.D. Thurmond and M. Kowalchik, Bell Syst. Tech. I 169 (1960)

15) S. Suriñach, M.D. Baró and F. Tejerina. To be published.

16) S.P. Yatsenko and V.N. Danilin, Inorg. Mater. 4 (1968) p. 758.

# Instrumentation

V. Krämer
University of Freiburg, FRG

H. G. Wiedemann
Mettler Instrumente AG
Greifensee/Switzerland

DEHYDRATION OF CRYSTALLIZATION WATER IN SALTS USING A NEW
DESIGNED CONTROLLED-WATER VAPOR MICRO DTA

M. Taniguchi, H. Moriguchi, S. Shimizu
Department of Chemical Engineering, Faculty of Engineering,
Tokyo Institute of Technology, Tokyo 152 Japan

## ABSTRACT

The effects of water vapor pressure on the equilibria and kinetics for the dehydration of crystallization water in salts have been studied using a new designed controlled-water vapor micro sample DTA instrument. The dehydrations of $CuSO_4 \cdot H_2O$, $CaC_2O_4 \cdot H_2O$, $ZnC_2O_4 \cdot 2H_2O$ and $CaSO_4 \cdot 1/2H_2O$ have been examined respectively. The partial pressures of $H_2O(g)$ in $N_2$ carrier gas flowing through the DTA chamber were kept constant within $\pm 0.001$ atm over the wide pressure range from 0.01 atm to 0.7 atm. A dumbbel type Pt detector and Pt pan were used. For the comparison of data, the usual Ni block cell was also used. The van't Hoff plots by using the initiation temperatures of DTA peaks and the fixed $P_{H_2O}$ values gave the reasonable enthalpy changes for the dehydration reactions. The activation energies were obtained by introducing the correction term $(1-P/P_e)$ in the kinetic equation.

## INTRODUCTION

One of the first controlled-atmosphere and -water vapor DTA instrument provided with gas-flow device was reported by Stone(1,2). By using this instrument, the mechanism of thermal decomposition of ammonium metavanadate was determined previously by one of the authors(3). For the further detailed clarification of the mechanisms of thermal decomposition of salts and dehydrations of hydrated salts, the precise controlled-pressure of atmosphere and water vapor micro DTA instrument have been expected. In general, the thermal analysis(4,5,6) and the equilibrium experimental technique(7) considering the precise pressure controlle of gas phase give us the valuable results of the kinetic(8) and the equilibrium studies(3, 9, 10) on the thermal deconpositions of solids or solid-gas heterogeneous reactions. In this study the enthalpy changes and the activation energies of the dehydrations of crystallization water of $CuSO_4 \cdot H_2O$, $CaC_2O_4 \cdot H_2O$, $ZnC_2O_4 \cdot 2H_2O$ and $CaSO_4 \cdot 1/2H_2O$ have been determined by using a new desinged controlled-water vapor micro DTA instrument. These values are discussed by comparing the transpiration data(9) and the general concept for endothermic reversible dehydration reaction.

## EXPERIMENTAL PROCEDURE

<u>Apparatus</u> The water vapor was generated in the flask (E) and $N_2$ carrier gas was bubbled into the hot water in flask(E) at fixed flow rate. The mixed gas of water vapor and $N_2$ carrier gas was flowing through the condenser(G) which was fixed at the constant temperature by means of the hot water circulation pump(I) keeping at the constant temperature as shown in Fig.1. Thus, the $N_2$ carrier gas with fixed partial pressure of vapor was flowing through the pipe I and (N) to DTA cell chamber(K) (Fig.2). To avoid the condensation of water vapor, the furnace and DTA cell chamber were kept in the air bath(L) at the constant temperature above 100°C and the pipe lines were wounded by ribbon heater. The heating rate was fixed at 5°C/min due to the limitation of the capacity of furnace.

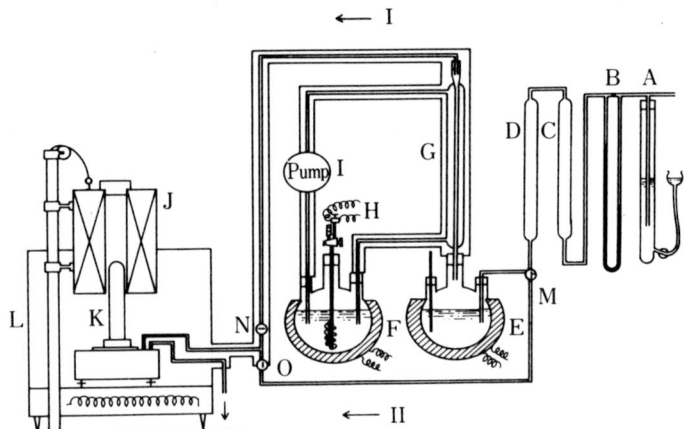

Fig.1 Controlled-water vapor micro DTA apparatus

DTA cell: A Ni block type was used as shown in Fig.3. A dumbbel type Pt detector(11) and a Pt pan type cell were used as shown in Fig.4.

<u>Materials</u> The analytical grade and 200-325 mesh powder samples were used. The samples with fixed hydrated degree were prepared by the preheat treatment on the basis of the TG data. The 50mg sample was taken in Ni block cell and the 10mg sample was taken in Pt pan cell.

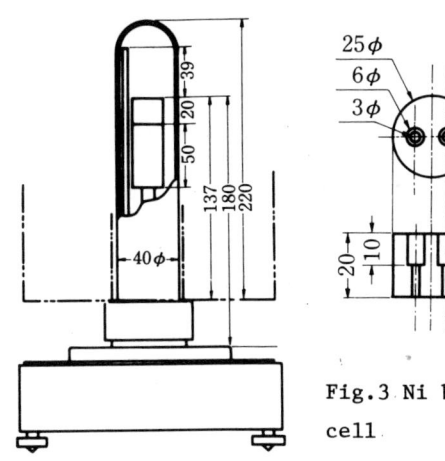
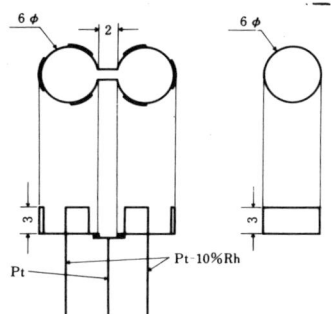

Fig.3 Ni block cell

Fig.4 Dumbbel type Pt detector and Pt pan cell

Fig.2 DTA cell holder and chamber

## RESULTS AND DISCUSSION

The dehydration reactions are as follows:

$$CuSO_4 \cdot H_2O = CuSO_4 + H_2O \quad (Kp = P_{H_2O}) \quad [1]$$

$$CaC_2O_4 \cdot H_2O = CaC_2O_4 + H_2O \quad (Kp = P_{H_2O}) \quad [2]$$

$$ZnC_2O_4 \cdot 2H_2O = ZnC_2O_4 + 2H_2O \quad (Kp = P^2_{H_2O}) \quad [3]$$

$$CaSO_4 \cdot 1/2H_2O = CaSO_4 + 1/2H_2O \quad (Kp = P^{1/2}_{H_2O}) \quad [4]$$

The initiation temperature, peak temperature and half width value of DTA peak were considerably influenced by the fixed water vapor pressure in the DTA chamber as shown in Fig.5. The magnitude of the influence of water vapor pressure varied with the dehydration reactions and the type of cell. Equilibria The plots of the fixed water vapor pressure in the chamber vs. the peak temperature scattered. However, the plots of the fixed water vapor pressure vs. the initiation temperature of peak were illustrated by the smooth curves as shown in Fig.6 and 7. It is well known that the equilibrium water vapor pressures for the dehydration reaction are generaly determined by the static method or the transpiration method. In present experiment, a initiation temperature of the DTA endothermic peak for dehydration was determined at the some fixed controlled water vapor pressure, and it was estimated that this vapor pressure value is respect to the apparent quasi equilibrium water vapor pressure for the dehydration reaction at this initiation temperature. Thus, van't Hoff plots were

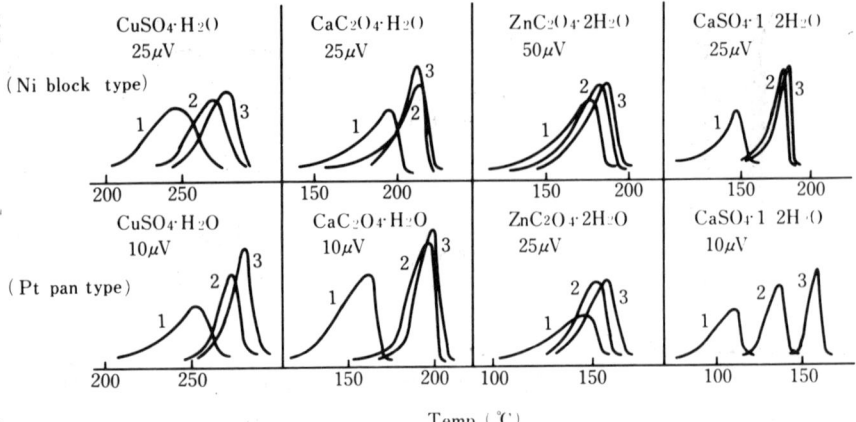

1. 0 Torr   2. 100 Torr   3. 200 Torr

Fig.5 Effects of water vapor pressure and type of cell on DTA peak

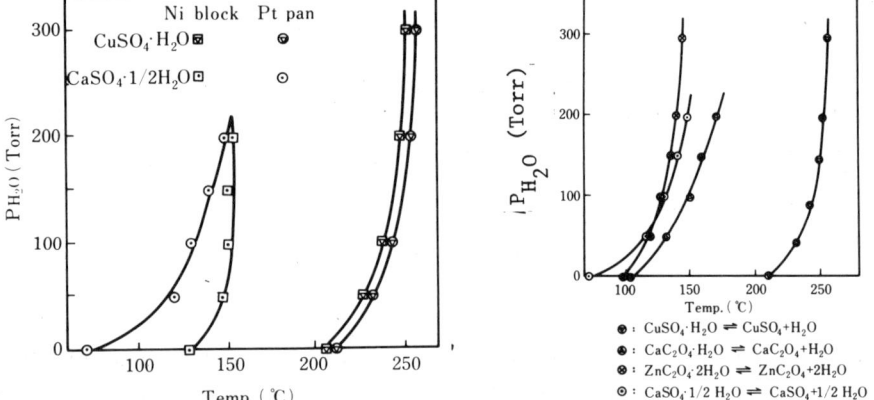

Fig.6 Plots of fixed water vapor pressure vs. initiation temperature of peak

Fig.7 Plots of fixed water vapor pressure vs. initiation temperature of peak by using Pt pan cell

tried by using the relation between Kp and $P_{H_2O}$ in Eq.[1]-[4] as shown in Fig.8. The van't Hoff plots gave H and S values for dehydration reactions as listed in Table 1 respectively. The van't Hoff plots by using the transpiration data are located at the portion of much lower Kp values or temperature range than those by the water vapor DTA data as shown in Fig.8. However, it is interesting that H or S values obtained by DTA method almost agreed with those by transpiration method except the few as shown in Table 1. The van't Hoff plots by using Ni block cell data showed slightly higher Kp values than those by using Pt pan cell data due to the difference in contact with solid and vapor phase as shown in Fig.8.

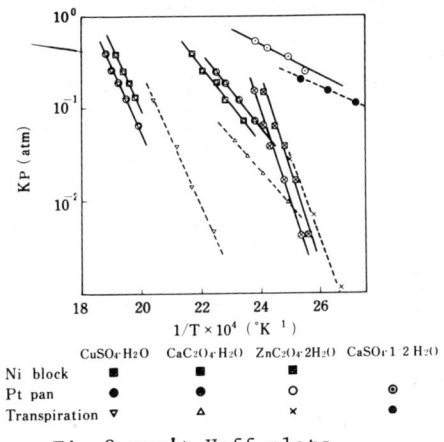
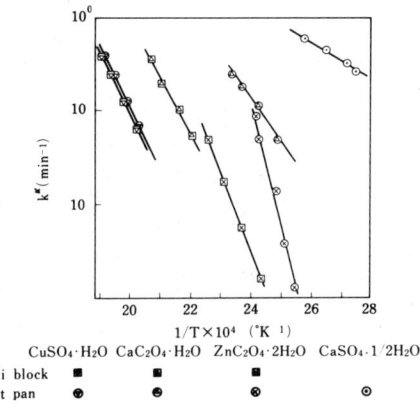

Fig.8 van't Hoff plots          Fig.9 Arrehenius plots

<u>Kinetics</u>  It was assumed that the dehydration rate can be explained by the equation [5] as follow:

$$d\alpha/dt = k(1 - \alpha)^n \quad [5]$$

The $\alpha$ is the fraction reacted. The $\alpha$ value was determined by the peak area and $d\alpha/dt$ was obtained from the peak hight at the some fixed time t. The n value was determined by Kissinger method. Considering the effect of the water vapor pressure on the dehydration rate the correction term (Pe - P)/Pe was introduced as follow:

$$d\alpha/dt = k*(1 - \alpha)^n (Pe - P)/Pe \quad [6]$$

Pe is the apparent quasi equilibrium water vapor pressure obtained by reading the plots of $P_{H_2O}$ vs. temperature as shown in Fig.7, and P is the controlled fixed water vapor pressure in the chamber. The Arrhenius plots by using k* values (Fig.9) gave the activation energies for the dehydration reactions as listed in Table 1. These results are almost consistent with the general concept for the endothermic process: $\triangle H \leq E$.

Table 1 $\Delta H$, $\Delta S$ and E values for the dehydration reactions

| System | Method | | Temperature range(°C) | Water vapor pressure range(atm) | $\Delta H$ (KJ/mol) | $\Delta S$ (e,u.) | E (KJ/mol) |
|---|---|---|---|---|---|---|---|
| $CuSO_4-H_2O$ | DTA | Pt pan | 230 - 263 | 0.0658 - 0.658 | 131.4 | 239.3 | 176.6 |
| | | Ni block | 228 - 265 | 0.0658 - 0.658 | 125.5 | 233.9 | 166.1 |
| | Transpiration | | 174 - 217 | 0.00679 - 0.133 | 123.8 | 235.6 | --- |
| $ZnC_2O_4-H_2O$ | DTA | Pt pan | 120 - 147 | 0.0658 - 0.395 | 180.3 | 412.5 | 204.6 |
| | | Ni block | 120 - 145 | 0.0658 - 0.395 | 210.5 | 493.3 | 160.2 |
| | Transpiration | | 94.5 - 128 | 0.0213 - 0.158 | 146.4 | 335.1 | --- |
| $CaC_2O_4-H_2O$ | DTA | Pt pan | 133 - 172 | 0.0658 - 0.395 | 74.8 | 159.4 | 119.7 |
| | | Ni block | 165 - 195 | 0.0658 - 0.395 | 72.8 | 150.2 | 82.0 |
| | Transpiration | | 129 - 161 | 0.0102 - 0.0459 | 69.5 | 134.3 | --- |
| $CaSO_4-H_2O$ | DTA Pt pan | | 120 - 150 | 0.0658 - 0.263 | 32.6 | 72.4 | 38.5 |
| | Transpiration | | 90.5 - 122 | 0.00922 - 0.0454 | 30.9 | 65.3 | --- |

(The transpiration data of Ref.(9) were used.)

## REFERENCES

(1) R.L. Stone and H.F. Rase, Anal. Chem. 29 (1957) 1273.

(2) R.L. Stone, Anal. Chem. 32 (1960) 1582.

(3) M. Taniguchi and T.R. Ingraham, Can, J. Chem. 42 (1969) 2467.

(4) P.D. Garn, Thermoanalytical Methods of Investigation, Academic Press (1965) p.224.

(5) R.C. Mackengie(ed.), Differential Thermal Analysis vol.1, Academic Press (1970) p.63, 101, 395.

(6) W.W. Wendlandt, Thermal Methods of Analysis (2nd ed.), John Wiley & Sons (1974) p.202.

(7) M. Taniguchi, Kagaku to Kogyo (The Chem. Soc. of Japan) 18 (1965) 645.

(8) M. Taniguchi and Y. Yamamoto, Proc. of the 5th ICTA, (1977) 497.

(9) S. Shimizu and M. Taniguchi, Nippon Kagaku Kaishi (The Chem. Soc. of Japan), 1977 (7) 957.

(10) M. Wakihara, T. Uchida and M. Taniguchi, Metal Trans. B 9B (1978) 29.

(11) A. Yamamoto, K. Yamada, M. Maruta and J. Akiyama, Proc. of the 2nd ICTA, Vol.1 (1969) 105.

"OPEN SYSTEM" DSC : A NEW APPROACH OF ISOTHERMAL INVESTIGATIONS

Pierre LE PARLOUËR

SETARAM, LYON, F

## ABSTRACT

The "open structure of SETARAM DSC 111 allows a new approach of isothermal method by using an automatic and symmetrical introducer of samples. By reducing the the thermal effect due to the introduction of the cells, more informations on the beginning of the reaction can be collected.

The interest of the device is shown through different classical isothermal applications polymerisation, crystallisation, decomposition.

## INTRODUCTION

Differential Scanning Calorimetry, is essentially used, as pointed out by its name, according to the mode of temperature programming. However the interest of isothermal investigations remains when kinetics of reaction are studied.

In such a test, the temperature of reaction must be reached very quickly. Whatever method employed (fast heating, direct introduction of sample at the temperature of the test), the data relative to the beginning of the reaction are hidden by the thermal disequilibrium of the calorimeter (1,3).

In order to improve the DSC isothermal method, the sample introduction must be symmetrical and reproducible. This can be performed by using an automatic device, directly matched on the DSC.

## AUTOMATIC INTRODUCER OF SAMPLES

SETARAM Differential Scanning calorimeter DSC 111 has an "open" structure which allows introduction of the sample simply sliding it in the tube into the sensitive zone of the DSC(2) In an isothermal test, the calorimeter is preset at a defined temperature, and the sample is introduced from ambiant temperature or after preheating in a little oven fitted on the calorimeter.

The thermal disequilibrium is very large if the introduction is done only in one tube. The symmetrical introduction of two identical cells allows to reduce this thermal effect.

In order to make the introduction reproducible, a symmetrical introducer has been developed (Figure 1).

It is constituted of a servo-mechanism which drives two identical rods by means of an articulation.

Temperature of the cells is controlled in the preheating oven before the introduction into the calorimeter.

The guide brings the cells in the sensitive zone and returns at its initial position.

The time of transfer from the oven to the calorimetric detector is about five seconds.

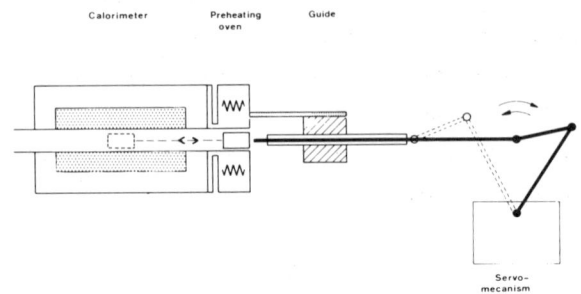

Figure 1 - Automatic introducer with preheating oven

## USE AND APPLICATIONS

The use and the advantages of the automatic introducer are shown in some characteristic isothermal applications : polymerisation, crystallisation, decomposition.

### Polycondensation of phenolic resin

Most studies of isothermal polymerisation by DSC have been run on epoxy resins (1, 3). The investigation of phenolic resin is more delicate because polycondensation releases water. The test must be carried out in a tightened cell. The stainless steel high pressure cell developed by SETARAM is well adapted for this application.

The resin sample and an identical previously polymerized one which acts as a reference are preheated at 60°C in the little oven.

The temperature of DSC is stabilized at 430 K.

The first introduction of both cells produces a short thermal effect before the peak corresponding to the polycondensation. When the reaction is finished, the first thermal effect corresponding to the disequilibrium is measured by running a second introduction of the cells in the same experimental conditions.

As the thermal effect corresponding to the introduction is reproducible, the difference of the two curves is performed by means of the Hewlett-Packerd 9825 A calculator, and drawn by a plotter (Figure 2).

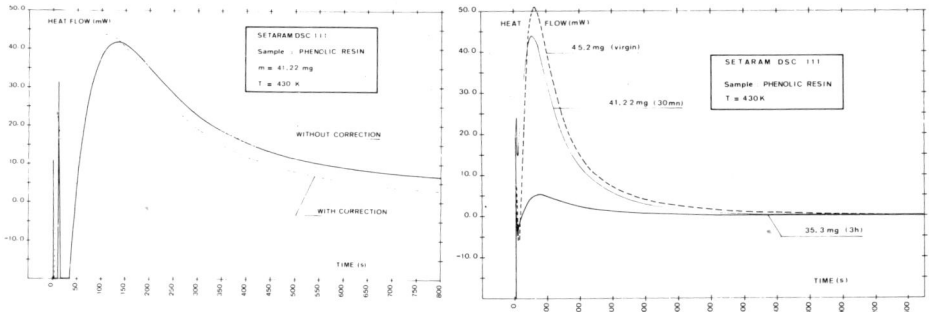

Figure 2 - Curves of phenolic resin polycondensation without and with correction of "blank".

Figure 3 - Polycondensation of phenolic resin differently preheated.

We notice, that the corrected curve gives data on the polymerization from 30 seconds, that is quite less than with the other methods.

According to this method of introduction, polycondensation of phenolic resins which have been previously preheated at 110°C during 30 mn and 3 h, is studied (Figure 3) The curves, obtained after "blank" correction corroborate that, with the automatic introducer, it is pratically possible to detect the beginning of the reaction, even in the case of fast reactions.

## Crystallisation of polypropylene

The study of crystallisation of polymers by DSC is a method frequently used (4,5). However the method is limited by the cooling rate and the response time of instruments when the difference between melting and crystallisation temperatures becomes large. In this case, the peak of crystallisation is not separated from the disequilibrium thermal effect.

The use of the automatic introducer allows to improve the method.

A sample of polypropylene is melted in the preheating oven at 200°C, and introduced, as well as a reference cell with an identical calorific capacity, in the calorimeter at the defined crystallisation temperature.

As the disequilibrium thermal effect is weak, there is no problem to detect the peak of crystallisation at every temperature (Figure 4). At low temperature, it is the first effect which is hidden by the crystallisation peak.

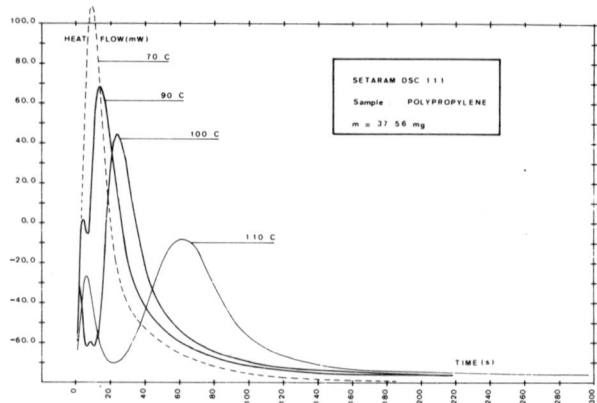

Figure 4 – Crystallisation of polypropylene (Melting at 200°C)

### Decomposition of azodicarbonamid

Decomposition investigations by DSC are mostly carried out in scanning mode and numerous methods of calculation have been developed in order to determine kinetics of reaction. They have all their limits, because thermal equilibrium in the sample is never reached during the test. The isothermal study partially solves this difficulty, but the problem of the data at the beginning of the reaction remains.

For some studies of decomposition (pyrotechnic compositions for example), only the isothermal test is available in order to have firing of the powder at a temperature above its decomposition temperature (6).

The automatic introduction with DSC 111 has been adapted for the study of azodicarbonamid decomposition.

The sample is preheated at 60°C, then introduced into the DSC at 200°C. When the decomposition is finished, a second introduction is carried out in order to measure the disequilibrium thermal effect. The difference of the two curves allows to have informations on the beginning of the decomposition (Figure 5).

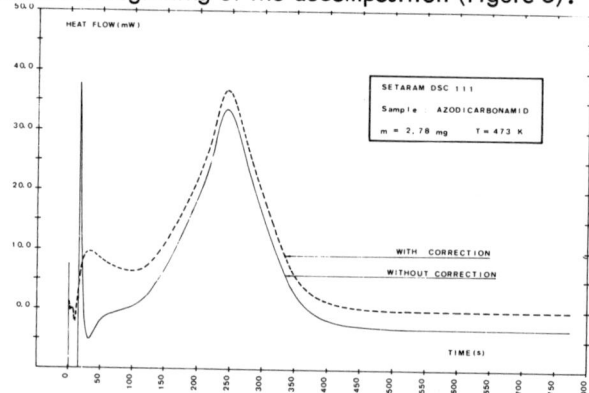

Figure 5 – Decomposition of azodicarbonamid

## CONCLUSION

The use of a symmetrical and automatic introducer, adapted to SETARAM DSC 111, allows to improve some isothermal methods. The device reduces the disequilibrium thermal effect at the beginning of the test, and gives rapidly data on the reaction studied.

## REFERENCES

1 - R.A. FAVA,    Polymer, 9 (1968) 137
2 - J. MERCIER,   J. Thermal Anal. 14 (1978) 161
3 - G. WIDMANN, Thermal Analysis Vol.3, Proceedings 4th ICTA, Budapest 1974, p. 359
4 - A.A. DUSWALT, Hercules Chemist, 57 (1968) 5
5 - J.R. KNOX,    Analytical Calorimetry (1968) p.45
6 - M. BOURGEON International Conference on explosifs of pyrotechnic, Spatial Applications, Toulouse, October 1979

# THERMOVOLTAIC DETECTION: A NEW TECHNIQUE FOR THE STUDY OF THERMAL DECOMPOSITION REACTIONS

W.W. Wendlandt

Thermal Analysis Laboratory
Department of Chemistry
University of Houston
Houston, Texas  77004   USA

The thermovoltaic detector (TVD) consists of two metal electrodes, such as aluminum and platinum, which when placed in contact with certain types of samples at elevated temperatures, spontaneously generates a voltage that is proportional to the decomposition reaction taking place. The compounds investigated, metal salt hydrates and various ammine type coordination compounds, yield characteristic TVD curves which are related to the thermal decomposition reactions observed by other thermal analysis techniques. The TVD output voltage is of the order of 0-1V over the temperature range of 25 to 500°C. The apparatus for the technique is extremely simple, consisting mainly of a digital multimeter with an analog voltage output (Keithley Model 160B), a X-Y plotter, and a furnace temperature programmer.

The TVD curve, like other thermal analysis techniques, is dependent on the furnace heating rate and other parameters such as concentration of the sample in the matrix, thickness of the sample disk, and so on. Some of the compounds studied by this technique include $NiSO_4 \cdot 6H_2O$, $CoSO_4 \cdot 7H_2O$, $CuSO_4 \cdot 5H_2O$, $Ni(py)_4Cl_2$, $[Co(en)_3]X_3$ (X = halide, nitrate etc.), $[Co(NH_3)_6]X_3$ complexes, and others. The technique was found to be very useful for the investigation of the thermal decomposition reactions of these compounds.

# THERMAL HAZARD EVALUATION BY AN ACCELERATING RATE CALORIMETER

J. C. Tou and L. F. Whiting
Analytical Laboratories
D. I. Townsend
Process and Development Laboratories
The Dow Chemical Company
Midland, MI  48640, USA

## ABSTRACT

An accelerating rate calorimeter was developed for thermal hazard evaluation to provide time-temperature-pressure data for chemical reactions taking place under adiabatic conditions. The data interpretation is illustrated with an nth-order reaction. The technique was applied to a study of the thermal decomposition of N-methyl-N-nitroso-p-toluene sulfonamide, commonly known as Diazald.

## INTRODUCTION

The hazards of most concern in the chemical industry are the thermal hazard and the pressure hazard. Many techniques have been developed for the purpose of studying the thermodynamic and/or kinetic aspects of these hazards (1).
In this paper, the principle of an accelerating rate calorimeter, based on the adiabatic calorimetric principle, is reported. With this technique, not only can the kinetic aspects of temperature and pressure associated with chemical reactions be evaluated, but also the heat of reaction can be estimated.

## EXPERIMENTAL PROCEDURE

The accelerating rate calorimeter (ARC) was initially developed in the laboratories of the Dow Chemical Company. Recently, Columbia Scientific Industries of Austin, Texas, added a microprocessor control and commercialized the unit under an abbreviated name CSI-ARC. A brief discussion of the instrumental design and the operational logic follows.

A spherical sample bomb is mounted inside a nickel plated copper jacket with a tube fitting on which is attached a pressure transducer and a sample thermocouple. The jacket is composed of three zones which are individually heated and controlled by the Nisil/Nicrosil type N thermocouples. During an exothermic reaction, the temperature differences between the bomb and three zones of the jacket are digitally controlled to maintain zero. The ARC follows the heat-wait-search logic until a self-heat rate greater than the preset rate is detected. The calorimeter will be maintained at adiabatic conditions until the completion of the experiment. The heat of decomposition of the Diazald solutions was also determined by DSC with a modified sealed glass ampoule microreactor (2).

## GENERAL PRINCIPLE

Under adiabatic conditions, the heat of reaction, $\Delta H_r$, can be evaluated from the average heat capacity, $\bar{C}_v$ over the experimental range, the mass of the sample M and the experimentally determined adiabatic temperature rise, $\Delta T_{AB}$, which is the difference between the final and initial temperatures, $T_f - T_o$,

$$\Delta H_r = M \bar{C}_v \Delta T_{AB}$$

The concentration of the reactant, C, at any temperature or time t can be approximately related to the temperature of the system.

$$C = \frac{T_t - T}{\Delta T_{AB}} C_0$$

Where $C_o$ is the initial concentration of reactant. Substituting this equation after differentiation with respect to T into the rate equation of an nth-order reaction, the following fundamental equations can be obtained in relating the thermally measured quantity, T, to a kinetic event.

$$m_T = \frac{dT}{dt} = k \left(\frac{T_t - T}{\Delta T_{AB}}\right)^n \Delta T_{AB} C_0^{n-1}$$

$$k^* = C_0^{n-1} k = \frac{m_T}{\left(\frac{T_t - T}{\Delta T_{AB}}\right)^n \Delta T_{AB}} \quad \text{or} \quad \ln k^* = \ln C_0^{n-1} A - \frac{E}{R}\left(\frac{1}{T}\right)$$

and $\quad \ln m_0 = \ln \Delta T_{AB} C_0^{n-1} A - \frac{E}{R}\left(\frac{1}{T_0}\right)$

where $m_T$ is the self-heat rate measured at temperature T, or time t, $k^*$ the psuedo zero-order rate constant at temperature T and $m_o$ the initial self-heat rate at temperature $T_o$.

It is also important to know the temperature-time relationship, which can be evaluated from the integration of $m_T dt$. The upper boundary of the integration is set at the temperature of the maximum self-heat rate. For a reaction with high activation energy, for example, 20 Kcal/mole$^{-1}$, the major portion of reaction time is the time to maximum rate (TMR), $\Theta_m$. Above the temperature at maximum rate, $T_m$, the reaction decelerates rather quickly until the completion of the reaction. The $\Theta_m$ can be evaluated as follows:

$$\theta_m = t_m - t = \int_t^{t_m} dt = \int_T^{T_m} \frac{dT}{k\left(\frac{T_t - T}{\Delta T_{AB}}\right)^n \Delta T_{AB} C_0^{n-1}}$$

An approximate analytical solution for the above integration is derived,

$$\theta_m \approx \frac{RT^2}{m_T E} - \frac{RT_m^2}{m_m E}$$

For a reaction with a high activation energy, the second term is relatively insignificant as compared to the first term. Then

$$\theta_m \simeq \frac{RT^2}{m_T E}$$

This is equivalent to the equation derived for the determination of homogeneous condensed explosives following a first-order reaction rate(3). By substituting the self-heat rate into the above equation, the following approximate expression can be obtained.

$$\ln \theta_m \simeq \frac{E}{R}\left(\frac{1}{T}\right) - \ln A$$

For a reaction vessel with a known heat transfer, it is important to estimate the temperature of no return, $T_{NR}$, below which the reaction temperature will never increase, since all reaction heat can be transferred from the system. Above $T_{NR}$, the reaction temperature will spiral upward and a runaway reaction will result. At thermal equilibrium, the following equation is derived

$$\theta_{m_0} \text{ at } T_{NR} \text{ or } \theta_{T_{NR}} = \frac{M\bar{C}_v}{Ua}$$

where U is the overall heat transfer coefficient and a the heat transfer area.

The quantity, $M\bar{C}_v/Ua$, represents the time line for the equipment, and the quantity, $\Theta_{T_{NR}}$, is the adiabatic zero-order time to maximum rate at $T_{NR}$. All the equations described above have been developed for a sample held at adiabatic conditions. However, in the present accelerating rate calorimetric system, part of the heat generated from the reaction is being used to heat up the sample bomb as well. Therefore, the following correction has to be made:

$$\Delta T_{AB} = \phi \Delta T_{AB\textit{a}}, \qquad T_t = T_0 + \phi \Delta T_{AB\textit{a}} \text{ where } \phi = 1 + M_b \bar{C}_{vb}/M\bar{C}_v.$$

where $M_o$ and $\bar{C}_{vb}$ are the mass and the heat capacity of the bomb and $\Delta T_{AB,S}$ the adiabatic temperature rise for the total system comprising the sample and the sample bomb.

The two fundamental equations now become

$$C = \frac{T_{t,s} - T}{\Delta T_{AB,s}} C_0 \quad \text{and} \quad m_{T,S} = k \left( \frac{T_{t,s} - T}{\Delta T_{AB,s}} \right) \Delta T_{AB,s} C_0^{n-1}$$

where the subscript reflects that the measurement is made for the system.

The psuedo zero-order rate constant, k*, and therefore the fundamental kinetic parameters, A and E, can be evaluated in an identical manner to the calculations based on isolated chemicals ($\phi=1$).

## RESULTS AND DISCUSSION

The performance of a CSI-ARC was characterized with the thermal decomposition of di-t-butyl peroxide in toluene and in its pure form. Different amounts of samples were loaded in various sample bombs, which cover a wide range of $\phi$. The kinetic parameters determined should not depend on $\phi$. This would be a stringent test of the kinetic performance of the instrument (4). The results are shown in Table 1, which compare well to the literature values ranging from 36 to 38 Kcal mole$^{-1}$ in various solvents systems (1).

The ARC principle was applied to a study of the thermal decomposition of N-methyl-N-nitroso-p-toluene sulfonamide, commonly known as Diazald. The compound is a widely utilized reactant for the preparation of diazomethane, one of the most versatile reactants in organic synthesis.

The ARC data are shown in Figures 1-6. Figure 6 shows the experimental TMR, $\Theta_{m,s}$, with $\phi=2$, the derived TMR $\Theta_m$ and the zero order TMR, $\Theta_{m_0}$ curves of the diethyl ether solution of Diazald. The zero-order TMR curve for pure Diazald was also calculated from the kinetic and thermodynamic data obtained by considering the solvent as being part of the bomb. With this consideration, thermal inertia is estimated to be 7.6, assuming $\bar{C}_v$ of diethyl ether and Diazald in solution to be 0.5 cal.°C-g$^{-1}$. For the two vessels studied, the $T_{NR}$ of the solution and pure Diazald were estimated to be ~85°C and ~73, respectively, in the case of the standard 500 ml three-neck flask. From this study, handling or storing this solution or pure Diazald in the laboratory vessels at ambient temperature does not present any thermal hazard. However, in the case of the 10M gal. well-insulated tank car, the $T_{NR}$'s are ~26 and 18°C for the solution and pure Diazald respectively. Under such conditions, a thermal runaway and a possible explosion due to the pressure build-up are predicted if the ambient temperature is greater than $T_{NR}$.

1) D. I. Townsend and J. C. Tou, Thermochimica Acta (in press) and the references therein.
2) J. C. Tou and L. F. Whiting, submitted to thermochimica Acta for publication.
3) S. H. Lin and H. Eyring, Annual Review of Physical Chemistry, Vol. 21, 1970, ed., H. Eyring, C. J. Christensen, and H. S. Johnston, P.225, Annual Review Inc., Palo Alto, California, and the references therein.
4) J. C. Tou, and L. F. Whiting, to be published.

TABLE 1: THE KINETIC PARAMETERS, E AND A AND THE HEAT OF THE THERMAL DECOMPOSITION REACTION OF di-t-butyl PEROXIDE IN TOLUENE AND IN ITS PURE FORM

| Bomb type | Conc. | $\phi^*$ solution | $\phi^*$ pure | E (Kcal/mole) set average | A (min$^{-1}$) set average | $\Delta H_r$ (Kcal/mole) ARC | $\Delta H_r$ (Kcal/mole) DSC |
|---|---|---|---|---|---|---|---|
| Ti | 30% | 1.7 | 5.6 | 38.077 | 1.16 × 10$^{18}$ | -41 | -42 |
|    | 30% | 1.6 | 5.4 |        |                  | -42 |     |
| Light weight Hastelloy C | 10% | 1.8 | 17 | 37.944 | 9.75 × 10$^{17}$ | -33 ($\dot{m}_m$=0.073°C/min) | -38 |
|  | 20% | 1.8 | 9.0 |  |  | -43 | -42 |
|  | 30% | 1.8 | 6.0 |  |  | -41 | -42 |
|  | 60% | 1.8 | 3.1 |  |  | -33 ($\dot{m}_m$=467°C/min) | -41 |
| Heavy weight Hastelloy C | 60% | 9.4 | 16 | 37.200 | 4.12 × 10$^{18}$ | -49 | -41 |
|  | 60% | 6.4 | 11 |  |  | -48 |  |
|  | 60% | 5.0 | 8.3 |  |  | -46 |  |
|  | 60% | 4.3 | 7.2 |  |  | -45 |  |
| Heavy weight Hastelloy C | pure |  | 8.3 | 37.278 | 5.23 × 10$^{17}$ | -37 | -38 |
|  | pure |  | 7.9 |  |  | -41 |  |

* $C_{vb}$ = 0.1 cal/°C gm , $C_{vs}$ = 0.5 cal/°C gm

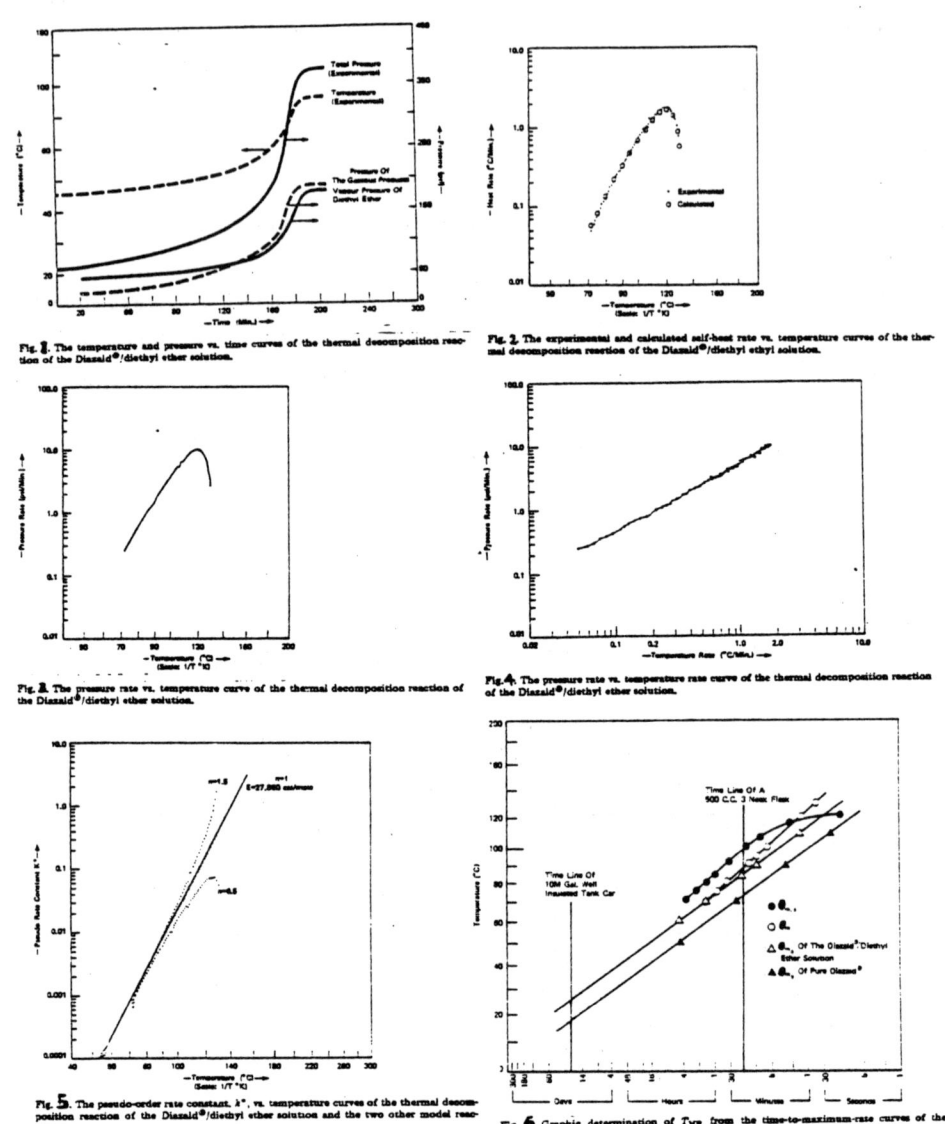

Fig. 1. The temperature and pressure vs. time curves of the thermal decomposition reaction of the Diazald®/diethyl ether solution.

Fig. 2. The experimental and calculated self-heat rate vs. temperature curves of the thermal decomposition reaction of the Diazald®/diethyl ethyl solution.

Fig. 3. The pressure rate vs. temperature curve of the thermal decomposition reaction of the Diazald®/diethyl ether solution.

Fig. 4. The pressure rate vs. temperature rate curve of the thermal decomposition reaction of the Diazald®/diethyl ether solution.

Fig. 5. The pseudo-order rate constant, $k^*$, vs. temperature curves of the thermal decomposition reaction of the Diazald®/diethyl ether solution and the two other model reactions.

Fig. 6. Graphic determination of $T_{MR}$ from the time-to-maximum-rate curves of the thermal decomposition reaction of Diazald® and the equipment time curve.

# ESTIMATION OF INSTANTANEOUS CALORIC EFFECT IN THE HEAT EXCHANGE TYPE OF CALORIMETRY

M. Nakanishi and S. Fujieda
Ochanomizu University, Bunkyo-ku, Tokyo 112, Japan

## ABSTRACT

The previously reported technique of the heat exchange type of titration calorimetry, which depends on operational amplifier circuitry for computation, has been improved to give much more precise values of instantaneous heat. Since double differentiation of the observed temperature is necessary for obtaining instantaneous heat and the analog computation is not suited for differentiation, advantage of a digital technique has been successfully taken for the on-line calculation of instantaneous heats.

## INTRODUCTION

In following chemical phenomena thermometrically, exact estimation of instantaneous heat, or heat evolution per unit time, is important as well as the corresponding total heat. The heat exchange type of titration calorimetry (1),(2) can provide sufficiently exact values of total heat when operated with the so far developed analog computation circuit which consists of operational amplifier net work. But this is not suited to obtain instantaneous values of heat for the following reason.

As will be described in a later paragraph, the expression for instantaneous heat (q in equation 3) involves double differentiation with respect to time of the obtained values of temperature. The experimentally obtained temperature output inevitably contains small noises and these noises are more amplified in the analog differentiation, to give erroneous results. Possibly the noises may be reduced or eliminated by introducing a capacitive component in the circuit, but the result may lose in accuracy.

Lately digital computation technique was introduced to obtain instantaneous as well as total heats from the same experiment as was carried out in the previous analog treatment.

In this investigation an attempt has been made to introduce a microcomputer for the rather complicated control and on-line calculation in the heat exchange type of titration calorimetry.

## HEAT EXCHANGE TYPE OF TITRATION CALORIMETRY

The essential part of the calorimeter is schematically shown in Figure 1. A sample solution to be titrated is placed in a glass vessel (S) for sample which is provided with a thermistor, an electrical heater and a capillary tip of an electrically driven piston buret. The solution in the vessel is agitated with a magnetic stirrer rod. Another glass vessel (R) for reference of the identical size and configuration with the sample vessel is placed side by side in a sufficiently large water bath.

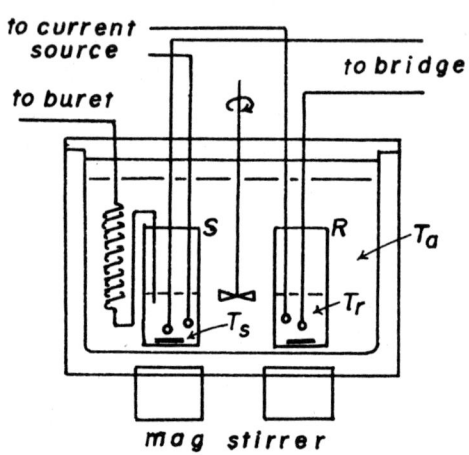

Fig.1. Schema of the titration calorimeter

The water bath is also agitated vigorously to maintain a temperature $T_a$ uniform throughout the bath. The actual temperatures in the vessels S and R are $T_s$ and $T_r$, respectively. A titrant solution is delivered usually at a constant rate. During long passage through a helical tube, the temperature of the titrant is equilibrated with that of the ambient water, $T_a$, so long as the latter temperature remains almost constant.

When heat is evolved at a rate q per unit time in the sample vessel either by any chemical or electrical process, the following equation holds:

$$\frac{dT_s}{dt} = \frac{q + q'}{C} - \alpha(T_s - T_a) \tag{1}$$

where $\alpha$ is a constant relating to the heat transfer and $q'$ is the heat effect per unit time due to the difference in temperature between titrant and titrand, $C$ is the effective heat capacity of the vessel. The corresponding equation is valid for the reference solution,

$$\frac{dT_r}{dt} = -\alpha(T_r - T_a) \tag{2}$$

If a small heat evolution exists as a result of stirring in the stand-by state, $T_s$ is slightly greater than $T_a$ and a stationary state is reached. A Wheatstone bridge in two arms of which the thermistors in S and R are incorporated is equilibrated in such a stationary state, to give a null output.

Equations 1 and 2 are combined to give the following equation:

$$\frac{dT}{dt} = \frac{q + q'}{C} - \alpha T \tag{3}$$

where $T = T_s - T_r$.

The response $\theta$ of thermistors involves delay of the first order to a change of temperature, thus:

$$\frac{d\theta_s}{dt} = \beta(T_s - \theta_s), \quad \frac{d\theta_r}{dt} = \beta(T_r - \theta_r) \tag{4}$$

where $\beta$ is a constant relating to delay of response. These are combined into following one:

$$\frac{d\theta}{dt} = \beta(T - \theta) \tag{5}$$

where $\theta = \theta_s - \theta_r$.

Our final aim is to derive $q$ from the knowledge of observed values of $\theta$'s.

First, $\theta$'s can be converted into T's using the relation in equation 5, thus:

$$T = \frac{1}{\beta} \frac{d\theta}{dt} + \theta \qquad (6)$$

Previously this process involving differentiation was performed in an analog computational way and therefore the process that follows had to contain an analog integration, to give a smooth and reliable curve. The analog technique is replaced by the digital computation in this paper, and it proved the differentiation could be repeated without losing in accuracy.

Second, $\alpha$ in equation 3 is made constant regardless of an increase of the heat exchange surface area by simply covering the base, plus a small portion of the side wall, of the cylindrical vessels with a thermally insulating material, so that the area on the wall through which heat is exchanged is proportional to the solution volume.

Third, the effective heat capacity C can be expressed by a sum of the initial heat capacity $C_0$ and a product $\rho cvt$, where $\rho$ is density, c is specific heat capacity and v is the delivery rate of titrant. q' is considered to be $\rho cvT$. Thus, equation 3 can be rewritten as follows and q obtained accordingly.

$$\frac{q}{C_0} = \left(\frac{dT}{dt} + \alpha T\right)\left(1 + \frac{\rho cvt}{C_0}\right) + \frac{\rho cvT}{C_0} \qquad (7)$$

## CONTROL AND COMPUTATION

A SORD microcomputer, model M223 mark II, provided with a Z-80 microprocessor, 64K byte RAM and a floppy disc for external storage of the program and the data files, was used for both control and computation. Analog signals, both input and output, were treated through an analog-digital interface HC-AIO, supplied by SORD, via an S-100 bus.

The program for data processing, display, and storage was composed in a BASIC language. It comprised setup of experimental conditions for both chemical and electrical, such as the number and interval of observations, delivery rate of titrant, initial volume of titrand, current strength and time duration of electrical heating, control of starting of titration or alternative process. The experimental conditions were set in a dialogic fashion by means of a key board and a CRT display.

The constant $\alpha$ can be determined from the data in an electrical calibration run in the free cooling stage where $\theta$ varies in an exponential fashion and $\alpha$ is the only factor to determine the rate of temperature fall. On the other hand, $\beta$ is more difficult to determine directly from the experimental data. It can be determined either from the time constant corresponding to $\beta$ in the analog circuit or by successive approximation to give an adequate result in digital calculation.

Values of temperature $\theta$ in a thermal phenomenon approximated within a small range of time by a quadratic expression and a linear combination of exponential terms. For the sake of simplicity the former will be mentioned below.

Observed points were assumed to be unequally distanced with respect to time, thus:

$$\theta(t) = a(t - t_n)(t - t_{n-1}) + b(t - t_n) + c \qquad (8)$$

The constants a, b, and c are expressed as follows:

$$a = \frac{1}{t_{n-1} - t_{n-2}} (b - \frac{\theta_n - \theta_{n-2}}{t_n - t_{n-2}})$$

$$b = \frac{\theta_n - \theta_{n-1}}{t_n - t_{n-1}} \qquad (9)$$

$$c = \theta_n$$

From relation 8 and 9, and 6, T's can be calculated from corresponding $\theta$'s. Then again, T's within a small range of time are expressed in a quadratic formula with respect to t and treated in the same way as in expressions 8 and 9, to give q in equation 7. Naturally values in term of a caloric unit are determined by comparison with known caloric quantities in electrical calibration runs.

## RESULTS

Compared with results obtained in an analog computation, those by digital calculation obtained in this investigation are in the same order of accuracy and precision, but far better with the instantaneous values of heat. Though the microcomputer control does not yet cover the complete process, the experimental procedures have been facilitated to a large extent.

## REFERENCES

(1) M.Nakanishi and S.Fujieda, Anal. Chem. $\underline{44}$ (1972) 574; Proceedings of the 4th ICTA 1974 (Budapest), Vol.3, (1975), p.929.
(2) M.Nakanishi and S.Fujieda, Proceedings of the 5th ICTA 1977 (Kyoto), (1977), p.393.

# THERMOANALITICAL INVESTIGATIONS ON SOLID - VAPOUR REACTIONS

S. Gál, J. Sztatisz and [+]L. Fodor

**Institute for General and Analytical Chemistry,**
[+]Institute of Agricultural Chemical Technology,
Technical University, Budapest, Hungary

## ABSTRACT

A divided sample holder has been worked out for thermoanalytical investigations on solid - vapour reactions, which often involve experimental difficulties. It can be used in DSC and TG apparatus conveniently. One part of the sample holder contains the solid reactant, the other is for the liquid or solid source of the vapour. The gaseous compounds cannot leave the surroundings of the sample below a certain selected pressure. Reactions of formic acid and acetic acid vapours with sodium carbonate and calcium carbonate are described, using the results obtained with the divided sample holder.

## INTRODUCTION

One of the difficulties arising in thermoanalytical studies on solid - vapour reactions is related to the maintenance of a suitable /partial/ pressure of components being a liquid at ambient temperature. The usual way to supply the reactive vapour into the sample chamber is saturating a flowing gas with the vapour. For higher partial pressures the vapour can be generated and led to the sample chamber at high temperatures. This method requires a very careful construction and operation of the apparatus to avoid condensation. For DTA investigations under constant pressure vapour atmospheres, Garn (1) worked out a system, which can be used easily provided the sample resists submerging under the liquid at lower temperatures.

The sample holder shown in Fig.1 was worked out for studies on
$$Solid_I + Vapour_I = Solid_{II} + \ldots$$

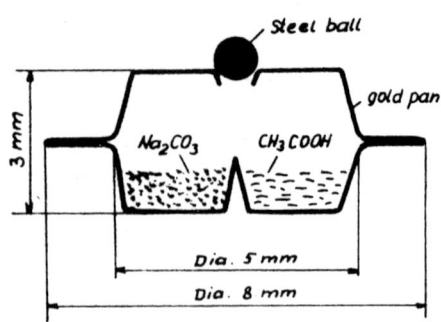

Fig.1. Divided sample holder for solid - vapour reactions

type processes. At first, the reactants are placed into the lower part of the sample holder, which is a divided pan, putting $Solid_I$ on one side, the solid or liquid source of $Vapour_I$ on the other. Then the pan is covered with the upper part and the two pieces are sealed hermetically. At last, the lid is punched, and a ball is placed on the opening. This design prevents diffusion between the inner space and the surroundings, and permits an outflow of the vapours and gases above a certain pressure difference, determined by parameters of the ball and the opening. Of course, the pressure outside the sample holder may be different from atmospheric.

In the present paper, the reactions of formic and acetic acid vapours with sodium carbonate, bicarbonate and calcium carbonate are taken as examples to demonstrate the use of the divided sample holder. These reactions are generally carried out with acid solutions.

## EXPERIMENTAL

The sample holder described in the preceding section was used in the TG and DSC devices of the DuPont 990 Thermal Analysis System. Analytical grade reagents were applied. The initial mass of the samples are included in the diagrams, the heating rate was 5 and $10°C/min$.

## RESULTS AND DISCUSSION

Fig 2 shows the reaction of sodium carbonate with acetic acid vapour. The amounts of glacial acetic acid and the carbonate were in stoichiometric relation. The reaction started at room temperature, and the sample reached a

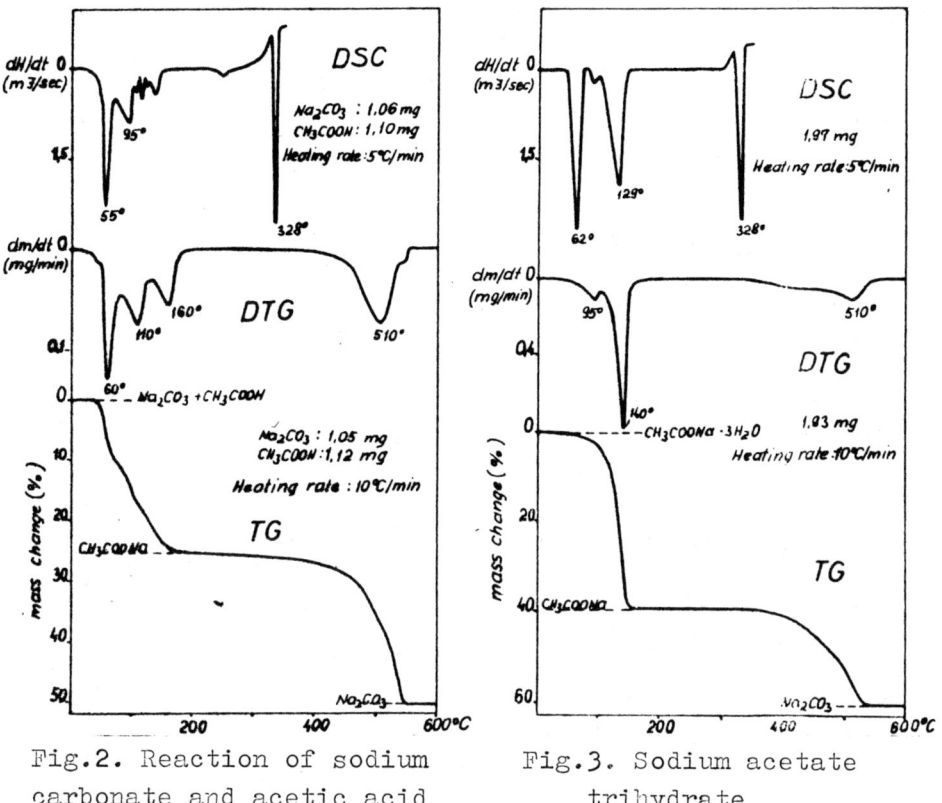

Fig.2. Reaction of sodium carbonate and acetic acid

Fig.3. Sodium acetate trihydrate

constant mass above 150°C. The mass change corresponds to the

$$2 CH_3COOH + Na_2CO_3 = 2 CH_3COONa + H_2O + CO_2$$

equation. The decomposition of the product - starting slowly at 250°C - was first endothermic, but it soon became exothermic. During this, the sample melted at 328°C, the decomposition ended at about 550°C /TG curve/. Thermoanalytical curves of sodium avetate trihydrate were recorded, too /Fig.3/. This material melted at 62°C /in the water of crystallization/, then the water was lost up to 150°C yielding a solid product. Anhydrous sodium acetate started to loose mass at 300°C; at 328°C, melting occurred during decomposition. The transformation of the sample to sodium carbonate was complete by 550°C. So, thermal behaviour of sodium acetate was found the same as reported by other

authors (2),(3). Comparison of the curves in Figs.2 and 3 proves that the
$$Na_2CO_3 + 2\ CH_3COOH\ /vapour/$$
reaction was stoichiometric.

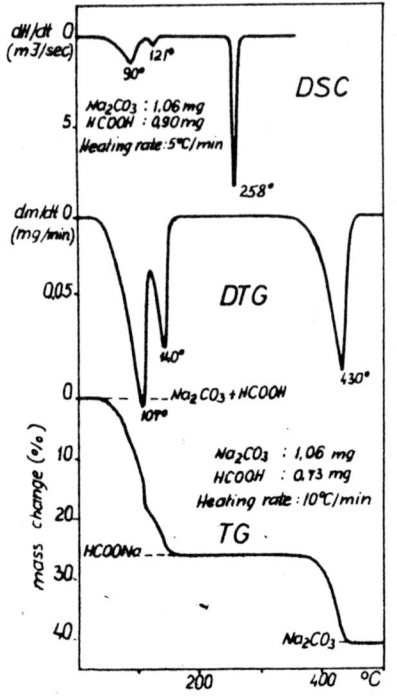

Fig. 4. Reaction of sodium carbonate and formic acid

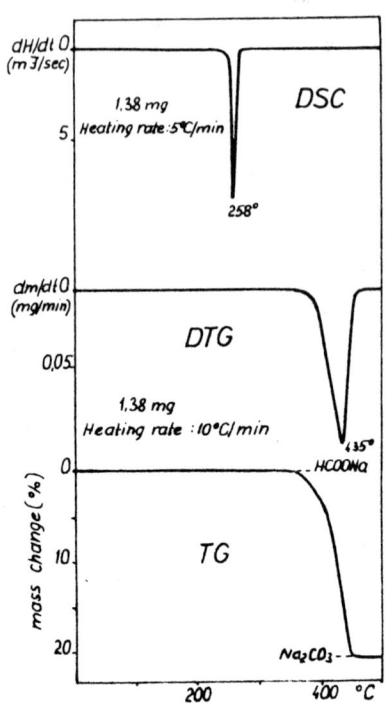

Fig 5. Sodium formate

Sodium carbonate can react with the vapour of formic acid as well. This process is characterized in Fig.4, while Fig.5 the thermoanalytical curves of pure sodium formate. The solid - vapour reaction was complete at about 170°C. By this point, formic acid reacted quantitatively; sodium carbonate - being present in an excessive amount - transformed partly to sodium formate, partly to sodium bicarbonate. The latter decomposed below 170°C, too, as shown by the DTG peak at 140°C. This proves the assumption of sodium bicarbonate as an intermediate. On further heating, the sample melted at 258°C; decomposition to carbonate became intensive near 350°C.

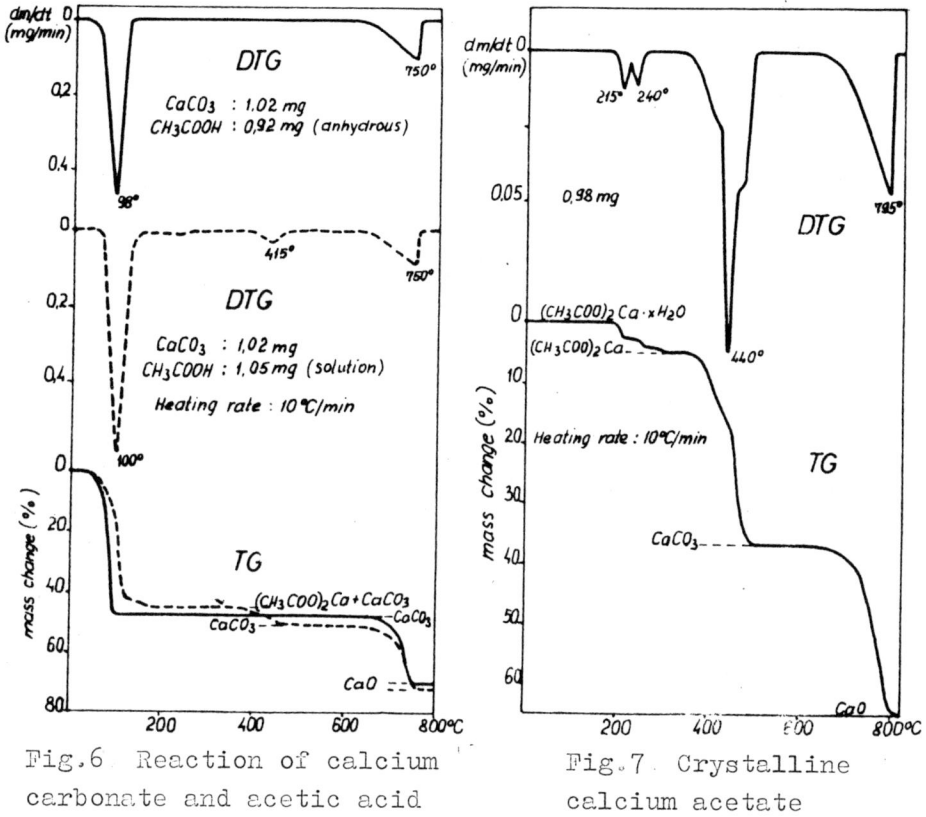

Fig.6 Reaction of calcium carbonate and acetic acid

Fig.7 Crystalline calcium acetate

In Fig.6, showing TG and DTG curves of calcium carbonate and acetic acid, an experiment with glacial acetic acid is represented by the full line; the dotted line belongs to a sample where acetic acid was applied in 85% aqueous solution. The solid-vapour reaction took place and the unreacted acetic acid left the sample holder below 150°C. Anhydrous acetate decomposed in the range of 350 to 500°C, calcium carbonate /consisting of the unchanged fraction of the initial sample and the product of calcium acetate decomposition/ transformed to the oxide above 650°C Comparing these results to the behaviour of calcium acetate - Fig.7 - suggests that water vapour is needed for the reaction of calcium carbonate and acetic acid. Using 85% solution a mere 22% of the acetic acid yielded calcium acetate; complete conversion can be achieved with a dilute solution.

Unlike acetic acid, vapours of formic acid reacted with calcium carbonate in the absence of water too. However, this process was not stoichiometric either: under our experimental conditions the conversion reached nearly 80% - related to the initial amount of the acid.

### REFERENCES

(1) P.D. Garn, Rev. Sci. Instr. <u>44</u> /1973/ 231
(2) C. Duval: Inorganic Thermogravimetric Analysis. Elsevier, Amsterdam 1963, p. 210
(3) G. Liptay ed.: Atlas of Thermoanalytical Curves, vol 1 Heyden and Son - Akadémiai, London - Budapest 1971, no. 8

ISOTHERMAL THERMOGRAVIMETRIC ANALYZER USING INFRARED IMAGE FURNACE AND MICROCOMPUTER SYSTEM

Akikazu Maesono, Masahiko Ichihasi, Koichiro Takaoka and Akira Kishi

SINKU RIKO CO., LTD.
Midori-ku, Yokohama 226, Japan

1. Introduction

A high speed thermobalance using an infrared image furnace has been developed by combining a microcomputer-controlled heating and data processing system. This new instrument includes a differential thermobalance, microcomputer-controlled temperature controller and data processing system, keyboard, floppy disk, CRT display and xy plotter.

The system can be used very effectively in both isochronel and isothermal thermogravimetry.

The conventional isothermal thermogravimetric technique, in which, for example, the sample is introduced in a superheated oven, involves a number of problems because of its instrument design, such as excessively long time to reach the specified temperature, instability of temperature control at the desired temperature, un-uniformity of temperature distribution within and around the sample cell, cell construction, responsivity of the detecting system, troublesome operation and others.

Thus the application of the conventional isothermal thermo-balance has been limited and its attainable temperature is not very high and the reaction rate is relatively low.

With this new system, it is much easier to obtain the isothermal condition, the TG sample can be heated up to 1300°C within one minute with no overshoot of temperature and the temperature can be controlled within $\pm 0.5°C$. Furthermore, a wide range of programmed heating rates from 1°C/hr to 50°C/sec are available.

Photo 1 shows the overall view of the system.

## 2. Differential Thermobalance using Infrared Image Furnace

For the balance is used a torsion ribbon type top loading balance.

The sample cell is made of platinum and the standard size is 5 mm in diameter and 5 mm deep (the internal volume is approximately 70 µℓ). The principle of operation and structure of the infrared image furnace are as reported previously.[1]

For improved stability of the DTA baseline and improved accuracy of TG measurement at high heating rates, the following improvements have been made on the previous model.

(1) An infrared image furnace with a parabolic reflecting surface is employed instead of the infrared image furnace with an elliptical reflecting surface. This has improved the radial temperature distribution in the proximity of the sample and has made the DTA baseline more stable.

(2) A platinum foil cap (e) has been added as shown in Fig. 1. This has decreased the temperature gradient in the inner atmosphere of the cap (d) and has eliminated the problem that the temperature of the atmosphere gas is lower than that of the sample cell. It also prevents the light of high color temperature emitted by the infrared lamp from irradiating the sample directly, thereby the measurement dispersion that may be caused by the sample color being eliminated.

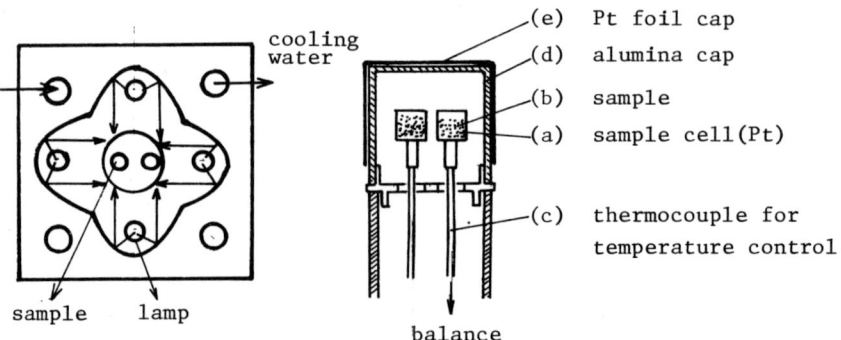

Sectional view of funace and sample assembly

Detail of sample assembly

(e) Pt foil cap
(d) alumina cap
(b) sample
(a) sample cell (Pt)
(c) thermocouple for temperature control

FIG.1 SAMPLE ASSEMBLY

3. Data Processing System (Model DPS-2)

The new system is provided with the data memory and arithmetic operation functions, besides the functions of the programmed temperature controller and recorder provided in the conventional thermobalance. Fig. 2 shows the block diagram of this system.

Features of this system include the following.

(1) The result can be graphically displayed on the CRT with the temperature or time on the abscissa and signals $Y_1$ and $Y_2$ and data $Y_3$ processed from $Y_1$ and $Y_2$ by on-line control (e.g., TG, DTA and DTG in thermogravimetry) on the ordinate. The data can also be printed out on the printer as a hard copy. It is also possible to display and tabulate digital data to the right of the graph (e.g., weight loss percent of 20∿100°C, 20∿200°C, 20∿300°C ...)

(2) It is possible to specify two points, A and B, on the curve by operating the cursor and to calculate the weight change between the two points for display and recording.

(3) Almost infinite number of combinations of temperature-time programs can be easily preset by means of the temperature program key. Improved PID-K control is employed for isothermal measurement and high rate heating. The EMF curve for two types of thermocouple is linearized with an error of 0.1% or less.

(4) Two floppy disks are provided for the memory and can be used in storing the measured data and programs necessary for the operation of the equipment.

4. Isothermal Thermogravimetry

The system is provided with conditions necessary for isothermal thermogravimetry.

o Rapid heating can be easily effected without overshoot up to 1300°C.

o The accuracy of holding temperature at a specified temperature is within ±0.5°C.

o Temperature control being effected with the temperature of the sample cell, the reaction temperature can be known accurately and the temperature can be maintained constant independently of the endo- or exothermal reaction of the sample.

o Measurement can be made with several milligrams of sample, so that thermal conductivity, gas diffusion and other effects in the sample can be held to a minimum.

o The stability of the balance is excellent even at a high rate of heating and the **responsivity** of the balance is also high.

An isothermal example of decarbolxylation reaction of calcium carbonate is given. Approximately four milligrams of calcium oxalate monohydride is put in a cell 5 mm in diameter and 3 mm in depth and is heated up to 700°C in the air at a rate of 200°C/min to make it $CaCO_3$, followed by rapid heating to a specified temperature in steps of 20 to 60°C, when the isothermal reaction starts. Fig. 3 shows one example.

On the other hand, the sample was heated at a constant rate from 1.25 to 320°C/min and the same reaction was subjected to kinetic analysis by the Ozawa's method[2] and compared with the isothermal measurement.

References:

1) M. Ichihashi et al., Proc. of the 5th ICTA, p. 554, Kyoto, 1977
2) T. Ozawa, Bull. Chem, Soc. Jpn., **38**, 1881 (1965)

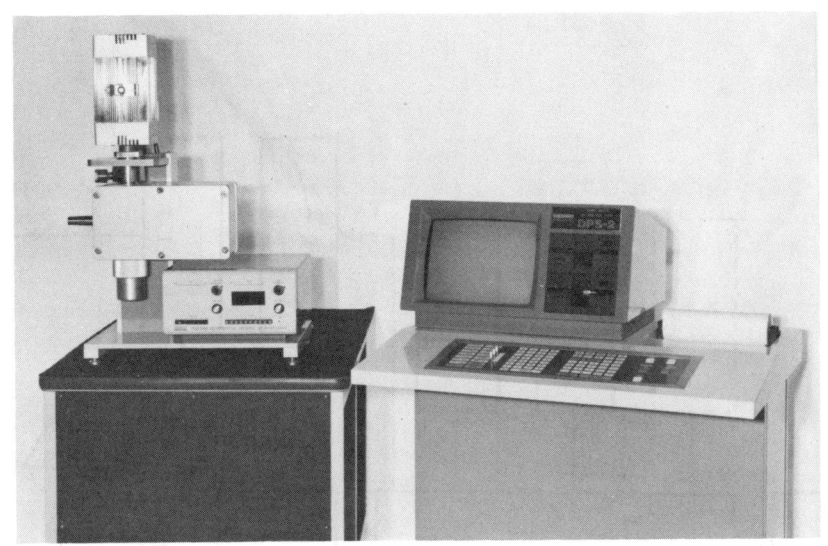

PHOTO 1.    OVERALL VIEW OF THE SYSTEM

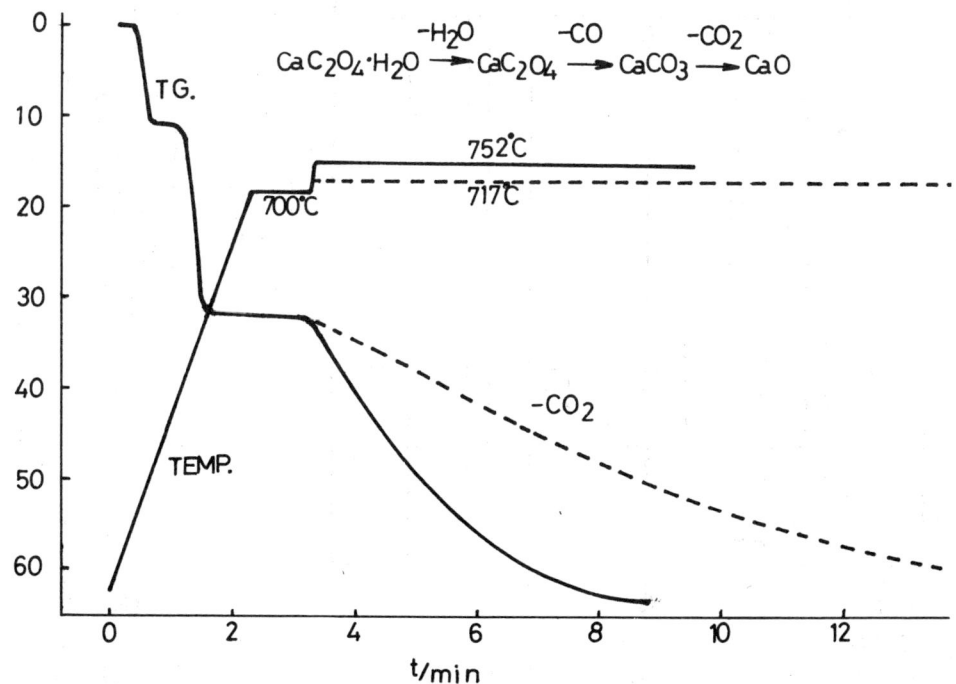

FIG.3   ISOTHERMAL TG ( DECARBOXYLATION OF CALCIUM CARBONATE )

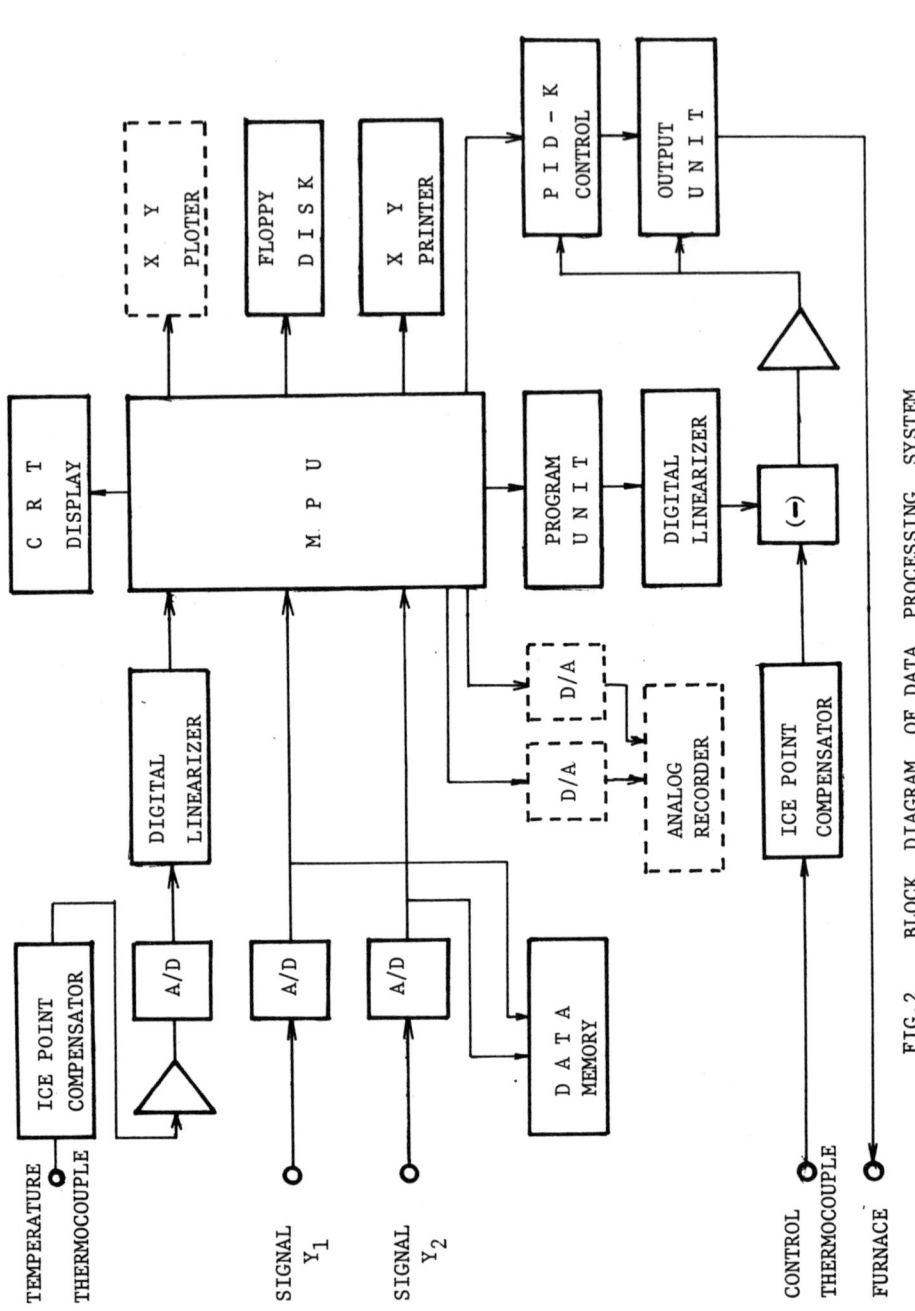

FIG.2  BLOCK DIAGRAM OF DATA PROCESSING SYSTEM

# TEMPERATURE CALIBRATION IN THERMOGRAVIMETRY

Paul D. Garn, The University of Akron
Akron, Ohio, USA
Oscar Menis, U. S. National Bureau of Standards
Washington, D. C., USA
Hans-Georg Wiedemann, Mettler Instruments
AG, Greifensee, Switzerland

## ABSTRACT

Five magnetic materials have been tested and certified as temperature reference materials for thermogravimetry. Use of these materials will enable the relating of data from thermobalances of different designs.

## INTRODUCTION

In most thermobalances the temperature sensor is not in contact with either the sample or the sample holder; the temperature of the sample is inferred from the measured temperature of sensor. Unless the temperature gradient within the furnace is known from experiment to be small, the operator has a considerable uncertainty in his estimate of the sample temperature. The actual error will vary with the distance and direction of the sensor with respect to the sample and with the geometry of the furnace-sample-sensor assembly. Particularly if the sensor is above or below the sample, the temperature difference between the sample and sensor will vary both with temperature and with heating rate. Still further, the heat transport through vacuum is different from that through air or nitrogen or from that through argon or carbon dioxide; the sample-sensor temperature difference will therefore be dependent upon the atmosphere and its pressure.

These considerations led the Committee on Standardization of the International Confederation for Thermal Analysis to seek reference materials for thermogravimetry. The Committee tested a variety of real weight-change processes such as dehydrations and decompositions and concluded that such processes were intrinsically not reproducible without specific conditions of atmosphere and pressure, hence unsuitable for general calibration of thermobalances. It then directed its attention to magnetic reference materials[1]. A test program was carried out to determine the utility of a number of materials and the practicality of the method with a variety of thermobalances. The results led to instituting a larger program -- The Sixth International Test Program -- and to the certification of five materials by the ICTA as Certified Reference Materials for Thermogravimetry.

The Certified Reference Material for Thermogravimetry set is available from the U.S. National Bureau of Standards, Washington, D.C. 20234, USA. It is listed as GM-761. The Certificate is available separately only by sending a check or money order payable to the ICTA for $5 US to the chairman of the Committee on Standardization. The present chairman is Prof. Paul D. Garn, The University of Akron, Akron, OH 44221.

THERMAL ANALYSIS . ICTA 80 . BIRKHAEUSER VERLAG, BASEL, BOSTON, STUTTGART

## THE SIXTH INTERNATIONAL TEST PROGRAM

Sets of five discs of each of the candidate materials were distributed to about thirty workers, along with a protocol, a report form and a questionnaire on the apparatus. Eighteen reports were returned, one having data from two instruments. Eleven different models of thermobalance were used; these were from eight manufacturers of complete assemblies and one investigator who had incorporated a commercial electrobalance into a thermobalance.

The protocol emphasized the importance of using ordinary operating conditions for the measurements, listed precautions and defined the points to be measured. Figure 1 shows the model curve with the defined points. The detailed description of the test program is given in the Certificate that accompanies the Certified Reference Materials.

The materials for these Certified Reference Materials were purchased from the Vacuumschmelze GMBH, Hanau, Federal Republic of Germany. They are, in ascending order of their magnetic transitions; Permanorn 3, Nickel, Mumetal, Permanorm 5, and Traforperm.

Typically, the magnetic transition temperature is highly susceptible to variations in composition such as might take place from batch to batch; nickel is well known to be highly susceptible to impurities. The Committee emphasizes that these *materials* are not being certified; only *these batches* of materials are certified.

a. Systematic bias

Examination of the unweighted raw data and comparison with the means disclosed immediately that systematic bias was the major source of deviation. This was expected because of the diverse methods chosen by instrument manufacturers to provide a temperature measuring point. No extensive statistical evaluation appeared appropriate. Instead, the data from each observer were examined in terms of their relation to the means.

b. Random error

The data on a given material from any one participant differed typically by 0-5° for any of the three points. Because there were not "standard" ways of arranging the magnet, comparison of identical instruments is less meaningful than in the previous test programs on DTA-DSC reference materials. It can be concluded, however, that data reproducible within a few degrees can be obtained on any one instrument.

c. Heating rate dependence

The data of individual participants were examined to learn whether or not a variation due to heating rate existed. In most cases the differences were small, 0-3°, much less than the systematic deviation discussed above. The differences were not even completely consistent in sign.

d. Unweighted means

With the exclusions noted above, the unweighted means and standard deviations were calculated from the participants' means. These are given in Table I. In only five of the 213 means did a participant's stndard deviation for a given data point equal or exceed the overall standard deviation. Each of these five data sets was from an instrument in which a wide range of adjustment of the thermocouple position is possible.

e. Significance of the means

The mean values of these data are useful as reference points from which to measure the deviations found in an individual apparatus. The reference points can thereby be used to relate measurements from laboratory to laboratory -- even though different instruments are used -- because common materials, tested for homogeneity, were used.

The mean values of these data cannot be taken as an accurate measure of the magnetic transition temperature. The defined points on the TG curve are necessarily arbitrary but are readily defined geometrically; they have no firm relationship in principle to the absolute value of the temperature at which the material loses its paramagnetism, even when that event occurs at a well-defined temperature. This does not detract in any way from their utility in dynamic measurements.

f. Breadth of deflection

A feature worth noting is the difference between the measured $T_1$ and $T_3$ which can be defined as the breadth of the deflection. Not only are there large differences in breadths but also these have some consistencies with respect to both material and apparatus.

Nickel has an extremely sharp transition which the small breadth reflects, whereas Permanorm 3 had the greatest breadth, nearly five times that for nickel.

The $\Sigma(T_3-T_1)$ for each participant discloses that some had characteristically large or small breadths. Five participants had small values for one or more materials, these data were from four different instruments.

g. Sample loading position

Three general types of balances are readily identifiable -- the top-loaded, the bottom-loaded, and the beam-loaded, in which the term identifies the position of the load (including sample) with respect to the beam. Even though there is no obvious direct effect arising from the load position, the question had been raised so a test of the data was indicated.

The spans, the differences between the high and low investigator means for each group, disclose some systematic errors. The data on beam-loading have smaller spans than the others partly because only one (commercial) balance is represented. The top-loading balances were five in number, two manufacturers each represented by two models. The bottom loading group represented six models, counting one particular model of balance separately for each different control and measuring system with which it is supplied. The separate counting is appropriate because manufacturers can position sensors differently in different models.

The deviations within a balance type can be attributed with confidence to differences in operator adjustment. When the data from top-loaded balances are arranged in numerical order the sequence of participants is precisely repeated for each of the five materials. For the beam-loaded data, the same participant was consistently high, neither of the other two being consistently lowest.

This suggests a systematic difference either in calibration, which can occur with either balance, or in placement of the measuring point, which can occur in one but not in the other. The evidence that this can occur emphasizes the importance of calibrating under programmed temperature as compared to an independent calibration of the thermocouple.

## CONCLUSIONS

The reproducibility demonstrated by the several participants indicates that the materials are suitable temperature reference standards. The variability between participants is largely due to instrument design, particularly with regard to the geometric relation between the sample and the temperature measuring point. In some instruments, variation of this relationship is possible from investigator to investigator or even from day to day in the same laboratory. These variations, avoidable or not, make the use of temperature reference standards necessary for correlation of data.

The recommendations below are provided to aid in the use of these Certified Reference Materials:

1. Position of Magnet

The optimum position of the magnet is directly above or below the sample holder so that the magnet flux is aligned with the gravitational field. Another possible arrangement is the use of a small magnet well out of the heated zone with the flux concentrated by a permeable rod leading closer to the sample.

2. Strength of Magnetic Field

No a priori values can be established. The magnetic flux for a given magnet decreases with the second power of the distance. The magnet need not be large because it needs to produce only an identifiable deflection, not a half- or quarter-scale deflection.

A variable field would be usedul to enable calibration during ordinary use of the thermobalance. This can be done by (a) using an electromagnet; (b) varying the position (proximity) of the magnet; or (c) if permeable rods are used, changing the length of the rod.

3. Multiple Calibrations

There is no reason why more than one reference material cannot be used in a single run. Difficulty in recording may arise from using an excessive portion of the range for calibration but re-zeroing can be used to enable full use of the balance range for the real weight loss.

4. Calibration for each set of conditions

The calibration should be performed for each atmosphere and pressure combination used. The differences in thermal conductivity of the possible atmospheres can lead to different temperature relationships between the sample and the sensor.

## THE CERTIFIED REFERENCE MATERIALS

Each of these Certified Reference Materials is supplied in the form of a thin metal strip or strips that can be cut to whatever size provides a suitable response with the magnet and sensitivity range used. Most participants have been able to place a magnet close enough that 10-20 mg provided an easily detectable signal with commercial apparatus.

[1] S.D. Norem, M. J. O'Neil, A. P. Gray, Thermochim. Acta 1, 29-38(1970).

## ACKNOWLEDGEMENTS

The Committee on Standardization is grateful to the several participants in the Sixth International Test Program and their organizations that enable their participations. The participants were V. Amicarelli (Italy), G.D'Ascenzo (Italy), P. A. Barnes (UK), M. Escoubes (France), C. R. Foltz (USA), P. K. Gallagher (USA), B. Haglund (Sweden), P. J. Haines (UK), M. Harmelin (France), K. Heide (DDR), J. M. Jervis (Canada), H. Kambe (Japan, J. P. Mathiew (Switzerland), H. G. McAdie (Canada), O. Menis (USA), Oshigama (Japan), H. R. Oswald (Switzerland), T. Ozawa (Japan), A. Quivy (France), D. Stewart (USA), E. Sturzenegger (Switzerland), Y. Takahashi (Japan) and H.-G Wiedemann (Switzerland).

Present members of the committee are P. D. Garn, Chairman; H.-G. Wiedemann, Vice Chairman; K. Heide; H. Kambe; G. Lombardi; R. C. Mackenzie; H. G. McAdie; H. R. Oswald; T. Ozawa; F. Paulik; J. P. Redfern; and O. T. Sørensen. Our co-author, Oscar Menis, died December 12, 1979.

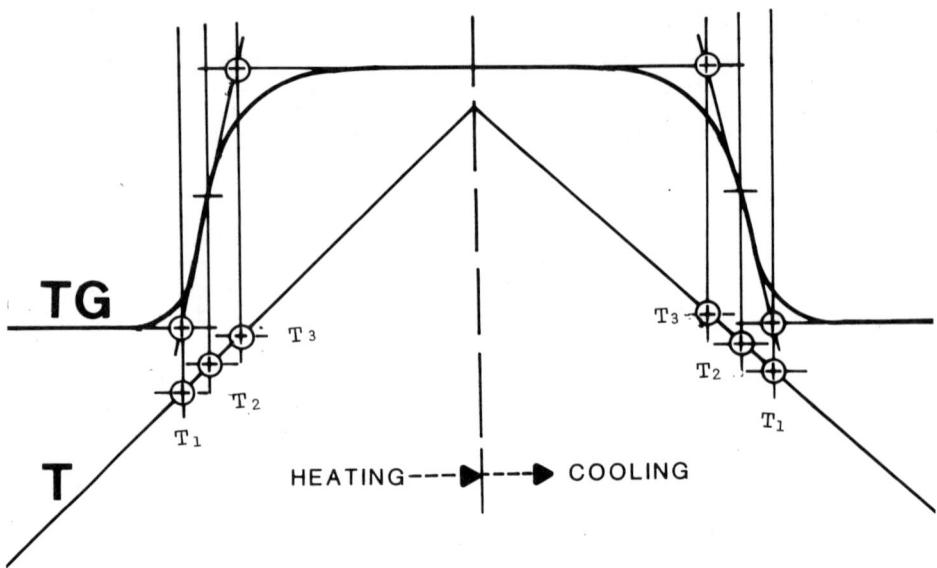

Figure 1. The defined points on the thermogravimetric curve.

TABLE I

The unweighted mean values of the defined points, $T_1$, $T_2$, and $T_3$, on the thermogravimetric curve for The Certified Reference Materials.

| Material | $T_1$ | $T_2$ | $T_3$ |
|---|---|---|---|
| Permanorm 3 | 253.1°C | 258.8°C | 266.3°C |
| Nickel | 351.4 | 352.8 | 354.4 |
| Mumetal | 377.0 | 381.4 | 386.2 |
| Permanorm 5 | 450.8 | 454.5 | 458.8 |
| Traforperm | 749.2 | 750.5 | 753.8 |

CONVECTION EFFECTS IN THERMOGRAVIMETRY:
SIGNIFICANCE AND CORRECTIONS

Bengt O. Haglund and Torsten Luks
Coromant Research Center
Sandvik AB, P.O. Box 42 056
S-126 12  Stockholm 42, Sweden

## ABSTRACT

During thermogravimetric measurements disturbances due to convection effects on the crucible are often encountered. Thermogravimetric blank tests were made using different furnaces, crucibles, atmospheres, gas flow rates and temperatures in a Mettler thermoanalyzer TA 1. The observed force on the crucible could be related to the friction forces calculated from previously known formulae for laminar gas flow. The contribution from this force is most pronounced at high gas flow rates and at high temperatures.

## INTRODUCTION

In thermogravimetric measurements disturbances of the weight recording from the surrounding atmosphere generally are encountered. The effects can be due to i.a. sample container air buoyancy, convection currents in the furnace atmosphere and the Knudsen effect (1). Often such effects are neglected, but for quantitative work corrections must be applied. A simple way to solve the correction problem is to run a blank test before or after each experiment under identical conditions, but this procedure is time-consuming and does not always give reproducible results. It would be most convenient and time-saving to correct the results directly, e.g. by means of a computer, once the theoretical background to the corrections is understood.

## CONVECTION CURRENTS IN THE FURNACE ATMOSPHERE

The influence of convection currents in the atmosphere on the weight recording has been discussed to some extent in the literature (1, 2, 3, 4). The studies of this effect have been mainly phenomenologic and no general conclusions can be drawn from the published results.

## BASIC RELATIONS

From the theory of a fluid in motion it is known (5) that the total energy of the fluid is composed of internal, potential, pressure and kinetic energies according to the formula

$$E = U + zg + \frac{P}{\rho} + \frac{w^2}{2}$$

From this equation the Bernoulli equation can be derived:

$$z_1 \cdot \rho_1 \cdot g + P_1 + \frac{\rho_1 w_1^2}{2} = z_2 \cdot \rho_2 \cdot g + P_2 + \frac{\rho_2 w_2^2}{2}$$

The first, second and third terms are known as the potential, pressure and velocity terms, respectively. In the velocity term, w relates to the linear velocity along a streamline. If the equation applies to steady flow in a tube the mean linear velocity v in the tube can be inserted instead of w, if a dimensionless correction factor $\alpha$ is inserted:

$$z_1 \cdot \rho_1 \cdot g + P_1 + \frac{\rho_1 \cdot v_1^2}{2\alpha} = z_2 \cdot \rho_2 \cdot g + P_2 + \frac{\rho_2 \cdot v_2^2}{2\alpha}$$

The correction factor $\alpha$ takes care of the velocity distribution across the tube. It can be shown that $\alpha$ is 1/2 for laminar flow and approximately 1 for turbulent flow in a tube of circular cross-section.

## THE FORCE ON A SOLID OBSTACLE

In the equations the velocity term is also known as the dynamic pressure. If the fluid flow hits a solid obstacle a small amount of fluid is brought to a standstill. The dynamic pressure then results in a force acting on the obstacle according to the formula

$$F_f^o = c_f \cdot A \cdot \frac{\rho \cdot v^2}{2\alpha}$$

The coefficient $c_f$ is known as the friction factor, which is dimensionless. It is, however, not constant, but is related to the fluid flow pattern and the shape of the solid obstacle. For laminar flow a typical relation for the coefficient $c_f$ is written in the form

$$c_f = k/Re^n$$

where Reynolds number Re is given by $Re = v \cdot d/\nu$

Different values of the constant k and exponent n are to be found in the literature. For spheres, hemispheres, ellipsoids and circular discs the friction coefficient $c_f$ at low Reynolds numbers (Re < 40) converges towards a common relation in the form:

$$c_f = 30 \, Re^{-0.81}$$

If the diameter d of the obstacle is inserted in the expression for Reynolds number the force on the solid obstacle is given by

$$F_f^o = 7.5 \cdot \pi \cdot \nu^{0.81} \cdot d^{1.19} \cdot v_\infty^{1.19} \cdot \varrho$$

The linear velocity $v_\infty$ has to be measured in the undisturbed flow far away from the obstacle.

## THE FORCE ON A CYLINDER

The friction of the gas against the cylinder surface, e.g. of a crucible, gives rise to a pressure drop between the ends of the cylinder, which is counterbalanced by the shear stress $R_w$ over the surface of the cylinder.

The total friction force is given by (5):

$$F_f^c = R_w \cdot \pi \cdot d \cdot L$$

The basic friction factor $j_f$ is defined by:

$$j_f = \frac{R_w}{\varrho \cdot v^2}$$

For laminar flow $j_f = \frac{8}{Re}$

and thus the friction force can be written

$$F_f^c = \frac{8}{Re} \cdot \pi \cdot d \cdot L \cdot \varrho \cdot v^2 = 8\pi \varrho \cdot L \cdot v \cdot \nu$$

In this case for v the mean linear velocity $v_c$ at the level of the crucible shall be used.

## THE TOTAL FORCE ON A CRUCIBLE

The total force can be calculated as the sum of forces acting on the bottom and the top of the crucible and the frictional forces acting on the cylinder surface:

$$F_f^{tot} = (7.5 \cdot \pi \cdot \nu^{0.81} \cdot d^{1.19} \cdot v_\infty^{1.19} + 8\pi\nu \cdot L \cdot v_c) \cdot \varrho$$

## EXPERIMENTAL PROCEDURE

Three different kinds of atmospheres (hydrogen, nitrogen and argon) were introduced in a Mettler Thermoanalyzer TA-1 at three or four different, but constant, flow rates. Two different types of crucibles and one flat, circular sample container were used alternatively, see Fig. 1. The crucibles and sample container pan were empty. Weight recordings were made at room temperature, 100, 400, 600 and 1200°C as isothermal tests under sufficient long time to obtain steady-state conditions and to allow corrections for the long-term drift. From the curves the apparent weight decrease was evaluated and corrected to force (in newtons) by multiplying by 9.81 m/s$^2$.

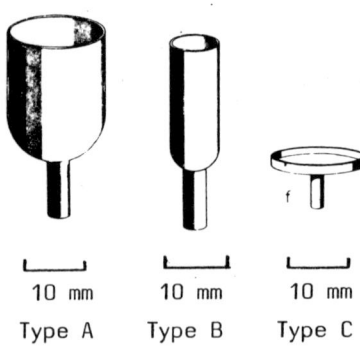

Fig. 1 Some of the various crucibles used. The pan type C was made of platinum, the others of alumina.

## RESULTS AND DISCUSSIONS

Data from the experiments regarding the measured and calculated friction forces are reported in Figure 2.

It is found that at low gas flow rates and at temperatures below approximately 400°C the necessary corrections are less than 0.1 mg, which often may be within the normal experimental error. However, for quantitative work corrections are to be made, especially at higher temperature and a gas flow exceeding 10-20 l/h NTP. In this range there is a reasonably good agreement between the calculated and measured values, which is indicated by the fact that the slopes of these lines are between 0.75 and

2.5. The lowest values are found for the large crucible in nitrogen and the highest values are found for the pan type of sample container. It seems that the friction factor has to be increased by a factor 2 for the pan type sample container. For the other crucible types the calculated corrections can be used with good results.

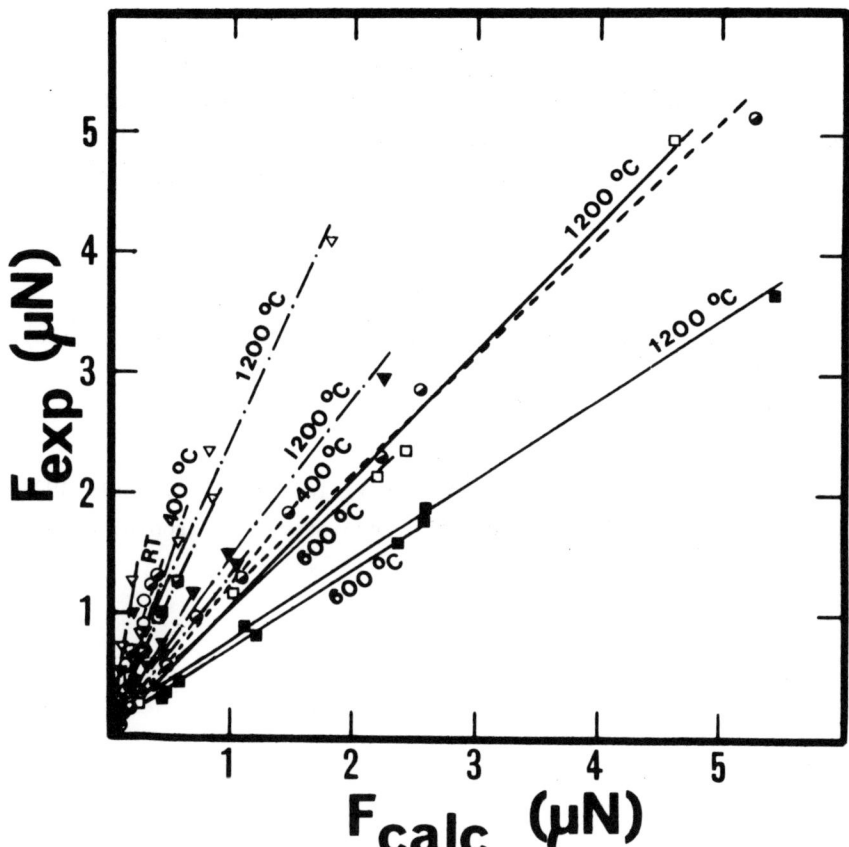

Fig. 2  Relation between measured and calculated friction forces

Legend:  Crucible        Hydrogen    Nitrogen    Argon
         Type A            □————————■————————◪
          "   B            ○--------●--------◓
          "   C            ▽—·—·—·—▼—·—·—·—▿

It has, however, to be pointed out that friction factors found in the literature are based on experiments under conditions which may differ considerably from those used in this investigation. Thus, for a certain experimental geometry it may be useful to perform calibration experiments. For routine work this seems unnecessary but the formulae can be applied directly.

## ACKNOWLEDGEMENT

Thanks are due to Dr Ulf Smith and Dr.-Ing. Rudolf Hiltscher for valuable discussions and to Dr Lars Aschan for correcting the English text. The authors are indebted to Sandvik AB for the permission to publish this work.

## REFERENCES

(1) W.W. Wendlandt, Thermal methods of analysis. Second edition. Wiley, New York 1974
(2) A.E. Newkirk, Anal. Chem 32 (1960) 1558
(3) G.M. Lukaszewski, Nature 194 (1962) 959
(4) P. Vallet, Thermogravimétrie. Gauthier-Villars Editeur, Paris 1972
(5) F.A. Holland, Fluid flow for chemical engineers. E. Arnold, London 1973

## ERRORS IN THERMOGRAVIMETRIC EXPERIMENTS RESULTING FROM ADSORPTION ON THE COUNTERWEIGHT

E. Robens
Battelle-Institut e.V.
Am Römerhof 35, D-6000 Frankfurt am Main, GFR

### ABSTRACT

The sorption behaviour of quartz, glass, aluminum, gold and PTFE was examined with respect to their use as counterweight materials. Using nitrogen, krypton, water, alcohol and neopentane, reversible S-shaped isotherms were observed with a monolayer cover at a relative pressure of about 0.1, and up to eight layers at the saturation pressure. When organic vapours are used chemisorption may occur. Near the saturation pressure condensation effects on porous surfaces and contaminations have to be taken into account.

### INTRODUCTION

When performing thermogravimetric experiments or gravimetric determination of gas density and pressure, adsorption on the moving parts of the balance has to be taken into account, in addition to buoyancy, thermal flow, insufficient heat transfer to the sample, etc., which are reviewed elsewhere (1). Whereas most of the adsorption effects can be cancelled out using a symmetrical balance design, there is still adsorption at the counterweight. Typical materials used for counterweights include quartz, glass, gold, and aluminum. By combination of these materials counterweights with densities between 0.002 and 0.019 kg m$^{-3}$ can be achieved; with hollow glass spheres even lower values are possible. Polytetrafluorethylene (PTFE) is also of interest as a material for counterweights because it is hardly wettable by any condensate.

In order to assess the errors resulting from adsorption we have reviewed the available literature and supplemented these results by experiments (2). For the experiments we used our Gravimat*) prototype (3) comprising two Gast balances, one for sorption measurements, the other one as as buoyancy gauge, pressure controller and thermostats and a vacuum aggregate with a turbo molecular pump (Fig. 1).

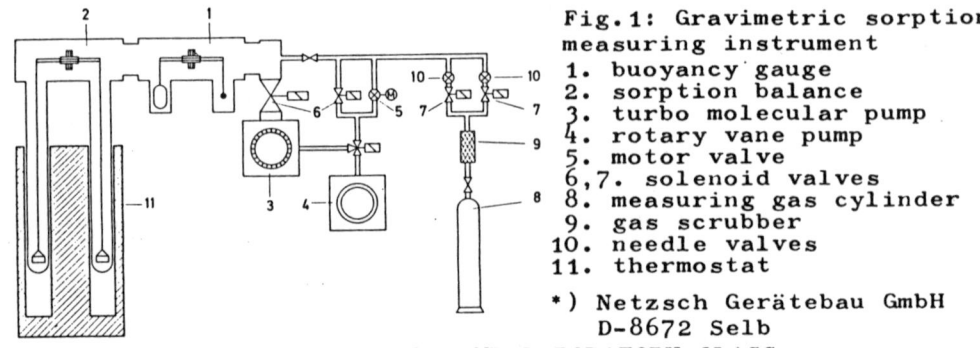

Fig.1: Gravimetric sorption measuring instrument
1. buoyancy gauge
2. sorption balance
3. turbo molecular pump
4. rotary vane pump
5. motor valve
6,7. solenoid valves
8. measuring gas cylinder
9. gas scrubber
10. needle valves
11. thermostat

*) Netzsch Gerätebau GmbH D-8672 Selb

## ADSORPTION ON QUARTZ AND LABORATORY GLASS

A newly generated quartz or glass surface e.g. created by breaking exhibits active free radicals. In air the surface anneals very quickly forming a gel-like surface which is covered in about half an hour by a water film in equilibrium.

Degassing experiments (Fig. 2) show that in practice at 130°C about 35 % of the water film evaporates independent of the kind of glass and the chemisorbed remnant steadily evaporates up to temperatures of 500°C (4). Fig. 3 shows reversible water isotherms on quartz glass (5), and nitrogen, water, neopentane and methanol isotherms on glass powder or glass wool (6). All isotherms are S-shaped with a monolayer cover at a relative pressure of about 0.1 and four layers at about 0.8. With methanol a certain proportion remains chemisorbed and can be removed only by heating.

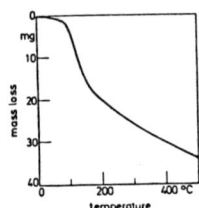

Fig.2: Water desorption from glass according to Korányi (4)

Fig.3: Adsorption isotherms on glass

## ADSORPTION ON GOLD

Although in the bulk of gold no oxide is stable, oxygen can be chemisorbed at the surface. Stable surfaces with stoichiometric compositions include Au, AuO and

$AuO_2$. Small amounts of impurities and lattice defects may explain the contradictory literature reports (2). Above 200°C gold functions as a catalyst in oxidation/reduction processes. Fig. 4 shows oxidation/reduction experiments with high purity gold spheres prepared by melting in hydrogen. After baking in a vacuum and surface area determination with krypton at 90 K, several oxidation/reduction experiments were performed which resulted in a final loss of about seven layers. Oxygen chemisorption was observed even at room temperature. In spite of the relatively clean conditions during handling, the surface was obviously contaminated. Most astonishing, however, is that the specific surface area did not change as a result of this treatment.

Fig. 5 shows nitrogen, water, methanol and neopentane adsorption on gold foils; all of these reactions being strongly reversible, as is also the case with krypton and xenon. Irreversible adsorption as reported by other authors (see (2)) may be due to impurities.

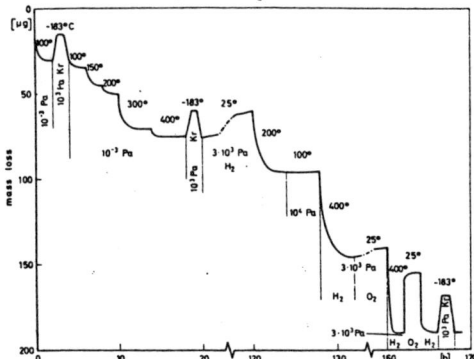

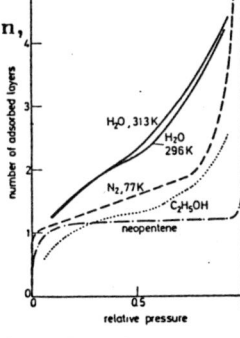

Fig.4: Oxidation, reduction and surface area determination using krypton on 1.5 g gold spheres of 10 to 60 µm in diameter

Fig.5: Adsorption isotherms on gold

Adsorption of acetic acid on a pure gold surface is reversible whereas in the case of a oxygen-covered surface one chemisorbed layer remains at the surface (Fig.6). For subsequent adsorption of acetic acid this chemisorption layer acts as a stable surface on which reversible isotherms have been observed. The specific surface area is practically the same

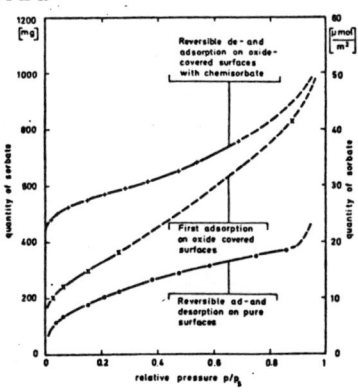

as that of the pure gold surface.
Gold can amalgamate with mercury vapour.

## ADSORPTION ON ALUMINUM

Aluminum is always covered with an oxide layer which sometimes exhibits a highly porous structure and consequently a large surface area.

Fig. 7 shows nitrogen isotherms on aluminum foils, on aluminum powder and on pure aluminum oxide. In contrast to the aluminum powder, the isotherm on alumina exhibits a hysteresis loop. Furthermore, two krypton isotherms on a foil and on powder are shown. Because krypton at 90 K condenses in solid form there is no sharp increase near the saturation pressure. With all isotherms, at a relative pressure of 0.1 about one layer is adsorbed and at the saturation point not more than eight layers.

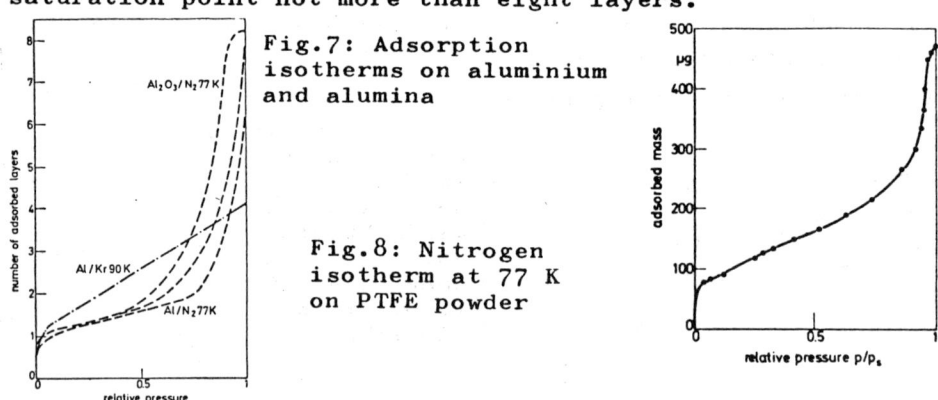

Fig.7: Adsorption isotherms on aluminium and alumina

Fig.8: Nitrogen isotherm at 77 K on PTFE powder

## ADSORPTION ON POLYTETRAFLUORETHYLENE

With PTFE powder (Hostaflon TF 14, Hoechst AG) the nitrogen isotherm shown in Fig. 8 was measured. Similar results are reported in the literature (7, 8) with nitrogen and with argon. We also obtained an S-shaped physisorption isotherm with propane, whereas a few percent of neopentane were absorbed in the bulk.

## CONDENSATION IN CAPILLARIES AND ON CONTAMINATIONS

Oil and grease vapours from the pump and from sealing materials as well as grease and salts from fingerprints form solutions with organic vapours and water vapour respectively. Fig. 9 shows the water isotherm on a fingerprint on an otherwise clean stainless steel foil.

Because under the microscope smooth surfaces reveal a rough and often porous structure, condensation takes place below the saturation pressure, which tends to result in a sharp increase in the isotherm near saturation pressure. If the saturation pressure has been exceeded for a short time only, an additional amount will condense. When reducing the pressure these layers remain on the surface.

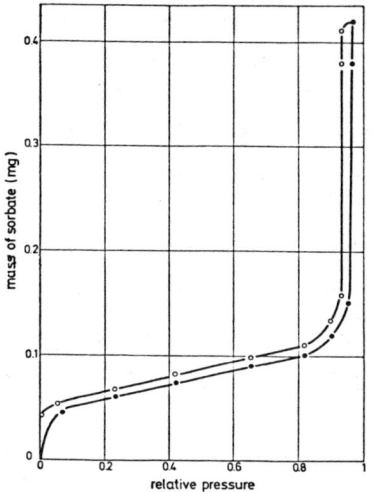

Fig.9: Sorption isotherm on a fingerprint on a stainless steel foil at 23°C

## CONCLUSIONS

With nitrogen, krypton, xenon, water, alcohol and neopentane on quartz, glass, gold and aluminum, S-shaped and mostly reversible physisorption isotherms of type II of the BDDT classification were observed. At a relative pressure of about 0.1 a monolayer and at saturation pressure four to eight layers are adsorbed. Thus, in most cases the error can be neglected. If a correction of the results appears necessary, the error is the geometric area of the counterweight multiplied by the roughness factor of the material (about 1.05 for quartz and glass, at least 1.3 for gold and aluminum if the latter surface is proved to be non-porous).

Since PTFE shows the same sorption behaviour but dissolves some vapours, there seems to be no advantage in the use of this material.

With gold, the effects of oxidation, reduction, chemisorption or amalgamation have to be borne in mind. Chemisorption with some vapours can also occur on quartz and glass. Because of the simple composition of quartz, this material should be preferred to glass. Due to its possibly porous surface aluminum should not be used as counterweight material.

If the saturation pressure is exceeded additional condensation will occur. At the counterweight this can impeded by keeping it at a somewhat higher temperature than the sample. When cooling with liquid nitrogen, this may be achieved by the addition of a few per mille of oxygen, without significantly influencing the buoyancy.

By careful cleaning with acetone and hot water it is possible to remove grease, oil and salts from fingerprints, etc.

## REFERENCES

A more complete reference list is included in paper (2).

(1) A.W. Czanderna, (ed.): Microweighing in Vacuum and Controlled Environments; Part I: Beam Balances; Amsterdam: Elsevier 1980

(2) E. Robens; Wägen und Dosieren, (1980) in print

(3) E. Robens, G. Walter, Thermochimica Acta 9 (1974) 23-28

(4) G. Korányi: Surface Properties of Silicate Glasses; Budapest: Akadémici Kiado 1963; p. 29 ff.

(5) N. Hackerman, A.C. Hall; J.Phys.Chem. 62 (1958) 1212-1216

(6) R.I. Razouk, A.S. Salem; J.Phys.Chem. 52 (1948) 1208-1227

(7) R.B. Perry, K.H. Svatek; J. Colloid Interface Sci. 68 (1979) No. 2, 393-395

(8) J.W. Whalen, W.H. Wade, J.J. Porter; J. Colloid Interface Sci. 24 (1967) No. 3, 379-383

MICROCOMPUTER-CONTROLLED DILATOMETER WITH INFRARED RAPID HEATING SYSTEM
AND ITS APPLICATION

Akikazu Maesono[*], Masahiko Ichihashi[*], Akira Kishi[*], Koichiro Takaoka[*],
Zenshiro Hara[**] and Kiyoaki Akechi[**]

[*]  SINKU RIKO CO., LTD.
     Midori-ku, Yokohama 226, Japan
[**] Institute of Industrial Science, Univ. of Tokyo
     Minato-ku, Tokyo 106, Japan

1. Introduction

A high speed dilatometer using an infrared image heater[1] has been developed by combining a microcomputer-controlled heating and data processing system. This new dilatometer system consists of a push rod dilatometer with an infrared furnace, microcomputer-controlled temperature controller and data processing system, keyboard, floppy disk, CRT display and xy plotter. This system has many features:

o Highly accurate and sharp heating and cooling control up to 1000°C/min.
o Thermal expansion and its derivative curves are instantaneously displayed on CRT.
o Mean coefficient of thermal expansion and characteristic temperatures such as transition temperatures can be displayed on CRT by using two cursor points.

With this dilatometer, the influence of transformation on sintering of titanium powder has been identified[2]. Compacts of high purity dehydrated titanium powder are sintered in argon at various heating conditions.

In a cyclic step heating between 843°C and 924°C, the shrinkage rate of each β-phase temperature cycle immediately after passing the α-β transformation temperature (882°C) is several times greater than that in the last sintering stage of the prior β-phase cycle.

Photo 1 shows the overall view of this system.

2. Dilatometer with Infrared Rapid Heating System

The system is a combination of a push rod type thermal dilatometer and an infrared image furnace and permits measurement up to 1200°C when a quartz push rod and supporter are used. It permits measurement of expansion by stepwise heating as shown in Fig. 1, besides by constant rate heating of 10°C/min or less as in normal dilatometry. The measurement error in stepwise heating is 2% or less as compared with that in the constant rate heating. The dilatometer can also be used effectively in the study of the initial stage of sintering. As seen in Fig. 2 (b) and (c), the sintering start time $\tau = 0$ and the shrinkage at each time $\Delta L$ can be measured more accurately than by the conventional method using the resistance furnace (a).

3. Data Processing System (Model DPS-2)

The system is provided with data memory and arithmetic operational functions, besides the functions of the program temperature controller and recorder provided in the conventional thermal dilatometer. The general description of the data processing system is given in the separate report "Isothermal Thermogravimetric Analyzer using Infrared Image Furnace and Microcomputer System".

Combined with the dilatometer, the system allows the total expansion curve and expansivity curve (dL/dt) to be displayed on the CRT. It also allows the mean coefficient of thermal expansion, e.g. for 20∿100°C, 20∿200°C, 20∿300°C ... and the percent of expansion between temperatures specified by two cursors for display and tabulation on the CRT and then printed out as a hard copy.

4. Application of This Thermal Dilatometer -- Influence of transformation on sintering of titanium powder --

On the influence of the phase transformation on sintering of metal powders, several studies using Ti, Fe and other metal powders have been reported. However, their conclusions do not agree with each other. The problem whether the sintering process is promoted by transformation or not has not yet been solved. In this study, Ti powder compacts were sintered and the influence of the Ti α-β transformation on the sintering rate was studied.

(1) Method of experiment

A quartz cell 5 mm in diameter and 7 mm in depth was packed with titanium powder of -350 mesh prepared by the hydride-dehydride method to make a sample. The sample temperature was detected by a PR13 thermocouple attached to the compact. Measurement was conducted in the following modes to identify the effects of the α-β transformation on the sintering rate.

Mode A : Isothermal sintering at 920°C in the β-phase (bcc)
Mode B : Cyclic sintering at 802°C and 869°C in the α-phase (hcp)
Mode C : Cyclic sintering at 924°C and 985°C in the β-phase
Mode D : Cyclic sintering in the α-phase (843°C) and β-phase (924°C)

We made a solid Ti pellet having the same height as used for the green compacts, recorded its dilatometric curves in the same heating modes as for the Ti powder compacts and, comparing both curves, determined the <u>net shrinkage</u> due to sintering ($-\Delta L$). In practice, through constant temperature periods, the solid pellet did not change in length. Thus, we could determine the "sintering rate" as $-\delta(\Delta L)/\delta t/L_0$ ($\delta t$: time interval between two measuring moments, $\delta(\Delta L)$: change of $\Delta L$ through the time interval $\delta t$). Taking $\delta t$ as small as possible, we could precisely determine the sintering rate at any moment.

Fig. 3 shows the measurement results of mode A and mode D and Fig. 4 shows those of mode C and mode D.

In the cyclic sintering in single phase, the sintering rate monotoneously decreases with time. On the other hand, in the α⇄β cyclic sintering, the sintering rate immediately after passing the α→β transformation point in any β-phase stage (A, B, C, ---) is several times greater than that at the final moment of the prior β-phase stage (A', B', C', ---). This proves clearly the sintering promotion effect of the α→β transformation. This effect rapidly decreased with time after transformation point and the time interval at each cycle is $\lesssim$ 2 min. On the contrary, the β→α transformation seems to have no effect on the sintering rate.

The titanium powder compact was further sintered with no external load in the following two modes, each for five hours, to measure the volume shrinkage.

Mode E : 920°C x 2 min → cooling 1000°C/min → 600°C x 0.5 sec
→ heating 1000°C/min → 920°C x 2 min → cyclic

Mode F : Isothermal sintering at 920°C

The volume shrinkage in mode E is approximately seven times as great as that in mode F.

References:

1) M. Ichihashi et al., Proc. of the 5th ICTA, p. 554, Kyoto, 1977
2) K. Akechi and Z. Hara , Proc. of the 4th International Conference on Titanium, Kyoto, 1980

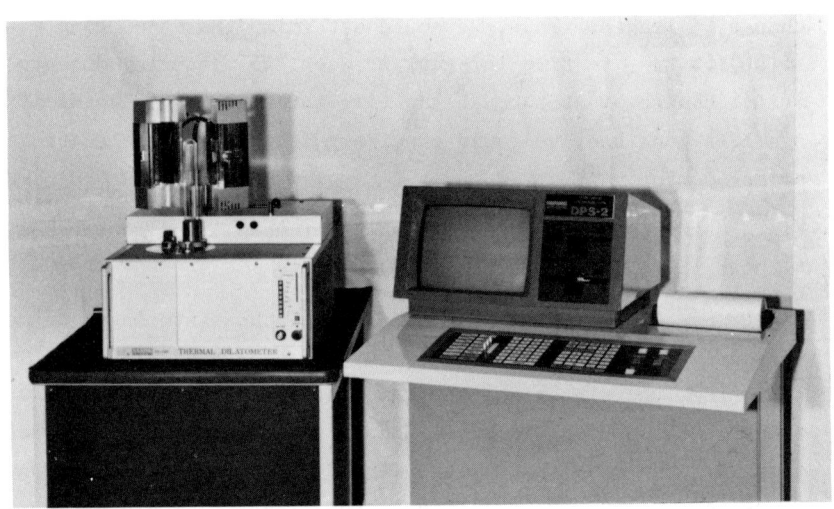

PHOTO 1.   OVERALL VIEW OF THE SYSTEM

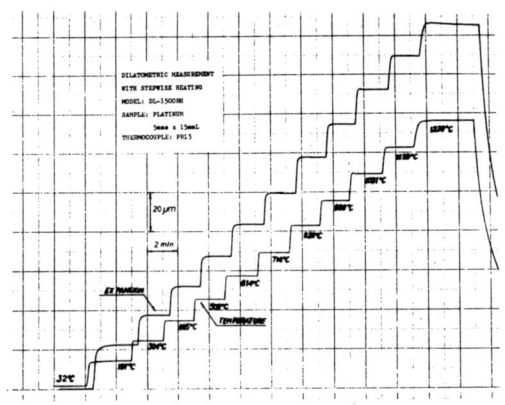

FIG.1

STEPWISE HEATING

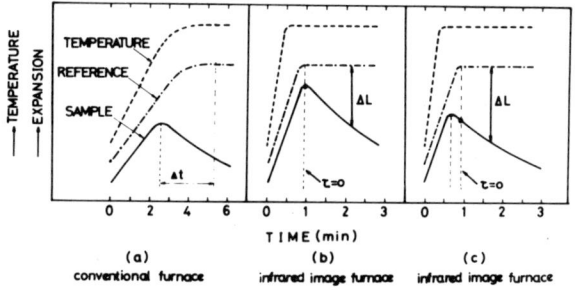

FIG.2

FIRST STAGE OF SINTERING

"reference" is expansion of well sintered sample

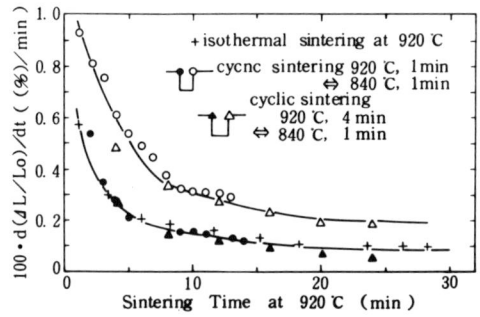

FIG.3

RELATION BETWEEN NET SINTERING TIME AND SHRINKAGE RATE

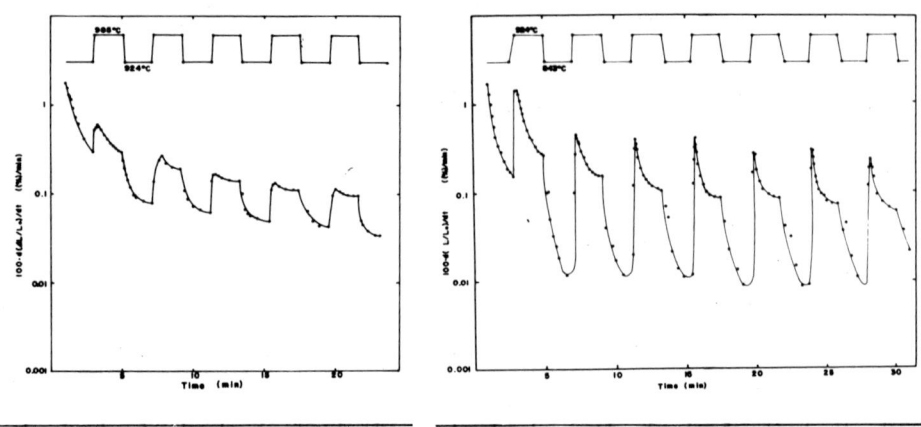

FIG.4　ELATIONS BETWEEN SINTERING RATE AND SINTERING TIME DURING CYCLIC SINTERING

LEFT;　$\beta \rightleftarrows \beta$　985 °C $\rightleftarrows$ 924 °C
RIGHT;　$\beta \rightleftarrows \alpha$　924 °C $\rightleftarrows$ 843 °C

# DYNAMIC ELASTICITY MEASUREMENTS ON PLASTICS

B. Andrejs, J.P. Schulz, E. Wappler

W.C. Heraeus GmbH, Hanau, FRG

## INTRODUCTION

It is especially to characterize plastics and control their properties, that thermoanalytical methods continue to be used increasingly. The user is interested not only in the processability of the material, but also, particularly, in its mechanical properties as a function of temperature. These mechanical properties are for instance: expansion behaviour, also under pressure or under tensile load, bending strength, flow behaviour, softening temperature as a function of pressure, swelling behaviour under the influence of gases or liquids, and flexion of multi-component systems. Thermal analysis is used for these measurements, because, on account of the high resolution, it also allows investigations to be carried out on thin foils or very thin films of lacquer.

It is important to both the manufacturer and the user, in most cases, not only to know these particular properties, but also to have an idea of the plastic and elastic behaviour of the material.

In this connection, Heraeus has developed an accessory unit for use with the Dilatometer TMA 500, allowing the temperature behaviour of a sample to be investigated under varying loads.

## MEASURING METHOD

This accessory unit consists of a coil into which the core is drawn with a linearly increasing force. In this way, a load between 0.2 and 15 p can be compensated. As previously, for the measurement, a probe is placed upon the sample. The probe shift is recorded by an inductive linearly variable differential transformer. Since the load varies, the probe is pressed into the sample, and is lifted again, by the restoring force of the sample, when the load is removed. This process can be recorded at constant temperature as well as under varying temperatures. The charging and discharging frequency can be chosen between 0.03 and 5 Hz, the load of the probe can be varied between 0.2 and 15 p. The contact surface of the probe is spherical, its diameter is 0.395 mm. This diameter has been chosen to enable these measurements to be compared with hardness measurements made on vulcanized rubber, in accordance with ISO-Standard 48-1975 (E.).

## RESULTS AND DISCUSSION

Measurements have been made on silicon rubber, fluoro-elastomer, NBR butadien-acrylonitrile rubber, polyester lacquer and other materials, giving a survey of the elastic and plastic behaviour of the material as a function of temperature. Additionally, however, the investigation reveals fine structures which may be due to relaxation phenomena.

Figure 1 shows the elastic behaviour of a polyester lacquer at constant temperature. The sample was charged linearly with a weight of 15 p, and then discharged.

The diagram shows that, on account to the load, the probe penetrates approx. 6 $\mu$m into the sample, and is pushed out again, almost completely, when discharged. Plastic deformation of the lacquer begins after 5 to 10 minutes. The glass temperature of the material is between 45 and 53 °C.

Figure 2 shows the charging and discharging of an NBR sample as a function of temperature. The investigation was carried out in the range between -100 and 240 °C. Below the glass temperature, approx. -40 °C, the probe penetrates only approx. 3 $\mu$m into the sample. When the glass transition begins, the plastic properties are increased and flow processes occur. An equilibrium is reached only at approx. +10 °C. The sample shows uniformly elastic properties up to approx. 180 °C. The depth of penetration of the probe is approx. 70 $\mu$m in this range.

When the charging and discharging frequency is increased, the depth of penetration of the probe is reduced, the material being less capable of following the charging frequency. Plastic deformation is increased. The time of relaxation, when the load is removed, is longer than the time when the load is applied, so that, under dynamic measuring conditions, the initial state cannot be restored.

The temperature range in which the behaviour under load is uniform is maintained; glass transition, however, is shifted to a higher temperature (figure 3). Figure 4 shows a higher charging and discharging frequency (5 Hz). The curve is quite different. The high frequency results in a curve where the individual penetration processes are no longer resolved by the recorder. The penetration depth is reduced in both the lower and the higher temperature range, the glass transition is shifted to a higher temperature, and fine structures of the material become visible above 120 °C. Due to the higher mechanical load caused by a higher frequency, a flow in steps occurs.

The measurements show that a rapidly varying mechanical load results in changes in the flow behaviour, glass transition, and hardness of the plastic part. Due to the varying load, the mechanical stability of the NBR is limited to a narrower temperature range.

Figures 5 and 6 show a comparison between two different rubber samples under like conditions - temperature range -100 to 360 °C, 0.14 Hz, load 10 p.

Above the glass temperature, the silicon rubber shows a distinct plastic flow. With rising temperature, the penetration depth increases. Above 275 °C, the material begins to soften and is deformed plastically. Above the glass temperature (-2 °C), the fluoro-elastomer sample shows a considerably reduced plastic flow, and up to 300 °C, the mechanical behaviour hardly changes, even under load. In the case of exacting requirements regarding mechanical properties, the material tested is better suitable in the large temperature range than the silicon rubber sample measured.

These simple comparisons of the plastic behaviour can be carried out on very small samples, down to thin films of lacquer. What is of interest, in addition to the selection of suitable materials for mechanically stressed parts, is, above all, investigations into the ageing properties of lacquers exposed to heavy atmospheric influences.

Examples of typical fields where dynamic elasticity measurements are carried out on plastics, are hardness and elasticity determinations, determination of plastic and elastic deformation, fatigue tests, and behaviour at the glass transition. In the case of vulcanized rubber, through hardness measurements, also the complex E-modulus (Young's Modulus) can be calculated (1).

## REFERENCES

1)   ISO-STANDARD 48-1975 (E)

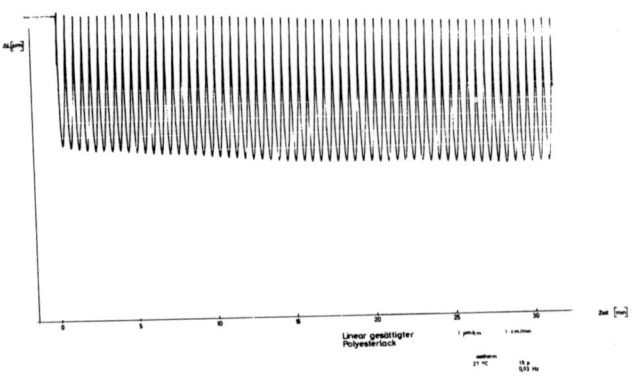

Fig. 1    Dynamic Elasticity Measurement on Polyester lacquer at 27 °C

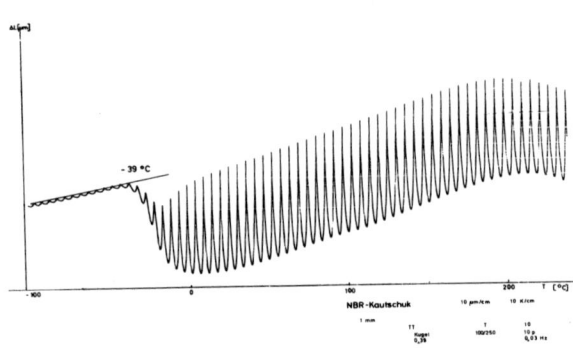

Fig. 2    Dynamic Elasticity Measurement on NBR from -100 to 250 °C at 0.03 Hz.

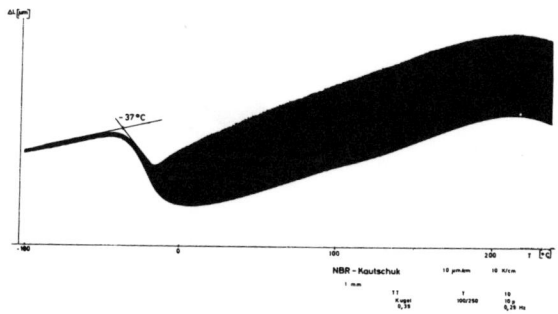

Fig. 3   Dynamic Elasticity Measurement on NBR from -100 to 250 °C at 0.29 Hz.

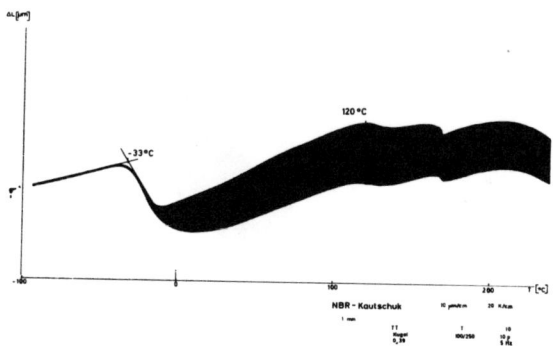

Fig. 4   Dynamic Elasticity Measurement on NBR from -100 to 250 °C at 5 Hz.

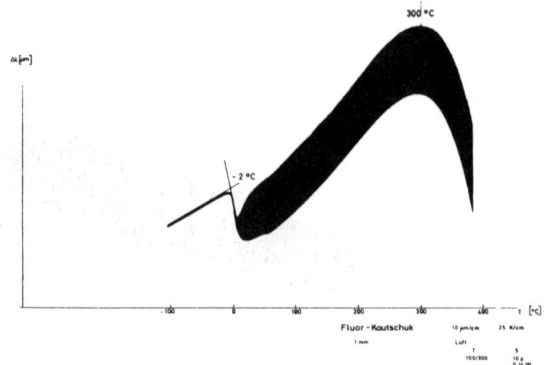

Fig. 5    Dynamic Elasticity Measurement on fluoro-elastomer from -100 to 360 °C at 0.14 Hz.

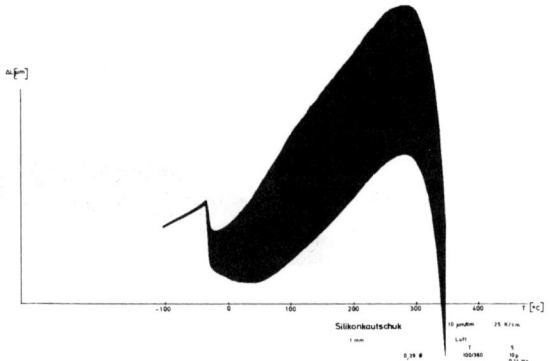

Fig. 6    Dynamic Elasticity Measurement on silicon rubber from -100 to 360 °C at 0.14 Hz.

231

Densification Studies of Ceramic Powder Compacts
by Quasi-Isothermal Dilatometry

O. Toft Sørensen

Risø National Laboratory, Roskilde, Denmark

ABSTRACT

Quasi-isothermal dilatometry (QID) is a new technique which can be used with great advantage in sintering studies of powder compacts. Contrary to conventional dilatometric measurements the optimum heating rate can be determined in a single run by this technique and from the measured shrinkage curve the controlling sintering mechanism and its activation energy can be determined. In this paper the advantage of this method is demonstrated and a method is given for calculating the activation energies from the QID curves. As an example the curves obtained in sintering studies of $CeO_2$-$Gd_2O_3$ compacts are analysed and the kinetic data for the different stages involved in the sintering are determined.

INTRODUCTION

Quasi-isothermal dilatometry (QID) is a new technique, which has proved to be very useful in sintering studies. It is based on the same principle as that used in quasi-isothermal thermogravimetric analysis (QIA) (see Paulik and Paulik (1-3) and Sørensen (4-5)). Compared to conventional dilatometric studies it has the following advantages: a) In this technique the overall heating rate is controlled by the densification rate which assures that the specimen is very closely sintered under equilibrium conditions; the optimum sintering can therefore be determined in a single run, and b) the sintering takes place in isothermal steps and the necessary data for a calculation of kinetic parameters (activation energy and pre-exponential factor) for the different processes involved in the sintering can be determined directly.

DESCRIPTION OF THE QID METHOD

The principle of the QID-method is shown in Fig. 1. The specimen (powder compact) is heated in the dilatometer at a constant rate until the sintering temperature is reached; the length of the specimen then begins to decrease and a dl/dt-signal proportional to the slope of the length vs time curve is produced. With a regulator, two limits are preset with two switches: switch 2, which sets an outer limit at the point at which the heating is stopped, and switch 1, which sets an inner limit where the heating is resumed when the signal becomes smaller than this limit. By this technique the sintering thus characteristically takes place in isothermal steps as also shown in the experimental curves in Fig. 2, and this technique is thus quite different from the quasi-isothermal thermodilatometry technique described by Paulik and Paulik (5) in which the measurements are carried out at a constant rate of change in length. In this latter method, the temperature increases continuously (but not at a constant rate), and it is thus not particularly useful in kinetic studies. The same drawback applies to the rate-controlled sintering methods described by Palmour III et al. (6,7).

## THEORY

Sintering is an activated process and according to Thümmler and Thomma (8) the densification can be described by the following general equation:

$$\frac{\Delta l}{l_o} = (Kt)^n = [(Z \exp{-Q/RT}) \cdot t]^n \tag{1}$$

where $\Delta l$ is the change in length, $l_o$ the initial length of the specimen, K the Arrhenius constant with the preexponential factor Z and the activation energy Q. T and t designate the temperature and time respectively, whereas n is an exponent which characterizes the sintering process.

According to this equation, isothermal plots of $\ln \frac{\Delta l}{l_o}$ vs. $\ln t$ should give a straight line with a slope equal to n. This was not the case, however, with the experimental data obtained in the present work, which only gave a straight line when $\Delta l$ was plotted against t and the equation to be used for QID data must be:

$$\Delta l = K \cdot l_o \cdot t = Z \exp(-Q/RT) \cdot l_o \cdot t \tag{2}$$

The slope of this line is $K \cdot l_o$ and according to

$$\ln(K \cdot l_o) = \ln Z + \ln l_o - Q/R \cdot T \tag{3}$$

the activation energy can thus be determined from the slope of the straight line obtained by plotting ln $K \cdot l_o$ versus $1/T$. From Equation (3) the preexponential factor can also be determined when $l_o$ and $Q$ are known.

## EXPERIMENTAL

### Equipment

The QID runs were performed in the Netzsch dilatometer equipped with an electronic regulator developed by Netzsch for their Simultaneous Thermoanalyser. In this dilatometer a horizontal sample holder of $Al_2O_3$ is used. The recorder reading, therefore, does not give the specimen length directly but instead the difference between the contraction of the specimen and the expansion of the sample holder. The corrections were determined in a blank run (without a specimen) and it appeared that these are negligible (ca. 30 μm at $1500°C$) compared to the large contractions observed during the sintering (ca. 3,000 μm at $1500°C$).

### Materials

The cylindrical compacts used in the present investigation were prepared by pressing a powder mixture of 65% $CeO_2$ - 35% $Gd_2O_3$, which was milled for 50 h in a ball mill ($ZrO_2$ balls) in order to improve the sintering activity.

### Experimental conditions

The compacts were sintered under the following conditions:

| | | | |
|---|---|---|---|
| atmosphere: | air | dl/dt: | 1 μm/min.cm |
| max. temp.: | $1500°C$ | switch 1: | 0.5 μm/min |
| heating rate | | switch 2: | 0.75 " |
| (between steps): | 2% min. | | |

## RESULTS AND DISCUSSION

The temperature and length curves recorded during the QID measurement are shown in Fig. 2. The shrinkage (Δl) determined at each isothermal step from these curves is shown as a function of time in Fig. 3 and obviously a straight-line relationship is obtained for all steps. In some cases, however, the shape of this line changes before the densification is complete at a step (4 and 11) indicating a change in the kinetic parameters for the sintering process. Finally, the log ($K \cdot l_o$) values derived from the slopes of the straight lines in Fig. 3 are shown in Fig. 4 as a function of 1/T whereas the activation energies and preexponential factors calculated from this figure are given in Table 1.

Both from Table 1 and Fig. 4 it is clear that the sintering takes place in several stages (A-F) each with a characteristic activation energy and preexponential factor. The rate-controlling process for these stages, however, cannot be established in the present analysis although the rather low activation energies observed can indicate that surface diffusion plays an important role, at least during the initial stages, and a very active powder was obviously obtained because of the extended milling time used.

Another interesting feature of this experiment is that the data are apparently described better with the exponent n=1, whereas values ranging from 0.2-0.6 normally are observed (See table in Thümmler and Thomma (8)). This indicates that a higher sintering rate is obtained by QID than in conventional sintering studies, perhaps because the sintering in QID takes place very closely to equilibrium conditions.

### ACKNOWLEDGEMENT

The author wishes to acknowledge the able assistance of H. Jensen and H. Frederiksen.

### REFERENCES

(1)  J. Paulik and F. Paulik, Anal. Chim. Acta, 56 (1971) 328.
(2)  F. Paulik and J. Paulik, Thermochim. Acta, 4 (1972) 189-198.
(3)  F. Paulik and J. Paulik, J. Thermal Anal., 5 (1973) 253-270.
(4)  O. Toft Sørensen, J. Thermal Anal., 13 (1978) 429-437.
(5)  O. Toft Sørensen, Thermochim. Acta, 29 (1979) 211-214.
(6)  F. Paulik and J. Paulik, J. Thermal Anal., 16 (1979) 399-406.
(7)  M.L. Huckabee, T.M. Hare and H. Palmour III.
     Materials Science Research: Processing of Crystalline Ceramics (H. Palmour III, R.T. Davis and T.M. Hare, Eds.) Vol. 11, 205-215. Plenum Press 1978.
(8)  F. Thümmler and W. Thomma, Metallurgical Reviews 12 (1967), 69-108.

Table I. Calculation of activation energies and preexponential factor for sintering stages

| Curve no.* | $Q/R \cdot 10^{-4}$ | Q(kcal/mole) | $\frac{Q}{R} \cdot \frac{1}{T}$ ** | | Z min$^{-1}$ |
|---|---|---|---|---|---|
| A | -0.489 | 22.37 | 3.531 | (1) | 0.247 |
| B | -0.857 | 39.22 | 6.016 | (4) | 88.51 |
| C | -0.613 | 28.03 | 4.181 | (8) | 1.33 |
| D | -1.083 | 49.57 | 7.148 | (11) | $1.58 \cdot 10^3$ |
| E | -0.720 | 32.95 | 4.644 | (13) | 4.71 |
| F | -0.29 | 13.26 | 1.757 | (17) | $5.42 \cdot 10^{-3}$ |

\* see Fig. 4, \*\* step no. in Fig. 2 and 3

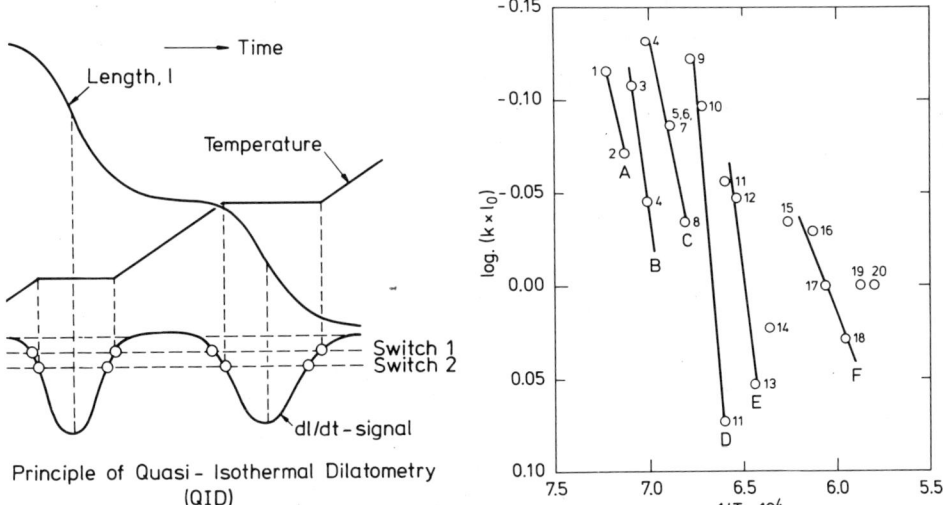

Fig. 1  Principle of Quasi-Isothermal Dilatometry (QID)

Fig. 4. Arrhenius plot of log $(K \cdot l_o)$ versus 1/T for sintering of a $CeO_2\text{-}Gd_2O_2$ (35%) compact. The number indicated at each point is the step number used in Figs. 2 and 3.

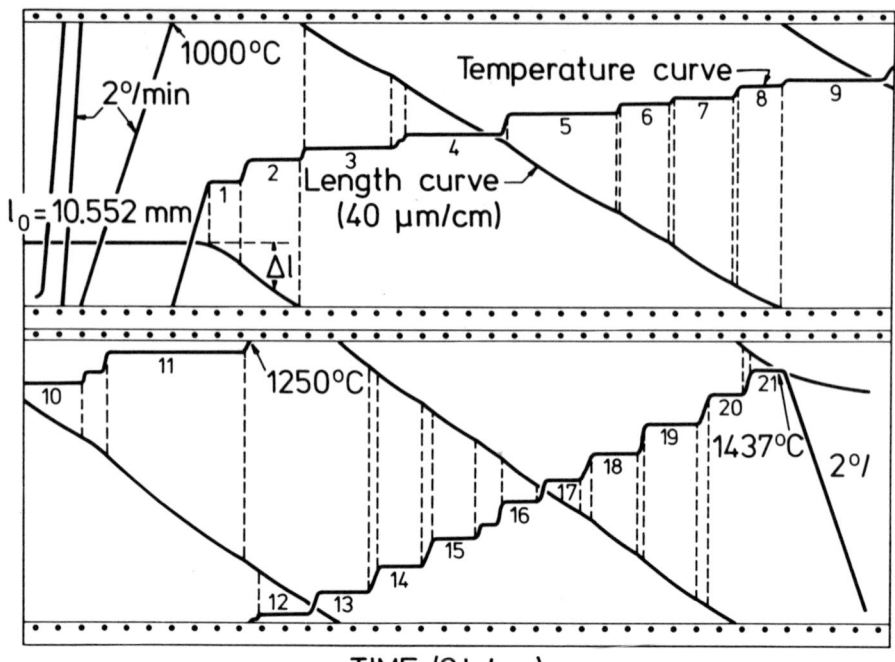

Fig. 2. QID curves (temperature and length) recorded during sintering of a $CeO_2$-$Gd_2O_3$ (35%) powder compact.

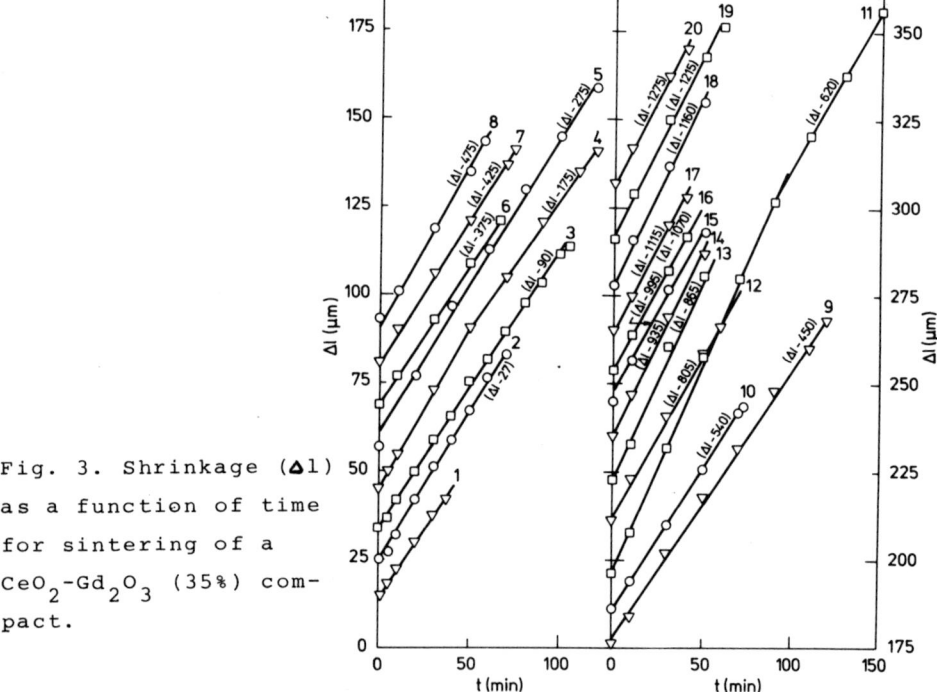

Fig. 3. Shrinkage ($\Delta l$) as a function of time for sintering of a $CeO_2$-$Gd_2O_3$ (35%) compact.

# A SIMULTANEOUS TG-DTA SYSTEM FOR OPERATION FROM -150°C TO 1500°C

E. L. Charsley, J. Joannou, A. C. F. Kamp,
M. R. Ottaway and J. P. Redfern

Stanton Redcroft, Copper Mill Lane, London, England

## ABSTRACT

A simultaneous TG-DTA instrument is described. Options are available for use to 1000°C, to 1500°C and from -150°C to 600°C, together with high capacity TG units to 1500°C. The performance and versatility of this new series of instruments is illustrated by reference to organic and inorganic systems.

## INTRODUCTION

The STA-780 series is a new range of equipment for TG and simultaneous TG-DTA measurements over the temperature range ambient to 1500°C. The modules available are:-

| | | |
|---|---|---|
| STA 780 | TG-DTG-DTA | ambient to 1000°C |
| STA 781 | TG-DTG-DTA | ambient to 1500°C |
| TG-782 | TG-DTG | ambient to 1000°C |
| TG-783 | TG-DTG | ambient to 1500°C |

A full account of the STA 780 has been given recently[1] and in this paper the other units in the series are discussed, together with a preliminary description of a prototype unit for simultaneous TG-DSC in the range -150°C to 600°C (STA 785).

## STA 781 Module

A cross-section of the furnace and head assembly is shown in Fig. 1. The sample and reference are housed in flat-bottomed 6mm dia. platinum crucibles D supported by plate-type platinum v platinum-13% rhodium thermocouples P. For work above 1000°C, alumina discs 0.1mm thick are inserted between crucibles and thermocouples to avoid high temperature welding. Alumina crucibles are available for materials that attack platinum. The head assembly supported by a four-bore alumina rod R

hangs within an alumina cup A, which has been electrically shielded for
high temperature work. The cup, which seats against the alumina baffles
B, ensures even heat distribution and enables very good atmosphere
control to be maintained around the sample, by passing the desired gas
in an upwards direction through the cup as shown, eliminating the need
for lengthy purging or evacuation of the balance system.

The sample is heated by a water-cooled furnace F, wound non-inductively
with a platinum-rhodium alloy W. The furnace is motor driven for ease
of handling and precision location. It seals against the water-cooled
finger C. Switch selected heating rates from 0.1 to 50degC/min are
available using a solid state digital programmer.

A photograph of the complete assembly is shown in Fig. 2. As with the
other models in the system, a 5 gram capacity electronic microbalance is
used with a digital control unit incorporating a microprocessor. Any
desired weight range from 2 to 200mg full scale deflection can be
selected with a resolution of 1µg in the range 2-20mg and 10µg in the
range 2-200mg. This facility makes it easy to display a selected
percentage of the sample weight as full scale. Full digital taring,
multiple inject and DTG facilities are incorporated.

Thermocouple connections are taken from the balance beam to measuring
circuits using very fine wires made of the thermocouple materials. The
$\Delta T$ signal is amplified by a low noise D.C. amplifier giving a maximum
sensitivity of 10µV full scale. Cold junction compensation and
temperature linearisation are available for the sample temperature
output.

The design of the gasflow system in conjunction with the water-cooled
finger enables runs to be carried in flowing atmospheres without having
to make a large weight correction, the magnitude of the latter generally
being below 0.1mg. Typical flow rates are normally in the range
25-75ml/min. To enable direct linkage to GC or MS equipment a heated
capillary system has been developed in which the probe enters directly
into the sample cup and is positioned above the sample. The resistance
of the system to blockage has been demonstrated by vaporising waxes in

an inert atmosphere and the efficiency demonstrated by a near
quantitative yield of HCl from the decomposition of PVC.

The furnace will cool from 1500°C to 600°C in approximately ten minutes
when the furnace can be lowered to remove the sample. With the furnace
in the up position a programme cooling rate of 10degC/min can be
maintained down to 120°C. Isothermal stability at 1500°C is better
than $\pm 1°C$ measured over a two hour period.

The performance of the instrument over the range a-1500°C is illustrated
by a run on calcium dichromate pentahydrate shown in Fig. 3. The
reactions involve dehydration in the range 50-150°C, decomposition to
calcium chromate in the range 460-520°C and formation of calcium
chromite above 1080°C.

## TG-782/3 Modules

For high capacity TG work the TG-DTA hangdown can be readily replaced by
a TG hangdown enabling a tapered platinum crucible 8mm high with a top
diameter of 12mm to be used. The crucible is supported by a ring-type
precious metal thermocouple, thus retaining the considerable advantage
of direct readout of sample temperature. The crucible is still housed
within the sample cup so that the benefits of the atmosphere control
system already described are maintained. The flanges shown in Fig. 1
are raised slightly to provide additional clearance. A maximum load
of approximately 1.2g can be housed in the crucible. This coupled with
a balance range of 2mg full scale enables small weight changes to be
measured with a high degree of accuracy. For many applications it is
convenient to set 1% of the sample weight as full scale deflection on
the recorder. The multiple inject facility eliminates the possibility
of going off-scale at these high sensitivities giving over 100 spans
of the chart automatically. An example of the use of the system for
measuring small weight changes is given in Fig. 4 which shows the
TG-DTG curve for a stearic acid coated, calcium carbonate filler used
in the polymer industry.

The head can also be used for studying non-homogeneous materials where
the user may prefer to use larger sample weights in order to obtain a

representative sample. It is also useful for preparing reaction products under carefully controlled conditions for analysis by other techniques.

## TG-DSC Module (STA 785)

Work is in progress on development of a simultaneous TG-heat flux DSC unit for operation from $-150°C$ to $600°C$. The high sensitivity head is based on chromel-alumel plate type thermocouples and is interchangeable with the standard TG-DTA head. The water-cooled furnace is replaced by a heated silver block, with integral liquid nitrogen cooling facilities. This assembly enables linear heating or cooling rates to be obtained over the complete range of the instrument, while maintaining full atmosphere control.

A preliminary run on the equipment using the ICTA Low Temperature DTA Standard 1,2-dichloroethane is shown in Fig. 5. The DSC curve shows the melting of this compound in the region of $-36°C$, followed by an endotherm due to volatilisation of the liquid. The TG curve shows that the latter starts in the region of $-5°C$.

The unit has considerable potential particularly in the polymer field, since it combines the advantages of a heat flux DSC with those of a high sensitivity thermobalance.

## REFERENCES

1. E. L. Charsley, M. V. Collins, J. Joannou, A. C. F. Kamp, J. P. Redfern and N. Virji, Thermochim. Acta, in press.

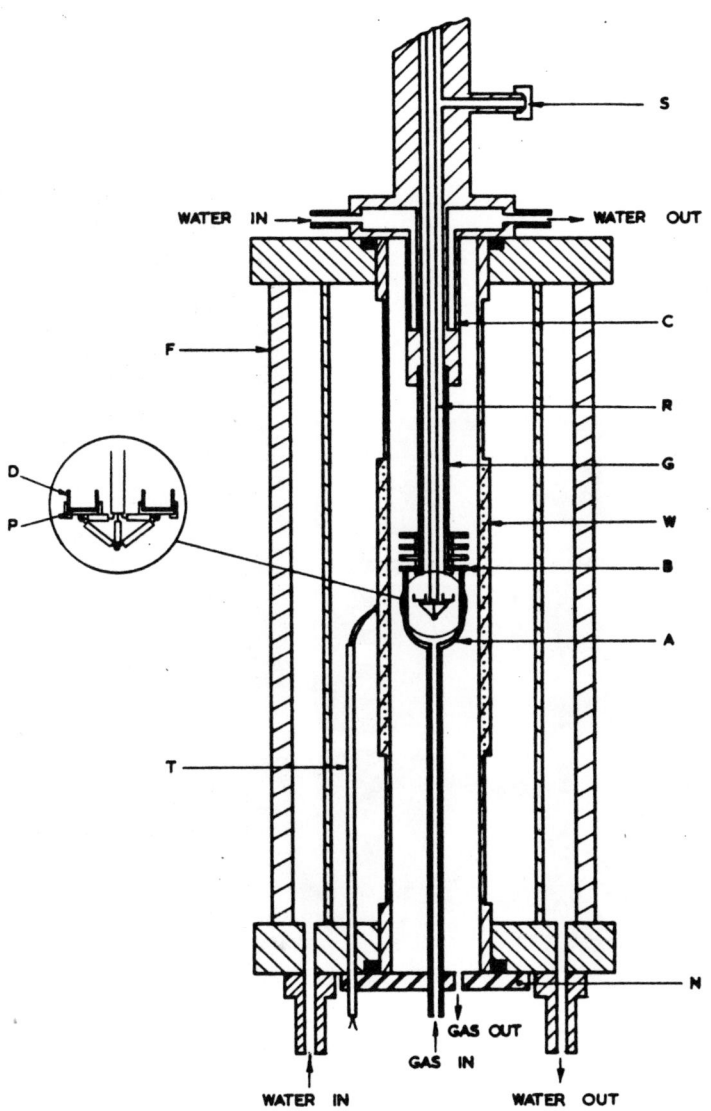

FIG. 1  Cross Section of 1500°C Furnace with TG-DTA Hangdown in Position

Fig. 2. Photograph of the STA 780

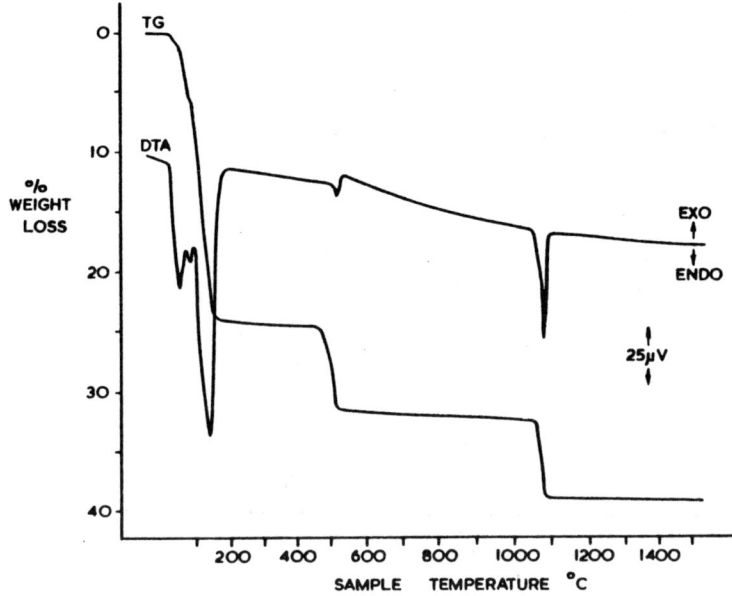

FIG. 3    Simultaneous TG-DTA Curve from Calcium Dichromate Pentahydrate (19.4mg heated at 10°C/min in Static Air

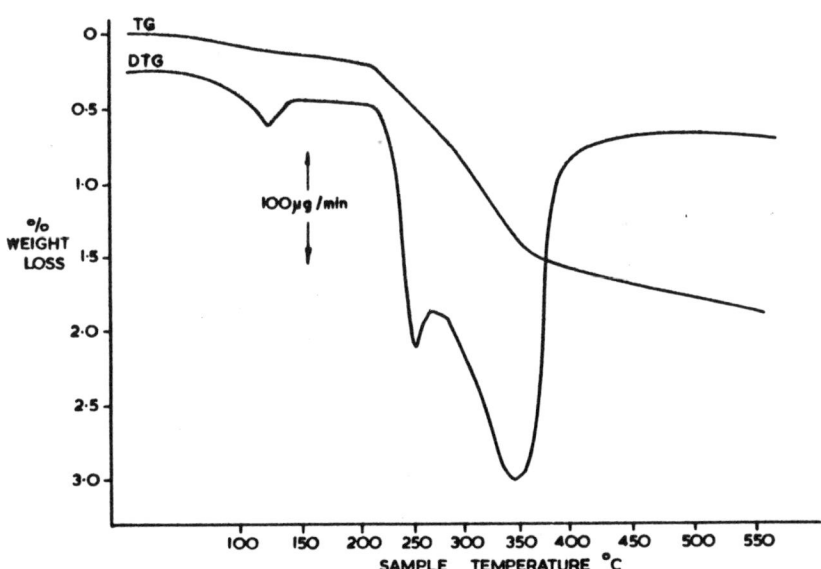

FIG. 4    Simultaneous TG-DTG Curve from a Stearic Acid Coated Calcium Carbonate Filler (463.9mg heated at 20°C/min in Air, 50ml/min)

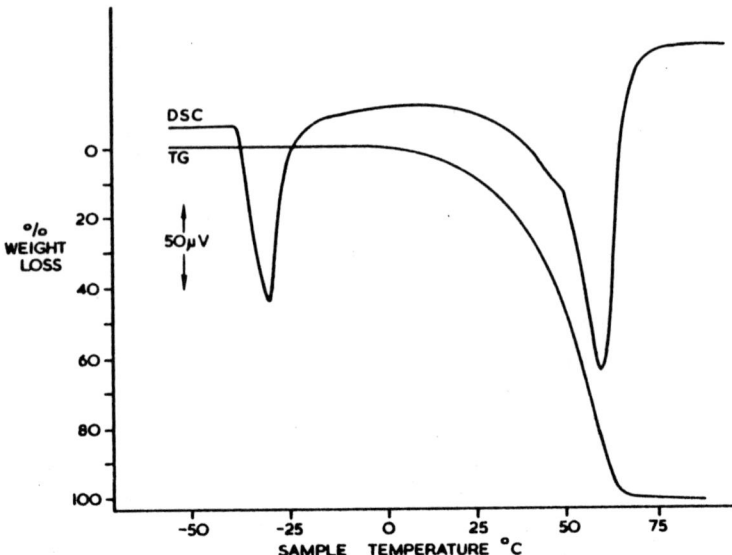

FIG. 5  Simultaneous TG-DSC Curve from 1,2-Dichloroethane (6.8mg heated at 10°C/min in Static Air)

# TECHNIQUES FOR COUPLING MASS SPECTROMETRY TO THERMOGRAVIMETRY

Jen Chiu and A. J. Beattie
Polymer Products Department
E. I. duPont de Nemours & Company
Experimental Station
Wilmington, Delaware 19898

## ABSTRACT

Techniques combining thermogravimetry (TG) and mass spectrometry (MS) are described whereby TG provides precise heating conditions and weight loss information and MS identifies volatiles evolved during the weight loss process. In the present work the DuPont 990 thermal analysis system, a DuPont 21-104 mass spectrometer, and a Hewlett-Packard 21-MX/Digital Equipment PDP-10 computer system are combined through simple, unique interfaces. The instrumentation features flexibility in thermal treatment of the sample and good atmospheric control. Tests with model substances and practical industrial problems have shown good enrichment and recovery of eluted components and no significant delay in TG operations.

## INTRODUCTION

Many attempts have been made to achieve TG-MS combination during the past decade. The main emphasis of research efforts has been centered around the development of an interfacing system to suit the particular TG and MS instruments being used. Techniques reported include direct connection under vacuum (1-5), direct connection under reagent gas (6), the use of metering valves (7-8), and the use of capillary-orifice (9-13). It appears to be a rather difficult task to develop an interface to meet the requirements of an ideally coupled TG-MS system such as free choice of vacuum or atmospheric pressures, no dilution of the sample of interest, no mass discrimination, no loss of highly volatile or non-volatile components, continuous monitoring, and easy adaptation to various instruments. The present paper describes several simple coupling techniques which meet most of the requirements.

## EXPERIMENTAL

A schematic diagram of the coupled TG-MS system is shown in Figure 1. In the first technique, the interface used is a 6-port microvalve connected to a U-shaped liquid nitrogen trap made of ca. 3 inch x 1/16 inch O.D. stainless steel or glass tubing. The substances evolved during a certain weight loss step are condensed in the liquid nitrogen trap and then introduced into the MS by properly switching the microvalve. No dilution of the sample results when helium is used as the carrier gas. This technique is highly sensitive to determine trace amounts of effluents from TG, and provides both qualitative and quantitative analyses of the off-gases during a weight loss step. However, it is not suitable for continuous monitoring during the weight change process.

The second technique replaces the microvalve-trap system by a glass tee as shown in the dashed portion of the diagram. Two arms of the tee are connected to the TG furnace tube and the inlet system of the MS, respectively. The third arm of the tee is connected to an oil pump. By properly controlling the carrier gas flow and the sample size, most components of interest are introduced into the MS. Thus, the differential pumping capability between the oil pump and the MS pumps functions to enrich the heavier molecules in the helium stream.

Both enrichment and recovery of the off-gases are further improved by using a tubing-tee combination instead of the glass tee interface. Still, both types reduce the atmospheric pressure somewhat in the sample chamber, and do not provide satisfactory control of a reactive atmosphere. To obtain overall balance of performance, a technique based on the tubing-tee interface in conjunction with a sampler to allow normal flow of carrier gas without evacuation of the TG sample chamber has been devised.

## RESULTS AND DISCUSSION

I. <u>Total Condensation Method</u> - The TG unit is operated by the standard procedure. The derivative TG (DTG) curve is used to better define the weight loss step, while the TG curve provides more convenient calculation of the absolute weight.

Figure 2 shows the TG and DTG scans of calcium acetate monohydrate. Volatiles evolved during the three distinct weight loss steps are condensed consecutively in the liquid nitrogen trap without interrupting the TG scan. The trap is then warmed and the trapped gases injected into the MS for identification. The main components identified as water, acetone, and carbon dioxide in cuts 1,2, and 3, respectively, are consistent with the reactions well established for decomposition of calcium acetate monohydrate. The quantitative information obtained from the TG scan also agrees with the stoichiometric values as expected.

A polyacetal resin part was analyzed by the TG-MS technique. TG showed clearly two weight loss steps involving 75% and 21%, respectively (Figure 3). MS analyses of the two cuts corresponding to the two weight loss steps showed mainly formaldehyde for the first cut and tetrafluoroethylene for the second cut. Since polyoxymethylene (POM) and polytetrafluoroethylene (PTFE) are known to depolymerize almost completely into their respective monomers, the resin part probably consists of 75% POM and 21% PTFE by weight.

II. <u>Continuous Monitoring Method</u> - By using the glass tee or tubing-tee combination as the interface, MS scans can be taken continuously. Thus, the MS scan follows the TG curve closely to provide interpretation of the thermal events.

The technique has been effectively used to study the curing of an experimental polyimide prepreg. Presumably the binder is in a diester-diammonium salt form, and further curing of the composition will eliminate water and ethanol to form the polyimide.

A continuous TG scan of the polyimide prepreg is shown (Figure 4) with MS scans taken sequentially every two minutes as numbered on the DTG curve.

The two major components, water and ethanol, are monitored by mass peaks 18 and 31, respectively, and their relative ion intensities are plotted as a function of temperature (Figure 5). The results showed that water and ethanol are involved in all weight loss steps, contrary to previous assumption that dehydration would be followed by ethanol elimination to form the polyimide.

## REFERENCES

1. H. G. Wiedemann, "Thermal Analysis", Vol. 1, R. F. Schwenker, P. D. Garn, Eds., Academic, New York, 1969, p. 229.

2. D. E. Wilson, F. M. Hamaker, "Thermal Analysis", Vol. 1, R. F. Schwenker, P. D. Garn, Eds., Academic, New York, 1969, p. 517.

3. E. K. Gibson, Jr., S. M. Johnson, Thermochim. Acta, $\underline{4}$, 49 (1972).

4. G. J. Mol, Thermochim. Acta, $\underline{10}$, 259 (1974).

5. H. Eppler, H. Selhofer, Thermochim. Acta, $\underline{20}$, 45 (1977).

6. E. Baumgartner, E. Nachbaur, Thermochim. Acta, $\underline{19}$, 3 (1977).

7. F. Zitomer, Anal. Chem., $\underline{40}$, 1091 (1968).

8. R. G. Beimer, Am. Chem. Soc., Div. Org. Coat. Plast. Chem. Pap., $\underline{35}$ (1), 428 (1975).

9. H. P. Vaughan, Proc. 17th Ann. Conf. Mass Spectr. Allied Topics, 1969, p. 223.

10. D. L. Geiger, G. A. Kleineberg, Proc. 20th Ann. Conf. Mass Spectr. Allied Topics, 1972, p. 125.

11. G. A. Kleineberg, D. L. Geiger, W. T. Gormley, Makromol. Chem., $\underline{175}$, 483 (1974).

12. W. Dunner, H. Eppler, "Thermal Analysis", Vol. 3, I Buzas, Ed., Heyden & Son, London, 1975, p. 1049.

13. W. D. Emmerich, E. Kaisersberger, "Thermal Analysis", H. Chihara, Ed., Kagaki Gijutsu-Sha, Tokyo, 1977, p. 67.

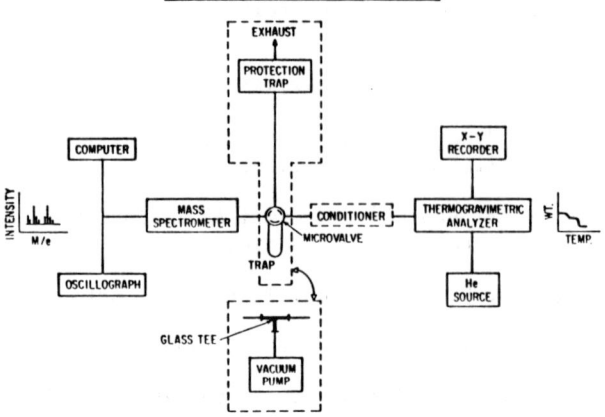

FIGURE 1
SCHEMATIC DIAGRAM OF TG-MS SYSTEM

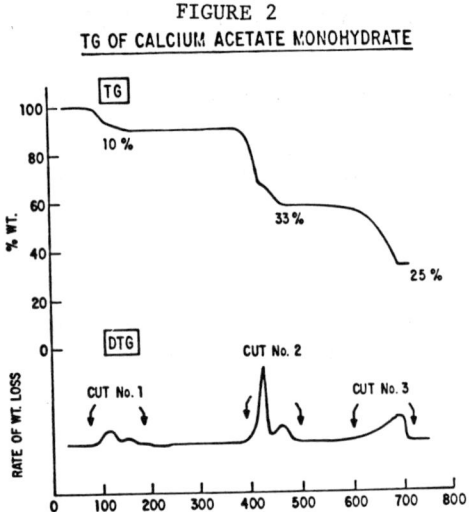

FIGURE 2
TG OF CALCIUM ACETATE MONOHYDRATE

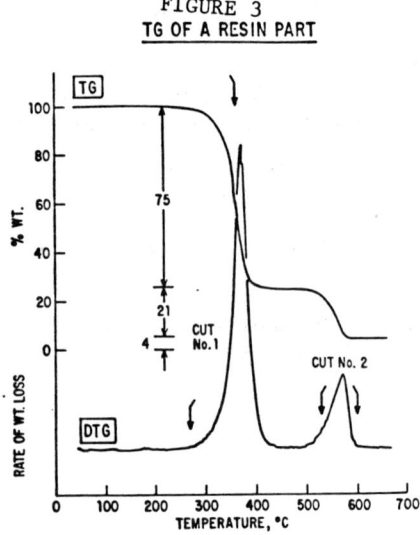

FIGURE 3
TG OF A RESIN PART

FIGURE 4
TG OF POLYIMIDE PREPREG

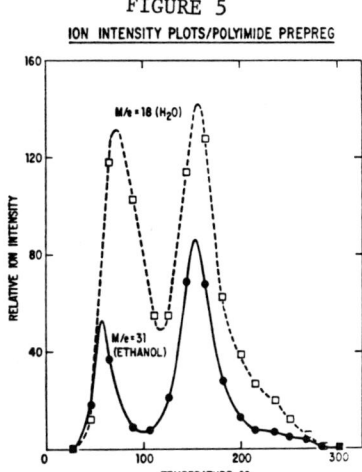

FIGURE 5
ION INTENSITY PLOTS/POLYIMIDE PREPREG

FURTHER DEVELOPMENT AND APPLICATION OF A COMBINED SYSTEM FOR
SIMULTANEOUS THERMAL ANALYSIS AND MASS SPECTROMETRY

Erwin Kaisersberger
NETZSCH-Gerätebau GmbH, Selb, FRG

ABSTRACT

The experiences obtained with a combined simultaneous thermal analysis (TG-DTA-DTG) apparatus and quadrupole mass spectrometer resulted in developments in the applied high-temperature coupling system. The geometric arrangement of the gas inlet orifices can be varied, the distance between samples and ion source has been reduced. These steps improved the sensitivity detecting condensable metal vapours.
The calculation of the heat of evaporation of cadmium from a silver-cadmium alloy is demonstrated by means of the observed increase in intensity for cadmium in the mass spectrum.

INTRODUCTION

A gas inlet system for a mass spectrometer, which is combined with a thermogravimetric apparatus working up to 1550°C, sets high requirements on material and mechanic precision. The majority of thermogravimetric tests is carried out at atmospheric pressure in the test space in air or other gases. The gas inlet system for the mass spectrometer should bring all gases and vapours released from the sample unaltered to detection and analysis in the mass spectrometer in the useful temperature range of the combined apparatus. A gas inlet system must be hardly judged by its effect on the demixturing with regard to time of multicomponent gas or vapour mixtures as well as fractionation and interferences.

During the decomposition of a sample in high-vacuum the distance and geometric arrangement of the ionization device of the mass spectrometer towards the vaporizing sample are decisive, in order to transform the arising molecular stream most effectively into a detectable ion beam.

THERMAL ANALYSIS . ICTA 80 . BIRKHAEUSER VERLAG,BASEL,BOSTON,STUTTGART

While during high vacuum operation the detection sensitivity of a mass spectrometer is scarcely used, the partial pressure sensitivity of the mass spectrometer specifies the detection limits for measurements under atmospheric pressure.

## EXPERIMENTAL

A two-stage gas inlet system combining a NETZSCH STA 429 with a Balzers QMG 511 mass spectrometer is the base of the present report. This coupling system works up to sample temperatures of 1550°C. (1, 2)

The pressure reduction of atmospheric pressure on the sample, which decomposes by emitting gas or vaporizes, up to the high vacuum required in the recipient of the mass spectrometer is accomplished by orifices arranged in $Al_2O_3$ tubes. The first orifice having a diameter of about 0,1 to 0.2 mm effects a pressure drop to about 1 to 5 mbar. It is exhausted by a rotation pump, whereat the dimensioning of the orifice and the suction capacity of the pump provide for viscous inlet into the intermediate vacuum (reynolds' number abt. 900 with dry air 20°C, Reynolds' number abt. 100 with dry air 1000°C). The Hagen-Poiseuille law for gases is applied: (3)

$$Q_V = \frac{\pi R^4}{8 \eta L} \cdot \frac{P_1^2 - P_2^2}{2 P_2} (1 + 4 \frac{\zeta}{R}) \cdot K_A$$

$Q_V$ = gas flow  
$R$ = orifice radius  
$L$ = length  
$\eta$ = dyn. viscosity  

$p_1$ = pressure in front of the orifice  
$p_2$ = pressure behind the orifice  
$\zeta$ = slip coefficient  
$K_A$ = correction for inlet stage  

The viscously streaming gas quantity is largely independent on the molecular weight, the composition of a gas mixture is not changed with time.

By a second orifice the pressure is reduced to $= 10^{-5}$ mbar. The mean free path length of the gas molecules exceeds the recipient dimensions, molecular streaming conditions prevail. The following definition is accepted: (4)

$$Q_m = k \cdot R^2 \sqrt{\frac{T}{M}} (P_1 - P_2) = L \cdot \Delta P$$

T = absolute temperature      K = constant for orifice
M = molecular weight          L = admittance for the orifice

The gas quantity streaming in at a fixed pressure gradient depends on the temperature, the molecular weight and on geometric factors. Due to the dependence on the molecular weight, there arises a demixturing with regard to time, more heavy components will concentrate in front of the orifice in the stationary case. By the gas exchange before the second orifice, which does not depend on the molecular weight, this effect is without influence.

For the analysis of the gases on the sample being under atmospheric pressure, which is to be carried out continuously, this two-stage orifice system offers the only possibility to convey to the ion source of the mass spectrometer a representative portion of the sample atmosphere without demixturing and without fractionation. The dimensioning of the orifices and the pressure conditions provide for a low time constant. The integration into the high-temperature furnace of the thermal analysis apparatus as well as a suitably arranged additional heater at the orifice tubes prevent condensation of vapours on the orifices, even beyond $1000°C$.

In order to achieve a high ion output, it has proved advantageously to construct the position of the second orifice adjustably. The exact alignment of the orifices and ion source is accomplished by a laser, whereat the cooling trap is removed. For this, the cross beam ion source mounted on the quadrupole rod system can be adjusted, too.

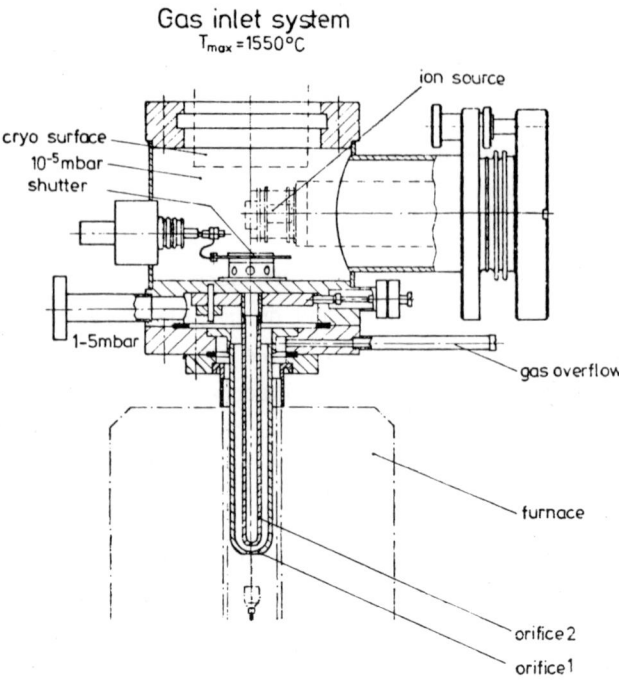

Fig. 1

As the apparatus can be largely used for the study of high condensable vapours, the ionization and analyzer system must be protected against contamination. A shutter scattering the molecular beam (with beam limitation in open position) is provided, as well as a cooling trap for the defined condensation of all non-ionizated molecules. The distance between sample and ion source was narrowed in order to reduce losses in intensity by wall contacts in the range of the molecular beam. This leads also to increased detection sensitivities during high vacuum operation, when the pressure reducing orifices are removed. By the shutter a control of the background spectrum of the apparatus by scattering the molecular beam is possible. The use of turbo-molecular pumps to obtain a vacuum free from hydrocarbon has proved in this units combination, no influence on the weighing by high-frequent vibration is notable.

## RESULTS AND DISCUSSION

Main application of a combined TG-DTA-MS system present besides the exact detection of permanent gas, the identification of high-molecular, polar and also unstable decomposition products. The capacity of the two-stage gas inlet system is discussed in publications (1, 2, 5) and has shown further confirmation studying glasses, mixtures of tobaco, metal-organic compounds, mineral salts and ceramic raw materials. Interesting use is to be found in the frame of environmental pollution stating carcinogenic polycyclic aromatic hydrocarbons in the soot of energy producing plants for fossil fuels, as well as in the study of recycling methods for fine metal residues.

It is reported about a high vacuum experiment with a silver cadmium alloy. The DTA curve shows the start of melting of the alloy at 900°C. Beginning at 360°C a peak with the isotope distribution of cadmium at m/e 106 to 116 is clearly seen in the mass spectrum.

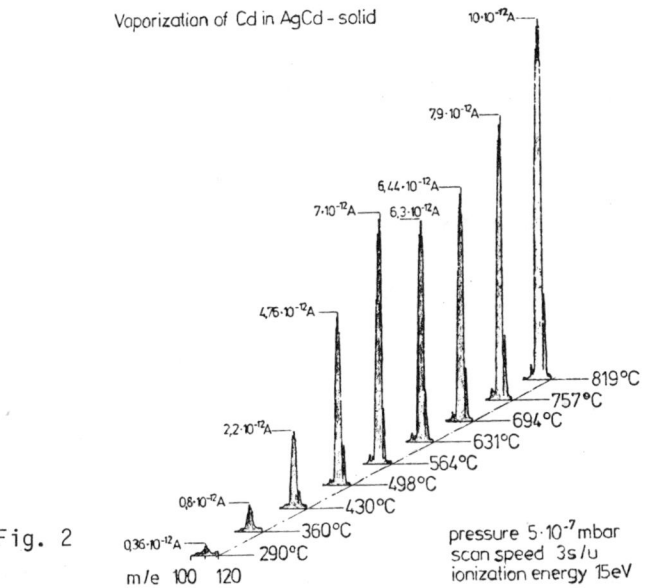

Fig. 2

The steady increase of intensities for cadmium is interrupted in the temperature range 600 to 700°C.

For the vaporization into vacuum ($5 \cdot 10^{-7}$ mbar) there is applied:

$$\frac{dQ}{dt} = \frac{F \cdot P_s}{2\pi MRT} \text{ moles /s}$$

F = surface of the sample, $P_s$ = vapour pressure

The pressure on the ion source and thus the vapour pressure for the vaporization of cadmium from the alloy can be determined by the course of intensities and the sensitivity of the mass spectrometer /A / mbar). Because in the discussed temperature range the vapour pressure of silver is smaller by several orders than the vapour pressure of cadmium, the problem is simplifying dealed as one-component system. For the solid-gaseous transition is applied:

$$\frac{d\ln k_p}{dT} = \frac{\Delta H}{RT^2} \quad \text{or} \quad \frac{d\ln k_p}{d(\frac{1}{T})} = -\frac{\Delta H}{R}$$

$K_p$ = equilibrium constant of the vaporization process

For the transition solid gas the equilibrium constant can be equated with the pressure of the gases. In this way, the average heat of vaporization $H_v$ in the studied temperature range can be calculated from the slope of the ln p dependence on 1/T (fig. 3)

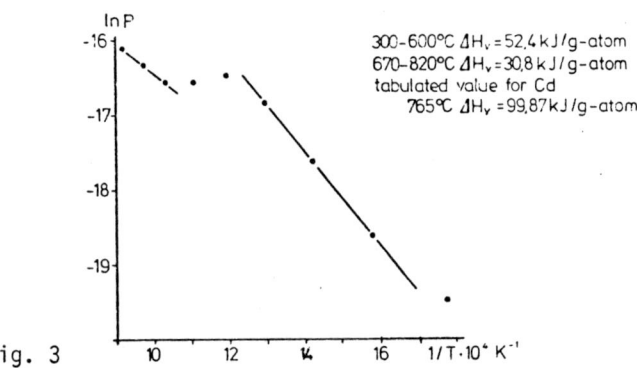

Fig. 3

The evaluation of fig. 3 shows for the heat of vaporization of cadmium from the silver-cadmium alloy (sensitivity $10^{-4}$ A/mbar):

300 - 600°C  $\Delta H_v$ = 52,4 kJ/g-atom
670 - 820°C  $\Delta H_v$ = 30,8 kJ/g-atom

The course of intensities for the vaporization of cadmium from the alloy runs similar to the curve for the vapour pressure of cadmium and is also shown in the changing slope of the plot ln p against 1/T and thus the calculated heat of vaporization.

This example of calculation shows that by means of the simultaneous TG-DTA-MS apparatus also metal vapour with higher condensation temperature can be detected quantitatively. In similar manner the vaporization of sodium from glasses as well as mineral salts could be demonstrated successfully.

## REFERENCES

(1) W.-D. Emmerich, E, Kaisersberger, Simultaneous TG-DTA-Mass Spectrometry to $1550^\circ C$, I. Thermal Analysis 17 (1979), 195-212
(2) E. Kaisersberger, Gas Analytical Methods of Thermal Analysis in Comparison, Thermoch. Acta 29 (1979), 215-220
(3) Kohlrausch F., Praktische Physik, Bd. 1, T.G. Teubner Stuttgart (1968), 178-179
(4) Kienitz H., Massenspektrometrie, Verlag Chemie Weinheim 1968 XV., 31
(5) H. Eppler, H. Selhofer, Thermoch. Acta 20 (1977), 45-52

# HIGH-SENSITIVITY MULTICHANNEL DTA APPARATUS

Michio Maruta, Yoshiomi Kunimatsu and Kiyotsugu Yamada
Scientific & Industrial Instrument Division, Shimadzu Corporation

## APPARATUS

A multichannel differential thermal analysis(DTA) apparatus has been developed, which permits simultaneous analysis of 3~5 samples under the same temperature and atmosphere conditions. The flow rates of the atmospheric gases can be separately controlled for each sample.

The multichannel DTA apparatus has 3~5 units of sample holders for single channel DTA, which are mechanically connected. Equipped with a heating block and a dumbbell type detector, each sample holder can be used as an independent unit.

The temperature of the sample holders is controlled with a thermocouple and a temperature program controller.

Since our experiments have shown that temperature is uniformly controlled through the sample holders, the temperature of one unit, instead of the temperatures of all the units, is recorded. A 4~6-pen recorder is effectively used, therefore, for 3~5 channel DTA.

The apparatus has a temperature range from ambient to 300°C.

## INTRODUCTION

The technique of differential thermal analysis(DTA) is now widely used for various purposes in various industrial and academic fields.

The most popular mode of DTA is to detect physical and chemical changes of a sample which is heated at a constnant rate through a selected temperature range. "Isothermal" DTA, however, is sometimes used to evaluate the thermal stability and to test the accelerating deterioration of electric insulators (1), pharmaceuticals, etc. In "Isothermal" DTA, the sample is rapidly heated up to, and then kept constant at, a selected temperature. Since this temperature is generally fairly lower than the decomposition temperature observed by the ordinary DTA, a run of analysis will take from a few hours to even more than ten hours. As presumed from the fact that in ordinary DTA, only a low sensitivity is provided when a low heating rate is selected, a higher sensitivity is required of the detector for "Isothermal" DTA. The following features are incorporated in our newly-developed multichannel DTA apparatus, so that a high analytical efficiency is provided, and that it is easy to compare concurrently the properties of more than one sample. Also, an excellent cost-performance ratio was accomplished.

The features of this multichannel apparatus are;
1) High sensitivity
2) Long-term high stability of base line
3) Ease of exchanging sample cells and of changing atmosphere conditions
4) Capability to increase or decrease the number (1~5) of the single-channel units combined and to freely change the atmosphere conditions separately for each unit.

## APPARATUS

a. Generals

Fig. 1 shows the schematic construction which permits simultaneous DTA of three samples. Fig. 2 is the external view of the multichannel DTA apparatus, including 3-channel DTA sample holder, 3-amplifiers and a temperature program controller.

The sample holder consists of three single-channel sample holders, which are connected mechanically. Each single-channel sample holder has a heating block, a dumbbell type detector, a base line stabilizing mechanism, and a flow controller for atmospheric gases. The heating blocks are connected to each other with screws and heated with two sheathed heaters of almost the same length as the total length of the blocks.

The temperature is controlled with a chromel-alumel thermocouple, inserted into the heating block, and a temperature program controller. The heating block is made of copper, because ①, the highest temperature necessary for accelerating-deterioration and screening test of pharmaceuticals and electric insulators made of high polymers is 300°C, and because ②, a good temperature uniformity is required.

Differential thermal signals and sample temperature signals are simultaneously recorded on a multi-pen recorder. Since the temperature is distributed with a uniformity better than ±0.1°C in the three single-channel units combined together, the sample temperature of only one of the units is recorded as being representative of all of them.

The atmospheric gas is changed simultaneously for all the units by means of solenoid valves. The flow rate can be seperately controlled at each unit.

b. High-sensitivity detector

The detector is the high-sensitivity dumbbell type which has been reported previously(2). It is a disc type detector with a thin platinum plate(6 dia x $0.2^t$mm ) welded to the hot junction of the thermocouple by an electron beam welder and which also serves as the holder of the sample cells.

Since the heat capacity of the detector and the sample cell is very small, a minute change of heat in sample can be detected as a great temperature change - extremely high sensitivity is ensured.

Since the holder plate of the sample cells is made of platinum, a good thermal-contact is ensured between the sample cell and the detector. The detector is a plug-in type so that it can be easily dismounted. Thus the contaminants often attributable to pyrolysis products in the DTA of organic compounds can be easily removed by burning off the dirt.

c.  Base line stabilization

The instability of base line is an important problem in high-sensitivity DTA. In our high-sensitivity DTA apparatus, this problem has been solved by placing the two hot junctions of the differential thermocouple on the hypothetical isothermal line in the furnace by adjusting the relative potions of the furnace and the detector. Since the temperatures of the two hot junctions are equal, the output of the differential thermocouple is zero in principle. This system, installed in a furnace in which the temperature distribution is fairly uniform, ensures a stable base line up to a high temperature region of 1500°C. This system also has the advantage that, since sample and reference are placed on the isothermal line, the amounts of heat leaking from them are almost equal and hence, the lineality between the sample weight and the peak area is as good as that provided by heat-flux type differential scanning calorimeters.

d.  Atmosphere conditions

The thermal stability of a sample differs greatly depending on the kind of atmospheric gases and the material of the sample cell, even under the identical temperature conditions.
It is desirable to use a copper cell for analysis of electric insulators made of high polymers, because they are always in contact with copper wires in practical use. Our appratus permits use of sample cells made of various materials.

Since vacuum pump is not used to displace the atmospheric gases, disruption around the sample is prevented. Instead, only switching solenoid valves are used. The sample holder is designed to be compact so that the gases in it can be displaced rapidly.

## RESULTS AND DISCUSSION

A blank test in which the flow rate of $N_2$ or $O_2$ as atmospheric gas was varied from 0 to 50 ml/min. showed that the flow rate has little effect on the data. This test, in which the heating blocks were kept at a constant temperature near 200°C, produced a stable base line for more than 24 hours.

Three 2.2mg indium samples were crimped in three aluminum cells which were them set on the detectors. Then, an ordinary DTA was performed using empty aluminum cells as the reference and setting the heating rate at 2, 5, and 10°C/min., subsequently. The melting peaks differed by atout 0.4, 0.8, and 1.8°C from each other at each heating rate. From these results, it was estimated that the temperature difference between the three detectors was less than ±0.1°C at isothermal conditions, and that the temperature signal of any one of the three thermocouples was reliable enough to indicate the temperature of the three.

Fig. 3 shows DTA curves of a polyethylene resin sample. The sample was heated up to 190°C at 5°C/min. in nitrogen atmosphere. Then it was kept at 190°C and when the base line was stable, the atmosphere was switched from nitrogen to oxygen.

The data shows that in analysis of the same sample under the same set of operational conditions, an aluminum cell and a copper cell give data singifficantly different in the length of time before the oxidation-exothermic peak appears.

Though this multichannel DTA apparatus has been developed to be used under a quasi-isothermal operational condition, it can be effectively used for ordinary DTA with a heating rate of 5°C/min. or less (see Fig. 4 which shows the three DTA curves of $CaSO_4 \cdot n\, H_2O$ samples).

## REFERENCES

(1) John B. Howard, Polymers Engineering and Science, 1(6) 429(1973).
(2) A. Yamamoto, K. Yamada, M. Maruta, J. Akiyama, Thermal Analysis 1, 105, (1969).

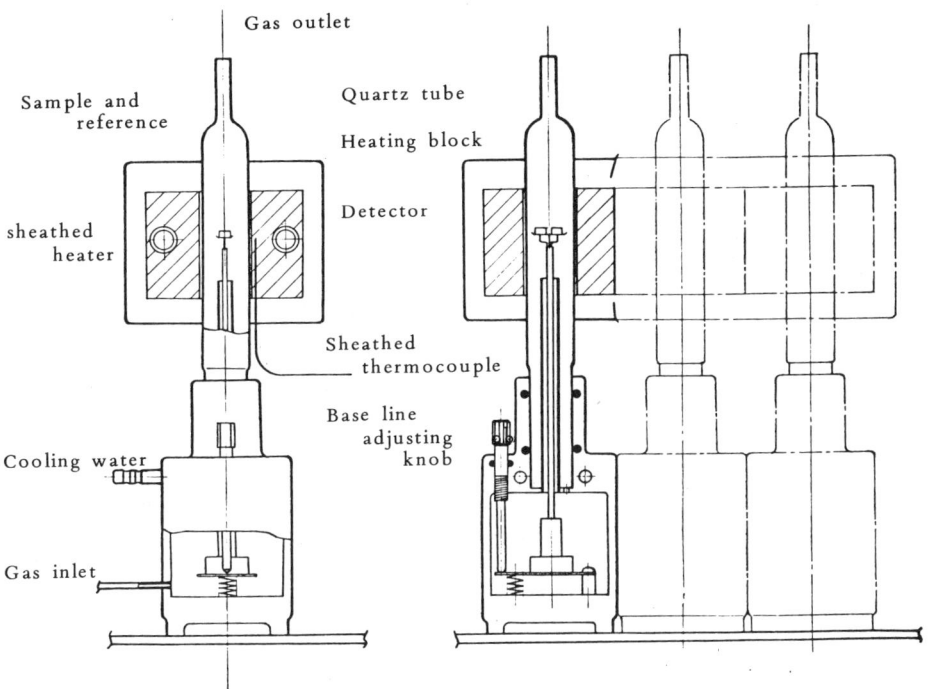

Fig.1  Schematic construction of 3-channel DTA sample holder

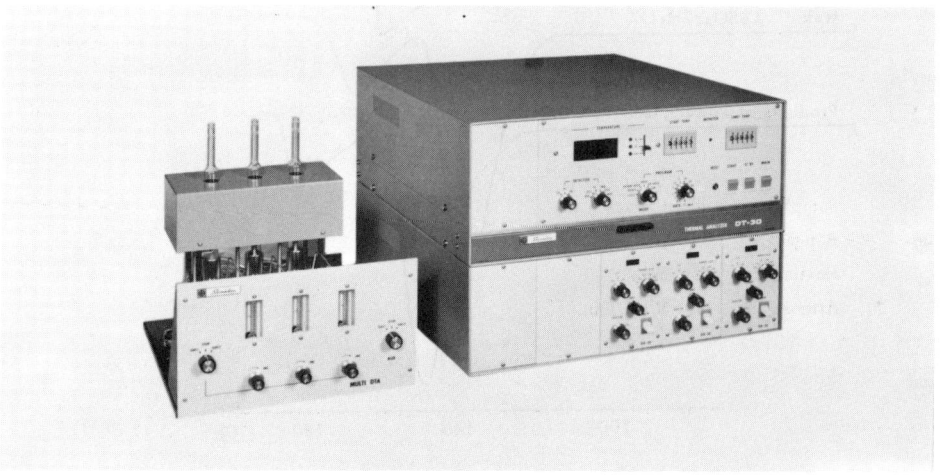

Fig.2  External view of 3-channel DTA apparatus

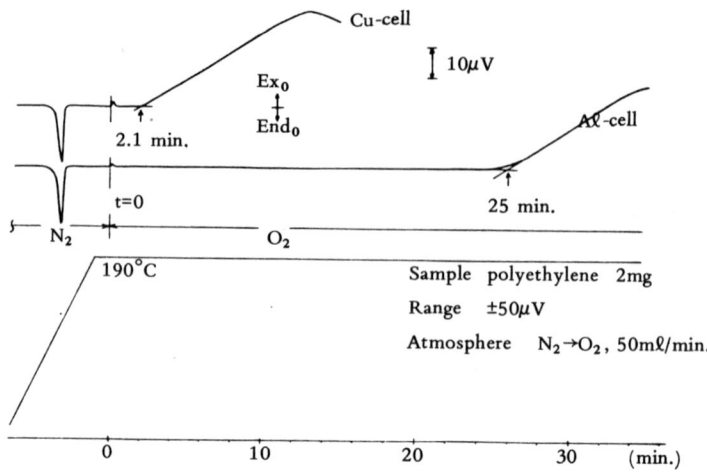

Fig.3 DTA Curves of polyethylene

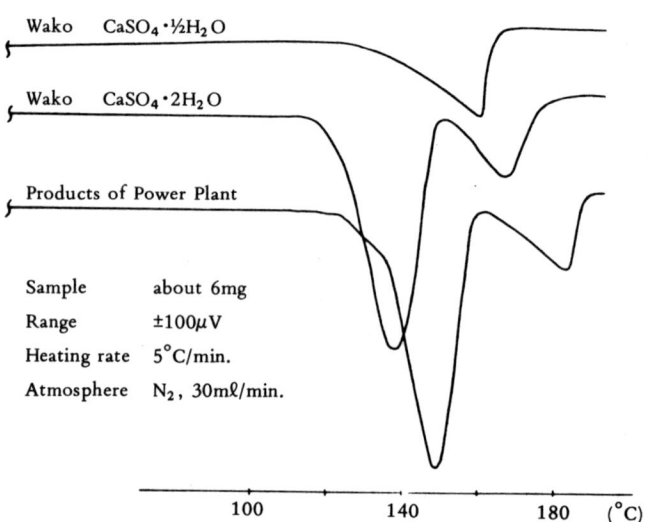

Fig.4  DTA curves with 3-channel DTA apparatus

# A MICROCOMPUTER BASED DTA

R.L. Fyans, J.S. Mayer and W.P. Brennan
Perkin-Elmer Corporation, USA

Most workers in thermal analysis are undoubtedly aware that Perkin-Elmer has been the exclusive manufacturer of thermal null Differential Scanning Calorimeters, which embody our patented power compensation principle, since 1963. Because of their unique construction and operating principle, thermal null DSC's inherently have unprecedented calorimetric accuracy, faster response, and exceptional resolution. However, as a direct consequence of their unique construction, it has not been possible to design a thermal null DSC that operates to high temperatures. For example, the upper temperature limit of the DSC-2C is $725^{\circ}C$. Consequently, at Perkin-Elmer we have been involved in designing a high temperature DTA, which will take full advantage of the available microcomputer technology and greatly help the advance of this important technique. The result of this project is the Model DTA 1700 High Temperature Differential Thermal Analyzer, shown in Figure 1. The Model DTA 1700 is a high temperature microcomputer based DTA which offers a highly sophisticated degree of instrument control and signal conditioning.

The cell of the DTA 1700 consists first of a ceramic support base with three V-grooves. Two of these V-grooves hold the sample and reference cup support tubes, which are hollow and also made of ceramic. These tubes not only support the platinum sample cups but also contain the platinum-platinum 10% rhodium thermocouples which are in good contact with the sample and reference cup liners. The cup liners are of either 60 or 100 cubic millimeter volume and available in both ceramic and platinum. The choice of liner depends on the nature of the sample to be analyzed. The third V-groove holds the purge gas tube.

In the normal mode of operation, the purge gas enters at the bottom of the purge tube and is, therefore, preheated before entering the sample area. This purge system will minimize vertical temperature gradients, which would result in thermal assymetry in the system. Since the

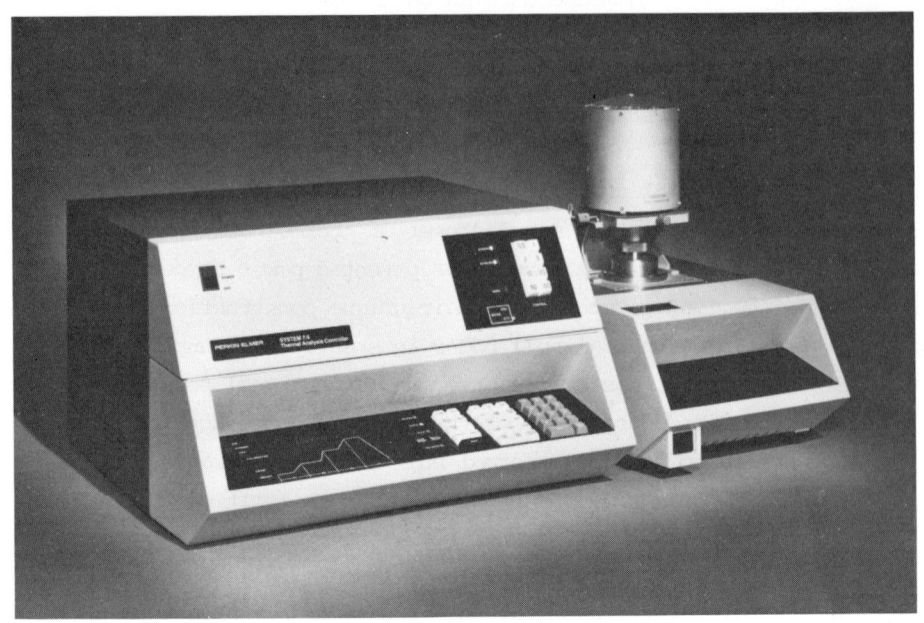

Fig. 1

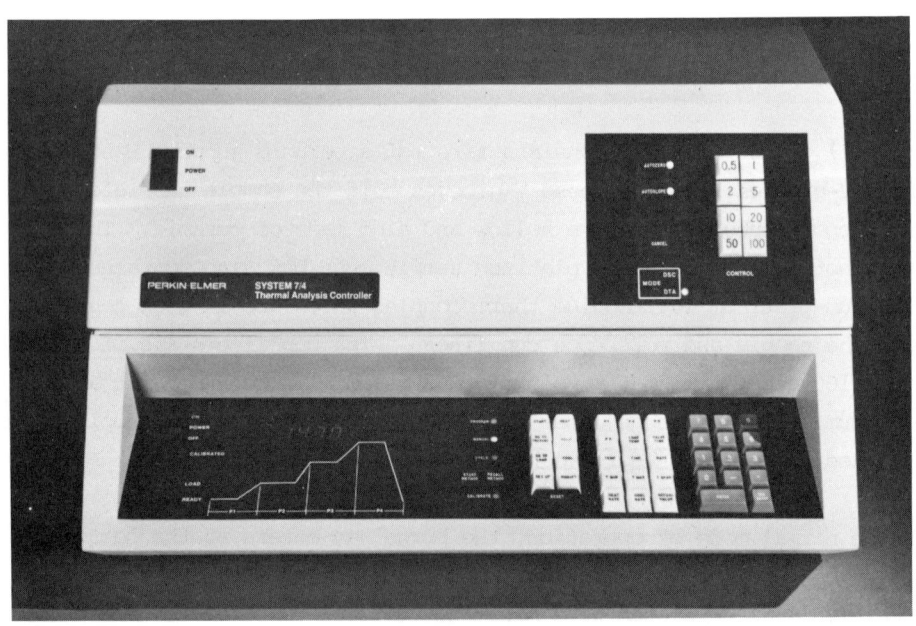

Fig. 2

microcomputer can only correct behavior that is predictable or repeatable, it is absolutely essential that those factors which influence the DTA curve be minimized and controlled. With this purge system, we have been able to vary the purge rate from 30 to 300 cc/min as well as change the flow gas from nitrogen to helium without noticeably affecting baseline performance.

To contain the purge, the cell is covered with a cylindrical, closed top sample tube, which is surrounded by, and well coupled to, the furnace. The furnace itself is demountable and provided with horizontal adjustments which move the furnace closer to one thermocouple or the other, again for baseline optimization. The furnace mandrel is symmetrically wound with a platinum-40% rhodium wire and sealed to provide good thermal coupling. These furnaces routinely operate to 1500°C. The furnace design has provision for an auxiliary purge to facilitate furnace cool down. Cooling from 1500°C back to 100°C is achieved in less than 45 min.

Having designed the DTA 1700 Analyzer with detail to baseline repeatability, its controller, the System 7/4 is capable of automatically performing several signal conditioning functions for optimizing and controlling the baseline. These functions, see Figure 2, include Scanning AutoZero, AutoZero and AutoSlope. The Scanning AutoZero function is initiated by a keyboard code to prevent inadvertant erasure. As the DTA 1700 scans through a temperature range during a calibration baseline run, the microcomputer stores the corrections required to bring the baseline to a zero signal level. At the conclusion of the scan, the microcomputer automatically switches to the run mode and during subsequent scans, applies the stored correction. Each correction accumulated by the System 7/4 corresponds to a specific temperature and is applied prior to the sensitivity attenuation. In this way, the sample can be run at scanning rates and sensitivities other than those used for the calibration scan. In addition, any temperature range within the calibrated range can be scanned.

The AutoZero function is activated by depressing the AutoZero key. If depressed prior to starting the scan, the microcomputer waits one minute, then applies sufficient correction to suppress the ordinate signal to zero. If depressed during the scan, the correction is immediately applied. Likewise, the AutoSlope function is activated by depressing the AutoSlope key. When depressed, the System 7/4 tracks the baseline slope for fifteen seconds and then applies this correction throughout the scan. Again, if the key is depressed prior to starting the scan, a one minute delay is observed before tracking the slope. Both functions can be cancelled by depressing the function key together with the Cancel key.

The microcomputer based in the DTA 1700 also provides linearization of the sample temperature signal as well as the ordinate signal. As shown, ordinate sensitivities are selected by eight keys ranging from 0.5 to 100. The DTA 1700 may be operated as either a conventional DTA or as a heat flux DSC. In the DTA mode, the selected sensitivity is in units of differential temperature ($\Delta T$) for full scale displacement. However, in the heat flux DSC mode, the $\Delta T$ signal is conditioned so that the output is calibrated in units of millicalories per second per inch. Mode selection is made on a rear panel switch and an illuminated lamp indicates mode of operation. This heat flux DSC should not be confused with Perkin-Elmer's exclusive thermal null DSC, the DSC-2C. Since the DSC-2C measures energy directly, it has unprecedented calorimetric accuracy and sensitivity. However, since the temperature dependence of the limiting thermal resistances of the DTA 1700 were made to be repeatable and could be determined, microcomputer optimization was possible. Actually, the quantitative capabilities of the DTA 1700 are quite good.

Figure 3 shows a scan of 16.55 milligrams of pure gold run in the DSC mode. For a series of scans of this type in which both the sample weight and heating rate were changed by at least a factor of four, as well as crystallizing the sample, the calorimetric response changed by less than 5%. This scan also illustrates the performance of the AutoZero and AutoSlope functions. Note that the slope is corrected and the baseline suppressed to recorder zero.

Figure 2 shows the keyboard of the System 7/4. On the right are the numerical entry keys, which are used to initialize or modify program parameters. For initialization an automatic set-up routine leads the operator through a complete set of the essential parameters. Modification of parameters can be accomplished at any time--before, during, or after the analysis. The program parameters are those necessary to direct a complete thermal analysis temperature program from a simple heat-cool program to a multi-step program consisting of several user-selected rates temperatures, and isothermal equilibration times. The next set of keys are those of the commands; for example, "heat", "hold", "cool", "go to the loading temperature", and "start the program". The "modify" function is noteworthy in that any of the program parameters can be changed at any time during the program without disrupting the progress of the program. Hence, scanning rates, equilibration times and temperatures, etc. can be modified even while that segment of the program is being executed.

The black keys at the left allow the user to select from among the operating modes of the microprocessor. These modes are:

First, a manual mode for simple heat-hold or heat-autocool methods.

Second, we have a program mode for more elaborate methods involving equilibration times, program cooling or multi-step programs. This mode provides for total automation from start to finish of a wide variety of programs.

Third, a cycle mode for continuous cycling of any designated program.

Next in this grouping of keys is a function whereby a complete set of values for the program parameters is transferred from the active registers of the microprocessor into memory. Once this set of values has been stored in the microprocessor memory, it is safe from erasure even if the instrument is turned off or there is a power outage. A stored set of parameters can be recalled into the active registers by simple keystrokes. Moreover, methods can be protected from alteration even when in use by activating a hidden keyboard-inhibit switch.

During the running of the DTA 1700 System, the display panel provides the operator with the relevant information on the status of the program. Indicator lights show when the temperature is equilibrated at the loading temperature, and allow the operator to follow each step of the program by illuminating the appropriate segment on a visual representation of the program ramp. There is also a four-place digital display of the actual sample temperature as monitored by the linearization thermocouple. This sample display is used during the set-up routine and during parameter modification to display the parameter being entered. Any number between 0.1 and 2000 may be entered as a parameter (heating rate, temperature, etc.). If the value selected is outside the safety limits of the analyzer, an error light will appear and the value will not be accepted. This allows the flexibility of allowing a wide choice of values for the parameters; but with instrument-specific safeguards against selection of values that would damage the analyzer.

The temperature calibration lamp indicates whether the analyzer has been temperature calibrated. If not, this can be readily done using the self-calibration function of the microprocessor. That is, by depressing the "calibration" key, the System 7/4 goes into a completely automated multi-step calibration routine, which checks and "trims-up" the program voltage to assure that the temperature called for will be accurately obtained.

In conclusion, microprocessor technology has provided an unprecedented opportunity to incorporate highly sophisticated means of instrument control and signal conditioning into a modern differential thermal analyzer. The incorporation of this technology along with an improved DTA cell design should greatly enhance the usefulness of high temperature Differential Thermal Analysis.

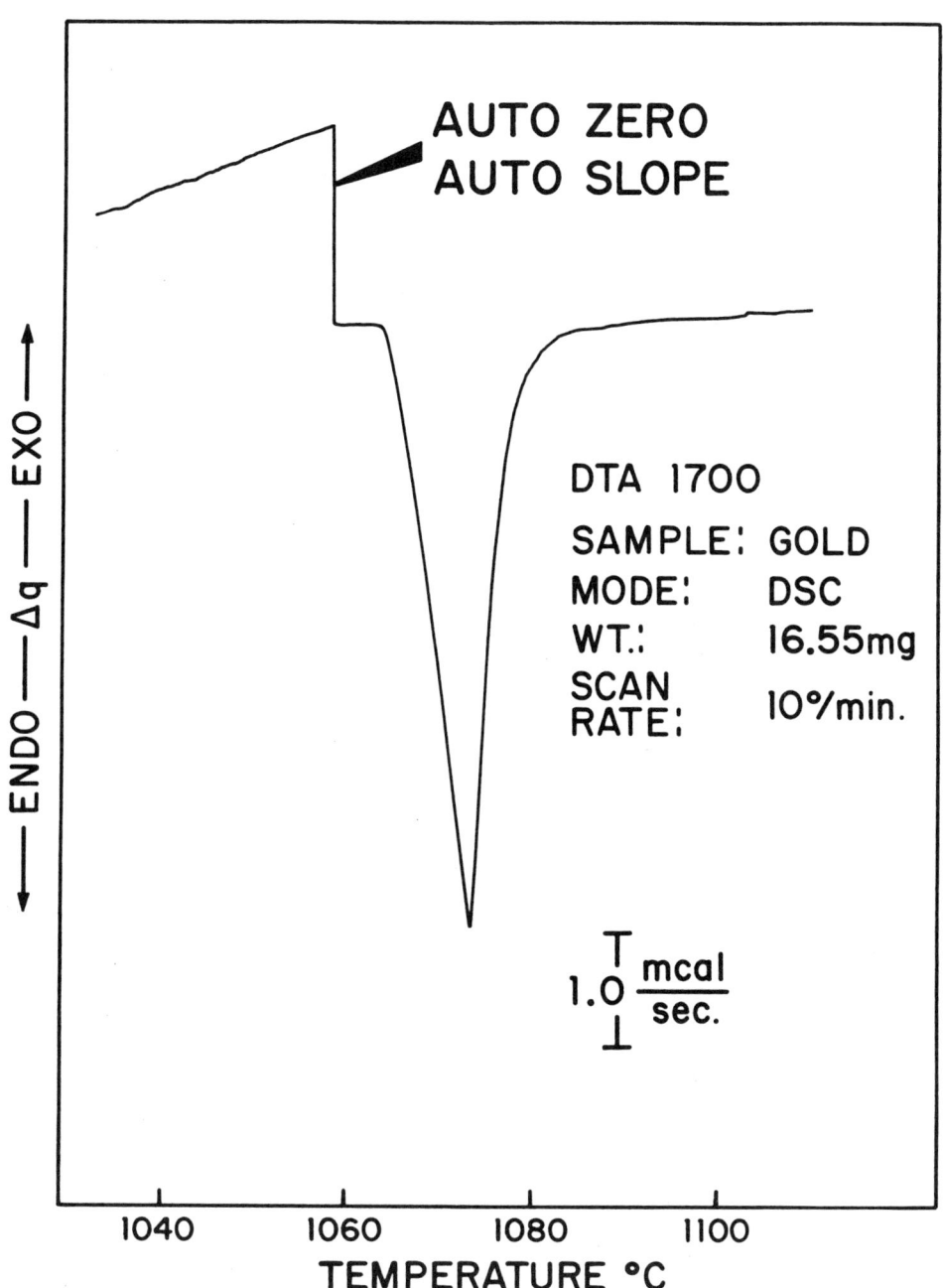

FIGURE 3

# NEW CONTROL MODULE FOR DSC, DTA, TMA AND TGA

E. Wappler, U. Kurpjuweit

W.C. Heraeus GmbH, Hanau, FRG

The HERAEUS Thermal Analysis System TA 500 has been extended by the addition of a new central control module, which permits simultaneous measurements to be made with up to three analyzers, in DSC, TMA and TGA.

The analyzers connected to this module are controlled by a microprocessor capable of handling even very complicated heating and cooling programmes at temperatures from 90 K to 1870 K, including control of cooling, inert, or reaction gases.
Eight programmes can be stored and recalled for routine analysis.
Recording and evaluation of curves are made possible by connection of XY- and Yt-recorders, suitable calculators, or computers.

## INTRODUCTION

Thermal analysis programmes were previously limited to the presetting of a starting temperature, a final temperature, and a linear heating rate. The measuring conditions for standardized routine investigations had to be adjusted for each individual analysis, and multistage analyses, in order to give reproducible results, required a considerable amount of work and time. These analyses necessitated continuous attendance by qualified personnel, and included many sources of error.

To provide a reliable instrument, easy to handle but offering many possibilities of programming, Heraeus has developed the Control and Recording Unit TA 500 SMC 3 (picture 1).

This unit permits simultaneous operation of three analysis modules, such as DSC, TMA and TGA, in any combination. These modules can be controlled according to temperature programmes consisting of a sequence of arbitrary starting and final temperatures as well as heating and cooling rates, including isothermal phases.

The desired values can be fed into the control and recording unit by means of keys resembling those of a calculator, and can be stored if necessary, so that, once established, the measuring conditions can be reproduced safely by pressing a key.

In this way, eight complete investigation programmes can be stored, which may be run as required. In addition to the temperature cycle, the atmosphere and the pressure can be controlled, and recording instruments are switched automatically.

This system, in conjunction with the Automatic Data Processing System TA 500 DV, provides all prerequisites for thermal analysis to take a firm place, in research and control, among the standard methods of analysis.

## DESIGN

The heart of the Control and Recording Unit TA 500 SMC is a mikroprocessor (CPU, Siemens 8085). This module coordinates the interaction of read-only memories (PROM), read-write memories (RAM), indicating module and programme emitter.

The microprocessor compares the actual temperature with the desired value and, if there is a difference, calculates the power input required. The properties of the microprocessor correspond to those of a PID controller.

## FUNCTION

The system described above permits the input, storage and operation of 8 programmes, each of which consists of up to 10 sections. The first section provides for the analysis modules to reach certain defined starting conditions, while all the other sections consist of the heating rate, or cooling rate, final temperature, duration of isothermal phase, and operating position of two relays. The limits within which these parameters may be varied are -180 $^{\circ}$C and +1600 $^{\circ}$C for the temperature, 0.1 and 199.9 K/min for the heating rate, and 0 and 1999 minutes adjustable in steps of 1 min. for the duration of the isothermal phase.

The possibility of forming a loop allows programming to be extended and simplified. This loop can be formed within any one programme. In this way, programmes, particularly those requiring repeated heating and cooling between two temperature limits, can be run most conveniently.

Two relays, to which any types of accessory units may be connected, serve to automate an investigation. In this way, programmed control of protective gases, or cooling gases, as well as vacuum pumps, recorders and other instruments is made possible.

When used with an accessory unit containing two power units, the Control and Recording Unit TA 500 SMC permits connection of three analysis modules. The control and recording unit itself incorporates one power unit for control of an analyzer. Moreover, the values measured by the analyzer, namely sample temperature and $\Delta T$-, or $\Delta L$-, or $\Delta G$-signal, are prepared for further processing, through linearization of the sample temperature and amplification of all signals. The signals are available at the connections for recorders and data processing systems.

The data of the two additional analyzers are processed in the accessory unit itself, which, too, has all necessary connections for recorders and calculators.

## HANDLING

The desired programme parameters are fed into the control and recording unit by means of alphanumeric keys. After a specific programme has been chosen by pressing the corresponding key (0 to 7), programming may begin with the first section, section 0. The light-emitting diode placed next to "Temperatur" lights up, i.e., the microprocessor interprets the numbers on the keys pressed next as the temperature which the analyzers are to reach by the beginning of the programme.

After the corresponding number has been fed into the control and recording unit, the indication of the set of light-emitting diodes changes to "Relais 1 0/1". The operator may now make his choice as to whether relay 1 is to be open during the run-in phase, or whether it is to be closed. After this choice has been made also in respect of relay 2, the microprocessor proceeds to the next section.

After the starting conditions for the reaction have been fixed, the programme proper is worked out. The microprocessor waits for information on the desired heating rate, or cooling rate, the final temperature, the duration of the isothermal phase, and the desired operating positions of the two relays, indicating, through the light-emitting diodes, when it wishes to receive the particular information. All the sections required are fed into the control and recording unit according to this pattern. The programme starts automatically when the key marked "Start" is pressed.

## EXAMPLES

The following example is to elucidate the way in which the Control and Recording Unit TA 500 SMC, with its many possibilities of programming, enables analyses to be carried out which were almost impracticable before.

For gelatinization of a PVC copolymer, it is sought to establish such optimal conditions as would render the material as resistant to disintegration by oxidation as possible. The glass transition temperature is to serve as a later criterion for the quality of the copolymer.

For this purpose, the process of gelatinization is carried out, and is recorded together with the glass transition temperature of the non-gelatinized product. In a second run, the glass transition temperature of the gelatinized product is recorded and the material tested for complete gelatinization. In the third run, the copolymer is heated in air. Again, the glass transition temperature is determined, and is compared with the values from the second run. The material is heated to 500 $^\circ$C, and the changes occurring due to disintegration by oxidation and temperature are recorded.

To solve this problem, 8 analyses were previously necessary:

1. Cool the non-gelatinized polymer to a temperature below glass transition temperature, Tg
2. Heat to gelatinizing temperature in an inert gas atmosphere
3. Hold at gelatinizing temperature
4. Cool to a temperature below glass transition temperature, Tg
5. Heat, in an inert gas atmosphere, to a temperature above gelatinizing temperature
6. Cool to a temperature below glass transition temperature, Tg
7. Heat to 500 °C in air
8. Hold at constant temperature for 20 minutes.

As, previously, each of these steps had to be specially set and the times checked by means of a watch, a high degree of concentration and accuracy was necessary to obtain reproducible results.

On the other hand, with the Control and Recording Unit TA 500 SMC, a corresponding programme, including control of recording, time of isothermal phases, and atmosphere, needs to be fed only once, and can then be run any number of times even by a semi-skilled operator, the accuracy remaining constant.

The programme for the investigation of the process of gelatinization consists of 5 sections. The sequence of the various steps of this programme is shown on picture 2, the corresponding parameters are given in the table 1, at the end of this paper.

Further examples where the Control and Recording Unit TA 500 SMC simplifies the investigation:

- Determination of the degree of brittleness of rubber, by dynamic elasticity measurement in a fatigue test involving several temperature cycles. The coolant and the DEM accessory unit are controlled directly by the control and recording unit.
- Routine determination of the purity of raw materials for drugs. A special programme, with different values for the heating rate and temperature range, is assigned to each of the various basic products. The evaluation takes place automatically, through a Data Processing System TA 500 DV and a calculator. The analysis can be carried out by semi-skilled personnel.

- Automatic goods entry control. A particular measuring programme is assigned to each product. The programme runs automatically, and the results are recorded by a calculator. The calculator compares the measurement with the results obtained from a former sample of the particular material. In the event of the permissible tolerances being exceede, notice is given automatically and the curve measured is plotted.

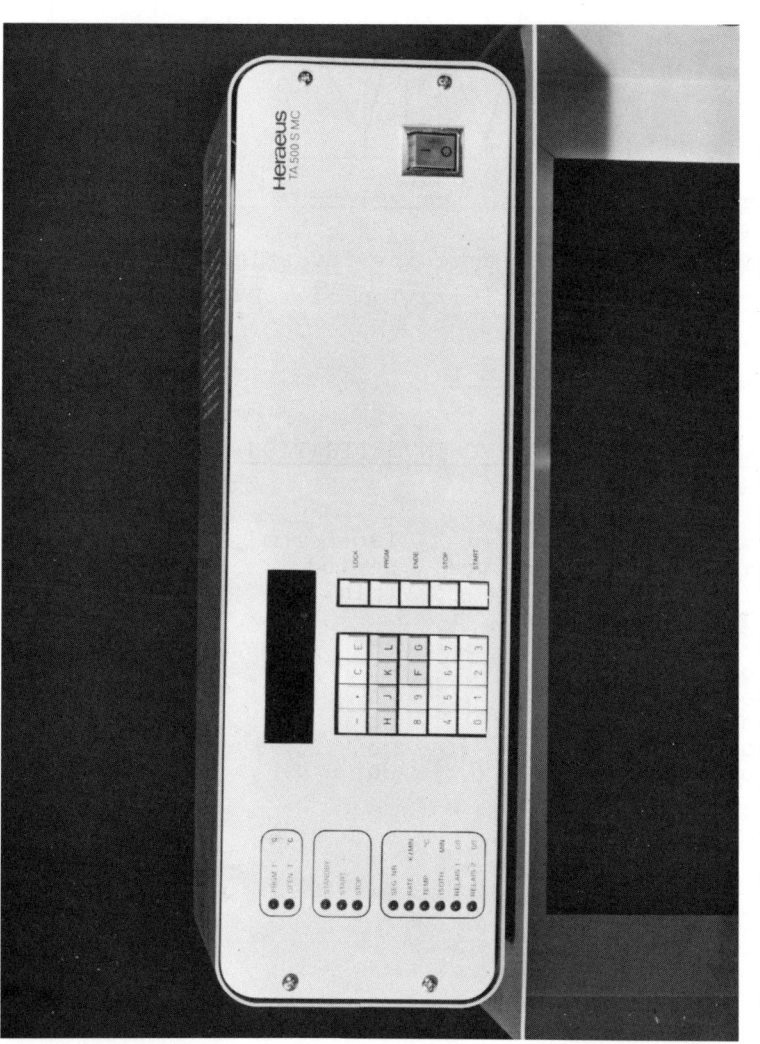

Heraeus Control Module TA 500 SMC 3    Picture 1

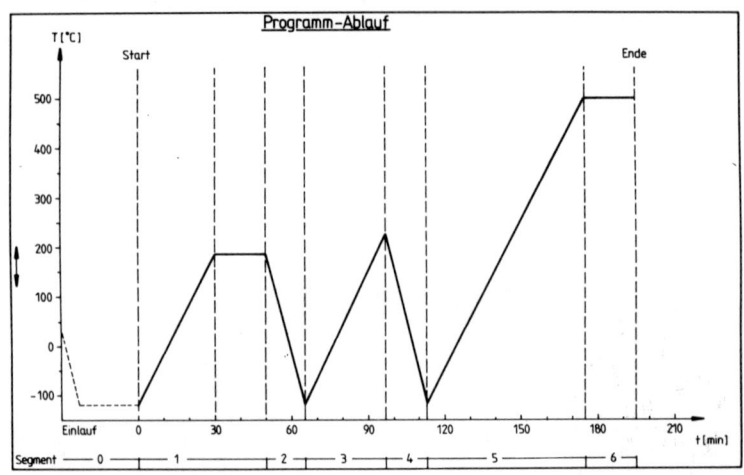

Picture 2  Programme for investigation of the gelatinization of a PVC copolymer

## TABLE 1

### PARAMETERS FOR RUNNING PVC INVESTIGATION

| Segment Nr. | Heating rate (K/min.) | Final temperature (°C) | Isothermal period (min.) | Relay 1 | 2 |
|---|---|---|---|---|---|
| 0 | -- | -120 | -- | 1 | 1 |
| 1 | 10 | 180 | 20 | 0 | 1 |
| 2 | 20 | -120 | 0 | 1 | 1 |
| 3 | 10 | 200 | 0 | 0 | 1 |
| 4 | 20 | -120 | 0 | 1 | 1 |
| 5 | 10 | 500 | 20 | 0 | 0 |

# A NEW INSTRUMENT FOR SIMULTANEOUS DIFFERENTIAL THERMAL ANALYSIS AND THERMOMICROSCOPY

W. Perron [1], G. Bayer [2] and H.G. Wiedemann [1]

1) Mettler Instrumente AG, CH-8606 Greifensee
2) Institute for Crystallography and Petrography, ETH, Zurich, Switzerland

## ABSTRACT

The correlation between morphological or structural changes and DTA-recordings can give valuable additional information on the mechanism of phase transformations. This fact gave the impulse for the development of a new instrument for simultaneous Differential Thermal Analysis (DTA) and Thermomicroscopy in transmitted light, where a DTA device is placed into a commercially available hot stage. A synopsis of the versatile application is given including the measurement of binary phase diagrams, the study of melting processes and nucleation in crystallisation.

## INTRODUCTION

Many methods of chemical analysis are unspecific and cover a broad field of application. Most of the thermoanalytical methods are unspecific e.g. Differential Thermal Analysis (DTA) and Differential Scanning Calorimetry (DSC). In Thermal Analysis the results, even if they are reproducible and quantitative, cannot always readily be correlated with the respective phase transformation or chemical reaction occuring in the substance. Scientists as well as instrument manufacturers have always tried to improve thermoanalytical methods by achieving a better correlation between results and physical or chemical changes of the substances beeing ivestigated.

Combination of methods has proved to be most successful. When performed simultaneously with DTA or DSC, Thermogravimetric Analysis (TG) is a very powerful tool for the study of chemical stability and decomposition of organics and

polymers. It is the purpose of this paper to demonstrate
the usefulness of simultaneous DTA and hot-stage-micro-
scopy. Effects which don't show any significant enthalpy
change in DTA often can be detected with microscopy. On
the other hand thermooptical investigations do not include
means to even estimate heats of transformation, fusion or
crystallisation.

### INSTRUMENTAL

The simultaneous measurement (Fig. 1) is based on a commercial microscope hot stage, the Mettler FP (FP5-Control
unit, FP52-Hot stage). The DTA-accessory can be added

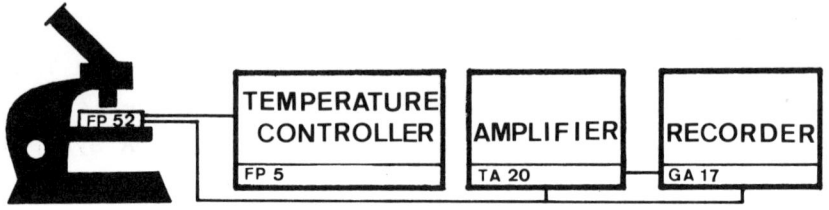

Fig. 1    Configuration of DTA-Thermomicroscopy


optionally to every Mettler hot stage FP5/52. Fig. 2 shows
the microscope hot stage and Fig. 3 the DTA-accessory
which has about the size of a microscope sample slide
and is mounted in it's place. The DTA-sensor is part
of it and consists of a special glass disc with thin-film-
deposited 5-fold thermopiles. The output of the thermopile
is amplified and recorded on a line recorder. The sapphire
crucible which contains the sample is positioned on the
DTA sensor together with an empty reference crucible.
Measurements are possible at constant heating or cooling
rate or isothermal. Microscopic observations can be made
in transmitting light. Simultaneously, changes of the
enthalpy of the same sample are recorded automatically.
Instrument specifications are as follows:

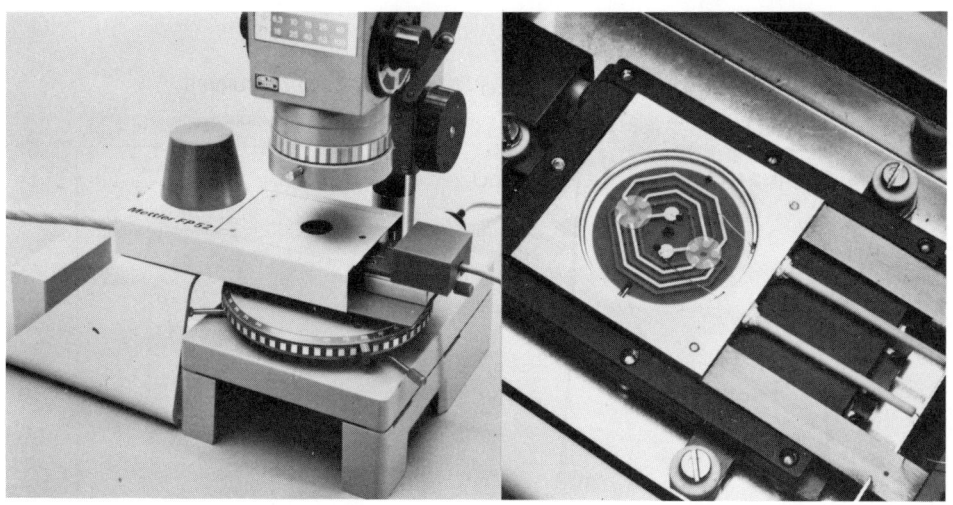

Fig. 2    Hot stage           Fig. 3    DTA accessory

| | |
|---|---|
| Temperature range | RT ... + 300°C |
| Temperature accuracy | ± 0,3°C |
| DTA-sensitivity | approx. 18 µV/mW |
| DTA-reproducibility | approx. ± 5% |
| Time constant of signal | approx. 5 s |

For high precision DTA-measurements the instrument can be calibrated, the microscope lamp should either be switched off or its influence on the DTA-sensor should be reduced using a heat filter.

## EXPERIMENTAL

From the variety of significant applications for DTA-Thermomicroscopy two examples will be discussed in the following:

### Phase diagram $KNO_3$-$NaNO_3$

According to previous micro-thermoanalytical investigations by A. Kofler (1) this system shows the formation of a continuous series of solid solutions. A special characteristic of these solid solutions however are peri-

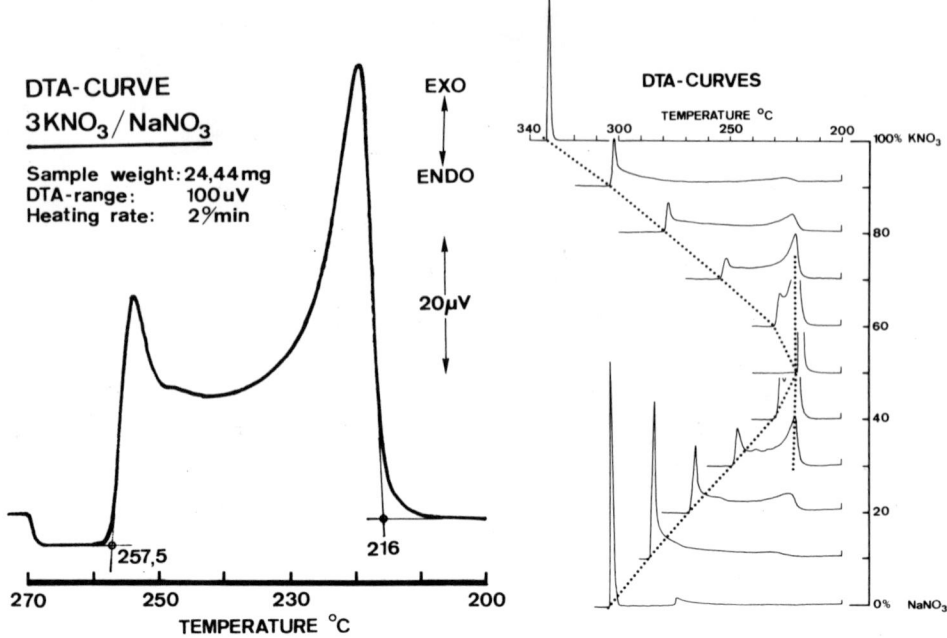

Fig. 4   DTA cooling curve         Fig. 5   Phase diagram

odical seggregation phenomena which occur during heating and cooling. In own experiments mixtures of $KNO_3$ and $NaNO_3$ were prepared in 5 mole % steps. These were molten and the corresponding cooling curves where recorded afterwards with DTA. Simultaneously the crystallization process was observed microscopically with the hot stage. As an example Fig. 4 shows the DTA cooling curve of a mixture 3 $KNO_3$/1 $NaNO_3$. The onset-point of the primary crystallization was 257.5°C. This is in good agreement with the microscopic observation, which shows the nucleation of the first crystals at 258°C (Fig. 6). The number of crystal nuclei increases up to about 248°C, further cooling

|   258°C   |   248°C   |   240°C   |   216°C   |

Fig. 6   Microphotographs of the crystallisation (160x) of 3 $KNO_3$ : 1 $NaNO_3$

to 240°C results in pronounced crystal growth with complete eutectic solidification at 216°C. Fig. 4 and Fig. 6 prove that the correlation between the DTA-curve and the microscopically observed crystallization phenomena is very good. The liquidus curve and the eutectic line for the system $KNO_3$-$NaNO_3$ as determined from DTA-curves are shown in Fig. 5. These results confirm that this system is a simple eutectic system. It should be pointed out that it is relatively difficult however to determine the exact temperature of the eutetic onset points for mixtures with low concentrations ( 5 mole %) of $KNO_3$ or $NaNO_3$ respectively. In these cases heating X-ray photographs proved to be very useful.

## Phase transitions of pure caffeine (2)

This application has been selected to demonstrate the practical value of the simultaneous DTA-Thermomicroscopy where both methods are mutually complementary. A DTA-curve of caffeine is shown in Fig. 7. The peak at 141°C corresponds to the first-order transition of waterfree $\beta$-caffeine into the $\alpha$-form. Melting occurs at 236,5° Both phase changes are superimposed by a drift of the

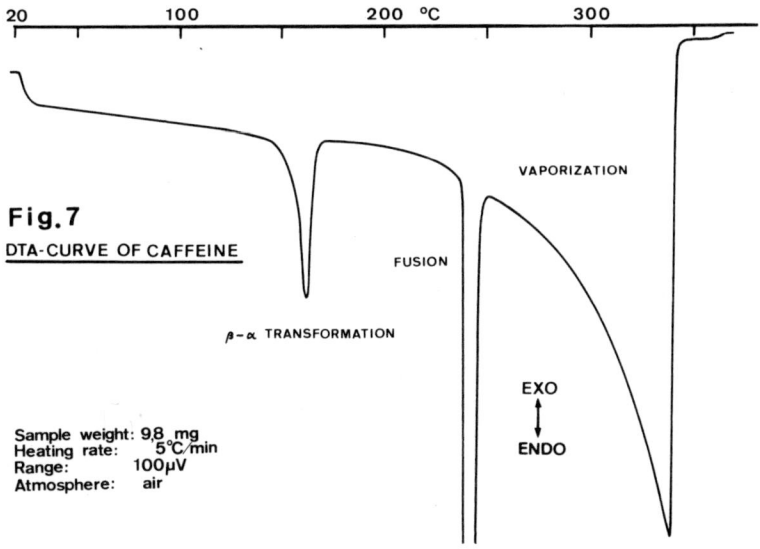

**Fig.7**
DTA-CURVE OF CAFFEINE

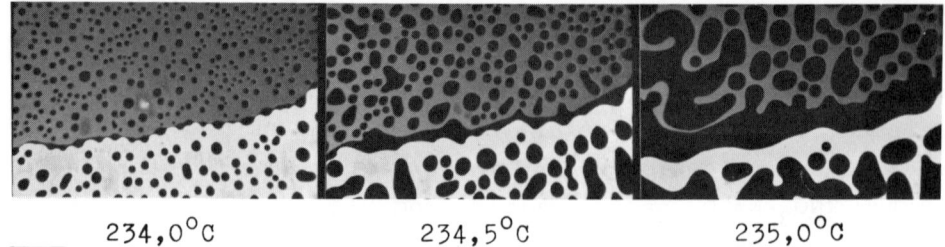

| 234,0°C | 234,5°C | 235,0°C |

Fig. 8 Microphotographs of melting caffeine (160x)

DTA-signal, which cannot be identified clearly by DTA only (evaporation, decomposition). Fig.8 reproduces the corresponding observations with the microscope at increasing temperatures. The phase transition clearly seen on the DTA trace cannot be detected with Thermomicroscopy but now the drift on the DTA-signal can be related to the evaporation of caffeine. The high vapor pressure of caffeine at it's melting temperature is mainly responsible for the bubble-formation.

## REFERENCES

(1) A. Kofler, Mh. Chem., Bd. 86, H. 4 (1955) 643-652
(2) H. Bothe and H. K. Cammenga, J. Thermal Anal. 16 (1979) 267-275

# TRANSMITTED LIGHT HOT STAGE MICROSCOPY IN THE TEMPERATURE RANGE −180°C TO 600°C

E. L. Charsley, A. C. F. Kamp and Jennifer A. Rumsey
Stanton Redcroft, Copper Mill Lane, London SW17 0BN, England

## ABSTRACT

A hot stage (Model TLHS) is described which allows direct observation of a sample by transmitted light, as it is heated or cooled at a controlled rate over the range from −180°C to 600°C. The sample can be viewed in normal or polarised light and the intensity of the transmitted light measured using a photocell and displayed on a recorder together with the linearised sample temperature. The performance of the unit is illustrated by reference to ICTA-NBS Temperature Standards.

## INTRODUCTION

The technique of transmitted light hot stage microscopy has been described in detail in the works of the Koflers(1) and of McCrone(2). The value of the technique to thermal analysts was considerably extended by adapting the apparatus to carry out depolarised light intensity measurements. This method, first applied to polymer systems by Magill in 1961(3), detects transitions in birefringent materials, by measuring the intensity of the polarised light transmitted by the sample as a function of sample temperature as the sample is heated or cooled at a linear rate. The technique has been shown to be complementary to DTA(4). The present work describes a hot stage which can be used over a much wider temperature range than conventional Kofler type units, under conditions of both controlled heating rate and atmosphere.

## DESCRIPTION OF THE APPARATUS

### The Hot Stage Unit

A cross-section of the hot stage is shown in Fig. 1. The body of the unit is made from anodised aluminium and is water-cooled. The sample, normally between two cover glasses, is heated by means of a silver block containing a nichrome heating element. Continuity of the heating surface and hence reduction of temperature gradients to a minimum is ensured by means of a sapphire window approximately 2mm in diameter which allows the sample to be viewed by means of transmitted light. The sample temperature is measured by a platinum resistor located in the block immediately adjacent to the sample. The resistor is also used to control the temperature programme.

Access to the sample chamber is by means of a removable lid housing a double window. Dry gas is passed through the latter at approximately 100ml/min to prevent condensation at sub-ambient temperatures and to protect the microscope optics at higher temperatures. The lid can be rotated while maintaining a gas-tight seal, thus if the viewing area above the sample becomes obscured by volatile products, a clean section of window can be readily obtained. Runs may be carried out under controlled atmosphere conditions by introduction of the desired gas via the connections shown. Flow rates of 10 to 25ml/min are normally used.

Using a Stanton Redcroft Model UTP/PR temperature programmer, switch selected heating rates of between 0.1 and 99°C/min are available. The programmer displays the sample temperature in °C in digital form and provides a linearised temperature output for a potentiometric recorder. A sensitivity of 100°C for full scale recorder deflection is normally used giving a temperature resolution of 0.2°C. In addition to heating and cooling, facilities are available for isotherming the sample at a preset temperature and for cycling between two set temperatures. Sub-ambient work is carried out using liquid nitrogen as a coolant. Using a controlled flow of the latter in conjunction with the temperature programmer enables linear heating or cooling rates to be obtained over the entire range of the apparatus.

A typical heating curve showing the wide operating range of the instrument is shown in Fig. 3. The low mass of the system enables rapid cooling rates to be obtained and these are markedly increased by the use of liquid nitrogen, thus the sample can be cooled from ambient to -100°C in less than 30 seconds. Equally, isothermal temperatures can readily be established and held to a high level of stability, e.g. within $\pm$ 0.2°C at 300°C for one hour.

## The Microscope Assembly

A microscope with a minimum working distance of 5mm and condenser focal length of at least 20mm is required for use with the hot stage (a model working with 16mm condenser lenses will be available shortly). In the present work an Olympus Series BHA polarising microscope was used with a 10X objective giving a total magnification of 100X. A photograph showing the hot stage in position on the microscope is shown in Fig. 2. Correct orientation of the sapphire window of the hot stage is ensured by working with crossed polars and rotating the hot stage until complete extinction is obtained. The standard microscope illumination system, fitted with a constant voltage transformer was used. The intensity of the transmitted light was measured using an IPL 16B silicon photo-detector, fitted with filters to approximate the human eye response. The detector was fitted in the camera tube of the trinocular head of the microscope and the output fed directly to one channel of a two pen recorder, the other channel being used to measure sample temperature. Photographic work was carried out using an Olympus PM-10A automatic 35mm system.

## APPLICATIONS

Samples are normally run between two cover slips, only a few milli-grams of material being required. Although solids can be run as crushed powders, superior results are obtained by using pre-melted samples. Liquids can also be run using a shallow dish made from a cover slip usually with a second slip as a lid to cut down evaporation. With volatile liquids it is necessary to avoid a fast flow of purge gas through the sample chamber and for sub-ambient work silica gel can be placed in the sample chamber to help reduce condensation effects.

For depolarised light intensity measurements samples were viewed using crossed polars so that the only light transmitted is due to rotation of the polarised light by the crystalline structure of the sample. Thus changes in the structure on heating will result in changes in the recorded transmitted intensity and melting of the sample (or formation of a cubic phase) will result in extinction of the light. Samples can also be viewed using white light by removing the analyser from the light path. Melting can be readily observed in this way.

Two examples of the depolarised light intensity technique are shown in Figs. 4 and 5, using ICTA-NBS Temperature Standards. Fig. 4 shows the melting and freezing of 1,2-dichloroethane. On heating the first change in the light intensity is seen in the region of $-38°C$. The major change took place in the range $-34.3°C$ to $-33.5°C$ compared with the mean ICTA values $35.7 \pm 2.0°C$ and $-31.4 \pm 4.3°C$ for the extrapolated onset and peak temperatures respectively. The curve also shows the magnitude of the supercooling given by this material, which makes it unsuitable as a standard in the cooling mode.

Fig. 5 shows the curve for a pre-melted sample of potassium nitrate, showing the phase change in the region of $128°C$, which is used as a temperature standard. This is followed by melting of the salt in the region of $335°C$ leading to complete extinction of the polarised light.

## REFERENCES

(1) L. Kofler and A. Kofler, Thermo-Mikromethoden, Verlag Wagner, Innsbruck (1954).
(2) W. McCrone, Fusion Methods in Chemical Microscopy, Interscience, New York (1957).
(3) J. H. Magill, Polymer, 2, 221 (1961).
(4) E. M. Barrall II, and E. J. Gallegos, J. Polymer Sci., A2, 5, 113 (1967).

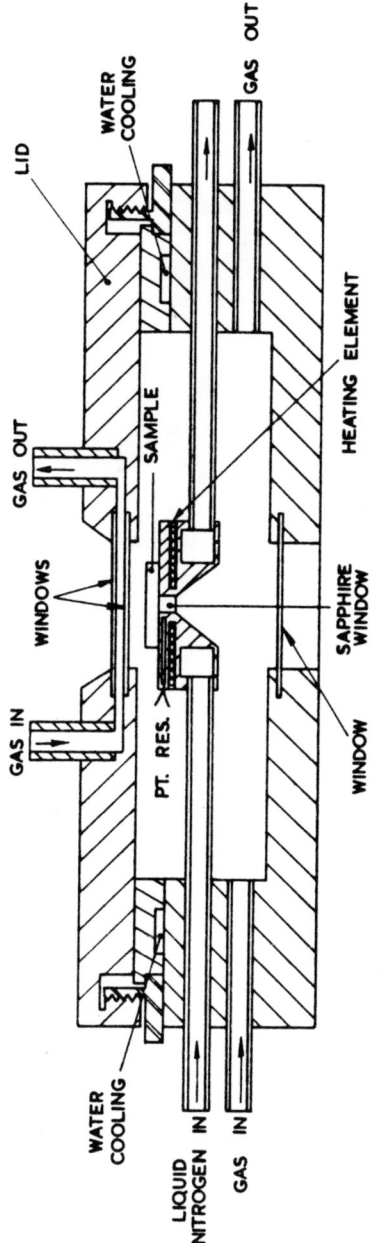

FIG. 1   Cross-section of hot stage unit

FIG. 2    Close up of hot stage

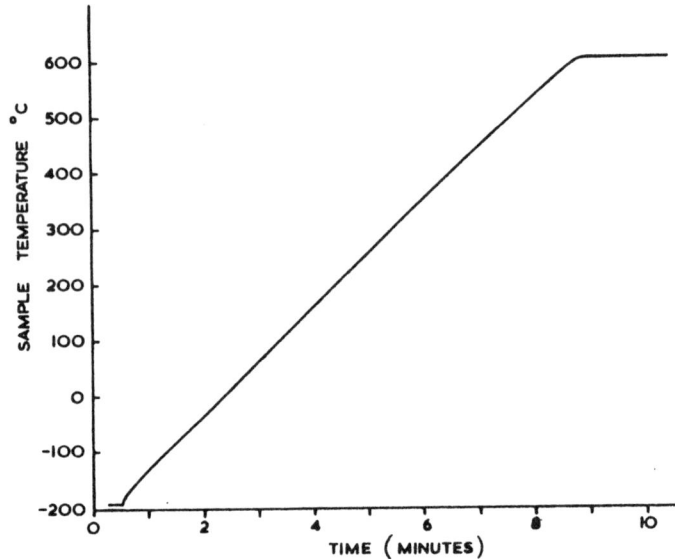

FIG. 3    Heating rate curve, 99°C/minute

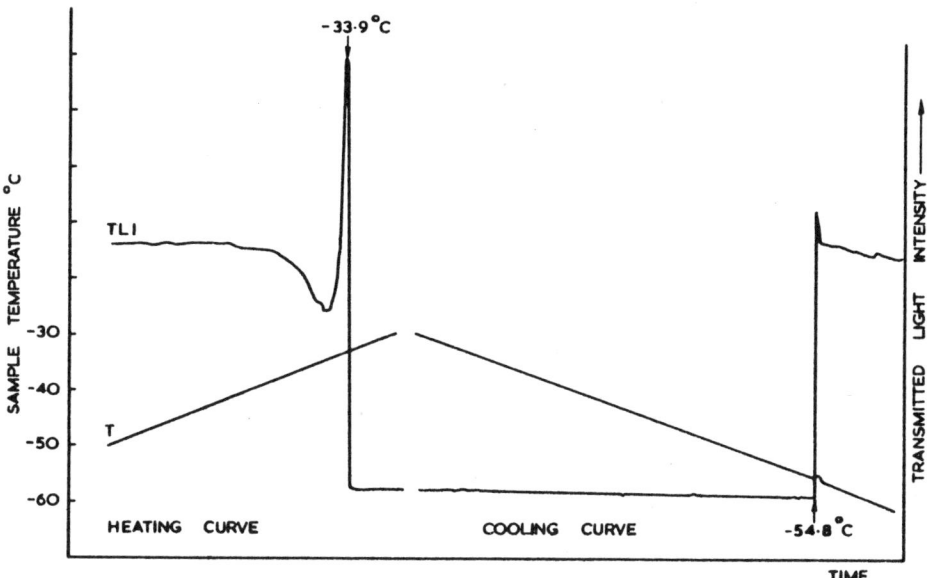

FIG. 4    Transmitted light intensity curve for
1.2-dichloroethane at 3°C/minute

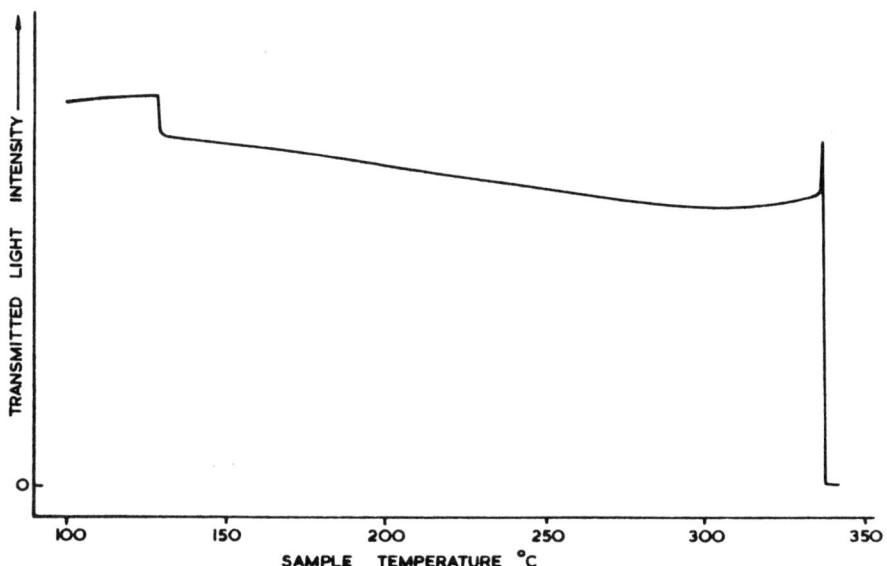

FIG. 5    Transmitted light intensity curve for
potassium nitrate at 10°C/minute

# APPLICATION OF AREA-THERMOCOUPLES IN THE THERMAL ANALYSIS OF AMORPHOUS THIN FILMS

W. Ludwig
Department of Chemistry, Friedrich Schiller University
69 Jena, DDR

## ABSTRACT

A technique is described in which the thermal behaviour of original thin layers was studied. By sputtering metallic films in the combinations Ni/Cr, Ni/Cu, Ni/Ag, Ni/Au, Cr/Cu and Cr/Ag on glass-substrates area-thermocouples were produced. Inspite of good thermocurrents they have a low mechanical stability. Stable thermocouples were prepared by electrolytic separation of Ni or Cr onto thin copper plates. After passivation by sputtering with SiO, the amorphous layer was sputtered in the diameters of 20 x 20 mm. The thermal investigation was undertaken in a special measuring bar by application of a high sensitivity DTA-arrangement under inert conditions.

Thin films of Se, $As_2Se_3$ and $Ge_2Se_3$ were investigated and compared with bulk melt samples or separated amorphous layers. Glass transition temperatures of the thin films are increased in comparison to glasses prepared by quenching melts. The rate of crystallization decreases significantly.

Comparative electrical measurements are included in the discussions.

## INTRODUCTION

Up till now the thermal characterization of vitreous systems by application of DTA has nearly only been carried out on bulk or powdered samples.

By the application of vapour deposition new possibilities are given for producing amorphous thin layers from substances, which are usually found as solid substances only in crystalline state.

With the application of the thin film technology for the

production of amorphous and crystalline opto-electronic devices, besides a series of another methods of investigation (1), experiments for the thermal characterization of thin films have been carried out.

However, in most cases, thin films were removed from the substrates or the foil with thin films was cut to pieces followed by thermal investigation (2,3,4). Important for a well defined assertion on the stability of crystallization of thin films is the thermal investigation of the original thin layer.

It is also of great interest to study the thermal behaviour of these amorphous thin layers in comparison to classically produced vitreous substances, that is by quenching a melt of the same composition (1,5).

## EXPERIMENTAL PROCEDURE

Thermocouples - By the areal shape of the specimens new methods had to be found for the sampleholder and the measuring of temperatures. Therefore it was necessary to produce area-thermocouples for separating amorphous layers on their surfaces.

Namba (6) prepared thermocouples in the combinations Au/Ni and Ag/Ni by sputtering the metallic films and realized layers of 500 Å.

In our experiments (7) metallic films in the combinations Ni/Cr, Ni/Cu, Ni/Ag, Ni/Au Cr/Cu and Cr/Ag were sputtered in a high vacuum-apparatus 'B 55' on glass-substrates in the diameters of 27 x 30 mm. The contact-area between the metals in each case was 20 x 20 mm. At the thickness of the layers between 100 and 500 Å these metallic combinations showed the following thermo-

| Thermo-couple | Thermo-current (mV) | Layer (Å) |
|---|---|---|
| Ni - Cr | 0,6 | Ni:100 Cr:280 |
| Ni - Cu | 0,75 | Ni:110 Cu:250 |
| Ni - Ag | 0,8 | Ni:110 Ag:500 |
| Ni - Au | 0,7 | Ni:110 Au:500 |
| Cr - Cu | 0,05 | Cr:280 Cu:300 |
| Cr - Ag | 0,1 | Cr:280 Ag:400 |

Table 1 : Thermocurrents of area-thermocouples

currents (Tab.1) in the temperatur range of 20-80°C. Further experiments were carried out with the combination Ni/Cu.

In spite of good thermocurrents the combinations Ni/Cu and Ni/Ag on the glass-substrates have a low mechanical stability. Stable area-thermocouples were prepared by electrolytic separation of Ni onto thin copper plates in the described diameters.

Fig. 1 shows the temperature dependence of the thermocurrent for a thermocouple in the combination Ni/Cu. The temperature calibration was performed using temperature standards $NH_4NO_3$, $KNO_3$ and $AgNO_3$. On this occasion the thermocouples were covered with the water solutions of these salts. After evaporation and carefully drying, temperature calibration is followed by DTA. In addition the melting temperatures of the metallic films of In and Sn were measured.

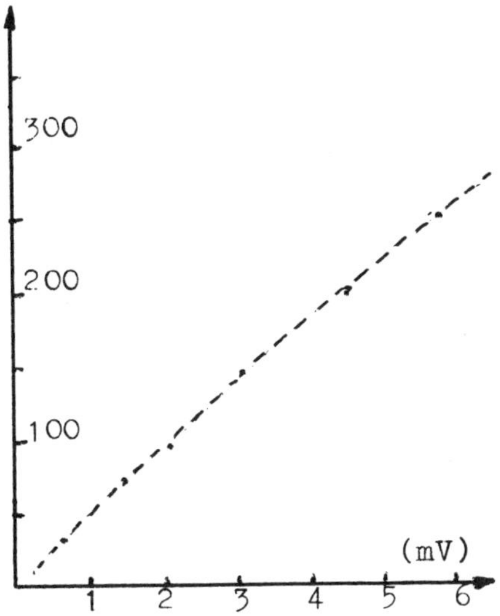

Fig. 1 Temperature calibration

<u>Differential Thermal Analysis</u> - The thermal investigation was undertaken in a special measuring bar shown in Fig.2. The sampleholder consists of two cylindrical aluminium bars (A,B) connected by a ring of aluminium (C). The thermocouples are symmetrically placed (D,E) in the sampleholder. The electrical contact was realized by in glass melt copper and nickel pins pressed by springs onto the thermoplates. The whole sampleholder was closed with a glass-shade

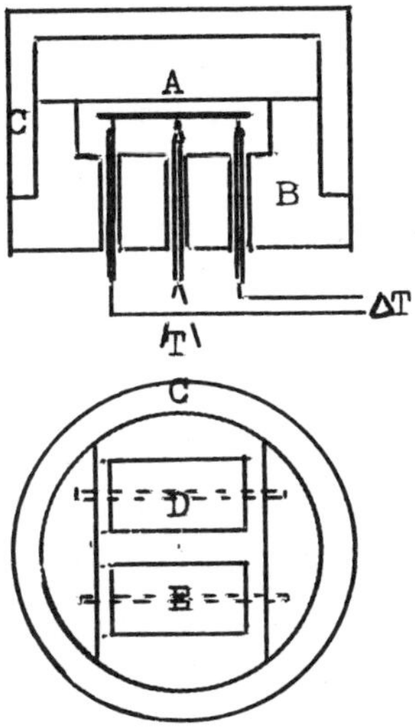

Fig. 2 Sampleholder

and could be evacuated and filled with dry nitrogen. The whole arrangement was surroundet by the furnace.

By using an amplifier of type F 116/1 (USSR) and a X,Y-recorder of type 'Endim 200' (GDR) the sensitivity of this DTA arrangement was 25 $\mu V$ $grd^{-1}$ (0.06 grd $cm^{-1}$ deflection). The heating rate in each cases was 4 K $min^{-1}$.

Passivation of the thermocouples - The reactivity of the investigated chalcogenides with the metals of the thermocouples also requires an adequate protection. After passivation by sputtering with SiO, the amorphous layers of 25-30 $\mu m$ of Se, $As_2Se_3$ and $Ge_2Se_3$ were sputtered in the above described diameters.

## RESULTS AND DISCUSSION

On the basis of the thermoanalytical arrangement it can be stated, that the original layer of Se, $As_2Se_3$ and $Ge_2Se_3$ under the given experimental conditions, in nitrogen, could be excellently characterized.

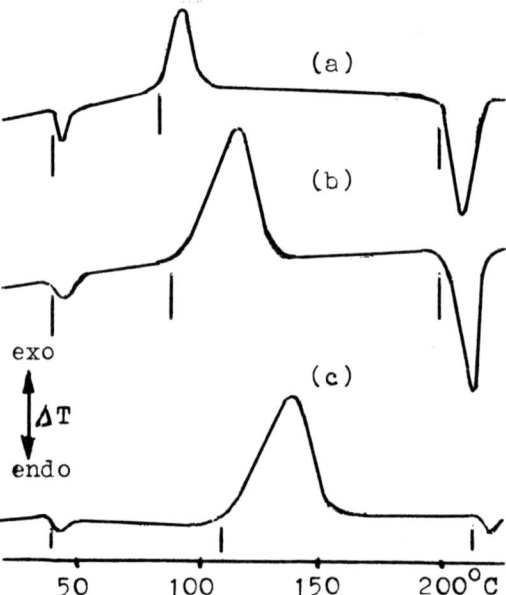

Fig.3 DTA curve of selenium, (a) thin film, (b) separated film, (c) bulk sample.

The DTA curves of a thin Se-film (a), compared with the thermal behaviour of a separated thin film (b) and a melt crushed sample (c) are shown in Fig.3.
At similar masses of the specimens the DTA curve of the amorphous selenium layer shows, in comparison to the thermal behaviour in the curves b and c, significantly a flat course of the glasstransition and increased temperature of crystallization. The process of crystallization seems to be retarded, obviously by the broad peak in the curve.

These differences of the thermal behaviour suggest lower degrees of order in the sputtered film.

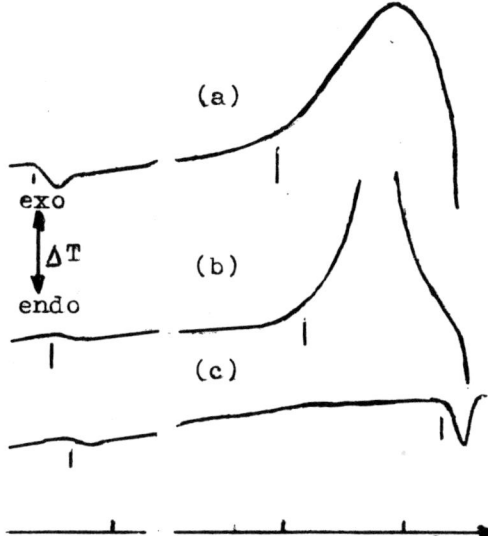

Fig.4 DTA curve of $As_2Se_3$, (a) thin film, (b) separated film, (c) bulk sample

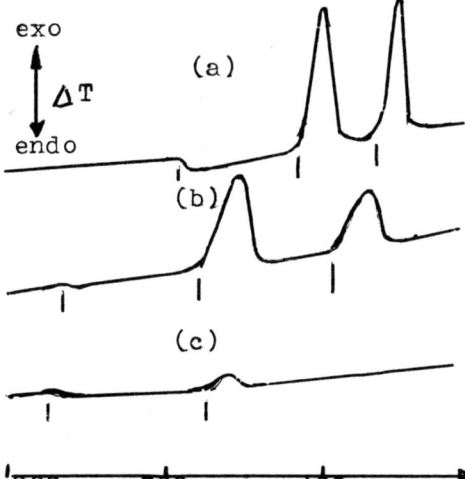

Fig.5 DTA curve of $Ge_2Se_3$, (a) thin film, (b) separated film, (c) bulk sample.

The DTA curves of Fig.4 demonstrate the different thermal behaviour of the original amorphous layer of $As_2Se_3$ (a), the separated layer (b) and a melt crushed sample of $As_2Se_3$ (c). Again at similar masses of the specimens there becomes evident a significant more flat course in the glasstransition and increased temperature at the original layer as well as at the separated layer opposite to the bulk sample. While the DTA curve of the separated thin layer shows a typical effect of crystallization, the original layer shows no sort of sign of a crystallization. Only a little slope in the baseline and a little effect of melting permits a partial crystallization. The lower degrees of order in the sputtered film caused this different thermal behaviour.

The situation of $Ge_2Se_3$, reproduced in Fig.5, seems to be not so well defined. The thermal behaviour of the original amorphous layer is shown in curve (a), the separated layer in (b) and the crushed bulk sample in (c). The glasstransition of the sputtered film is scarcely noticable and only at far lower temperatures than the bulk sample. Significant differences are seen at the crystallization. While the crystallization of GeSe and $GeSe_2$ take place successively at the bulk sample, the crystallization of $GeSe_2$ at the thin film remains quite at the same temperature, but the crystallization temperature of GeSe decreases by more than 50 K. From this it can be deduced, that in bulk specimens an interaction exists between structure units of GeSe and $GeSe_2$, while in the sputtered layers the units of GeSe and $GeSe_2$ exist independent of each other.

Comparative electrical measurements showed for the amorphous layers of $Ge_2Se_3$ an electrical resistance of two orders higher than at bulk samples.

The author wish to express his gratitude to Professor A. Feltz for his interest in this work and for valuable discussions.

## REFERENCES

(1) I. Gutzow, S. Kitova, M. Marinov, L. Tzakin; Glastechn. Ber. 49, (1976) Nr.6, 144.
(2) R. Messier, R. Roy; Mat. Research Bull. 6, Nr.8 (1971) 748
(3) D.D. Thornburg; Thin solid films, 37, (1976) 215
(4) D.D. Thornburg, R.J. Johnson; J. Non-Cryst. Solids, 17, (1975) 2
(5) A. Feltz; Conf. about amorphous, liquid and vitreous semiconductors, Sofia, 8.-12.5.1972.
(6) J. Namba, T. Mori; Oyo Butsuri (Jpn), 38, (1969) 1037
(7) J. Hopfe, Diplomarbeit 1972, Sektion Chemie der Friedrich Schiller Universität Jena.

# A NEW APPARATUS FOR STUDY OF STABILITIES IN HORIZONTAL AIR LAYERS HEATED FROM BELOW

B. Krstic

Royal Melbourne Institute of Technology

## ABSTRACT

An apparatus has been designed and built for studies and experimental evaluation of heat transfer by simultaneous conduction, convection and radiation. It consists of a rigid enclosure with two horizontal parallel plates separated by an air gap which was heated from below. Any preselected temperature difference, of up to $100°C$ across the plates could be maintained at a steady state. The apparatus was operated as a conduction-radiation cell. Using an energy balance and a calibration curve, "net" radiative heat transfer was evaluated and "emittance" of plain surfaces and "apparent emittance" of surfaces with V-notches was calculated. The results obtained were compared with these measured independently on the same samples by C.S.I.R.O. using optical techniques. It was found that a maximum difference of ±0.2% exists. To ensure that conduction and radiation only were present, a method similar to that used [1] to determine the onset of convection, was applied. Results obtained for "critical pressure" and "critical Rayleigh number" were of greater accuracy then theirs, due to the use of a small thin and very sensitive heat flux sensor of low thermal capacity.

## INTRODUCTION

The heat transfer across an air enclosure made up of two parallel plates heated from below is a complex phenomenon. It is known as Bernard-Rayleigh convection problem [2], [3], it has been with us for eighty years and has accumulated over 1000 relevant contributions. The heat transfer across it occurs by conduction, convection and radiation. It is known that convection enhances heat transfer [4], [5] and hence when convection is absent, (i.e. conduction and radiation are present on their own) the heat transfer is at a minimum. A further reduction in the heat transfer across an air enclosure can be achieved by filling it with insulating material. However, there are situations where introduction of insulating material is unacceptable, because it would impede the operation of the equipment. An example is the absorber plate and the cover system of a flat plate solar collector. The progress in this field is lagging behind, compared with that achieved in the development of selective surfaces. The cause for this is, in the author's opinion, lack of appropriate apparatus for evaluation of cover system performances with or without convection and radiation suppressing devices. There are no reliable experimental data on which the design could be based and no experimental set-up for their proper evaluation and optimisation. The apparatus has been designed for this type of evaluation and investigation. The apparatus is also suitable for numerous experimental investigations of simultaneous heat transfer by conduction and radiation with participating fluids which up to now have been carried out only by numerical analysis. To gain confidence in experimental work and in validity of obtained results the apparatus was used to determine "emittance" of plain surfaces and "apparent emittance" of surfaces with V-notches was evaluated from net radiative heat transfer.

## DESCRIPTION OF EQUIPMENT

The conduction-radiation cell shown in Fig.1 is mounted in a vacuum vessel whose pressure can be varied from atmosphere down to a few mm Hg. The air enclosure in the cell is bounded by two sets of square aluminium plates 200 mm wide, one set on the bottom and the other set on the top, separated by an air gap of D = 16.38 mm. The bottom set are the hot plates, the sample plate with unknown emittance being on the top. A cover plate, with a machined recess in which a heat flux sensor is fitted is screwed to it and the bottom side of the cover plate is heated by a flat electrical heater. A rigid insulation, sandwiched between two copper sheets with a thermocouple welded in the middle of each separates the main heater from the guard heater. The surface temperature of the heated sample plate could be kept constant within $\pm 0.016^\circ C$ over 15 minutes. To achieve a similar temperature control of the cold plate this was attached to a "boil-off" vessel filled with Freon F.11. The evaporated F.11 was condensed at atmospheric pressure. The side walls of the air enclosure are made of fibre-glass insulating board with a thermal conductivity $k = 0.0327 \frac{W}{mK}$ which is very similar to that of the air. This ensures minimal distortion in the temperature field of air close to the side walls. A copper coil soldered to a copper sheet, which is wrapped around the fiber insulation provides an external envelope which is kept at the same temperature as the cold plate by cooling water. Four nylon spacers placed in corners of copper wrapping maintain a constant distance between the plates. The temperature measurements were performed with calibrated chromel-constantan thermocouples inserted in a well drilled from the back to within 0.3 mm of the plate surface, or drilled to the root of the fin as in the case of the Blackbody Model Plate.

## ENERGY EQUATION AND CALIBRATION OF THE APPARATUS

To be able to apply the energy equation the number of unknown terms has to be reduced by one, the simplest way being to eliminate the convection term. A quantitative evaluation of the onset of convection is determined by the value of the dimensionless group called Rayleigh number - Ra which is defined, when air is the fluid, as

$$Ra = \frac{g p^2 \Delta T D^3 C_p}{R^2 T_m^3 \mu k} \quad \ldots \ldots \ldots (1)$$

where all the symbols have the usual meaning. The onset of convection is characterised by a "critical Rayleigh number" - $Ra_c$ which for enclosures bounded by plane rigid surfaces has a theoretically predicted [6] [7] and experimentally confirmed [8] [9] value of

$$Ra_c = 1707.7 \quad \ldots \ldots \ldots (2)$$

As long as $Ra < Ra_c$, instability will not occur, meaning that in an enclosure with a non-participating medium (air) the heat transfer will occur by simultaneous conduction and radiation only. Although simultaneous heat transfer by conduction and radiation in an enclosure is a complex phenomenon, because of the modification of the conductive temperature field by the presence of the radiative field, it has been shown [10] that the energy equation for Heater Element and Hot Plates can be reduced to

$$Q_{el} = K_1(T_1 - T_2) + \bar{F} \frac{A \sigma (T_1^4 - T_2^4)}{\frac{1}{\varepsilon} + \frac{1}{\varepsilon} - 1} \quad \ldots \ldots (3)$$

where $K_1$ is a constant representing physical dimensions, and properties of material which are independent of temperature. Although $K_1$ could be evaluated numerically, for accuracy reasons it was preferred to determine it experimentally. The second term in equation (3) represents a simplified expression for radiative heat transfer between parallel plates. $\bar{F}$ is "the total Interchange Area Factor" which is approximately equal to one [11]. The expression for radiative term is valid only when surfaces are gray and diffused emitter or diffused reflector. The usual errors in the application of the second term of equation (3) occurs because the original assumptions for its development are not fulfilled. The radiative energy emitted from the sample is reflected from the absorber plate back to the sample. To prevent this, the absorber plate should be a Blackbody model or have a high value of absorptance and a low value of reflectance. To this end a Blackbody plate with V-notches was developed [10], the dimensions of which are given in Fig.2 and used for calibration of the apparatus.

## EXPERIMENTAL PROCEDURE

The method used to ascertain that heat transfer across the enclosure occurs by conduction and radiation only is similar to that used [1] for determination of "$P_c$" and "$Ra_c$". It consists of monitoring the heat flux sensor reading across the air gap, while the air pressure is varied and the temperature difference of $\Delta T = 40.5°C$ and an effective distance between the plates $D_{eff} = 16.38$ where $D_{eff} = D + L_{Fin}$. The sought-after pressure is $P_{crit} = 286$ mm Hg which gives $Ra_c = 1708$ when substituted in equation (1). The pressure was further reduced to app. 150 mm Hg to allow for experiments in which a higher $\Delta T$ is used. Only after the surface temperature of the sample plate and the temperature between main and guard heater were within $\pm 0.016°C(\pm 1\mu V)$ for 15 minutes, the readings from a digital voltmeter were so averaged so that heat input could be evaluated. To obtain an equilibrium surface temperature of cold plate, it was necessary to bring F.11 in both "the boil-off" and the guard vessel to the boiling point.

## RESULTS AND DISCUSSION

It is convenient to plot the two right hand side terms of equation (3) against the temperature difference $\Delta T$: the conductive term - $Q_{cond}$ above and the radiative term - $Q_{rad}$ below the abscissa (Fig.3). When the conduction term in equation (3) was evaluated and plotted against $\Delta T$ a straight line was obtained. Additionally to the calibration curve, only

three curves are shown (for clarity): Aℓ plate "as received" $\varepsilon = 0.097$; Aℓ plate with 2 coats of black matt "pressure-pack" paint $\varepsilon = 0.715$ and the Blackbody Model plates anodized and 3 coats of same paint $\varepsilon_{app} = 0.968$
The results were compared with those independently obtained by C.S.I.R.O. using optical techniques and were found to be within the limits of ±0.2%. The results point out that with the apparatus and the method described in this paper, the radiative heat transfer between surfaces and the emittance can be evaluated more accurately than by any other calorimetric method using high vacuum, where an accuracy of between 2-4% can be achieved.

Note: This paper is taken from the author's M.Eng. Thesis. The continuation of this research program which envisaged the use of the apparatus for evaluation of cover systems for flat plate solar collectors by laboratory modelling, and also of the development of a technique for measuring of the onset of convection in flat plate solar collectors in the field. The project was however discontinued in 1978 due to lack of financial support.

## REFERENCES

[1] H.A. Thompson and H.H. Sogin, J. Fluid Mech., 24 (1966) 451
[2] E.L. Koschmieder, Adv. Chem. Phys., 26 (1974) 177
[3] J.A. Whithead Jr., "Fluctuations, Instabilities and Phase Transistor", edited by Tormod Riste (1975)
[4] M. Mull and H. Reicher, Beih. Gesund. Ing. 28 (1930)
[5] De Graaf and Von der Held, App. Sci. Res., A3 (1953) 393
[6] A.R. Low, Proc. Roy. Soc. (London), 125 (1929) 180
[7] A. Pellew and R.V. Southwell, Proc. Roy. Soc., A 176 (1940) 312
[8] R.J. Schmidt and S.M. Milverton, Proc. Roy. Soc. (Lond) A 152 (1935) 586
[9] P.L. Silveston, Forsh. Ing. Wes., 24 (1958) 29, 59
[10] B. Krstic, M.Eng., Thesis (V.I.C.) (1979)
[11] H.C. Hottel and A.F. Sarofin, "Radiative Transfer" McGraw Hill.

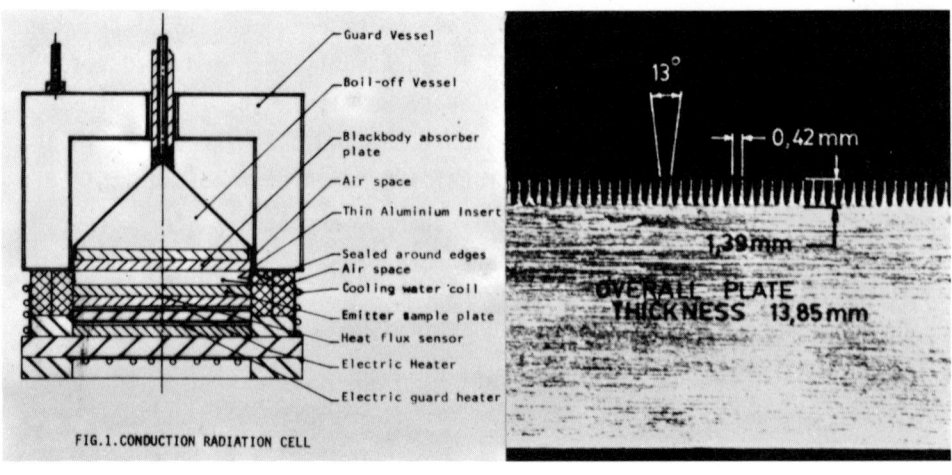

FIG.1. CONDUCTION RADIATION CELL

FIG.2. SAMPLE CROSS-SECTION OF BLACKBODY MODEL

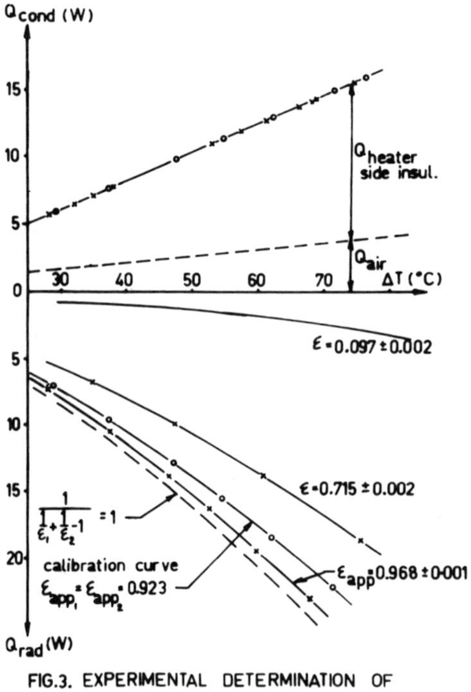

FIG.3. EXPERIMENTAL DETERMINATION OF THE $Q_{rad}$ AND CALCULATED $\varepsilon$ OR $\varepsilon_{app}$

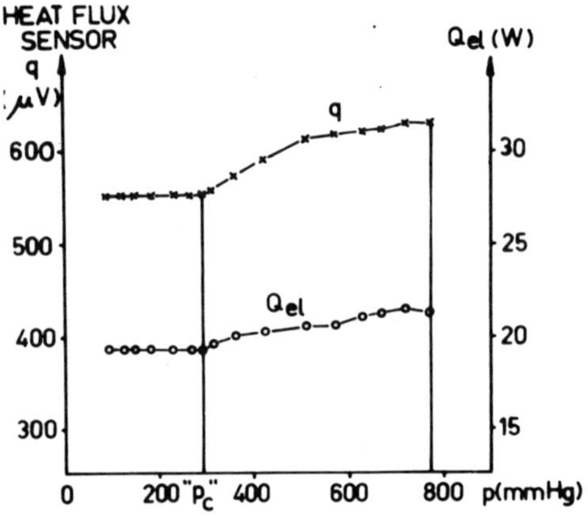

FIG.4. CRITICAL PRESSURE-"$p_c$" DETERMINATION FROM HEAT FLUX READINGS BY PRESSURE REDUCTION

$\Delta T = 40.5°C$   $D = 16.38 mm$   $p_c = 286 mmHg$

# PARTIAL PRESSURE MEASUREMENT BY THE FLOW METHOD

E. Marti, A. Geoffroy, B. F. Rordorf and M. Szelagiewicz
Central Function Research, CIBA-GEIGY Ltd.,
4002 Basle, Switzerland

## ABSTRACT

Accurate vapor pressures are determined by the gas flow method. The technique is reviewed in light of industrial application and important experimental aspects are discussed. A typical gas flow experiment is analyzed by transport equations. The results are illustrated on anthracene in a study of the dependence of transported mass on the inert gas flow rate. The measured vapor pressure data is analyzed in Rankine-Kirchhoff three parameter fits. A detailed comparison of our results on anthracene with literature shows good agreement with average literature values for heats of evaporation and entropies of evaporation. Comparison of the observed $\Delta c_p$'s on the other hand show that the present vapor pressure results are in much better agreement with expectations for $\Delta c_p$ than the compounded or the individual literature values.

## INTRODUCTION

The gas flow method, often called gas current, entrainement or transpiration method, is one of the oldest techniques to measure vapor pressures (1) to (4). Features which make this method particularly interesting for industrial application include the possibility to accurately measure partial pressures of active ingredients in the presence of large amounts of impurities. The method is furthermore characterized by a high experimental and analytical versatility, by small equipment costs and the possibility to automation.

A typical gas entrainement experiment for a solid sample is shown in Figure 1. Purified nitrogen is used as a carrier gas and is saturated with the sample. The sample is recondensed after flow through the excit capillary, which serves to reduce mass transfer by direct diffusion. Any quantitative analytic procedure which is selective for the main component can be used (Table 1).

Observables in the experiment are the inert gas flow rate, the sample transport, the saturation temperature, and the pressure and temperature at the flow meter. The calculation of the sample vapor pressure ($P_{2i}$) is straightforward (5) after assumption of Dalton's law and neglect of effects due to inert gas solubility in the sample. The computer programs developed in our laboratory for this purpose start out directly from the raw data of the analytical technique best suited to the problem.

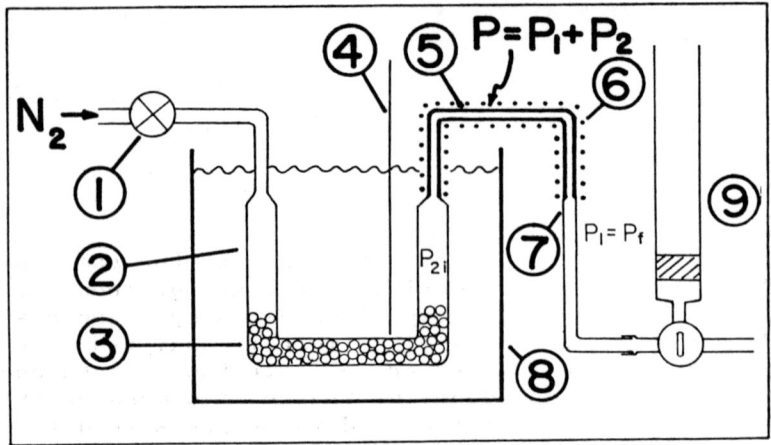

① NEEDLEVALVE ($\dot{n}_1$), ② CARRIER GAS PREHEATING, ③ SAMPLE COMPARTMENT CONTAINING GLASS PEARLS (∅ 2mm) COATED WITH EVAPORATING SAMPLE ($\dot{n}_2$), ④ THERMOCOUPLE ($T_1$), ⑤ EXCIT CAPILLARY ($\ell$, $r^2$) ⑥ CAPILLARY HEATER ($T \gtrsim T_1$), ⑦ CONDENSER, ⑧ THERMOSTATED OIL BATH, ⑨ FLOW METER.

Figure 1

Table 1

| Analytic Detection in the Gas Flow Method | |
|---|---|
| External | Gravimetry, Titration <br> UV, VIS, IR Spectroscopy <br> LC, TLC, GC (Chromatography) <br> any selective quantitative method known for the substance |
| On line | Real time detectors (FID, GC, laser induced fluorescence etc.) <br> Cold trapping gas chromatography <br> Mass spectroscopy <br> Laser ind. fluorescence, Raman, CARS, etc. |

## ANALYSIS OF THE TRANSPORT SYSTEM

Detailed insight into the transpiration experiment can be gained by solving the transport equations for a binary gas flow system:

$$J_\nu = \frac{\dot{n}_\nu}{r^2 \pi} = - Dc \frac{dx_\nu}{dz} + cx_\nu v \qquad (\nu = 1,2)$$

$J_\nu$ is the mass flux of species $\nu$ in the z direction, $x_\nu$ the mole fraction of $\nu$, c the total gas concentration, D the binary diffusion coefficient and v the sum of viscous and diffusion velocities. Solving this system of equations one arrives at the following expression (6):

$$P_{2i}^2 = x_{2i}^2 \left(P_f^2 + \frac{\dot{n}_1 + \dot{n}_2}{C} + \frac{A_{Ti}}{C} \ln(1 - (1-\gamma)x_{2i})\right) \quad (1)$$

with

$$x_{2i} = \frac{\dot{n}_2}{\dot{n}_1 + \dot{n}_2}(1 - e^{-\frac{\dot{n}_1 + \dot{n}_2}{A_{Ti}}}) \quad (1.1)$$

$$A_{Ti} = \left(\frac{\pi r^2}{\ell}\right) \frac{P \, D_{To}}{R} \left(\frac{Ti}{To}\right)^{0.75} \quad (1.2) \qquad D_{Ti} = D_{To}\left(\frac{Ti}{To}\right)^{1.75} \quad (1.3)$$

$$C = \frac{r^4 \pi}{16 \, R \, Ti \, \ell \eta} \quad (1.4)$$

The various symbols are explained in Figure 1 and Table 2. Index 1 stands for the inert gas, 2 for the substance, i for the saturation compartment and f for the location of condensation. $P_{2i}/x_{2i}$ is equal the total pressure ($P_i$) in the sample compartment and equation (1) can therefore be interpreted as a description of the pressure drop over the capillary region of the apparatus. C, the constant for mass transport by viscous flow, determines the pressure drop due to the viscous resistance. The expression $\dot{n}_1 + \dot{n}_2 + A_{Ti} \ln(1 - (1-\gamma)x_{2i})$ represents the total mass flow and its last term gives the flow of sample by diffusion. Evaluation of equation 1 for anthracene at 80 °C and $P_f$ = 101.3 kPa (Table 2) indicates a pressure drop over the capillary of less than 13 pascal even at the highest nitrogen flow rates used ($\dot{n}_1 \leqslant 10^{-4}$ mol/sec). A simplified equation $P_{2i} = x_{2i} P_f$ is therefore appropriate to evaluate experiments conducted at atmospheric pressures.

The exponential term of $x_{2i}$ (equation 1.1) is a correction to the observed vapor pressure, which takes account of diffusion contributing to the sample transfer. The constant for transport by diffusion, $A_{Ti}$, can be determined for substances of known vapor pressures in simple sublimation experiments (inert gas flow rate $\dot{n}_1$ = 0). Equation 1.1 rearranges for this case to

$$A_{Ti} = \frac{\dot{n}_2}{-\ln(1 - P_2/P)} \quad (2)$$

With an observed sublimation rate of $1.85 \times 10^{-13}$ mol/sec for anthracene at 80 °C and a vapor pressure of $4.64 \times 10^{-1}$ Pa (Table 3, selected literature plus gas flow) we calculate an experimental $A_{Ti}$ value of $4.0 \times 10^{-8}$ mol/sec. This value compares satisfactorily with the calculated

## Table 2

| Parameters for Anthracene in $N_2$ at 353.15 °K | | |
|---|---|---|
| $\ell$ | length of capillary | 28 cm |
| r | radius of capillary | 0.175 cm |
| $P_f$ | total pressure | 101.3 kPa |
| R | gas constant | 8.3143 Pa m³/°K mol |
| $\dot{n}_1$ | flow rate of inert gas ($N_2$) | $10^{-6}$ to $10^{-4}$ mol/sec |
| $\dot{n}_2$ | flow rate of sample | $10^{-11}$ to $10^{-9}$ mol/sec |
| $\eta$ | viscosity of inert gas | 211 μ poise |
| $D_{Ti}$ | binary diffusion coeff. | 0.071 cm²/sec (15) |
| $\gamma = \sqrt{\dfrac{\text{mol weight (sample)}}{\text{mol weight (inert gas)}}}$ | | 2.48 |
| $C_{Ti}$ | constant for mass transport by viscous flow ($\dot{n}_{visc} = C(P_i^2 - P_f^2)$) | $1.06 \times 10^{-10}$ mol/Pa² sec |
| $A_{Ti}$ | constant for mass transport by diffusion ($\dot{n}_\nu = A \cdot X_\nu$) | $8.46 \times 10^{-9}$ mol/sec |

value of $0.85 \times 10^{-8}$ mol/sec (Table 2), considering the simplifications used in the transport model (6). The effect of diffusion on the transfer of mass can now be predicted from the constant $A_{Ti}$. The relative contribution due to the diffusion can be represented by the quotient $q = \dot{n}_{2D}/(\dot{n}_2 - \dot{n}_{2D})$, where $n_{2D}$ is the mass transport due to diffusion only. Using equation 1.1 one arrives at the following expression for the inert gas rate

$$\dot{n}_1 = A_{Ti} \ln\left(\frac{1+q}{q}\right) \tag{3}$$

where the mole fraction $x_{2i}$ has been assumed to be much smaller than 1. For inert gas flow rates greater than $4.6 \times 10^{-8}$ mol/sec ($4.8 \times 10^{-3}$ l/h) one expects thus less than 1 % mass transfer ($n_{2D}$) by diffusion). In agreement with this prediction we find a plateau region in the saturation control curve on anthracene (Figure 2), which extends to very low inert gas flow rates. The apparent vapor pressure which is proportional to $\dot{n}_2/\dot{n}_1$ starts to raise indeed only at the lowest practicable nitrogen flow rate of $2 \times 10^{-6}$ mol/sec (0.23 l/h). The width of the plateau region, which is typical for a solid substance finely dispersed over the glass pearls as shown in Figure 1 (contact surface 600 cm²), allows for comfortably high gas flow rates. Complete saturation of the inert gas with anthracene is assured to flow rates of up to $1.4 \times 10^{-5}$ mol/sec (15 l/h).

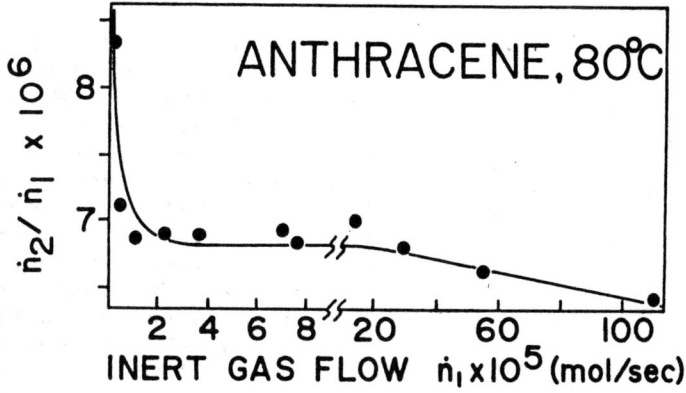

Figure 2

## VAPOR PRESSURE OF ANTHRACENE

Anthracene has an unusually high UV extinction coefficient ( (254 nm) = 10500), dissolved in dichloromethane it is therefore well suited to quantitative analysis by UV spectroscopy. The vapor pressures determined in our flow experiments where analyzed for their temperature behaviour by the well known Rankine-Kirchhoff equation, (eq. 4). The parameters can be interpreted in terms of the molar heat of evaporation $\Delta h(T)$, the molar entropy $\Delta s^+(T)$ and the change in heat capacity $\Delta c_p$ upon phase transition. Equations (4,3) shows that $\Delta c_p$ is assumed to be temperature independent.

$$\ln P (T) = A + B/T + C \ln T \qquad (4)$$
$$\Delta h (T) = R (TC - B) \qquad (4.1)$$
$$\Delta s^+(T) = R (A + C (1 + \ln T)) \qquad (4.2)$$
$$\Delta c_p = c_p^{gas} - c_p^{solid} = RC \qquad (4.3)$$

The three parameter fit was excecuted with a BMDP 77 library computer program (7) and the results for anthracene are shown in Figure 3A and Table 3.

A comparison of our data to literature is complicated by the fact, that no single very reliable vapor pressure measurement can be identified (8). The various literature values (9) to (12) were therefore pooled and analyzed collectively in a Rankine-Kirchhoff fit* (Figure 3C and Table 3). The fit was repeated after removing data points deviating by more than 4 standard deviations (8) and the results are shown in Figure 3D and Table 3. A combined analysis of this selected literature and our flow results is finaly shown in Figure 3B. It is not surprising to find a much tighter fit for the flow data than for the pooled literature values as is apparent in the residuals plots.

*Rankine-Kirchhoff fit = R-K fit

Identification of the residuals of our flow data for anthracene in the combined analysis (indicated with ■ in Figure 3B) shows that the measured vapor pressures fall right into the midst of the literature values. A tendency is clearly visible, however, for the residuals of our data indicating a systematic variance between the averadge literature and the present results. The slope of the indicated residuals points to a slightly smaller enthalpy of evaporation in the region of measurement (Table 3). The curvature of these residuals indicates a hefty difference between $\Delta c_p$ values for the pooled vapor pressures (literature and flow) and the flow data (see eq. 4).

Table 3 shows a comparison of $\Delta c_p$ values determined from the R-K fits. It can be seen that the change in heat capacity, $\Delta c_p$, for the gas flow experiment lies close to the calorimetrically determined value (9). Large misfits hold on the other hand for the pooled as well as for the individual (8) literature values, signifying wrong temperature dependences of the respective enthalpies of evaporation.

A negative $\Delta c_p$ value was already expected from the simple rule of Dulong and Petit which states that $c_p$ (solid) = $c_p$ (internal vibration) + 6R:
$\Delta c_p = (C_{vib} + 4R) - (C_{vib} + 6R) = -16.6$ J/mol.
The heat capacity of the solid was determined in a careful experiment (13) and this value compares well to expectations from solid state theory (4). Using statistical thermodynamics for the gas phase one arrives therefore at an independent $\Delta c_p$ value (8), which is in excellent agreement with the calorimetric value indeed (Table 3).

Table 3

| Anthracene | R - K fit (Eq. 4) A | B | C | Δh° kJ/mol | Δh(353.15°C) kJ/mol | ΔCp J/mol °C | ΔS(353.15°C) J/mol °K | P(353.15°C) Pa |
|---|---|---|---|---|---|---|---|---|
| All. lierterature | -17.71 | -6534 | 15.26 | 54.57 | 99.39 | 126.9 | 275.2 | 4.73 x 10⁻¹ |
| Selected literature | -12.97 | -9834 | 6.82 | 81.76 | 101.79 | 56.7 | 281.7 | 4.54 x 10⁻¹ |
| Sel. lit. + gas flow | -0.616 | -10434 | 5.01 | 86.75 | 101.46 | 41.7 | 280.9 | 4.64 x 10⁻¹ |
| gas flow | 52.91 | -12903 | -2.89 | 107.28 | 98.78 | -24.1 | 274.7 | 5.47 x 10⁻¹ |
| calorimetry (9) | | | | | | -31.8 | | |
| ΔCp (calc.)* | | | | | | -31.4 | | |

* $\Delta Cp$ (calc.) = $C_p^{25°C}$ (stat. thermodyn.) - $C_{p\ solid}^{25°C}$ (literature (13,14))

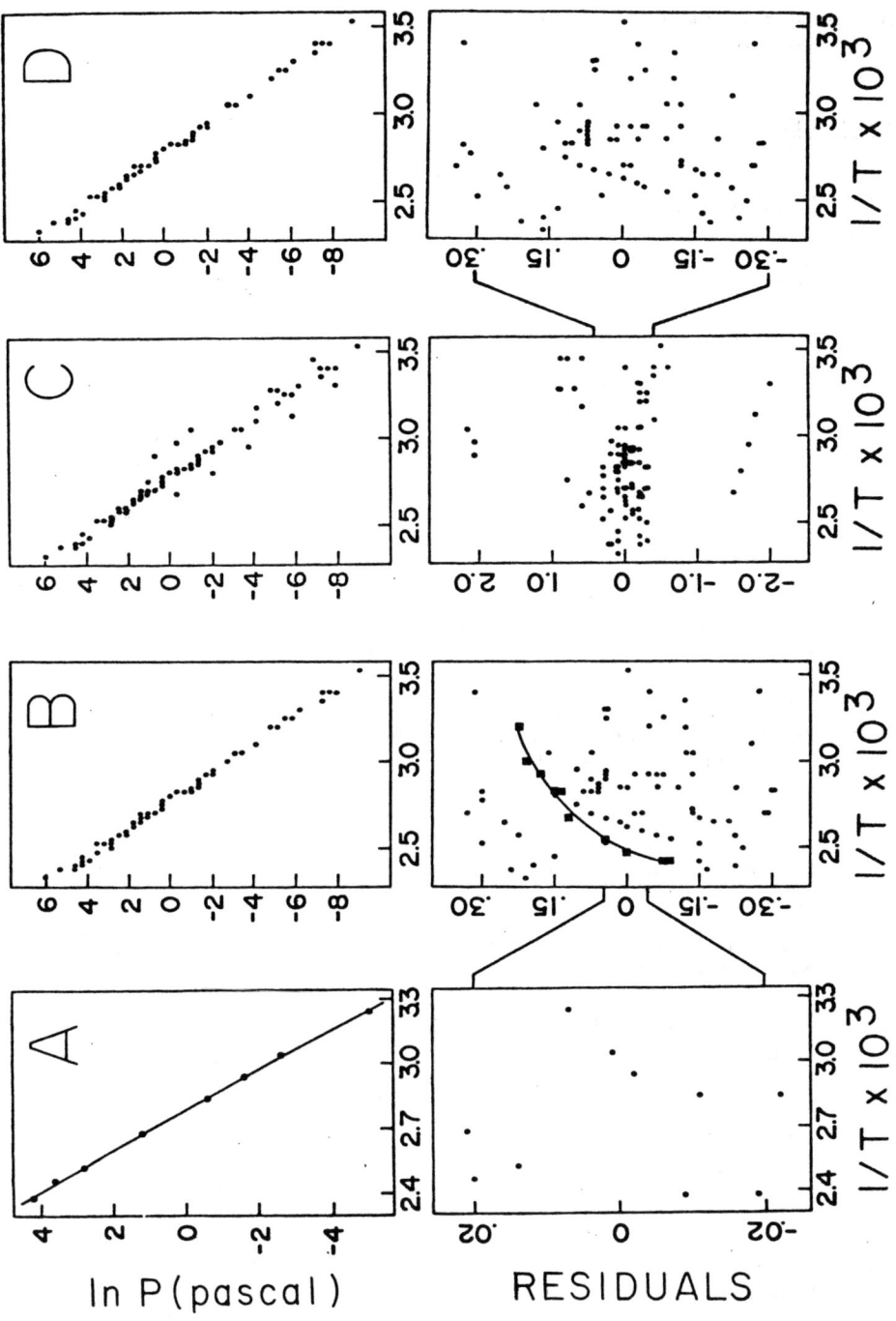

Figure 3

The authors acknowledge the contributions by O. Heiber and
E. Zimmermann.

RERFENCES

(1)  H.V. Regnault, Ann. Chim. 15 (1845) 129.

(2)  H.T. Gerry and L.J. Gillespie, Phys. Rev. 40 (1932) 269.

(3)  U. Merten and W.F. Bell, "The Characterisation of High Temperature Vapors", J.L. Margrave, editor, John Wiley & Sons: New York" (1967) 91.

(4)  G.W. Thomson and D.R. Douslin, "Determination of Pressure and Volume" in Physical Methods of Chemistry, Vol. 1, A. Weissberger, editor, Rossiter (1971).

(5)  B.F. Rordorf, A. Geoffroy, M. Szelagiewicz and E. Marti, this proceedings.

(6)  H. Kvande and P.G. Wahlbeck, Acta Chem. Scand. A30 (1979) 297.

(7)  BMDP (Biochemical Computer Programs, P-Series), UCLA Health Sciences Computing Facility, University of California Press (1977).

(8)  B.F. Rordorf and E. Marti, to be published.

(9)  L. Malaspina, R. Gigli and A. Bardi, J. Chem. Phys. 59 (1973) 387, and references therein.

(10) A.B. Macknick and J.M. Prausnitz, J. Chem. Eng. Data 24 (1979) 175.

(11) J.W. Taylor and R.J. Taylor, J. Chem. Soc, Faraday Trans. 72 (1976) 723.

(12) D.M. McEachern and 0.1 Sandovoal, J. Physics E: Sci. Instr. 6 (1972) 155.

(13) P. Goursot, H.L. Girdhar and E.F. Westrum, Jr. J. Phys. Chem. 74 (1970) 2538.

(14) G. Filippini, C.M. Gramaccioli, M. Simonetta and G.B. Suffriti, Chem. Phys. 8 (1975) 136.

(15) E. Mack, Jr., J. Am. Chem. Soc. 47 (1925) 2468.

# VAPOR PRESSURE METHODS FOR INDUSTRIAL APPLICATIONS

B.F. Rordorf, A. Geoffroy, M. Szelagiewicz, and E. Marti
Central Function Research
CIBA-GEIGY Ltd., CH-4002 Basel, Switzerland

## SUMMARY

Several vapor pressure methods have been developed under consideration of industrial feasibility. The principles of measurement range from equilibrium to non-stationary conditions. Methods such as static techniques, TG, DSC and gas saturation are described and vapor pressure results are compared for anthracene as a standard substance and found to be in excellent agreement with literature values.

## INTRODUCTION

Vapor pressure data of active ingredients and intermediate products are increasingly important in the chemical industry. Interest for vapor pressure ranges from development of chemical processes and formulations to registration requirements and toxicology related problems in manufacturing and handling of chemicals. Each of these applications need vapor pressure data in quite different temperature regions and the data asked for may range from relative pressure values for screening purposes of active ingredients to precise partial pressure measurement for chemical production, toxicology and registration. Table 1 serves to illustrate on the large variety of vapor pressures and substances encountered in industrial application.

Table 1

| Method | Instr. | Substance | Experimental Temp. Range $^\circ$C | Vapor[✱] Pressure mbar [+] | Appr. Vapor Pressure mbar ($20^\circ$C) |
|---|---|---|---|---|---|
| Boiling Point | DSC-1B* | 2-Phenylimidazole | 208 to 275 | 13.3 | $10^{-5}$ |
| " | " | " 1,2-Propanediamine | 42 to 118 | 40.0 | 12 |
| Evap. Rate | TGS-1* | Propyphenazone | 123 | 0.1 | $10^{-5}$ |
| Boiling Point | TGS-2* | Copper phthalocyanine | 540 to 600 | 13.3 | $10^{-25}$ |
| Flow | non commercial instr. | 1-Nitro-anthraquinone | 110 to 171.5 | $10^{-4}$ | $10^{-9}$ |
| Flow | non commercial instr. | 2-Chloro-4,6-di-fluoro-s-triazine | -18.2 to 10.1 | 0.2 | 20 |

* Perkin-Elmer Corp.   + 1 mbar = $10^2$ Pascal   ✱ low limit of temp. range

A huge body of methodical vapor pressure work exists in the literature and many different measuring principles are known (1) to (3). A selection of suitable methods for industrial use has to base on quite different arguments as in the choice for purely scientific work. A good industrial technique should cover a large pressure and temperature range. Depending on the industrial vapor pressure problem, further criteria include: selectivity and accuracy, low measuring time requirements, easy sample handling, low equipment costs and the possibility of automated or on-line data collection and processing.

Table 2 lists a selection of methods developed in our laboratory in efforts dating back to 1968.

Table 2

| Method | Determined Parameter | Technique | Pressure Range mbar | Temp. Range °C | Comments |
|---|---|---|---|---|---|
| Static | Total Vapor Pressure | -Capacitance Manometer<br>-Piezoelectric Transducer | $10^{-2}-10^{2}$<br>$10^{0}-5 \cdot 10^{3}$ | -50 to 70 | Solvents, low boiling substances and mixtures |
| DSC<br>DTA | Vapor Pressure | -Boiling Point Method | $10 - 10^{3}$ | 30 to 500 | Versatile method, accurate temp. determination, approx. determination of enthalpies of evaporation or sublimation |
| TG | Vapor Pressure | -Boiling Point Method<br>-Diffusion Controlled Evaporation in Air or $N_2$ at $10-10^3$ mbar<br>--Scanned Temp. Programm<br>--Isothermal | $10 - 10^{3}$<br>$10^{-4}-10$ | 30 to 1000 | Inexpensive, fast and versatile. With consecutive scans of the same sample proof for volatile impurities, decompositions, resonable accuracy |
| Flow or Transpiration Method | Partial Pressure | -Direct Partial Pressure Measurements upon Appropriate Choice of Analytical Procedure | $10^{-6}-10$ | -40 to 300 | High accuracy method, determination of enthalpies of evaporation or sublimation, analytical proof of impurities and decomposition products |

## STATIC METHOD

Static methods are applied where total pressures of mixtures in the range of over one mbar need to be known. Partial pressures are accessible only if simultaneous information on the gas phase composition is available (MS, spectroscopy, GC, etc.).

## DIFFERENTIAL SCANNING CALORIMETRY (DSC) AS BOILING POINT METHOD

Boiling points can be determined over a broad temperature range with DSC or DTA instruments by nonequilibrium experiments. The heat uptake of the sample is measured as a function of temperature and the enthalpy of evaporation can be measured directly if the whole sample is evaporated. Such an evaporation curve is shown in the insert of Figure 1. Point A is, after suitable calibration, a good approximation for the boiling temperature of a substance. If the evaporation of the sample is limited by diffusion only or, in case of Langmuir, by the evaporation rate into vacuum, the observable sample loss starts far below the boiling temperature. The evaporation curves are much better defined, if the sample is enclosed and evaporates through a pinhole of appropriate size. Viscous flow through the pinhole becomes rate limiting thus leading to a pressure buildup inside the capsule. The mass transport through the pinhole is approximately given by (4): $\dot{n} = C(P_{inside}^2 - P_{outside}^2)$. C, a constant for the viscous flow, is proportional to the forth power of the pinhole radius and inversely proportional to the wall thickness of the sample pan (7). It is clear that the relation between the boiling point ($P_{sample} = P_{outside}$) and the observed point A obtained from a dynamic DSC measurement is a complicated function of sample pan geometry and gas viscosity, temperature scanning rate and the resulting pressure difference between inside and outside of the sample pan.

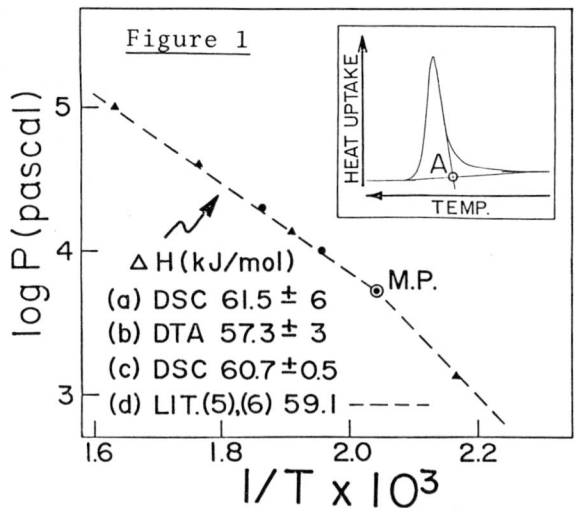

Figure 1

ΔH(kJ/mol)
(a) DSC 61.5 ± 6
(b) DTA 57.3 ± 3
(c) DSC 60.7 ± 0.5
(d) LIT.(5),(6) 59.1 ----

The linear part of the evaporation curve ($d^2n/dt^2$ = const, see insert Figure 1) is given by the restriction of the heat flow to the sample and the fact that the temperature of the sample remains at the boiling temperature while the instrument temperature continues to increase linearly. Boiling points can indeed be determined with an accuracy of ± 0.5 K by dynamic DSC and DTA methods. The reproducibility of the mechanically drilled pinholes (⌀ 0.1 mm), thus the sample fluctuation of the constant C, limits the accuracy of the method. We are presently engaged in developing an optimal procedure for punching pinholes.

In our experiments a constant carrier gas flow is maintained in the vicinity of the sample capsule to prevent recondensation of the sample vapor on the sample holder assembly. Outside pressures with our instruments range from 10 to $10^3$ mbar. The measured vapor pressures are plotted in the usual way in lnP vs 1/T curves to obtain the evaporation enthalpies. DSC (▲) and DTA (●) results for anthracene are compared in Figure 1 with literature values. Good agreement holds for the vapor pressures and also for the enthalpy of evaporation obtained by integration of (a) the DSC or (b) the DTA evaporation curves, and (c) from the vapor pressure plot. The errors of the mean values are given in terms of confidence intervals on a 65 % level.

## THERMOGRAVIMETRY

Boiling point experiments analogous to the DSC measurements just described can also be done by monitoring the mass of the evaporating sample instead of the heat uptake. Figure 2 compares the results for boiling point experiments by thermogravimetry (x) on anthracene with literature vapor pressures (---). Complete evaporation of the sample is in general inevitable in such boiling point experiments.

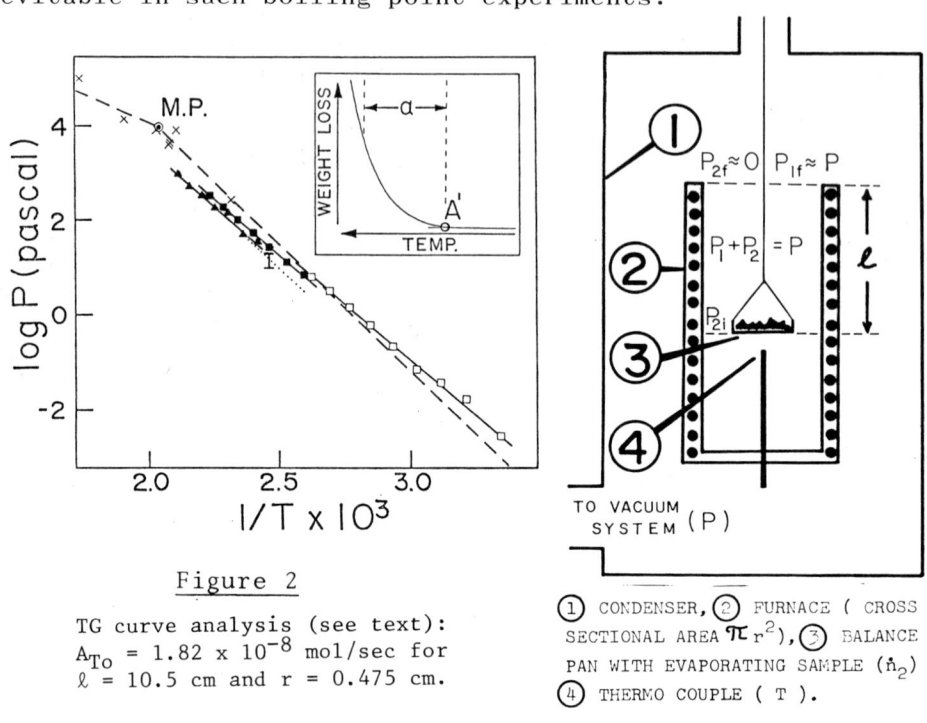

Figure 2

TG curve analysis (see text):
$A_{T_0}$ = 1.82 x $10^{-8}$ mol/sec for
$\ell$ = 10.5 cm and r = 0.475 cm.

① CONDENSER, ② FURNACE ( CROSS SECTIONAL AREA $\pi r^2$), ③ BALANCE PAN WITH EVAPORATING SAMPLE ($\dot{n}_2$) ④ THERMO COUPLE ( T ).

Figure 3

The possibility to obtain several thermo-gravimetry scans on one and the same sample is desirable, however. It allows to get information on volatile impurities, sample decomposition and helps to improve the experimental accuracy. Multiple scans are possible in TG vapor pressure experiments on samples placed on the open balance pan as shown in Figure 3. The elongated geometry of the oven insures control of the evaporation rate by diffusion of the sample vapor through the inert gas. The temperature inside the furnace is practically constant but drops very rapidly between the boundary of the oven and the condenser. Sample molecules reaching the top of the oven are thus rapidly removed, leaving the sample partial pressure ($P_{2f}$) close to zero. The resulting composition gradient inside the oven is responsible for sample diffusion.

Solving the transport equations for the binary system for diminishing inert gas flow ($\dot{n}_1 = 0$) one finds the following relation between the vapor pressure at the surface of the sample ($P_{2i}$) and the evaporation rate ($\dot{n}_2$) of the sample (see Eq. (2) in reference (7)):

$$P_{2i} = P(1 - \exp(-\dot{n}_2/A_T)) \text{ with } A_T = \left(\frac{\pi r^2}{\ell}\right) \frac{P\, D_{To}}{R} \left(\frac{T}{To}\right)^{0.75} \quad (1)$$

P is the total pressure adjustable on our instrument between 10 and $10^3$ mbar. $A_T$ is a constant for mass transport by diffusion, R is the gas constant and $D_{To}$ the binary diffusion coefficient ($D_{To} = 0.078$ cm$^3$ (8) for anthracene at 99.2°C). The evaporation rate ($\dot{n}_2$) is simply the first derivative of the weight loss curve.

A complete vapor pressure curve can be obtained from a single thermogravimetry scan and the temperature interval suitable for evaluation (indicated by a in Figure 2, insert) can be shifted according to the inert gas pressure. The vapor pressure results from a TG curve analysis using Eq. (1) are shown in Figure 2 (▲) for anthracene in air at atmospheric pressure. The agreement with literature vapor pressures (---) is good considering that these results were obtained in a dynamic and fast screening method.

The same analysis (Eq. 1) applies to isothermal TG scans. Vapor pressure curves are obtained for selected temperatures. Results of such isothermal TG experiments for anthracene are shown in Figure 2 for inert gas pressures of 101.3 mbar (■) and 8 mbar (□).

This discussion justifies an empirical rule, which has been used in our laboratory for some time to estimate vapor pressures: Point A' as shown in Figure 2 insert, corresponds to a certain evaporation rate and, by Eq. 1, relates to a constant vapor pressure of the sample. This pressure was empirically found to be in the range of 0.1 to 0.2 mbar for our experimental conditions (Figure 2 (I)). The temperature $T_{A'}$ correlates to the boiling point of the substance at these empirically established vapor pressures. A modified Trouton rule, $\Delta H = 40\, T_{A'}$, is used to estimate the evaporation enthalpy (see Figure 2 (...) Trouton).

## FLOW METHOD

The flow or transpiration method is especially attractive for industrial applications as a high precision partial pressure method. A detailed discussion is given in (7). The partial pressure of a substance $P_{2i}$ is calculated from the mass transfer of the substance $n_2$ at constant temperature $T_i$ by volume of nitrogen $V_i$ (symbols see in ref. (7)).

$$P_{2i} = n_2 RT_i/V_i = n_2 RT_f/V_f$$

These equations are derived assuming the ideal gas law, Avogadro's law, and a gas volume of the sample which can be neglected with respect to the volume of the carrier gas.

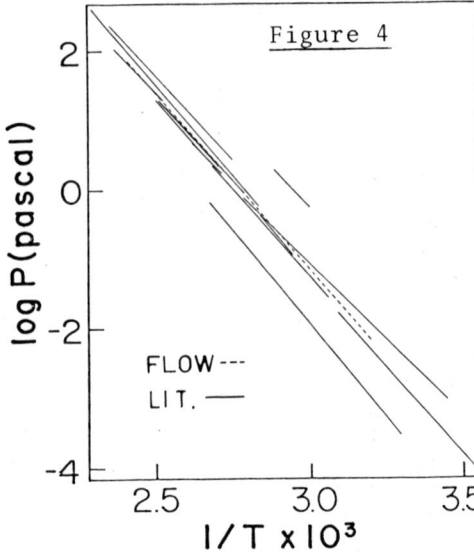

Figure 4

Fig. 4 presents results for anthracene obtained by the flow method together with several literature data quoted in (7). Our results fall right in the midst of the values reported from various vapor pressure methods. The difference of the heat capacity in the gas and solid phase, $\Delta c_p$, is calculated from the temperature dependence of the enthalpy of sublimation. The $\Delta c_p$ obtained from our vapor pressure curve revealed the best fit with $\Delta c_p$ values obtained calorimetrically or from theory (9).

High purity anthracene was chosen as a substance ideally suited to compare the different vapor pressure methods and, in case of the flow method, as a most suitable substance for testing partial pressure methods under idealized conditions with respect to chemical and physical stability and analytical assays.

REFERENCES

(1)  A.N. Nesmeyanov, "Vapor Pressure of the Chemical Elements", Elsevier Publ. Comp. (1963)

(2)  J.L. Margrave, editor, "The Characterisation of High Temperature Vapors", John Wiley & Sons, New York (1967)

(3)  G.W. Thomson and D.R. Douslin, "Determination of Pressure and Volume" in Physical Methods of Chemistry, Vol. 1, A. Weissberger, editor, Rossiter (1971)

(4)  K. Motzfeldt, H. Kvande, and Ph.G. Wahlbeck, Acta Chem. Scand. A31 (1977) 444

(5)  R.C. Reid, J.M. Prausnitz, and T.K. Sherwood, "The Properties of Gases and Liquids", McGraw Hill, New York (1977)

(6)  F.S. Mortimer and R.V. Murphy, Ind. Eng. Chem. 15, (1923) 1140

(7)  E. Marti, A. Geoffroy, B.F. Rordorf, and M. Szelagiewicz, these proceedings

(8)  E. Mack, Jr., J. Am. Chem. Soc. 47 (1925) 2468

(9)  B.F. Rordorf and E. Marti, to be published

THE USE OF SUBAMBIENT THERMAL VOLATILISATION ANALYSIS
TO STUDY VOLATILE PRODUCTS OF POLYMER DEGRADATION

I.C. McNeill

Chemistry Department, University of Glasgow,
Glasgow G12 8QQ, Scotland

ABSTRACT

Subambient TVA may be carried out on the collected products of heating a sample in a TVA system. The relationship to other TVA methods is considered and the technique is illustrated by several examples from polymer degradation studies.

INTRODUCTION

The variables which may be used to follow the progress of a polymer degradation reaction, as the sample temperature is raised, fall broadly into three groups - changes associated with volatilisation, heat changes, reflected in differences between the sample and an inert reference material, and changes in mechanical properties. There are also other variables, however, which it is important to recognise and to control. These include the sample size and form (it is primarily the thickness of sample which matters), atmosphere, heating rate and flow rate of carrier gas if used. Degradation mechanism can depend on the amount of heat supplied, or to be dissipated, on the rate of transfer of heat to or from the sample and on ease of product removal. If products are not removed efficiently from the reaction zone, secondary reactions may occur. The most efficient removal of products occurs under high vacuum conditions. Very high reaction temperatures or rapid rates of heating usually lead to temperature gradients in the sample and are therefore not to be preferred for serious studies of mechanism. Since both temperature gradients and secondary reactions involving products are more likely in thick samples, thinly distributed samples are desirable in degradation experiments.

The optimum conditions for degradation studies therefore involve thin samples, slow/moderate heating rates and high vacuum conditions. Given these, the next feature required is a method of collection, separation and identification of the reaction products.

## TVA METHODS

Thermal Volatilisation Analysis (TVA) is a particularly useful thermal analysis technique, since it meets all of the criteria discussed above.

All TVA experiments involve the measurement of pressure of substances undergoing transfer from one point to another in an initially evacuated system which is continuously pumped and give curves indicating rate of volatilisation versus temperature (time). The basic layout is shown schematically in Fig. 1.

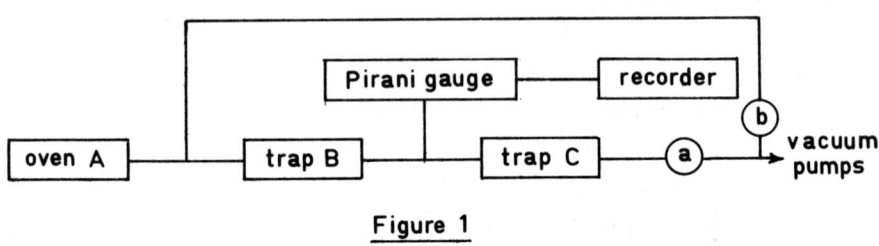

**Figure 1**

Three main types of experiment may be carried out, as follows, comprising TVA (1), (2), (3), differential condensation TVA (2), (3), (4), and subambient TVA (SATVA) (3), (5):

|  | I | II | III |
|---|---|---|---|
| Technique | TVA | TVA (diff. cond.) | SATVA |
| Situation | degradation | degradation | prod. separation |
| Sample at A | polymer | polymer | none |
| Heating prog. at A | 0 - 500°C | 0 - 500°C | none |
| Temperature, trap B | not in use | 0°>T>-100°C | -196°C |
| Temperature, trap C | -196°C | -196°C | -196°C → 0°C |
| Stopcock a | open | open | closed |
| Stopcock b | closed | closed | open |

TVA (column I) provides information about threshold and maximum rate temperatures for degradation reactions. With the additional facility of differential condensation of products (column II), it is also possible to deduce information about the volatility of the products at any stage and whether the composition of the products changes. An illustration is provided in Fig. 2, in which the differential

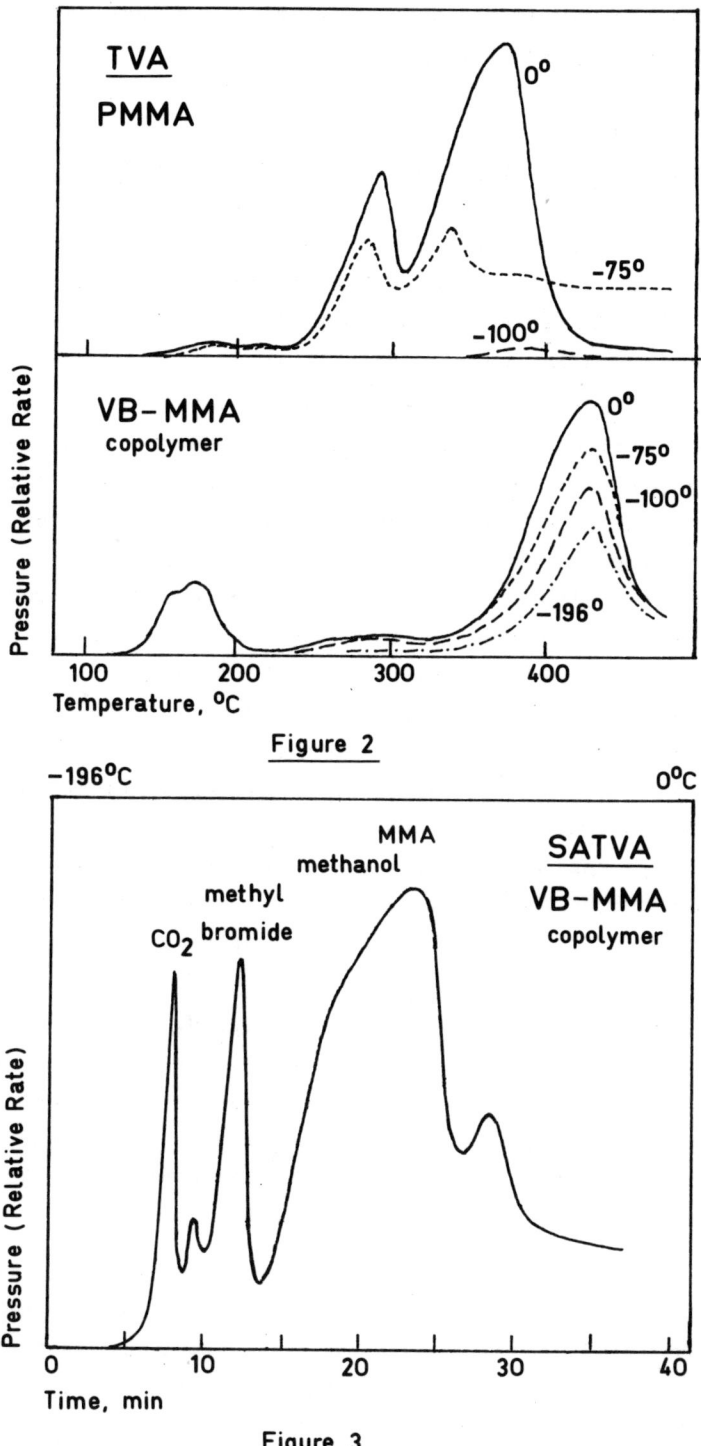

Figure 2

Figure 3

condensation TVA curves are compared for poly(methyl methacrylate) (PMMA) and a copolymer of vinyl bromide (VB) and MMA containing 19 mole % VB. It may be deduced from these data (6) that the homopolymer breaks down in two stages, essentially without change in product composition; the product, of moderate volatility, can be shown to be MMA. The copolymer breaks down in a totally different manner in three well-defined stages, with different products at each stage. The first and second stage products are much more volatile than MMA monomer and there is a mixture of products at the final stage, including non-condensable gases.

SATVA allows direct examination of product volatility and separation into fractions. The study of the same VB-MMA copolymer provides an example (Fig. 3). As the product trap C in Fig. 1 is allowed to warm up in a controlled manner over about 45 minutes, each product vaporises in turn according to its volatility. The volatilisation may be monitored to give a SATVA trace and the product(s) corresponding to each peak may be collected separately for further analysis. The three main products, carbon dioxide, methyl bromide and MMA, identified in this case by gas phase infrared spectroscopy using gas cells attached directly to the TVA system to collect the fractions, are well separated. Analysis by degradation stage (temperature ranges indicated by the data of Fig. 2) revealed also that the first stage consists exclusively of methyl bromide evolution and that the second stage involves elimination of small amounts of hydrogen bromide. MMA and carbon dioxide arise at the final stage of reaction.

Another example of SATVA is provided in Fig. 4, which shows the effect on the nature of the degradation products when the PVC chain is modified by the presence of 25% of copolymerised vinyl acetate units. The most interesting feature revealed in the SATVA results is that, in addition to the expected additional degradation product, acetic acid, significant amounts of acetyl chloride are formed.

An example of more immediate practical relevance concerns the degradation products of a polyurethane, derived from 1,4-butane diol and methylene bis-(4-phenyl diisocyanate), alone and in the presence of the fire retardant, ammonium polyphosphate (7). Product toxicity is of great concern in a fire situation and it is not only

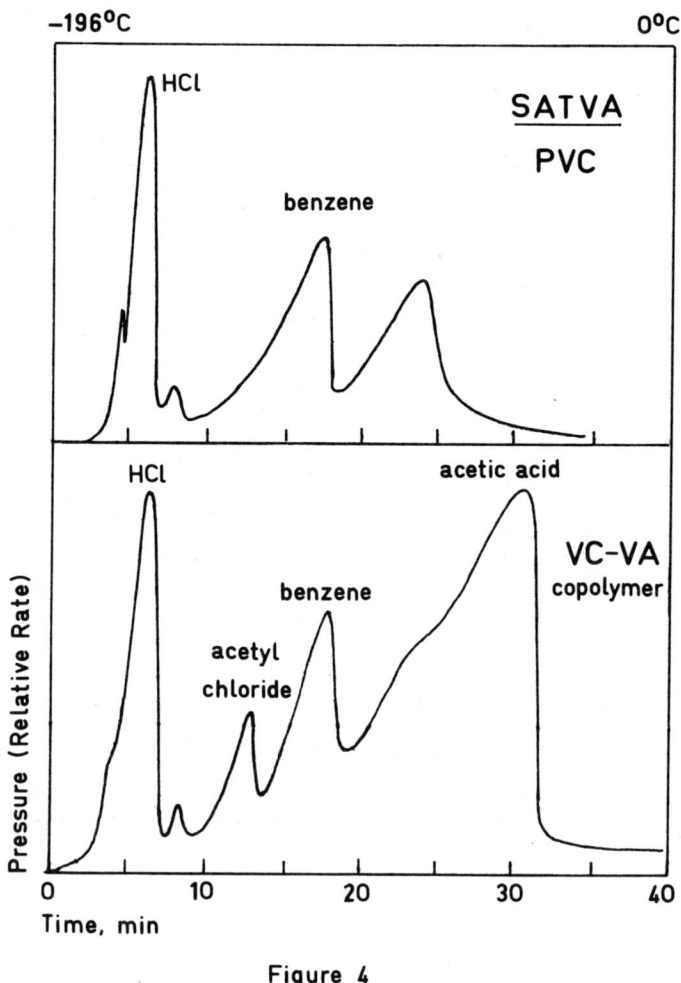

Figure 4

combustion products which matter, but also those generated purely by heat, ahead of the advancing fire. SATVA gives a useful picture of the distribution of products and shows that the sample with the fire retardant additive releases two additional, dangerously toxic products - formaldehyde and aniline (Fig. 5).

SATVA experiments such as those illustrated may be carried out consecutively after TVA degradations, in the same apparatus. Products from heating other involatile solids, as well as polymers, may be examined. If gas phase infrared spectroscopy or mass spectrometry is used for final identification, products need not be handled in the atmosphere, which is a considerable advantage when very small amounts of materials are involved.

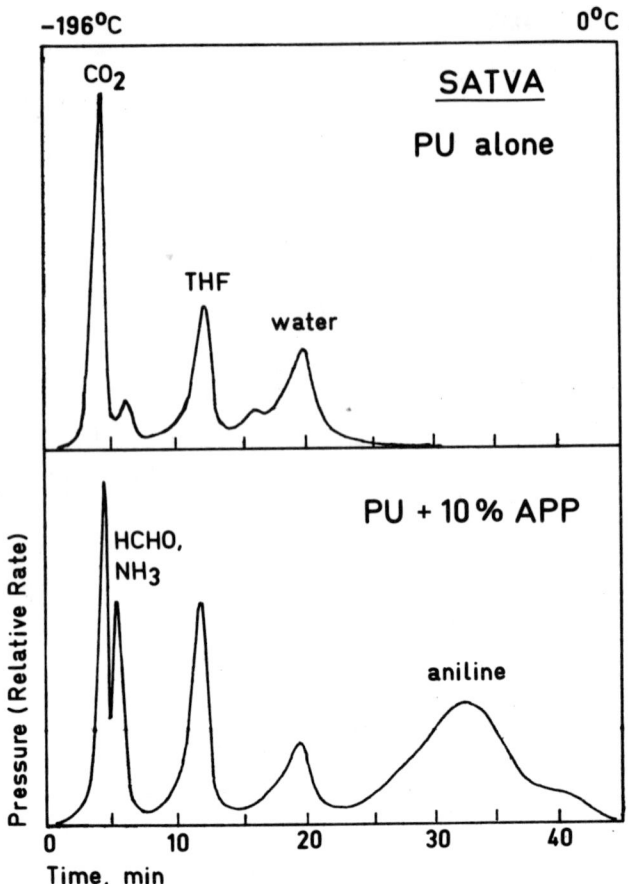

Figure 5

REFERENCES

(1). I.C. McNeill, Eur. Polym. J. *3*, 409 (1967).
(2). I.C. McNeill, in Thermal Analysis, Vol. 1, R.F. Schwenker and P.D. Garn (eds.), Academic Press, N.Y., ,969, p. 353, 417.
(3). I.C. McNeill, in Developments in Polymer Degradation - 1, N. Grassie (ed.), Applied Science, London, 1977, p.43.
(4). I.C. McNeill, Eur. Polym. J. *6*, 373 (1970).
(5). I.C. McNeill, L. Ackerman, S.N. Gupta, M. Zulfiqar and S. Zulfiqar, J. Polym. Sci., Polym. Chem. Ed. *15*, 2381 (1977).
(6). I.C. McNeill, J. Polym. Sci., Polym. Chem. Ed., in press.
(7). N. Grassie and M. Zulfiqar, in Developments in Polymer Stabilisation - 1, G. Scott (ed.), Applied Science, London, 1979, p. 197.

# Applied Sciences

W. Krajewski
T.H. Aachen, FRG

E. Lugscheider
T.H. Aachen, FRG

DTA/EGA USING A SPECIFIC DETECTOR

S. B. Warrington and P. A. Barnes
School of Health and Applied Sciences
The Polytechnic, Leeds 1, England

ABSTRACT

A highly sensitive, continuous-reading electrolytic hygrometer capable of measuring water concentrations in a gas stream down to ppm levels has been coupled to a Stanton-Redcroft DTA 671B Differential Thermal Analyser. The method of coupling and operating precautions are described. Integration of the EGA peaks obtained shows the combination to be capable of measuring sub-milligram quantities of evolved water to an accuracy of better than 4%. Applications discussed include hydrate decomposition and degradation of plant acids used in the food industry.

INTRODUCTION

The difficulties of interpreting an isolated DTA curve are well-known. Peak assignment, even in apparently simple systems can be erroneous and supplementary information is often required. Performing parallel experiments on another portion of material may introduce further confusion, as the conditions of the two tests can never be matched. Because of this, simultaneous methods have been pursued, in which the recording of various sample properties as a function of temperature or time refer indisputably to the same sample conditions.

EGD found early favour as a supplementary technique, allowing discrimination between DTA peaks associated with gas-loss and those which are not. A refinement, EGA, allows positive identification of the evolved gas, sometimes quantitatively.

Water is evolved from a wide variety of systems, by a wide variety of mechanisms, yet is notoriously difficult to measure quantitatively in small amounts. Most methods used have drawbacks, whether of accuracy, complexity or expense. The need for a sensitive, cheap, continuous-reading water detector appeared to be fulfilled by an electrolytic hygrometer.

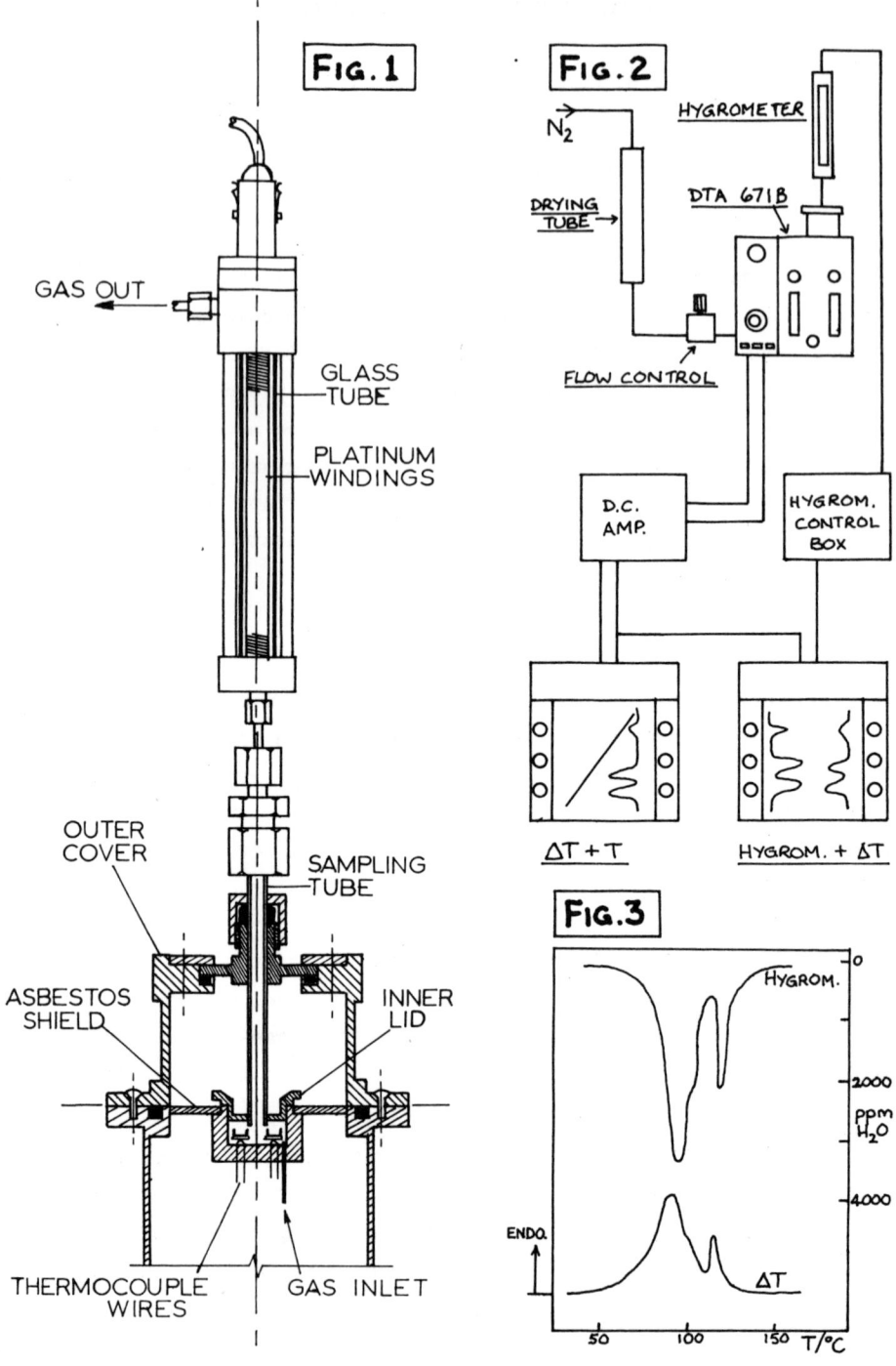

## THE HYGROMETER

This type of instrument was first described by Keidel (1), but the form used here is that devised by Still and Cluley (2) and manufactured by Salford Electrical Instruments Ltd.

It consists essentially of two fine platinum wires wound closely, but not in contact, on a PTFE former (Fig 1). The wires are coated with phosphoric acid and the element enclosed in a glass tube supported by a nickel plated frame. Before use the acid is electrolysed to dryness by a standing voltage of 100 V after which a negligible residual current flows. The acid coating is an efficient dessicator and when a gas is passed through the narrow annular space between the tube and wires, any water present is absorbed and simultaneously electrolysed. At equilibrium, with a fixed gas flowrate, the electrolysis current is proportional to water concentration in the gas.

The absorption efficiency with a fresh coating is high (2) and under the wide range of conditions in which our cell operates, recoating is found necessary every 4 to 6 weeks. Recoating takes about 24 hours. The control unit of the instrument has 6 switched ranges corresponding to 10 to 3000 ppm when the gas flow is 100 $cm^3$ $min^{-1}$, at S.T.P. Even though our work is on samples of a few milligrams, the highest range is normally used, which, with a flowrate of 50 $cm^3$ $min^{-1}$, becomes equivalent to 6000 ppm. A rheostat on the control box allows matching of output to a Phillips PM 8010 twin-pen recorder, so that 6000 ppm results in a displacement of 250 mm.

## THE DTA/HYGROMETER COMBINATION - OPERATIONAL

Certain features of the Stanton Redcroft DTA 671B were incompatible with accurate hygrometer work. The original two gas inputs, each with small rotameter and needle valve had polythene tubing in circuit, which was replaced with copper. The internal rotameter/valve combinations were by-passed and flow control effected by a finer needle valve/diaphragm-type controller (Fig 2). The asbestos shield (Fig 1) was cemented in place with silicone rubber to reduce the dead space into which vapour could diffuse and subsequently desorb. The inner surface of the upper cover was "silanised" with dimethyldichlorosilane to minimise adsorption. The upper cover (Fig 1) was fitted with a bulkhead union, allowing the $1/4"$ O.D. sampling tube to be raised or lowered through the inner lid. In the cavity in which the sample and reference materials rest in pans on their respective T/C beads, gas enters through a $^1/16$" O.D. tube. This is disposed asymmetrically with respect to the two pans and the cooling effect resulted in a displacement of the $\Delta T$ trace. A baffle cut from a bronze turning eradicated this problem.

Dry gas is obtained by passing "white spot" $N_2$ through a tube of molecular sieve. The resulting gas has a water content below 2 ppm. The apparatus is left with a purge of dry gas with the sampling tube raised, so that the inner cover is swept.

In use, the outer cover is removed and the sample quickly introduced. On replacing the cover the hygrometer reading rises to ca. 4000 ppm due to inclusion of ambient air, but in 5 minutes drops to ca. 300 ppm. The sampling tube is lowered and in 2 minutes the working background of ca. 40 ppm is achieved. A baseline of ca. 60 ppm can be maintained up to $300°C$ after which curvature results from desorption in the instrument.

## CALIBRATION

The meter on the control unit is scaled on the basis of Faraday's laws, relating electrolysis current to the amount of water. A practical calibration was felt to be more appropriate however and a series of 15 samples of A.R. grade copper sulphate pentahydrate were heated until the first four waters had evolved (between 60 and $150°C$). The relevant operating conditions were:

Heating rate: 10 K $min^{-1}$; Chart speed: 1 cm $min^{-1}$

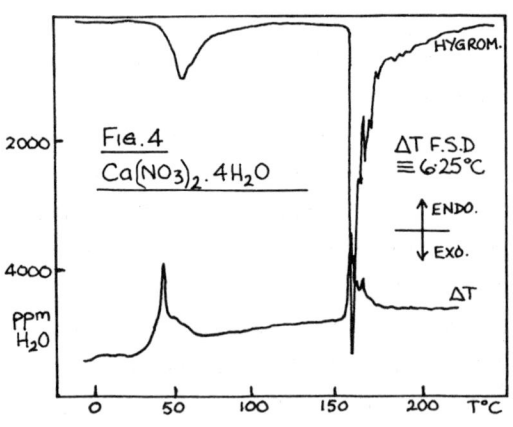

Fig.4 $Ca(NO_3)_2 \cdot 4H_2O$

Gas flow: 50 $cm^3$ $min^{-1}$
$\Delta T$: F.S.D. $6.25°C$
Hygrometer: 6000 ppm F.S.D.
A typical DTA/EGA trace is shown in Fig 3.
A plot of theoretical weight of water from the chosen sample weights against EGA peak weight (after tracing and cutting out) showed a correlation

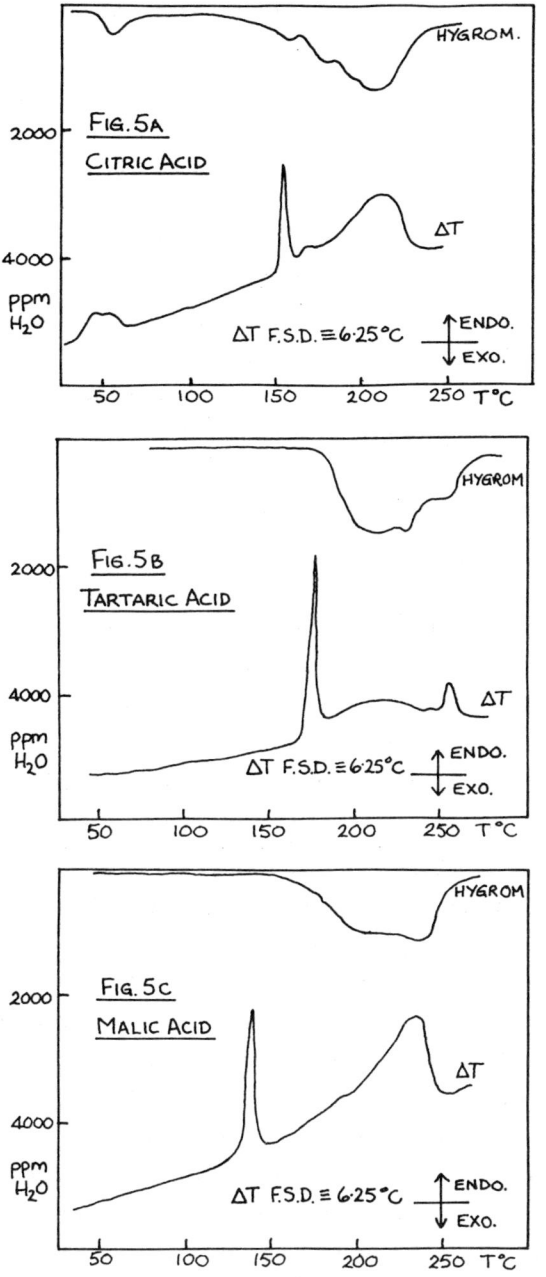

Fig. 5A Citric Acid
Fig. 5B Tartaric Acid
Fig. 5C Malic Acid

coefficient of 0.993, with no point lying more than 5% away from the least squares regression line. If the potential errors are considered, including tracing paper inconsistency and flowrate variations, this result may be thought encouraging.

A theoretical calibration, accepting the hygrometer control unit meter as correct, was less good, being about 12% too low; the chosen calibration procedure is thus vindicated. The calibration was checked with ca. 2 mg samples of calcium butyrate, which had been shown by TG to be a monohydrate. 3 samples analysed by the hygrometer arrangement produced peaks corresponding to the theoretical amount of water to within 4%.

An observation arising from the calibration runs is the wide variation in shape of the DTA/EGA curves obtained for $CuSO_4 \cdot 5H_2O$. Despite the fact that single well-formed crystals of similar history were used, under identical conditions, evolution of the first four waters produced either one EGA/DTA peak, or two. Although no attempt is made to explain this here, the observation may act as a reinforcement to the cautionary note by Pope[3] in which the need for precise specification of sample parameters is stressed.

## MISCELLANEOUS APPLICATIONS

As part of an investigation into nitrate decomposition some calcium nitrate tetrahydrate was prepared. This is an awkward material to crystallise, having a m.p. of 42.7°C and being highly soluble and deliquescent. Eventually, crystals were obtained by evaporating under vacuum at 20°C. A DTA run on these and on A.R. grade $Ca(NO_3)_2 \cdot 4H_2O$ showed the same features (Fig 4); a melting peak superimposed on a broader endotherm presumably resulting from water loss. When the two materials were heated with the hygrometer in series the A.R. grade material gave an EGA peak roughly coincident with the melting peak corresponding to the loss of two waters, whereas the laboratory-prepared sample showed a loss of 1.6 to 1.7 waters. This has been confirmed by TG.

Organic acids of plant origin have complex decomposition schemes involving release of $H_2$, $O_2$, CO, $CO_2$ or $H_2O$, the latter being the result of inter- or intra-molecular condensations. With the hygrometer interpretation of the DTA curves is simplified. At a basic level, the DTA/hygrometer combination may be used for more precise "fingerprinting" of such compounds. Figs 5A, 5B and 5C show the DTA/EGA traces for 3 common hydroxyacids - citric, tartaric and malic acids. All the figures (except citric, which loses some water of hydration at ca. 50°C) show the acids to be stable up to their melting points, after which a slow decomposition occurs. The structure of the EGA peaks suggests that the decomposition is a complex multi-step process, which would not be deduced from the DTA curves alone. In the case of tartaric acid, a sharp endotherm at ca. 260°C is not reflected on the EGA curve and is probably the result of $CO_2$ evolution. Data on the decomposition of these compounds is scanty, but the temperature ranges of the decompositions observed here do not agree with those quoted by e.g. Lorant (4). These materials yield unpleasant products on decomposition and a more precise knowledge of their fate in food-processing would seem to be urgently needed.

DTA has been applied to the determination of impurities in minerals. Although no work has been performed in this laboratory on such topics, it is felt that the determination of a mineral which evolved water at a characteristic temperature would be facilitated by the great sensitivity of the hygrometer.

## CONCLUSIONS

The hygrometer is reliable, robust, compact and cheap. It has the advantage of showing a linear relationship between water concentration and output. Its sensitivity is such that even with the milligram samples used here, the least sensitive range is normally used. The DTA/hygrometer combination produces EGA curves of excellent resolution, as may be seen in Fig 4, where the sharp peaks on the DTA curve, due to bubbling from a molten sample, are closely followed by the EGA curve. Sharp, multiple peaks of this type are not accurately integrated by tracing and weighing and the use of an electronic integrator with fast sampling rate is indicated. Nevertheless, determinations of quantities of water of about 100 $\mu$g are already routinely performed to an accuracy of better than 4% and the hygrometer has a potential increase in sensitivity of 2 orders of magnitude. The instrument cannot be used with any gas which reacts with $HPO_3$, such as ammonia. The upper temperature limit for the present assembly is 350°C, because of the silicone rubber sealant used.

Future development will include the incorporation of other gas detectors in parallel with the hygrometer to allow extraction of more information from a single DTA run.

## REFERENCES

1. H. A. Keidel    Anal. Chem. 31, 2043 (1959)
2. J. E. Still and H. J. Cluley    SAC Conf. (1965) p.405
3. M. I. Pope and D. I. Sutton    Thermochim. Acta 23, 188 (1978)
4. B. Lorant    Mitt.Geb. Lebensmittelunters u. Hyg. 57, 231 (1966)

## ACKNOWLEDGEMENTS

One of the authors (SBW) acknowledges the support of the Science Research Council.

# QUANTITATIVE DIFFERENTIAL THERMAL ANALYSIS AT ELEVATED PRESSURES

Prakash C. Jain and Deoraj Chaubey
Department of Physics and Astrophysics, University of Delhi
Delhi-110007, INDIA

## ABSTRACT

A simple, yet sensitive, form of apparatus for quantitative differential thermal analysis at elevated pressures is described. It has been calibrated and operated satisfactorily at pressures upto 6 kb and temperatures upto 600 K. It has been used to measure the melting point of indium as a function of pressure upto 6 kb. The measured values are in good agreement with earlier measurements. For quantitative measurements, the system has been used to construct phase diagrams for $KNO_3$ and $NH_4NO_3$. The dependence of change in enthalpy for a phase transition on pressure has been investigated. The results are in agreement with the results computed from the PVT data.

## INTRODUCTION

Application of DTA at elevated pressures has been rather limited. It has been used for detection of phase transitions in inorganic materials (1,2) and study of melting and crystallization of polymers (3-6) under pressure. These studies were basically qualitative in nature, no effort was made to determine the associated changes in enthalpy directly from the DTA curves. These were, however, determined indirectly by making use of Clausius-Clapeyron equation. This method requires data on the rate of change of the transition temperature with pressure and the associated changes in volume. This information may not be always available. In this paper we describe a simple DTA system which has been successfully used for determining changes in enthalpy associated with phase transitions at pressures upto 6 kbar directly from the peak area. Determination of the change in enthalpy from peak area is a widely accepted procedure in DTA. The change

in enthalpy $\Delta H$ can in general be represented by
$$\Delta H = f\ (A/m) \qquad (i)$$
where $f(A/m)$ is some function of the peak area, A and the mass m of the sample. In most of the cases, this relation is linear. The nature of the function $f(A/m)$ is related to the geometry and thermal conductivity of the sample holder. For a given system the function (i) can be determined by using a set of reference materials. Over a limited range of temperatures, the nature of the function (i) is nearly independent of the temperature. In our measurements we have made an additional assumption that it is also independent of pressure.

### EXPERIMENTAL PROCEDURE

The high pressure differential thermal analyser (HPDTA) developed consisted of a high pressure apparatus, DTA cell and a recorder for recording the $\Delta T$ and T signals.

The high pressure apparatus used was a piston-cylinder system. The pressure cylinder and the piston were made from hardened and tempered EN24 steel, details regarding their dimensions are given in Fig.1. Pressure communicating medium used was silicon fluid (DM 200, Metroark Ltd.,). The fluid was compressed by the action of the piston which was driven by an hydraulic ram. A metal sealing plug was screwed at the bottom of the cylinder. Different pressure seals, used in the system, were made from teflon O-rings supported on phosphor-bronze rings.

The fluid pressure was calculated from the ratio of the areas of the hydraulic ram and the piston. The necessary correction for the cylinder was estimated from the piston displacements in the forward and backward directions and the corresponding recorded ram forces.

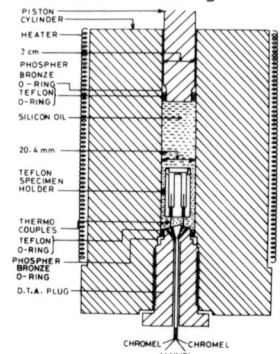

Fig.1:Design of high pressure apparatus and the DTA block.

The pressure chamber could be heated upto 600 K, the maximum heating rate being 4 K/min. Thermocouples for recording temperatures were introduced into the pressure chamber through the bottom plug. Twin-core mineral insulated chromel-alumel thermocouples in a stainless steel sheath passed through the conical holes drilled in the plug. These thermocouple were used to record the sample temperature and the difference $\Delta T$ between the sample and the reference temperatures. The sample temperature could be recorded to an accuracy of $\pm 0.25$ K. The DTA block was made from a teflon (PTFE) rod, 1.5 cm in diameter. Two holes drilled symmetrically in this block accommodated the sample and the reference materials. The two thermocouple junctions entered the sample and reference holders from the bottom.

The temperature (T) and the difference ($\Delta T$) signals were recorded on a two-channel recorder. The $\Delta T$ channel had a sensitivity of 10 $\mu$V/div. corresponding to 0.25 K/div. The required correction to the observed transition temperatures was computed from the observed and quoted values of the transition temperatures for the standard samples.

To measure heats of transition, the system was calibrated by using a set of inorganic and organic materials as standards. The materials used were ammonium nitrate, diphenylamine, magnesium nitrate, magnesium chloride, benzoic acid, urea, indium and tin. The samples used were of very high purity. DTA curves for varying weights of these samples were obtained at atmospheric pressure. The transition temperatures and the peak areas were determined in the standard manner.

The melting behaviour of indium as a function of pressure was investigated. Similarly the solid-solid phase transitions in $KNO_3$ and $NH_4NO_3$ were also studied as a function of pressure.

In all DTA measurements a constant heating rate of 4 K/min was maintained.

## RESULTS AND DISCUSSIONS

The variation of the peak area, corresponding to a transition, with the sample weight, for all the calibration standards used was always found to be linear. The variation of peak area per unit sample weight as a function of the change in enthalpy ($\Delta H$) for these samples as observed in the present experiment is shown in Fig. 2. The variation is more or less linear, however, it is better represented by the relation

$$\Delta H(cal/g) = 0.058 + 6.473\ A - 0.180\ A^2 + 0.006\ A^3 \qquad (ii)$$

where A is the peak area per gram of the sample expressed in square inches. The solid curve in Fig. 2 represents this variation. In estimating heats of transition from peak areas this relation has been used.

The variation of melting temperature of indium with pressure as obtained in the present experiment is shown in Fig. 3. The continuous line shows a fit to the Simon equation as obtained by Mc Daniel et al. (8). Our measured values are in good agreement with this curve. Similar results obtained by Turner (7) are also shown in this figure.

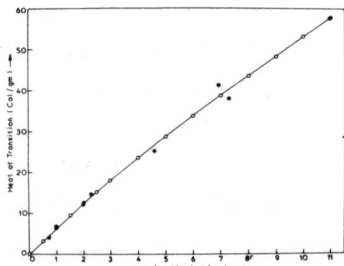
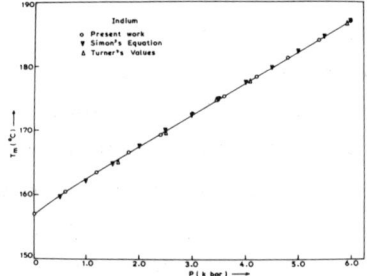

Fig. 2: Variation of change in enthalpy with the peak area.

Fig. 3: Variation of melting temperature of indium with pressure:

Complete phase diagrams for $KNO_3$ and $NH_4NO_3$ systems as obtained in the present experiment are shown in Figs. 4 and 5. In these figures the results obtained from PVT measurements by Bridgman (9) are also shown. The agreement between our results and those of Bridgman is very good.

A determination of heat of transition from the peak area

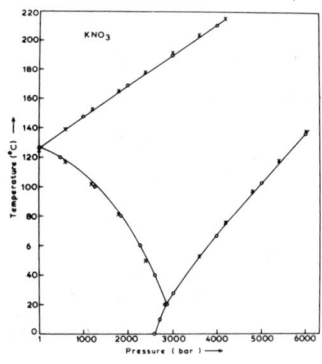

Fig. 4: Phase diagram for KNO$_3$

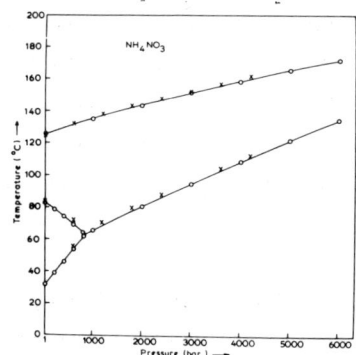

Fig. 5: Phase diagram for NH$_4$NO$_3$

involves the knowledge of a relationship between these two quantities for a given system. For our system such a relationship at atmospheric pressure is represented by eq. (ii) The values of the coefficients in this equation are related to the geometry and thermal conductivity of the sample holder. When the experiment is conducted at elevated pressures, the DTA block is subjected to a hydrostatic pressure and therefore no relative geometrical deformation between the sample and reference holders is expected. Furthermore the thermal conductivity of the DTA block material also does not change appreciably with pressure. Therefore, from these

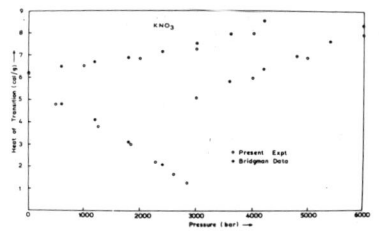

Fig. 6: Variation of change in enthalpy with pressure for solid-solid transitions in KNO$_3$

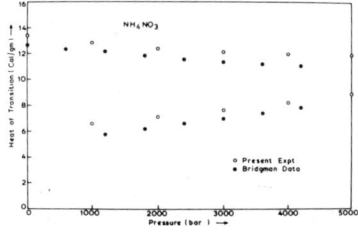

Fig. 7: Variation of change in enthalpy with pressure for solid-solid transitions in NH$_4$NO$_3$.

considerations the form of relationship between $\Delta H$ and the peak area could also be considered to be independent of pressure. Based on this assumption, the heats of transition for various solid-solid transitions in $KNO_3$ and $NH_4NO_3$ have been determined as a function of pressure. The results are shown in Figs. 6 and 7. In these figures the corresponding heats of transition computed from Bridgman's PVT data (9) making used of Clausius-Clapeyron equation, are also shown. There is good agreement between the values obtained directly from the peak areas and those computed from PVT data. Small departures could however be attributed to an implied assumption that the thermodynamic functions of equilibrium material parallel those of the actually studied metastable material.

## REFERENCES

(1) A. Jayaraman, W. Klement (Jr.), R.C. Newton and G.C. Kennedy, J. Phys. Chem. Solids 24 (1963) 7.

(2) G.C. Kennedy and R.C. Newton, Solids under Pressure Eds. W. Paul, D.M. Warschauer (McGraw Hill Book Co. New York, 1963).

(3) D.C. Bessett and B. Turner, Nature Phys. Sci., 240 (1972) 146 ibid Philos. Mag., 29 (1974) 925.

(4) S.K. Bhateja and K.D. Pal, J. Macromol Sci. Revs. Macromol. Chem. C13(1) (1975) 77.

(5) R.N. Gupta, P.C. Jain, V.S. Nanda and A.S. Reshamwala, J. Appl. Polym. Sci., 21, (1977) 2621.

(6) P.C. Jain, W. Wunderlich and D.R. Chaubey, J. Polym. Sci. A-2, 15 (1977) 2271.

(7) B. Turner, High Temp. High Press, 5 (1973) 273.

(8) M.L. Mc Daniel, S.E. Babb (Jr.) and G.J. Scott, J. Chem. Phys. 37 (1962) 822.

(9) P.W. Bridgman, Int. critical Tables, Chief Ed. E.W. Washburn (McGraw Hill Book Co., New York and London, (1928) Vol. 4, p. 9.

THE TREATMENT OF THERMOANALYTICAL DATA FOR EFFECTIVE PRESENTATION

J. S. Crighton and K. M. Li
University of Bradford, Bradford, West Yorkshire, U.K.

ABSTRACT

Techniques to facilitate extraction of the maximum meaningful information from thermoanalytical experiments are considered. Attention was given to the collection, manipulation and presentation of observations from DTA and TG. Off-line manipulative routines are assessed to provide smoothing, temperature linearisation, scaling, interpolation and differentiation of digitised analogue signals. The presentation of data to facilitate recognition and interpretation curve features is examined. The form of the manipulated data allows for the averaging of observations from repeated experiments to reflect representative behaviour. DTA curve corrections to eliminate the distorting influences of base lines are also facilitated.

INTRODUCTION

Thermoanalytical measurements generate analogue current or voltage signals which are displayed on galvanometric or potentiometric recorders. Despite the relative ease of measurement and recording, frequently significant amounts of the contained information are not used. From DTA curves often only the major enthalpic peaks or entropic shifts are reported. Although the position shape and size of enthalpic peaks form the basis of interpretations, justifiable caution is shown on the recognition of factors which can distort peak features. The use of apparent shoulders or peak shifts in DTA or of small overlapping mass changes in TG is affected by these influences. Many factors are inherent in the specific sample and instrumental conditions and are outside direct control (e.g. differing and variable thermoelectric responses by individual thermocouples). The collection, manipulation and presentation of thermoanalytical data can be achieved overcoming many of these problems and significantly improving the quality and potential of the information as well as the reliability of and confidence in the interpretations made. Collection of the data in the digital form is followed by computational transformation and presentation. Wendlandt has reviewed the use of non-interactive computational methods in thermal analysis (1). Reports of computational methods in

DTA/DSC have been concerned largely with reaction kinetics (2), polymer crystallisation (3) and purity estimations (4). Efforts have concentrated on procedures from the DTA curve. The quality and form of the DTA/TG curve data has received comparatively little attention. The acquisition of data and its reduction have been examined and described by Doelman in both DSC (5) and TG (6) applications. Procedures associated with data collection and manipulation are here examined to guide the establishment of effective presentation. The basic principles and the instrumentation have been reported previously (7).

## EXPERIMENTAL

A DuPont 900 with a 950 thermobalance were linked to a "Dart" data logger (Electronic Associates). The floating analogue signal inputs to the 900 console X-Y recorder were diverted and fed through screened leads to two channels of the logger. Repetitive sequential logging of these signals was achieved by bipassing the manual "operate" function with a Eurotherm cyclic timer. The timer determined the interval between each logged signal. This time can be digitally set at from 1 to 999 seconds. With this system digitisation to 1 in $10^4$ on ranges from 10mv to 100v is achieved with output through a teletype as punched tape. Smoothing, transformation and presentation were performed off-line on an ICL 1904S computer.

## DATA HANDLING PROCEDURES

The raw data was as a sequence of alternating weight and temperature digital values recorded at equal intervals determined by the cyclic timer. Meaningful manipulation including numerical differentiation, requires that the data was as free from noise as was practically possible. Precautions were taken to isolate the electrobalance from ambient vibrations and to stabilise the thermal environment around the equipment. The raw temperature and weight data were given an initial 5 point cubic smooth (7). The thermo electric temperature values were interpolated to generate values coincident in time with those of mass or temperature difference. For the transformation of emf to sample temperature secant iteration proved efficient (7). The DTA/TG curves are recorded directly as a function of sample temperature it is essential to confirm linearity/stability of programming prior to interpretation. A monitor of the "actual heating" rate as a function of temperature was constructed.

The (temperature difference/sample mass; temperature) values were concurrently smoothed and interpolated to provide temperature difference/sample mass values at defined one degree temperature intervals in the range of interest. By interpolating all data to the same temperatures and intervals averaging procedures and comparisons are facilitated. By averaging multiple analyses improvements in the quality of the final curves are achieved. With small DTA and TG samples the ability to sum observations from repeat samples provides an average curve more representative of the bulk. With the (sample mass, "sample" temperature) data the rate of mass change as a function of temperature differential at the defined temperatures can be concurrently generated within the routine. This derivative information facilitates the recognition of overlapping mass changes. Flexibility of scaling of all or sections from the DTA, TG and DTG curves is provided in the graphical presentation. With DTA the effect of base line movement and overlap on peak shape and position can restrict interpretation. If base line movement can be defined in regions where thermal transitions are absent, a function can be fitted to the data in this region and interpolated across the active portions of the curve. Subtraction of the generated base line to provide a compensated DTA curve which reflects only the thermally induced transitions in the active substrate.

## CONCLUSIONS

The transformation of digital data in DTA/TG has been shown as useful for maximising the information available and increasing its reliability. The reduced data provides for the averaging of repeated analyses which will improve the quality and representitiveness of the presentation.

## REFERENCES

(1) W.W. Wendlandt, Thermochim. Acta. $\underline{5}$ (1973) 225
(2) R.W. Crossley et al., in "Analytical Calorimetry" (1970), 429
(3) F. Gornick, J. Polym. Sci. $\underline{C25}$ (1968) 131
(4) E.F. Joy et al., Thermochim. Acta $\underline{2}$ (1971), 31
(5) A. Doelman et al., in "Analytical Calorimetry" $\underline{4}$ (1977), 1
(6) A. Doelman et al., IBM Journal of R & D $\underline{22}$ (1978) 81
(7) J.S. Crighton and A. Das submitted for publication
(8) J.S. Crighton and P.N. Hole Thermochim. Acta $\underline{24}$ (1978) 327

# EQUILIBRIUM SHRINKAGE-FORCE MEASUREMENT - A METHOD DESCRIBING THE STATE OF ORDER OF PET-FIBRES

Hans-Joachim Berndt and Gerhard Heidemann
Deutsches Textilforschungszentrum Nord-West e.V.,
Textilforschungsanstalt, D-4150 Krefeld, FRG

ABSTRACT

A new modified technique of measuring shrinkage forces provides in certain intervals of temperature an alteration of the clamped lengths by a minute stretching or shrinking of the fibres in a manner resulting a zero force deviation. So at each temperature we have an equilibrium of the released inherent stresses of the fibres and the recorded ones. These so called equilibrium-shrinkage-force-curves are capable of describing the state of order of PET-fibres in a better way than other methods do.


The properties of textiles from PET fibres are decisively depending on the stresses set within the material. Depending on the thermal stability of molecular interactions these stresses can be realesed by thermal treatments at different temperatures. As a consequence of stress release the fibres tend to shrink. A series of special thermal treatments is involved in the production of a textile fabric with a predetermined appearance and stable dimensions. Insofar nor shrinkage is desirable or possible in every case, and new stresses of different thermal stability may be set up within the fibres.

The thermal stability of stresses in semicrystalline polymers like synthetic fibres is dependent on the molecular mobility of the chain segments in the noncrystalline regions. This mobility, however, may be hindered by a physical network of crystallites. As can be shown by DTA crystalline networks of different thermal and/or mechanical stabilities may be created during processing of textiles.

Informations on the stresses built into synthetic fibres can be achieved by following up the contraction force (shrinkage force) as a function of temperature when heating the fibres under constant length. The resulting diagrams are called shrinkage-force curves. Providing comparable "effective temperatures" [1] of heating of the samples under investigation these curves are exactly reproducible. By variation of heating rate and/or heat transfer as it is usually the case with different instruments, retarded development of shrinkage force and stress relaxation may be observed.

A typical shrinkage-force curve of a commercial PET-filament yarn is shown in curve "A" of <u>figure 1</u>.

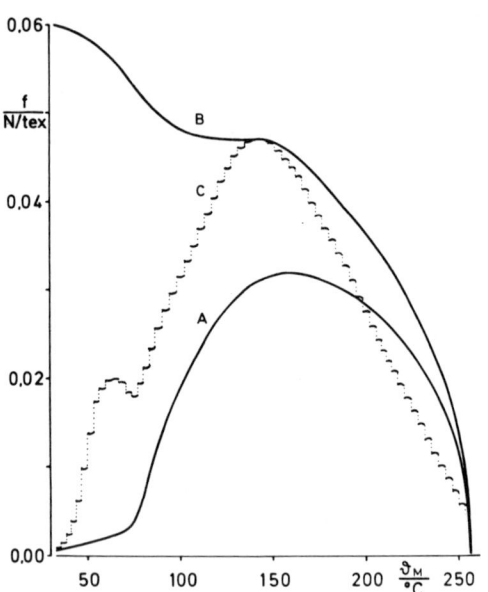

Fig. 1:
Temperature dependent shrinkage-forces of PET-filament yarns
A: Conventional measurement
B: Determination of the effective tension of pretreatment by a relaxation experiment
C: Equilibrium shrinkage-force measurement

Although one gets a quantitative information on tensions applied during the pretreatment of the fibre, there is no direct correlation between that tension and the shrinkage force observed. When increasing the preload of the specimen at the starting point of the thermomechanical investigation - this is done in going from curve "A" to curve "B" - a succession from stress retardation (development of shrinkage force) to stress relaxation will be observed with increasing temperature. During further heating an equilibrium of relaxation and retardation is reached in a certain temperature range just below the effective temperature of a definite pretreatment [2,3].

The stress level of this temperature dependent equilibrium corresponds to that having been set up during prior thermal treatment of the sample. The investigation of effective tensions of unkwon pretreatments by this method involves the measurement of some shrinkage-force curves with different preloads of the samples and requires a certain idea of the effective temperature of the unknown pretreatment which can be obtained by a DTA-analysis of the sample [4].

An equilibrium of stress retardation and relaxation at temperatures higher than the glass transition temperature of the sample will be reached within a few seconds if the shrinkage forces do not differ far from these equilibrium conditions. Therefore, in a thermomechanical analysis it is possible to correct the load during measurement at a constant heating rate by a stepwise alteration of strain realizing an equilibrium of retardation and relaxation. The graph of this experiment is shown in curve "C" of figure 1. In this curve intervals of constant strain during heating steps of about 4 K - corresponding to about 24 s of time - are performed to decide whether in case of an increase in shrinkage force the strain has to be increased or in case of a decrease of the shrinkage force the strain has to decreased. Within the intercepts of this curve with a constant strain very small extrema can be registered at a temperature and shrinkage force which cause an equilibrium of retardation and relaxation within the fibre. Therefore, connecting these extrema we get the so called equilibrium shrinkage-force curve showing the real stresses which are released at each temperature and are depending on the structure of the fibre itself. This curve is in a more or less extended intercept congruent with curve "B. This congruence appears in a temperature range beneath the effective temperature of the pretreatment of the sample in which the effective tension of this pretreatment had been blocked within the material for example by recrystallization during cooling.

As a practical example in figure 2 the equilibrium shrinkage-force curves of two PET-filament yarns are shown which both have been previously treated up to an effective temperature of 190 $^{o}$C at an effective tension of 0,02 N/tex but had been cooled down at different cooling rates.

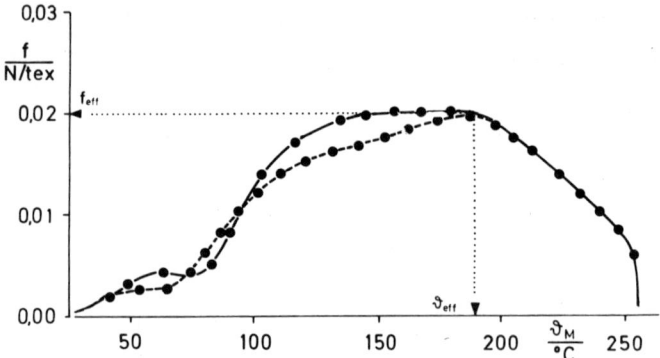

Fig. 2: Equilibrium shrinkage-forces of PET-filament yarns with a pretreatment at $\vartheta_{eff}$ = 190 °C and $f_{eff}$ = 0,02 N/tex and cooled down at different cooling rates of 20 K/min (●——●) and 20.000 K/min (●---●)

The slowly cooled sample delivers an inherent blocked stress of 0,02 N/tex in the temperature range from about 135 °C up to the effective temperature of the pretreatment of the sample. This is due to the stabilizing effect of the crystalline network that has been developed perfectly under the choosen conditions of cooling. Beneath a temperature of about 135 °C no more "perfect" crystallization of the PET material is possible.

On the other hand, in the sample quickly cooled down the recrystallization is imperfect and as a consequence the effective tension of pretreatment ist found only at temperature corresponding to the effective temperature of the pretreatment.

In a next step of investigation we looked at the correlation of the actual tension applied to the yarns during pretreatment and the effective tension measured by the described technique. As can be seen from figure 3 a direct correlation is found up to a tension of about 0,04 N/tex. By this almost the complete range of tension applied during textile processing is covered. The deviation from this correlation at higher tensions than 0,04 N/tex is due to the elastic recovery of the samples after pretreatment. As can be seen from figure 3 and is already well known this elastic recovery is dependent on the stability of the crystalline network within the structure of the fibres.

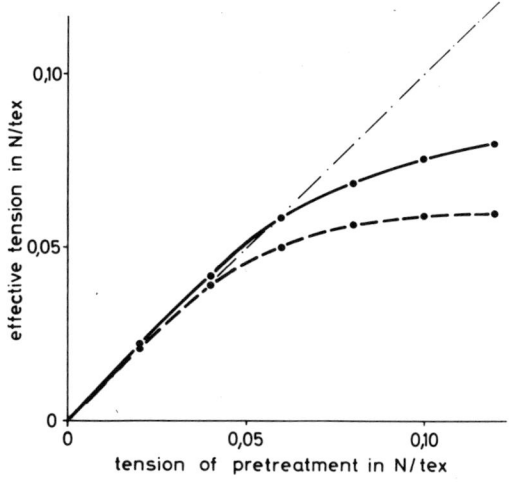

Fig. 3:
Correlation of effective and applied tension of pretreatment. The treatments had been performed at a temperature of $\vartheta_v$ = 190 $^{o}$C during $t_v$ = 60 s. The cooling rates have been 20 K/min (●——●) and 20.000 K/min (●---●)

The experiments have been performed by Mrs. M. Claessen und Mrs. G. Götz on a thermo-mechanical analyser developed in our institute [5] and built as well as distributed under the brand "Thermofil" by Textechno, D-4050 Mönchengladbach.

We are grateful to the Forschungskuratorium Gesamttextil for financial support of this project (AIF-Nr. 4655). The support was granted from resources of the Federal Ministry of Economics through a subsidy of the Association of Industrial Research Organizations (AIF).

REFERENCES

[1]   G. Heidemann and H.-J. Berndt, Melliand Textilber. _57_, 485 (1976)
[2]   H.-J. Berndt and G. Heidemann, Melliand Textilber. _58_, 83 (1977)
[3]   H.-J. Berndt and G. Heidemann, Melliand Textilber. _55_, 548 (1974)
[4]   H.-J. Berndt and A. Bossmann, Polymer _17_, 241 (1976)
[5]   H.-J. Berndt, Melliand Textilber. _56_, 928 (1975)

# THERMOMETRIC TITRIMETRY - A SUITABLE WAY INTO THERMOCHEMISTRY

L. Stäudel, G. Thiel and H. Wöhrmann

Chemische Institute der Gesamthochschule Kassel, FB 19
D 3500 Kassel, Heinrich-Plett-Str. 40

## ABSTRACT

A simple apparatus is described for the thermometric determination of the endpoint of reactions with analytical interest. Examples are the redox reaction $Fe^{2+}/MnO_4^-$, the neutralisation of $H_3PO_4$ and the determination of $K^+$, $NH_4^+$ ($Rb^+$, $Cs^+$) with $B\emptyset_4^-$ with an accuracy of about 2%.

## INTRODUCTION

Many chemical processes are not only accompanied by a change of color or other visible phenomena but rather by a change of energy. Unfortunately only large changes in energy are directly detected by the chemists fingers (forinstance the heat of dilution of sulfuric acid with water) and therefore enthalpy changes are not commonly used as methods to determine the endpoint of an analytical process or the grade of a reagent. Although there are a lot of commercial instruments in the field of calorimetry we decided to build a simple and not too expensive apparatus for use in the beginners analytical courses at universities as well as in schools.

## Theory

Unfortunately the amount of heat evolved or adsorbed within a reaction cannot be measured directely with any kind of an "energy-meter" but only the effect is measurable. If you look at a reaction

$$A + B \xrightarrow{\Delta H} AB$$

you have two ways to use the heat of reaction for analytical applications.
1) The two Substances A and B are mixed in one step and the measured temperature difference $\Delta T$, multiplied with the

heat canacity of the whole system, gives us the molarity n.
$$-n \cdot \Delta H = \Delta T \cdot C_p$$
The analytical method employing this relation (where the molar heat of reaction must be known) is called enthalpimetric titrimetry.

2) In usual titrimetry the molarity of an unknown substance is determined white the ammount and the concentration of the reagent are known. Sometimes it is difficult to find a suitable method for the endpoint determination but the heat of reaction may be used to determine the endpoint of many kinds of reactions. Unfortunately in this technique the heat capacity of the system is changing during the experiment and therefore $\Delta H$ cannot be determined very easily; the method is called thermometric titrimetry.

To get curves as shown in fig 4 it is necessary to have a quick response of the temperature measuring system, a constant flow of the relative by concentrated reagent (to avoid large changes in heat capacity), good insolation, short reaction periods and a quick and reproucibly working stirring system.

## APPARATUS

The temperature measuring element is a thermistor (1o K$\Omega$, 25°C, time constant o,2 sec, dissipation constant o,8 mW °C$^{-1}$ from Siemens) in a constant voltage driven Wheatstone bridge. For small temperature changes the output of the bridge, which is recorded with a strip chart recorder, is proportional the temperature difference $\Delta T$.

Fig 1 shows a simple home made system for the continuous reagent supply which is driven by a "grill motor". This system is usable for schools, the error is not bigger than 5%. In our experiments we used a n electronically controlled piston burette, which was build in our laboratories from a hand driven "Dosimat" from Metrohm". Good hose pumps are suitable too, but rather expensive and one must make sure, that they work pulse free.

fig 1
piston burette

Careful design of the reaction vessel is very important. We use thin walled test tubes (Ø 3cm) with a magnetic stirrer in a foam isolated box as shown in fig 2

fig 2
Reaction vessel

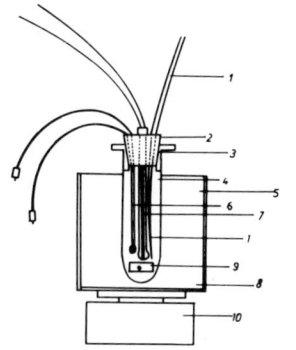

1 PE capillary for the reagent
2 cork stopper
3 PE stopper
4 test tube
5 box
6 thermistor
7 conductivity elektrodes
8 foam isolation
9 spin ball
1o magnetic stirrer

Fig 3 shows the whole assembly.

fig 3

set for thermo-
metric titration

## EXAMPLES

A variety of reactions have been studied by other authors[1] and ourselves[2]. We show here only three different types of reaction.

1) Redox titration of $Fe^{2+}/MnO_4^-$.
   We used 15 ml 0,005 m $Fe^{2+}$ and added 1,5 ml 0,02 m $MnO_4^-$ within a time of ca 20 sec (fig 4)

2) Neutralisation of $H_3PO_4$ with NaOH.
   We used 15 ml 0,052 m $H_3PO_4$ and added 2,35 ml 1 m NaOH within a time of ca 45 sec. Fig 5 shows the result, the dotted line shows the change of conductivity which was simultaneously measured with the electrodes 7.
   This example shows that the change of conductivity, which is often used as endpoint determination for neutralisation reactions yields the first and second step only.

3) Determination of $K^+$ and $NH_4^+$ simultaneously.
   By adding $B\emptyset_4^-$ to a solution of $K^+$ and $NH_4^+$ both ions form unsoluable precipitations (pH =6).
   With formaldehyde which forms Hexamethylentetramin with the $NH_3$ in alkaline solution.

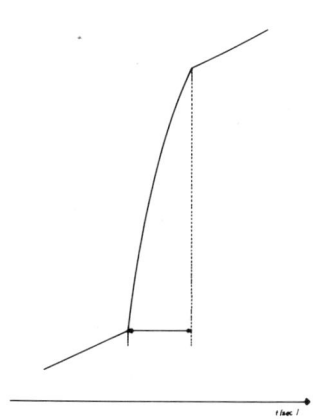 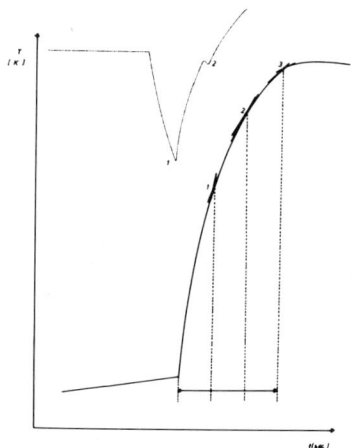

fig 4        fig 5

We used 15 ml o,o1 to o,1 m $(K^+ + NH_4^+)$ and titrated with o,1 m $BØ_4^-$. In the second run only $K^+$ is determined, an exess of formaldehyd is used. This method is of great use in a quick determination of $K^+$ and $NH_4^+$ especially in fertilizers. An accuracy of about 2% can be reached[3]. These few examples may demonstrate how thermometric titrimetry can be used in analytic chemistry. Very good results are obtained by using solvents and reagents with low heat capacity.

Enthalpimetric titrimetry can be used not only for analytical purpose, but also for a quick determination of ΔH by using a submerged burette with a known ammount of reagent and an excess of solution in the reaction vessel. The result is a jump of temperature. For simple calibration the stored energy of a capacitor should be used which generates a known amount of heat in a little heating element within the vessel leading to a second temperature jump. In this way to exotherm heats can be directy compared. Without used of knowing amy calibration of the temperature measuring unit and the heat capacity of the whole system the unkown heat is measurable.

## LITERATURE REFERENCE

(1) G.A. Vaughan, Thermometric and Enthalpimetric Titrimetry, Van Nostrand Reinhold Company, London 1973
(2) H.J. Morgret, G. Thiel u. H. Wöhrmann, Thermometrische Titrationen, MNU 32 (1979), S. 478
(3) L. Stäudel, A. Stille u. H. Wöhrmann, Thermometrische Titrationen von Alkalimetall- und Ammoniumionen mit Natriumtetraphenylborat, GIT Fachz.Lab., 23 (1979), S.291

# SINTERING OF CORUNDUM WITH AMMONIUM FLUORIDE

## A. M. ABDEL REHIM
Faculty of Science, Alexandria University, Egypt.

## ABSTRACT

Using a derivatograph, fluorination of corundum with ammonium fluoride was found to take place in three distinct steps with the formation of aluminium fluoride. The DTA curves indicate the formation of ammonium aluminium hexafluoride at 180°C. The endothermic peak at 225°C represents the dissociation of the resulted ammonium bifluoride & some decomposition of ammonium aluminium hexafluoride. The endothermic peak at 300°C represents the intensive formation of ammonium aluminium tetrafluoride, which dissociates at 360°C to aluminium fluoride.

## INTRODUCTION

Corundum or alpha-aluminium oxide is thermally the most stable form, crystallizing in trigonal system. Hexagonal beta-form & cubic gamma-form are known from synthetic & experimental work; on heating, these forms are both converted to corundum ( 4,7 ).
Ammonium fluoride has paid much attention as important fluorinating agent (1-3,5,6). Its DTA curve two large & sharp endothermic peaks. The first, at 158°- 170°C corresponds to the liberation of ammonia & formation of ammonium bifluoride. The second, at 225°- 240°C represents the dissociation of of the resulted ammonium bifluoride hydrogen fluoride and ammonia vapours.
The reaction of aluminium with ammonium bifluoride was reported to take place at 120°C. The reaction products indicate the presence of ammonium tetrafluoroaluminate & aluminium fluoride & ammonium aluminium hexafluoride (2). The present work represents a derivatographic study of sintering of corundum with ammonium fluoride.

## EXPERIMENTAL PROCEDURE

This research was carried out with corundum, separated from Egyptian black sands. Its chemical composition : 98.80% $Al_2O_3$, 0.56% $SiO_2$, 0.20% $TiO_2$, 0.15% $Fe_2O_3$, 0.22% MgO, 0.06% CaO. The X-ray diffraction data of the processed corundum agreed with those given in the ASTM index. In thin sections, corundum is bluish white, has a high relief, high birefringence, no cleavage & lamellar twinning on (10$\bar{1}$1). Mixes of corundum & ammonium fluoride were processed by repeated grinding followed by sieving. The thermal investigation of sintering was studied by using MOM derivatograph (8). Sensetivity of DTA & DTG circuits was 1/10 & heating rate was $10^{\circ}C\ min^{-1}$. The products of fluorination of corundum were identified both microscopically & by X-ray analysis using a Siemens Crystalloflex diffractometer. Nickel filtered copper radiation was used.

## RESULTS AND DISCUSSION

The thermal analysis data of sintering of corundum with ammonium fluoride mixes of ratios 1:1 and 1:0.8 are shown in Fig. 1 & 2 respectively. The endothermic peak at 180°C represents the formation of ammonium aluminium hexafluoride. The endothermic peak at 225°C may represent the decomposition of the resulted ammonium bifluoride & to some extent decomposition of ammonium aluminium hexafluoride. The sharp endothermic peak at 300°C represents the intensive dissociation of ammonium aluminium hexafluoride with the formation of ammonium aluminium tetrafluoride. These processes are connected with remarkable decrease in weight (TG curve) due to the liberation of ammonia & hydrogen fluoride.

The products of experiments at 110°, 250°, 300°, 400°, & 600°C were identified both microscopically & by X-ray diffraction analysis. It is observed that corundum grains appear in the run at 110°C & 30 min with $(NH_4)_3Al\ F_6$

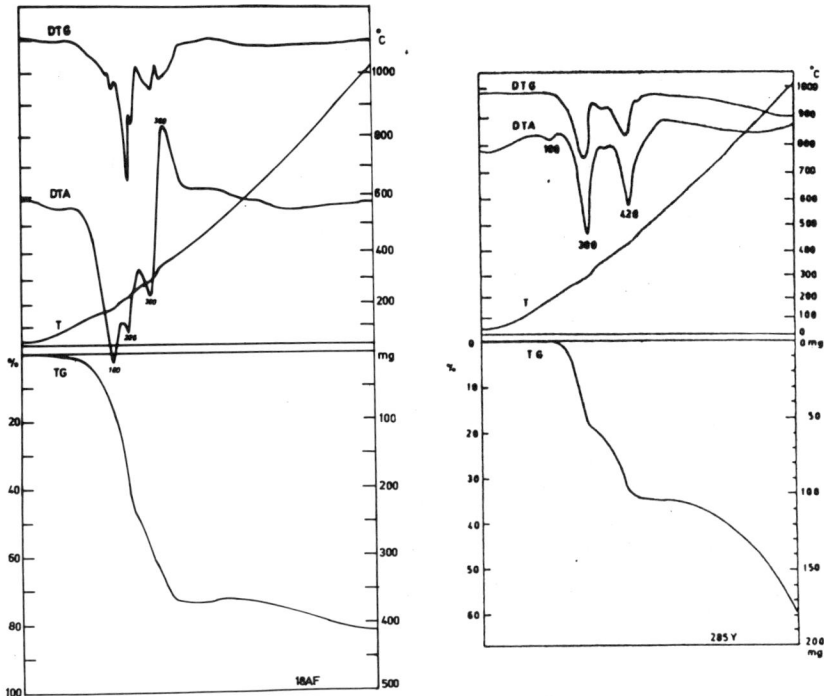

Fig. 1 & 2 : DTA Curves of sintering of corundum & ammonium fluoride mixes of ratios 1:1 & 1:0.8 respectively. indicating incomplete fluorination of corundum (Fig.3A). At 110° & 200°C during 2 hours (Fig.3B), the product is composed completely of ammonium aluminium hexafluoride phase & corundum grains are not observed. At 250°C & 2 hours (Fig.3C), ammonium aluminium tetrafluoride appears with relict grains of corundum. At 300°C & 1 h (Fig.3D), ammonium aluminium tetrafluoride constitutes the total composition of the product. The peaks of corundum & ammonium aluminium hexafluoride completely disappeared. At 400°C & 20 min (Fig.3E), the product consists of undissociated ammonium aluminium tetrafluoride & tetragonal gamma-aluminium fluoride with relict grains of corundum. The product of fluorination of corundum at 110°C & 2 h, was ignited at 600°C during 30 min. Its X-ray diffraction

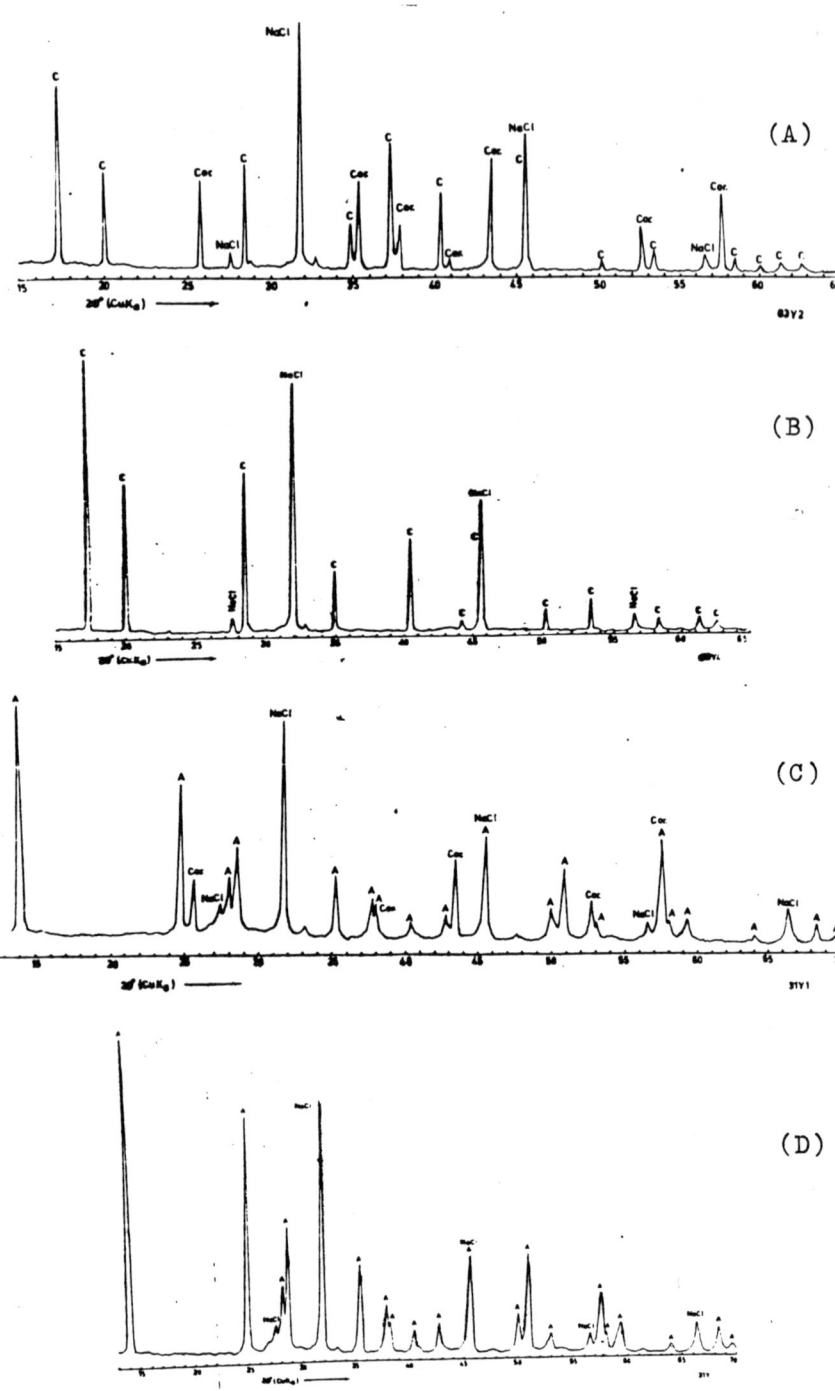

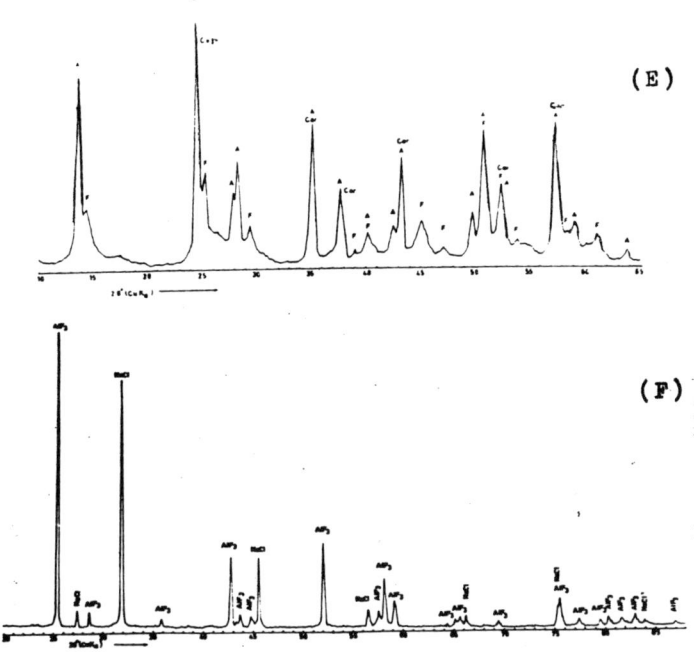

Fig.3 : X-ray powder diffraction patterns of the products of corundum sintering with ammonium fluoride of ratio 1:1 A, B, C, D, E, and F at temperatures & times (110°C, 30 min) (110° or 200°C, 2 h), (250°C, 2 h), (300°C, 1h), (400°C, 20 min) & product of B ignited at 600°C & 30 min respectively. $A = NH_4AlF_4$, $C = (NH_4)_3AlF_6$, Cor = Corundum and F = Gamma-aluminium fluoride.

pattern (Fig.3F) shows the charecteristic & distinct peaks of trigonal aluminium fluoride, which constitutes the total composition of the end product of fluorination of corundum. The peaks are sharp & intense, suggesting good crystallinity.

The obtained results show that fluorination of corundum with ammonium fluoride leads to different products, depending on the temperature. The reaction mechanism can be considered as the following :

At 110°- 180°C : Threaction of corundum with ammonium fluoride takes place with the formation of $(NH_4)_3AlF_6$.

$$Al_2O_3 + 12\ NH_4F \longrightarrow 2\ (NH_4)_3AlF_6 + 6\ NH_3 + 3\ H_2O$$

At 300°C : Ammonium aluminium hexafluoride is unstable and begins to decompose at such temperature, yielding the more stable ammonium aluminium tetrafluoride,

$$(NH_4)_3AlF_6 \longrightarrow NH_4AlF_4 + 2\ NH_3 + 2\ HF$$

or the reaction of corundum with ammonium fluoride takes place at 300°C, according to

$$Al_2O_3 + 8\ NH_4F \longrightarrow 2\ NH_4AlF_4 + 6\ NH_3 + 3\ H_2O$$

At 360°C and above : Dissociation of ammonium aluminium tetrafluoride takes place with the formation of $AlF_3$.

$$NH_4AlF_4 \longrightarrow AlF_3 + NH_3 + HF$$

or

$$Al_2O_3 + 6\ NH_4F \longrightarrow 2\ AlF_3 + 6\ NH_3 + 3\ H_2O$$

## CONCLUSIONS

Sintering of corundum with ammonium fluoride at 110° - 180°C leads to the formation of ammonium aluminium hexafluoride, which dissociates at 300°C to ammonium aluminium tetrafluoride. Ammonium aluminium tetrafluoride decomposes at 360°C with the formation of aluminium fluoride.

## REFERENCES

(1) L. Erdey, S. Gal & G. Liptay, Talanta, 11, (1964) 913
(2) A.A. Opalovsky, V.E. Fedorov & T.D. Fedotova, J.Ther. Anal., 5, (1973) 475
(3) W. Marshal, J. Am. Chem. Soc., 73, (1951) 1867
(4) W.A. Deer, R.A. Howie and J. Zussman, An Introduction to Rock forming Minerals, Longman, London, 1971
(5) A.M. Abdel Rehim, Thermochimica Acta, 30, (1979) 127
(6) A.M. Abdel Rehim, Proceed.5th Int. Conf. Therm. Anal. August 1977, Kyoto, Japan, (1977) 522
(7) R.C. Mackenzie, "Scifax", Differential Thermal Analysis Data Index, Cleaver-Hume Press, London, 1962
(8) F. Paulik, J. Paulik and Erdey L., Talanta, 13, (1966) 1405

# THERMAL ANALYSIS OF CONCRETE

A. Jarmontowicz and R. Krzywobłocka-Laurow
Building Research Institute, Poland

## ABSTRACT

The method of assessing composition of hardened concrete elaborated in Building Research Institute is relatively simple. The method is designed for determination of Portland cement, aggregate - limestone, gravel and sand contents in concrete. The error in the method does not exceed 2% in relation with volume density of concrete in case when components comply with government standards.

## INTRODUCTION

It is frequently necessary to assess in practise a composition of the concrete in building constructions and units.

Thermal analysis /DTA, DTG and TG/ is useful as a method making possible quick determination of water and $CO_2$ content in the concrete combined with cement during setting and hardening. The results obtained from thermal analysis connected with the results of determination of volume density and unsolved parts in HCl form the base to determine the composition of the investigated concrete (1).

## EXPERIMENTAL PROCEDURE

Sampling for the investigations of the concrete is made by cracking off or split off irregular lumps from the building unit or the construction. The samples of the concrete used in destructive tests of strength also can be utilised. The weight of sample depends on size of the aggregate grains. Minimum weight of the sample is not less than 3 kg and the size of aggregate grains not exceed 40 mm. The concrete samples with the coarser aggregate not less than 5 kg are used. Unless there is lack of data the used aggregate a petrographic examination should be performed by using a few representative grains of aggregate selected from concrete.

The next a volume density of concrete should be determined by using a direct method in case of regular shape of samples and a hydrostatic one when samples are irregular. Regardless of shape the determination should be tested on 3 samples. In case of limestone and sand used in concrete the total content of aggregate should be calculated by formula :

$$K = K_w + K_p \qquad (1)$$

where : $K_w$ - limestone content , %
$K_p$ - sand content , %

The determination of limestone aggregate bases on following assumption that limestone decomposes to calcium oxide and carbon dioxide at temperature range 580 to 1000°C. The loss of weight caused by this decomposition after being multi - plied by coefficient 2.27 means calcium carbonate content in the concrete under consideration.

The sample of concrete is ground fine and passes a 1 mm sieve for thermal analysis. The next by quartering of sample 50 - 100 g is weighted and is drying at 105° C until con - stant weight is obtained. After drying and careful mixing 5- 10 g is weighted from the prepared sample for thermal analysis /DTG, DTA and TG/. Clay minerals and quartz impurities are generally contented in limestone. When content of impurities exceed 1% the quantity of aggregate is fixed by suitable correction. When limestone is given the value of correction - P should be estimated in two different ways : by the thermal analysis method - $P_1$ or by the determination of unsolved parts in HCl - $P_2$ according to formulas :

$$P_1 = 100\% - 2.27A \qquad (2)$$
$$P_2 = K_{cz.n.} \qquad (3)$$

where : A - loss of weight of aggregate sample caused by decomposition of $CaCO_3$ , %
$K_{cz.n.}$ - unsolved parts in HCl content of limestone, %

The correction determined by thermal analysis is more accurate than the second one because both clay minerals and silica are partly soluble in HCl.

It should be pointed out that calcium carbonate is formed

during the process of carbonation of cement paste. Considering of that the aggregate content should be reduced by quantity of formed carbonate during carbonation. This value determined by experiments equals 2.5 for hardened concrete from 0.5 to 3 years. Limestone aggregate content in concrete should be obtaned by formula :

$$K_w = 2.27 / A - 2.5 / + P_1 \qquad (4)$$

where all the values are as in 2.

Value $P_1$ may be replaced with $P_2$ according the formula 4. Cement content in 1 $m^3$ of concrete should be obtained by formula :

$$C = \gamma_b - K - S \qquad (5)$$

where : K - aggregate content in concrete , kg
C - cement content in concrete , kg
S - quantity of compounds combined with cement during setting and hardening , kg
$\gamma_b$ - volume density of concrete

A bound water in hydration products and $CO_2$ in carbonation products of cement paste in concrete with limestone aggregate is determined by thermal analysis as loss of weight of sample at temperature range from 20 to $580^\circ$ C.

Described method can be applied to concrete with other aggregate chosing the suitable range of temperature determing loss of weight assumed as bounded water and $CO_2$ content in sample.

## RESULTS AND DISCUSSION

There is an example for determination of composition of hardened concrete with limestone aggregate /see Table 1 /. On the base of the numerous investigations carried out in Building Research Institute this method for testing laboratory samples and also samples taken on building site can be stated that this method is characterized by :

(i) good reproducebility of results under conditions of suitable homogenization of samples,

(ii) greater accuracy and easy performing of tests in comparison with classical chemical analysis.

Results of determination of aggregate and cement content in concrete    Table 1

| Characteristic | | Denotation result, unit | performance, result, unit variant I /without correction/ | Calculation variant II /with correction/ |
|---|---|---|---|---|
| Volume density of concrete $\gamma_b$ | | 2275 kg/m³ | | |
| Unsolved contents in HCl in concrete | | 25,7 % | | |
| Content | impurities in limestone | 5,0 % | | |
| | soluble substances in HCl in sand | 0,44 % | | |
| Content in concrete | limestone $K_w$ | 55,8 % | | $55,8 + 2,79 = 58,59$ % |
| | sand $K_p$ | 25,7 % | | $25,7/1 + \frac{0,44}{100}/ = 25,8$ % |
| Total aggregate content in concrete | $K_\%$ | | $55,8 + 25,7 = 81,5$ % | $58,59 + 25,81 = 84,4$ % |
| | $K_{kg}$ | | $\frac{81,5}{100} \cdot 2275 = 1854$ kg | $\frac{84,4}{100} \cdot 2275 = 1920,0$ kg |
| Components combined during setting and hardening | $S_\%$ | 5,2 | | |
| | $S_{kg}$ | | $\frac{5,2}{100} \cdot 2275 = 118$ kg | |
| Cement content in concrete | $C_{kg}$ | | $2275 - 1854 - 118 = 303$ kg | $2275 - 1920 - 118 = 237$ kg |

Moreover using thermal analysis the additional data concerning hydration and carbonation of cement paste in concrete are obtained.

## REFERENCES
(1) A. Jarmontowicz, R. Krzywobłocka-Laurow, Instruction for determination of composition of hardened concrete, ITB 1978 / in polish /

# APPLICATION OF THERMAL ANALYSIS TO THE INVESTIGATION OF PHASE COMPOSITION OF AUTOCLAVED CEMENT PASTES AND MORTARS

Irena Stebnicka - Kalicka
Building Research Institute, Warszawa, Poland

## ABSTRACT

Thermal analysis has shown that autoclaving of cement pastes and mortars generates products with lower content of bounded $H_2O$ than that obtained on curing under other conditions /air - dry, water, steam/. As a result an increased porosity of autoclaved materials is observed. Basing on the data obtained by thermal analysis, X-ray diffraction and mercury porosimetry the following conclusions can be drawn: lower strength of autoclaved cement materials, particulary in comparison with specimens cured in steam or in water, is caused by unfavourable effect of increased porosity and by the presence of hydrogarnets $C_3AH_6-C_3ASH_4$[1]. Absence of $\alpha-C_2SH$[1] has been proved. Besides, the increased porosity of autoclaved cement pastes and mortars caused an intensive carbonation with more quantity of cryptocrystalline $CaCO_3$ than in other specimens.

## INTRODUCTION

Experience of many building prefabrication factories and results obtained in numerous works /1/, /2/, /8/ have shown that curing under high pressure that means autoclaving is advisable when a part of cement, about 30-70% by weight, is replaced with some part of fine-grained substance containing $SiO_2$ /i.e. grunded sand, fly-ashes etc/. Then the proper strength of material is obtained because low-basic calcium hydrosilicates of tobermorite and C-S-H /I/[1] group are produced. The strength of cement materials autoclaved without the mentioned addition is generally lower than that of steam-cured or cured in normal conditions materials /except

---

[1] These chemical formulas are in conformity with notation used in chemistry of cement e.g. $CaO-C$, $SiO_2-S$, $Al_2O_3-A$, $H_2O-H$.

short period after autoclaving/. Some authors attribute this lower strength to the presence of high-basic calcium hydrosilicate $\alpha$-$C_2SH$ which forms as the consequence of the low $SiO_2$ content /1/, /3/, /8/. The other /6/, /9/ assume this worse strength is caused by increased porosity which is a result of formation the hydration products with the content of bounded $H_2O$ lower than that of products formed under other curing conditions. Our investigations support the latter and have additionally gave other information about the phase composition of autoclaved cement materials.

## EXPERIMENTAL PROCEDURE

The cement pastes and mortars of cement-sand proportion 1:3 were investigated. Water-cement ratio was of 0,28 for pastes and 0,5 for mortars with the same plasticity of fresh mixtures. The following curing conditions of cylindrical samples of diameter = height = 3 cm were used: 1/ on the air at $20°C$, 2/ in water at $20°C$, 3/ steam-curing at $80°C$ during 16 hrs, 4/ autoclaving at $140°C$ during 13 hrs. Series 1 and 2 were taken in moulds during the first 24 hrs at RH = 95% and $20°C$, the next they were demoulded. Hydrothermic treatment was applied in moulds after 3 hrs of precuring. After treatment demoulded samples were kept on the air at $20°C$. In this experiment Portland cement 350 consisting of /% by weight/: CaO-64.7, $SiO_2$-23.63, $Fe_2O_3$-1.73, MgO-2.44, $Na_2O$-0.36, $K_2O$--0.46, loss of ignition-1.43, was used. Phase composition was examined by thermal analysis /MOM derivatograph, Hungary/, X-ray analysis /diffractometer TUR M61, GDR/, mercury porosimetry /C.Erb's model 1520, Italy/.

## RESULTS AND DISCUSSION

The results of experiments have proved that compressive strength of autoclaved pastes and mortars is undoubtedly lower than that of steam-cured or stored in water specimens except paste cured in water till 28 days /Fig.1/. On the contrary the strength of cement materials cured on the air is lower than of autoclaved ones.

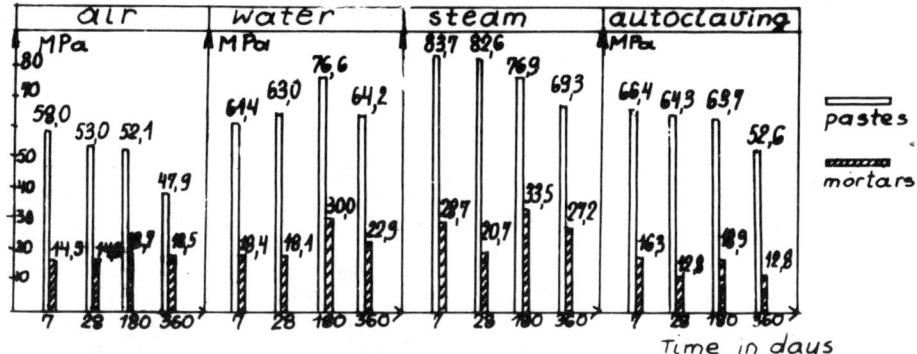

Fig.1. Compressive strength of cement materials cured in different conditions

Thermal and X-ray analysis have shown that phase composition of autoclaved materials is different from that of other samples. There are endothermic peaks on DTA and DTG curves of mortars cured on the air, in water or by steam, which are absent on the autoclaved samples thermograms /Fig.2/. In our opinion these peaks are connected with the AFm phases i.e. $C_4AH_{13}$ and/or $C_3A \cdot C_{\bar{s}} \cdot H_{12}$ /4/. The presence of these components is also proved by lines 8.2 Å /$C_4AH_{13}$/ and/or 8.9 and 1.66 Å /$C_3A \cdot C_{\bar{s}} \cdot H_{12}$/ on diffractograms of all specimens except autoclaved ones. On the contrary only on the thermograms of autoclaved pastes are wide, well-marked endothermic peaks with extremum at 400°C /7-days sample/ or at 370°C /1 year sample/. We attribute these peaks to dehydroxilation of calcium hydrogarnets of $C_3AH_6-C_4ASH_4$, composition which was supported by presence of following lines: 5.05, 4.39, 2.753, 2.036 Å. We could not attribute the peak at 400°C to the dehydration of $\alpha-C_2SH$ because the lines of this phase /4.2, 3.27, 2.41 Å/ were absent. The whole quantity of bounded $H_2O$ registrated at TG curve as the the loss of weight up to 550°C /Fig.3/ is in autoclaved pastes unquestionably lower than in other samples. Before measurements the unbounded water was eliminated by drying in gaseus flux of nitrogen at 30°C. We assumed that lower bounded $H_2O$ content in autoclaved pastes should result their higher porosity than that of other samples and this was supported by the data of porosimetry examinations /Table 1/. These statements refer to both pastes and mortars although the evidence for the latter is not so strong because of the sand in preparations.

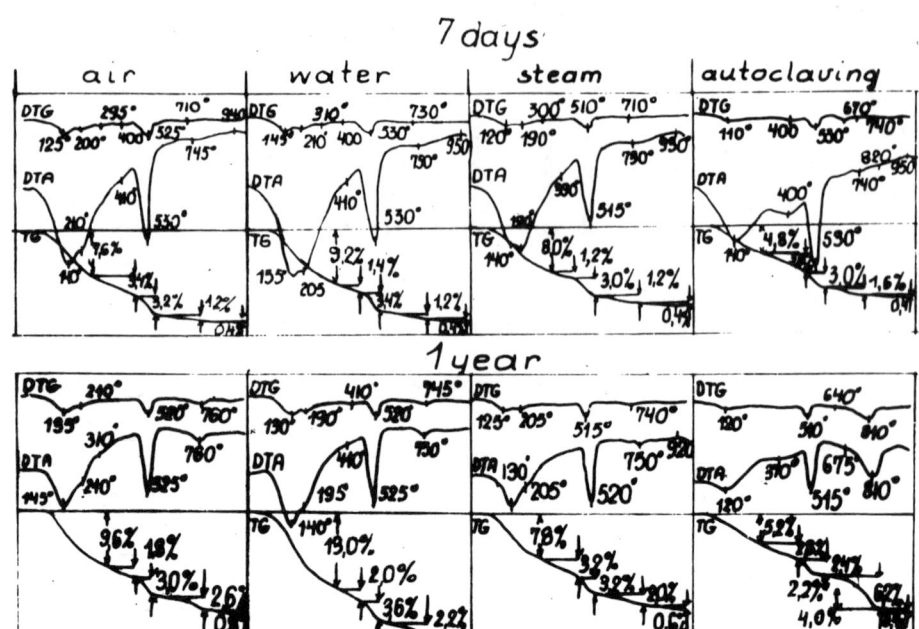

Fig.2. DTA, TG and DTG curves of cement pastes cured in different conditions /after 7 days and one year/

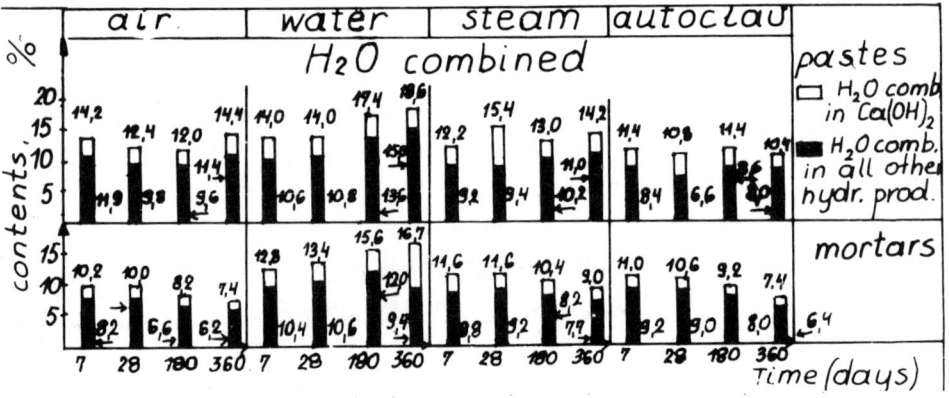

Fig.3. Contents of bounded H$_2$O in cement materials cured in different conditions

The more intensive carbonation of autoclaved samples was, undoubtedly, caused by their higher porosity /Fig.2, 4, and 5/. This was especially observed in mortars which are more porous than pastes /Fig.4 and 5/.

Table 1
Porosity of pastes

| Curing condition | Pore volume /cm$^3$/g/ | Specific surface /m$^2$/g/ |
|---|---|---|
| air | 0.095 - 0.099 | 3.75 - 3.950 |
| water | 0.009 - 0.016 | 1.671 - 2.730 |
| steam | 0.097 - 0.103 | 6.413 - 7.170 |
| autoclaving | 0.110 - 0.129 | 7.974 - 10.270 |

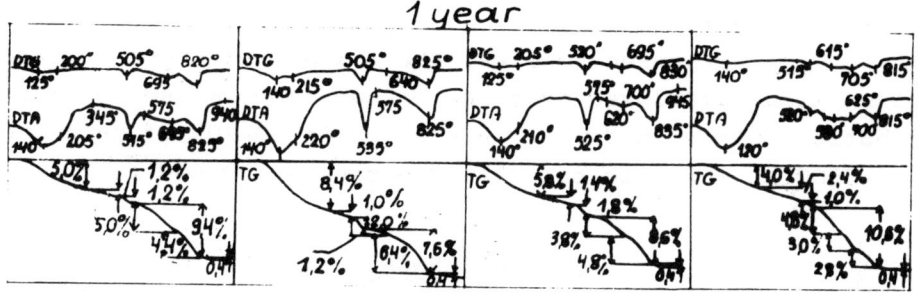

Fig.4. Thermograms of mortars cured in different conditions /after one year/

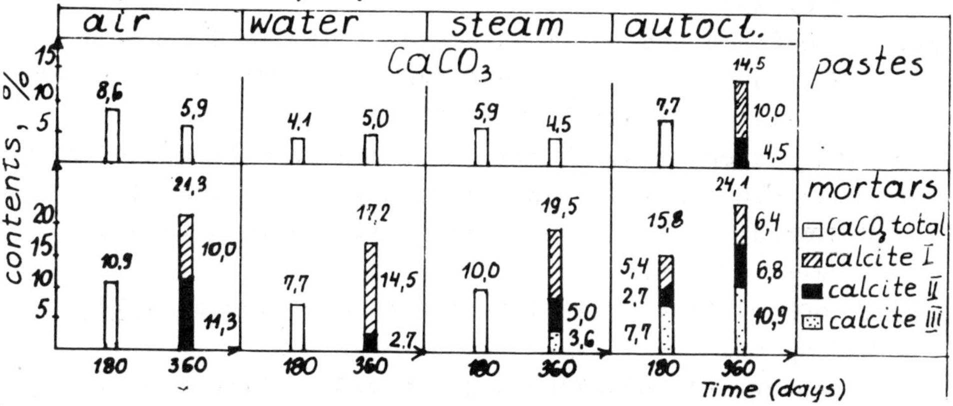

Fig.5. CaCO$_3$ content in cement materials cured in different conditions

CaCO$_3$ formed is mostly cryptocrystalline vaterite which decomposes in such a low temperature as 580-600°C /Fig.4/. The coarse-crystalline CaCO$_3$ dissociating at higher temperature i.e. at 800-835°C we have called, according to Šauman /7/

calcite I and the calcium carbonate dissociating at 700°C
and 620°C respectively calcite II and III.

The obtained results permit the following conclusions: lower strength of autoclaved cement materials in comparison with steam and water cured ones were caused by higher porosity and presence of calcium hydrogarnets of $C_3AH_6-C_3ASH_4$ series. The latter is supported by 2 facts: /i/ the $\alpha-C_2SH$ phase was absent in tested autoclaved specimens although the treatment temperature $120 < t < 175°C$ should have favorised formation of this phase, /ii/ the strength of calcium hydrogarnets is very low. According to Jambor /5/ it is only about 4% of that of tobermorite.

As for the air-cured samples, their lower strength in comparison with autoclaved ones may be caused by worse pore distribution. Since in dry-air pastes radii 60-70% of pores were 500-3000 Å, while in autoclaved pastes - 100-500 Å.

## REFERENCES

/1/ J.J.Beaudoin and R.F.Feldman. Cem.and Concr. Res. v.5 /1975/ 103

/2/ P.I.Boženov and V.I.Kavalerova. Trudy Mezd.Konf. po Probl.Uskor.Tverd.Bet. pri Izgot.Sbor.Železob. Konstr. Izd.Lit. po Stroit. /1968/ 232

/3/ Y.M.Butt and L.N.Raškovič. Tverdenije vjažuščich pri povyšennych temperaturach. Izd.Lit. po Stroit. /1965/

/4/ V.S.Gorškov. Termografia stroitelnych materialov. Stroizd. /1968/.

/5/ J.Jambor. Proc. V Intl.Symp.Chem.Cem. Tokyo. v.III /1968/ 541

/6/ J.Jambor. VI Meżd.Kongr. po Chim.Cem. Strojizd. t.II. kn.I. /1976/

/7/ Z.Sauman. Cem. /jug/ 7-9. /1971/ 135

/8/ H.F.W.Taylor. IV Meżd.Kongr. po Chim.Cem. Strojizd. /1964/ 140

/9/ A.V.Volženskij. VI Meżd.Kongr. po Chim.Cem. Strojizd. t.2. kn.2 /1976/ 91

# EFFECT OF ELEVATED TEMPERATURES ON THE HYDRATION OF CEMENT INVESTIGATED BY EMANATION THERMAL ANALYSIS

Vladimír Balek
Nuclear Research Institute, Řež, Czechoslovakia
Jiří Dohnálek
Building Research Institute, Technical University, Prague
Czechoslovakia
Wolf-Dieter Emmerich
NETZSCH-Gerätebau, Ltd., SELB, F.R.G.

## ABSTRACT

Emanation thermal analysis (ETA) has been applied to the investigation of hydration of cement at temperatures of 20, 35, 45, 65 and 85 $^\circ$C. ETA measurements enabled us to assigne the time intervals where different stages of hydration process take place. The structure of intermediate products, controlling the mechanical properties of the cement paste during hardening can be tested by ETA. ETA results are compared to measurements of penetration resistance. Using NETZSCH-equipment, ETA can be carried out automatically at required temperature and in gas medium with defined humidity, i.e. it can be used for testing cements during setting and hardening directly in technological conditions.

## INTRODUCTION

Among the most important problems of the modern technology of concrete are processes taking place during hydration of cement at elevated temperatures. In spite of the number of experimental works dealing with this topic the knowledge of this field is not sufficient enough with regard to the needs of technology. This paper aims to show the possibilities of emanation thermal analysis in investigating the hydration of cement at temperatures up to 85 $^\circ$C.

## EXPERIMENTAL

Emanation thermal analysis (ETA) is based on the measurement

of radioactive inert gases released from the materials previously labeled with radioactive isotopes. The labeling can be achieved by a number of methods (1), e.g. by the incorporation of the trace amounts of parent isotopes of radon ($^{228}$Th and $^{224}$Ra) into the sample, which represent a constant source of radon in the sample. Radon atoms are produced in the solid by disintegration: $^{228}$Th $\xrightarrow{\alpha}$ $^{224}$Ra $\xrightarrow{\alpha}$ $^{220}$Rn $\xrightarrow{\alpha}$ .
The simplified expression for the release rate of radon from the solid grain is valid

$$E = S \left[ K + (D/\lambda)^{1/2} \right] \rho \qquad /1/$$

where E is the radon release rate called emanation power, S - is the specific surface area, K - is a constant independent of temperature, D - is the diffusion coefficient of radon in the solid, $\lambda$ - is the decay constant of radon, $\rho$ - is the density of solid. Hence by measuring the rate of radon release we can obtain an information on changes of surface area and diffusion coefficient of radon in the solid. Penetration resistance (R) was determined as the ratio of the strength needed for the penetration of a cylindrical indentor into the setting cement paste to the cross-sectional surface of this indentor (2).

Preparation of samples. Samples of cement suspensions (about 1.0 g) were prepared by mixing slag cement (Čížkovice, ČSSR, surface area of 0.315 $m^2g^{-1}$ after Blaine) with water in ratio w/c = 0.3. For ETA measurements the cement powder was labeled by impregnation with a non-aqueous solution containing trace amounts of $^{228}$Th and $^{224}$Ra. The radioactivity of the labeled samples was 5 x $10^5$ Bq per 1 gram of the sample. For measurements of penetration resistance non-labeled cement was used. ETA apparatus used (which is a part of a series of equipment for simultaneous thermal analysis ETA-DTA-TG/DTG by NETZSCH (3)) allowed to perform the measurements under various temperatures maintained automatically. The cement suspensions were immediately after mixing homogenized for 20 s and put into the measuring cell of the ETA apparatus and thermostatically controlled at 20, 35, 45, 65 and 85 $^o$C. The air saturated by water vapour at the given temperature (flow-rate of 0.7 cm$^3$s$^{-1}$) passed over the sample and carried the radon released into the measuring chamber. ETA results are presented as time dependences of relative emanation power E, calculated from experimental data as E = A /A , where A is alpha radioactivity of sample measured in dynamic conditions of experiment, A is beta radioactivity of the sample. The measurements of penetration resistance (R) were carried out at standard temperatures (4) of 20 $\pm$2 $^{\circ}$C. During measuring time (60 s) the sample had to be removed from the thermostat.

## RESULTS AND DISCUSSION

When describing the setting and hardening of cement suspensions we usually distinguish several stages:
(i) primary hydration of cement on the surface of the grains which takes place immediately after mixing cement with water. In this stage a layer of hydration products is obviously formed on the surface of the grains which causes the slowing down of water transport and consequently brakes the hydration reaction; (ii) incubation of reaction characterized by a minimum hydration rate; (iii) the stage where the layer formed on the surface of cement grains begins to be destructed and the rate of hydration increases again. In this stage the decrease of deformability of the cement suspension takes place followed by an increase of its strength.

Fig 1 shows time dependences of relative emanation power E of cement suspensions at temperatures of 20, 35, 45, 65 and 85 $^{\circ}$C. The primary hydration (i) is indicated on every ETA curve analogically, by a steep increase of E, which brings about the formation of hydration products of relatively large surface area (most probably ettringite of large surface area and non-crystalline hydrated calcium silicate). The decrease of E indicates on all ETA curves the formation of relatively dense layers on the surface of cement grains where radon diffusion rate significantly decreases. Consequently, minimum permeability of this layer towards water molecules can be supposed.

The incubation stage (ii) is characterized in Fig. 1 by constant, relatively low values of E. The duration of the incubation period is shortened when hydration takes place at elevated temperatures. Values of $t_i$-time corresponding to the end of incubation period (indicated in curves 1 - 5 in Fig.1) fit the relation

$$t_i = 125 \, e^{-0.039 \, T} \qquad /2/$$

where T - is temperature in $^\circ$C at which the hydration process is carried out. Correlation coefficient r = 0.996.
Stage (iii) of hydration process has two periods. In the first period of this stage the increasing rate of the structure transition due to the cement hydration is indicated by ETA, followed by more or less intense decrease of E. After time $t_s$ (see Fig. 1) the rate of hydration has slowed down in such a degree that the second period of the stage (iii), where the hydration of remaining cement takes place, can last several months or even years.

The increase of E indicates the intense increase of the rate of hydration, being caused by the destruction of primarily formed surface layer, non-permeable for water (4,5). The following decrease of E is caused mainly by the decrease of cement hydration rate and also by the densification and recrystallization of the structure of hydrated cement paste. When comparing values of E of the samples studied corresponding to times $t_s$ (see Fig. 1) and greater, the structures of the cement paste hardened at temperatures of 35, 45, 65 and 85 $^\circ$C differ sufficiently. As it follows from Fig. 1, value of $t_s$ - the time of the end of the first period of this hydration stage (iii) - is shortened when hydration occurs at elevated temperatures. Following temperature dependence of $t_s$ has been obtained:

$$t_s = 780 \, e^{-0.039 \, T} \qquad /3/$$

correlation coefficient being 0.999.

Time dependences of penetration resistance (R) of samples cement suspensions hardened at various temperatures are given together with ETA curves at the same temperatures in Fig. 2.

The penetration resistance (R) measured at various temperatures begins to increase from the end of the incubation period ($t_i$), as indicated by ETA curves. In the time period of the intense increase of E, the increase of R is relatively lower, whereas in the period of decreasing E, R increases rapidly. This is in agreement with the modern conception of the hydration of cement (6,7).

The time dependence of R corresponding to 65 and 85 $^{\circ}$C (curves 4´ and 5´, Fig. 2) is slightly retarded with regard to the end of incubation period, which is due to the cooling of the sample caused by measuring R at 20 $^{\circ}$C.

## CONCLUSIONS

ETA sensitively indicates changes of surface and structure of cement paste during setting and hardening and enables us to distinguish different stages of hydration process and to determine their duration. It was proved that structure changes indicated by ETA during cement hydration at temperatures between 20 and 85 $^{\circ}$C are related to changes of mechanical properties.

The advantage of ETA is the ability to investigate the cement hydration continuously at given temperatures and in gas medium of required humidity, consequently, it can be used for the control of the technological process of the production of precast concrete.

## ACKNOWLEDGMENT

The authors express their gratitude to Prof.Dr.V.Šatava, Institute of Chemical Technology, Prague, for valuable discussion.

## REFERENCES

(1) V. Balek, Thermochim. Acta 22 (1977), 1-156
(2) V. Šatava, Silikattechnik 6 (1955), 338
(3) W.D. Emmerich, V. Balek, High Temp.-High Pressures 5 (1973), 67
(4) G.M. Idorn, Proceedings of 5th int. symposium on the chemistry of cement, Part III (1968), p. 311
(5) M. Venuat, Proceedings of the 6th int. congress on the chemistry of cement, Vol. II, Book 2 (1974), p. 109

(6) V. Balek, J. Dohnálek, Zement-Kalk-Gips, in print
(7) V. Balek, V. Šatava, J. Dohnálek, Proceedings of the 7$^{th}$ int. congress on the chemistry of cement (1980),in print

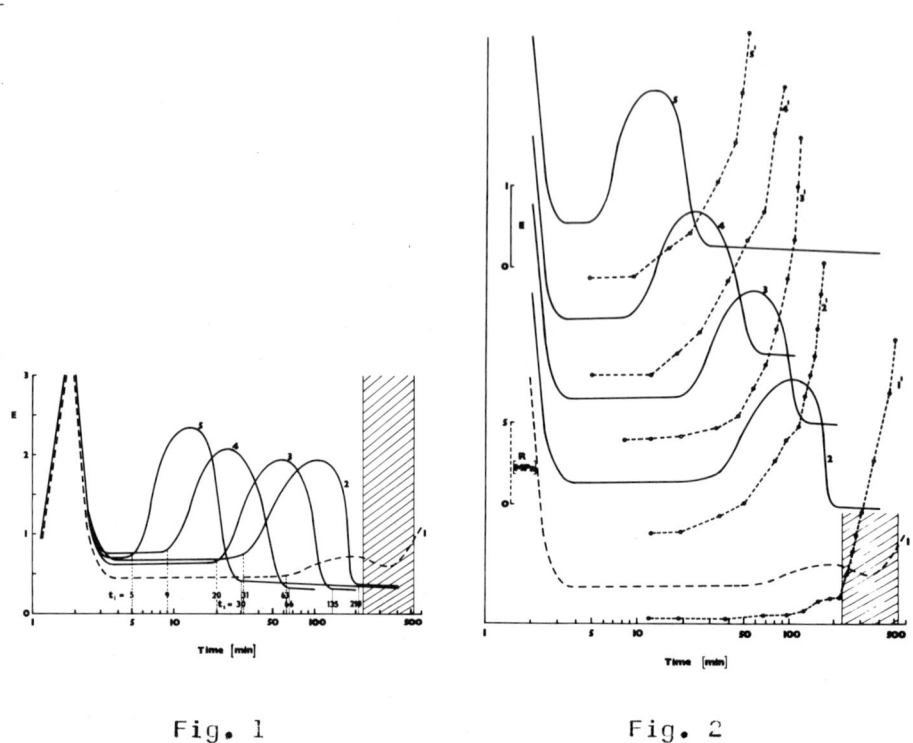

Fig. 1    Fig. 2

Fig. 1: Time dependences of relative emanation power E during the hydration of cement at temperatures of 20, 35, 45, 65 and 85 °C (curves 1, 2, 3, 4 and 5, resp.). $t_i$ - is the time of the end of incubation period (in min.), $t_s$ - is the time of the end of first period of hydration stage(iii).

Fig. 2: Time dependences of E (curves 1, 2, 3, 4, 5) and penetration resistance R (curves 1´, 2´, 3´, 4´and 5´) during hydration at temperatures: 20, 35, 45, 65 and 85 °C (corresponding curves are noted as in Fig. 1). The hatched area indicates the onset and the end of the setting period of cement paste determined by standard Vicat test at 20 °C.

A THERMOGRAVIMETRIC METHOD FOR STUDYING THE REACTION BETWEEN FLY-ASH AND
THE CALCIUM HYDROXIDE LIBERATED ON HYDRATION OF PORTLAND CEMENT.

F.G. Buttler and S.R. Morgan

Chemistry Department, Teesside Polytechnic, Middlesbrough, U.K.

ABSTRACT

The method of analysis involves the conversion of calcium hydroxide into calcium carbonate using gaseous carbon dioxide, followed by the decomposition of the calcium carbonate formed. The amount of the latter is proportional to the amount of calcium hydroxide in the samples provided they are heated to 200°C before exposure to the carbon dioxide. In this way the carbonation of hydrated calcium silicates and aluminates in the samples is prevented. The method has been used to monitor the amount of calcium hydroxide in set Portland cement pastes and to study the amount of reaction between the calcium hydroxide liberated from Portland cement with fly ash in mortars.

INTRODUCTION

A number of methods have been used to study the reactions between water and the compounds present in Portland cement, and many of the methods involve the estimation of the amount of calcium hydroxide produced. These methods include quantitative X-ray diffraction, thermal analysis and the extraction of the calcium hydroxide with solvents (1). Another method for studying these systems is that of trimethylsilyation which gives information with respect to the silicates produced (2). The methods used to estimate the amount of calcium hydroxide present tend to give widely differing results. Quantitative X-ray diffraction only measures the amount of crystalline material present and amorphous phases will not be detected. Solvent extraction suffers from the disadvantage that the amount of $Ca^{+2}$ ions removed is dependent on the extraction procedure and on the compounds present. The interpretation of the results from thermogravimetry can be difficult due to the overlap of steps corresponding to the loss of water from different compounds, and similar problems arise with differential thermal analysis (3).

This analytical method is based on the fact that, as it dehydroxylates, calcium hydroxide reacts rapidly with gaseous carbon dioxide to form calcium carbonate, and provided there is sufficient time of exposure the reaction can be used quantitatively. The amount of calcium hydroxide originally present can then be determined by using the high temperature decomposition of the calcium carbonate formed. It has been found that provided the bulk of the water present in the samples is removed before exposure to carbon dioxide, there is little or no carbonation of the other compounds arising from the hydration of the Portland cement. This procedure overcomes the problems associated with the concurrent loss of water from hydrated calcium silicates and aluminates and from calcium hydroxide.

EXPERIMENTAL PROCEDURE

Suitable portions of calcium oxide, prepared by ignition of 'Analar' calcium carbonate were weighed into the crucibles to be used for the thermal analysis, mixed with finely divided quartz, and the calcium oxide converted into hydroxide by adding carbon dioxide free water immediately before analysis.

Pastes of Portland cement and water with a water : cement ratio of

0.45 were prepared and stored at 20°C in closed tubes to minimise loss of water. Samples for analysis were obtained by drilling the pastes after different periods of time.

Six series of mortars were prepared as 5.0 cm cubes. These mortars ranged from 1:3 mixtures of Portland cement : sand (BS 4550 Part 6 quality) to ones in which up to 75% of the cement was replaced by fly ash. The mortars were machine mixed with sufficient water to attain a measured ASTM flow consistency of 100-110%. The amount of gauging water required decreased as the amount of fly ash in the mortars increased. After compaction by vibration the cubes were sealed and stored immediately, still in their moulds, in water at 20°C. All the cubes were removed from the moulds after 22 hours, sealed in polythene air evacuated envelopes and then taken through individual curing cycles by immersing in water baths at fixed temperatures. After 11 days the mortars were drilled and the calcium hydroxide content determined by the same thermoanalytical procedure. The details of the analyses of the cement and fly ash used, and of the curing regimes of the mortars, are given in Table 3. Details of the compositions of the mortars are shown in the Figures.

The thermogravimetric analyses were carried out using Stanton Automatic thermo-recording balances, models $HT_5$ and $TR_1$. The samples were contained in conical fused alumina crucibles and the mass changes determined in dynamic atmospheres of nitrogen and carbon dioxide with flow rates of 100 $cm^3$/min. Various heating rates and sample and crucible sizes were used in the experiments starting with calcium hydroxide. For the samples involving cement pastes and for the cement:fly ash:sand mortars a heating rate of 3°C/min was used.

### RESULTS AND DISCUSSION

The results from the carbonation of calcium hydroxide are shown in Table 1. The samples were heated in nitrogen and the temperature held at 200°C until the evolution of water was negligible. The temperature was then raised again but the atmosphere was changed to 80% nitrogen and 20% carbon dioxide. The carbonation reactions were completed by about 650°C but the same purging gas was used until about 750°C in order to obtain a good reference plateau on the TG curves before the decomposition of the calcium carbonate. Above 750°C the purging gas was changed again to nitrogen in order to allow the calcium carbonate formed to decompose easily. The same technique was used for all the other samples.

The results show that, starting with calcium hydroxide, and using a variety of heating rates, sample weights and crucible sizes, and different concentrations of calcium hydroxide in the samples, over 90% of the calcium hydroxide was converted into calcium carbonate.

The results obtained on the rate of hydration of the Portland cement paste are shown in Table 2. The results are expressed with respect to the anhydrous weight of the cement used and there was little change in the calcium hydroxide content after 64 days at 20°C.

Fig. 1 shows the thermoanalytical results obtained from the mortar samples. The amount of calcium hydroxide present is expressed as a percentage of the anhydrous weights of the samples. The calcium hydroxide content was determined by the method described above but it was necessary to make slight corrections to the loss in the final step in order to correct for the loss in mass from the fly ash under the same conditions and over the same temperature range. This was done both by studying the fly ash separately and by extrapolating the results from the mortars cured at 20°C to zero cement concentration. The latter was possible since with such mortars after 11 days there was no observable reaction between the calcium hydroxide liberated from the cement with the fly ash. Complete replacement of all the cement by fly ash gave an 'apparent' calcium hydroxide content of 0.63%.

The following is apparent.

(a) With mortars cured for 11 days at 20°C there is a straight line relationship between the percentage calcium hydroxide and the percentage cement in the mortars.

(b) For all the other curing regimes, in which both the maximum temperature attained and the time at temperatures in excess of 20°C have been increased the replacement of the cement by fly ash decreased the percentage calcium hydroxide remaining from the cement when compared with the mortars cured only at 20°C. The changes were most marked with increasing temperature of curing.

(c) For those mortars containing no fly ash there was a small increase in the percentage calcium hydroxide present on curing to higher temperatures.

In order to correlate the amount of calcium hydroxide in the mortars with their curing regimes and maturities, the maturities have been calculated in °C hours taking -10°C as the base temperature (4). Thus curing for one hour at 20°C is equivalent to a maturity of 30 °C hours.

In this way each set of mortars have been given maturities based on the sum of the total times they were held at each temperature (Table 3). The differences between the calcium hydroxide contents of the samples from curing regime 1 and those in samples containing the same cement: fly ash composition from the other curing regimes corresponds to the amount of calcium hydroxide removed from the mortars by reaction with the fly ash. By determining these differences using the results shown in Fig. 1 and combining them with the maturities of the mortars from Table 3, a map has been constructed in which the amount of calcium hydroxide removed is related to both the composition of the mortars and to their maturities (Fig. 2). This figure shows that the greatest amount of calcium hydroxide was removed in those mortars containing 17.5% cement and 7.5% fly ash with respect to the weight of dry solids taken. This corresponds to a 30% replacement of the cement by fly ash in the 1:3 cement:sand mortars. Because the calcium hydroxide removed must have reacted with the surface of the fly ash, and on the assumption that such a reaction will increase the strength of the mortars, this result implies that the greatest increase in strength should occur over this range of replacement. This result is in very good agreement with the crushing strength of similar mortars after the same curing treatment (5).

It is important that the environment of steel reinforcement in structural concrete should have a high pH in order that good corrosion protection for the steel can be maintained (6). The amount of calcium hydroxide present in a concrete is therefore of considerable importance, and it is of interest to examine what percentage of the calcium hydroxide available from the cement has been removed by reaction with the fly ash. This is shown as a map in Fig. 3 which has been constructed in a similar way to that used for Fig. 2. With a high degree of replacement very little free calcium hydroxide remains in the mortars. However it should be noted that, although some 50% of the available calcium hydroxide has been removed when using a 30% replacement of the cement by fly ash , the amount of calcium hydroxide remaining is more than sufficient to maintain good corrosion protection of iron.

## CONCLUSION

Although 100% carbonation of pure calcium hydroxide has not been achieved the analytical method gives results which are probably as accurate as those given by any other method on mortars and concretes, and with this method no pre-treatment of the samples is necessary. The method relies on the unreactivity of hydrated calcium silicates and aluminates to gaseous carbon dioxide under the experimental conditions used. Further work on these aspects is in progress, and the results so far indicate that the amount of reaction of such compounds is very small provided most of the water of hydration is removed first.

## ACKNOWLEDGEMENT

The authors would like to thank Pozzolanic Ltd., Chester for preparing the mortar samples used in this study.

## REFERENCES

1. F.M. Lea, The Chemistry of Cement and Concrete, Edward Arnold, London, (1970).
2. E.E. Lachowski, Cement concr. Res. 9, 111, (1979).
3. H.G. Midgley, Cement concr. Res. 9, 77 (1979).
4. V.M. Malhotra, Department of Energy (Ottawa Canada, Nov. 1971, Cat. No. M38-3/277).
5. P.L. Owens and F.G. Buttler, Proceedings of 7th International Congress on Chemistry of Cement, Paris, 1980.
6. M. Pourbaix, Atlas of electrochemical equilibria in aqueous solutions, Pergamon Press Ltd. (Oxford) 1966.
7. R.H. Bogue, Ind. Eng. Chem. analyt. Edn., 1, (4), 192, (1929).

## TABLE 1
### The carbonation of $Ca(OH)_2$

| Heating rate °C/min | Crucible | Wt. of CaO, g | Wt. of $SiO_2$, g | Wt. of $H_2O$, g | Final loss g | Wt. of $Ca(OH)_2$ calculated from final loss g | % carbonation of $Ca(OH)_2$ |
|---|---|---|---|---|---|---|---|
| 1.25 | A | 0.4378 |        | 0.3371 | 0.327  | 0.5505 | 95.2 |
| 1.25 | B | 0.0869 | 0.0996 | 0.1176 | 0.0632 | 0.1064 | 92.7 |
| 3    | A | 0.5920 |        | 0.8362 | 0.439  | 0.7383 | 94.4 |
| 3    | A | 0.4241 | 0.4380 | 0.5592 | 0.314  | 0.5287 | 94.3 |
| 3    | B | 0.0781 | 0.1533 | 0.1153 | 0.0557 | 0.0938 | 90.9 |
| 3    | A | 0.3843 | 1.1760 | 0.5181 | 0.277  | 0.4664 | 91.8 |
| 3    | B | 0.0788 | 0.4937 | 0.2901 | 0.0583 | 0.0982 | 94.3 |
| 3    | B | 0.3380 | 0.2540 | 0.4235 | 0.2485 | 0.4179 | 93.6 |
| 4.5  | A | 0.3988 | 0.4434 | 0.4297 | 0.279  | 0.4697 | 89.1 |

Crucible      A   height 23 mm diameter(base) 14 mm   diameter(top) 19 mm
Dimensions    B   height 18 mm diameter(base) 10 mm   diameter(top) 14 mm

## TABLE 2
### Calcium hydroxide content of hydrated Portland cement pastes

| Age (days) | 1 | 2 | 3 | 8 | 11 | 18 | 64 | 156 | 276 |
|---|---|---|---|---|---|---|---|---|---|
| % Ca(OH)$_2$ | 15.2 | 17.9 | 21.9 | 24.3 | 24.8 | 26.9 | 28.4 | 28.4 | 29.5 |

## TABLE 3
### Number of hours the mortars were held at each temperature

| Temp. sequence °C | 20 | 30 | 40 | 50 | 60 | 70 | 80 | 70 | 60 | 50 | 40 | 30 | 20 | maturity °C hours |
|---|---|---|---|---|---|---|---|---|---|---|---|---|---|---|
| regime 1 | 267 | | | | | | | | | | | | | 8010 |
| regime 2 | 6 | 4 | 36 | | | | | | | | | 24 | 197 | 9010 |
| regime 3 | 6 | 4 | 4 | 5 | 36 | | | | | 20 | 36 | 41 | 101 | 11030 |
| regime 4 | 6 | 4 | 4 | 5 | 50 | | | | | 20 | 36 | 41 | 101 | 12010 |
| regime 5 | 6 | 4 | 4 | 5 | 5 | 6 | 55 | 14 | 25 | 30 | 41 | 51 | 21 | 16010 |

|  | SiO$_2$ | Al$_2$O$_3$ | Fe$_2$O$_3$ | CaO | MgO | Na$_2$O | K$_2$O | SO$_3$ |
|---|---|---|---|---|---|---|---|---|
| Analysis of Portland cement | 20.13 | 5.60 | 2.56 | 63.10 | 2.25 | 0.18 | 0.78 | 2.92% |
| Analysis of fly ash | 52.5 | 27.5 | 9.5 | 1.8 | 1.9 | 0.9 | 3.8 | 0.7% |

Bogue (7) composition of Portland cement. $C_3S=56.3\%$, $C_2S=15.2\%$, $C_3A=10.5\%$, $C_4AF=7.8\%$.

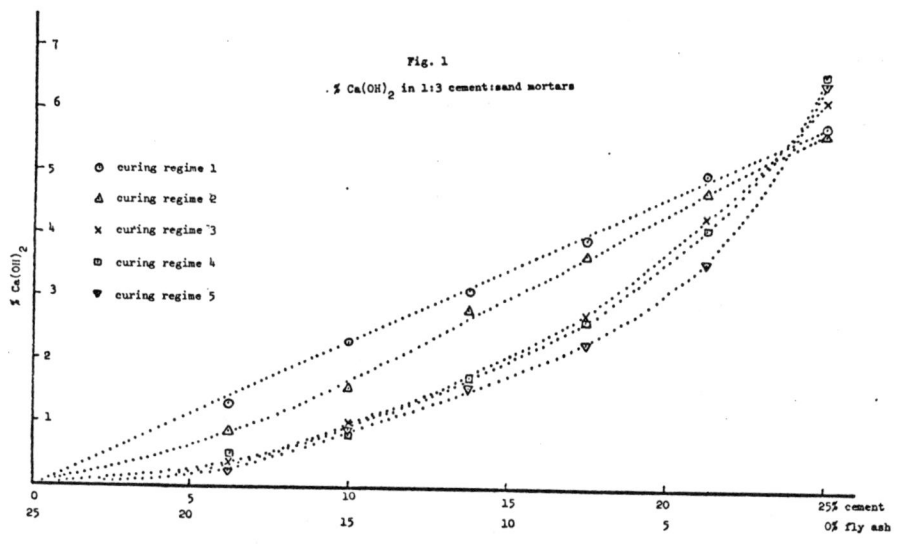

Fig. 1 — % Ca(OH)$_2$ in 1:3 cement:sand mortars

○ curing regime 1
△ curing regime 2
× curing regime 3
□ curing regime 4
▼ curing regime 5

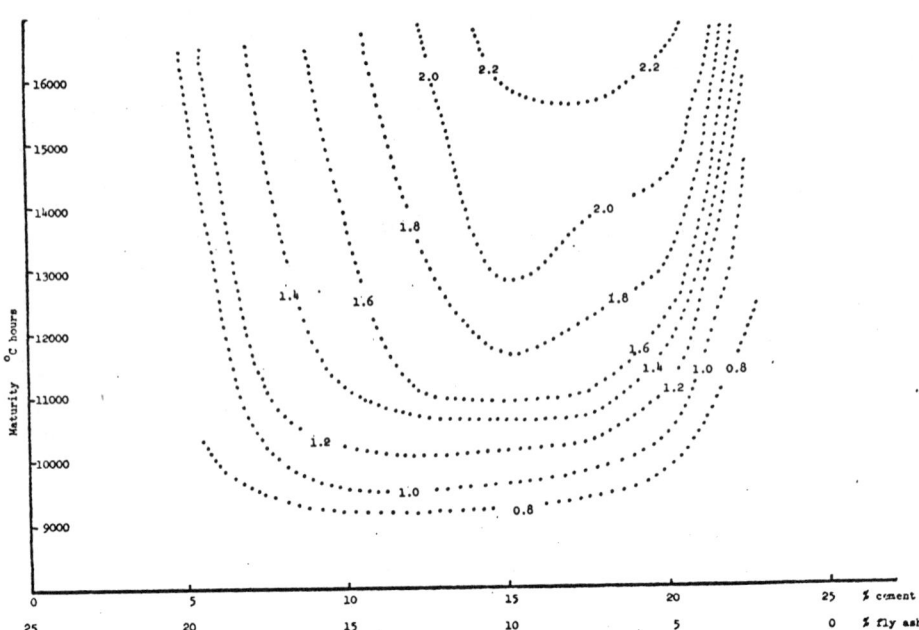

Fig. 2  The numbers in this figure refer to the percentage of calcium hydroxide removed

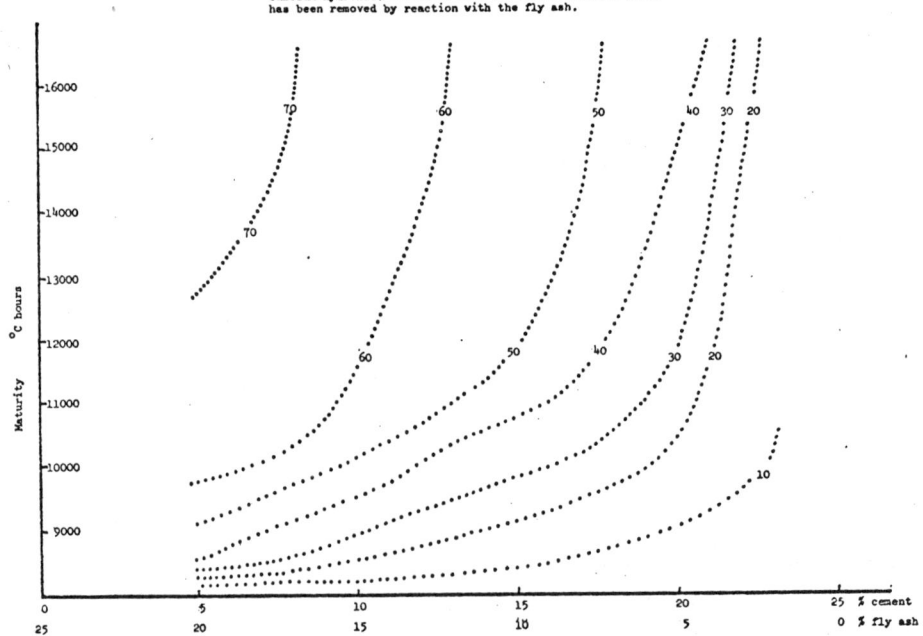

Fig. 3  The numbers in this figure refer to the percentage of the calcium hydroxide available from the Portland cement which has been removed by reaction with the fly ash.

# APPLICATION OF TEMPERATURE PROGRAMMED SORPTION AND REACTION AND OF DTA DATA TO THE CHARACTERIZATION OF BISMUTH MOLYBDATE CATALYSTS

E. Alsdorf[1], K. Habersberger[2], K.-H. Schnabel[1], W. Walkov[1]

[1] Central Institute of Physical Chemistry, Academy of Sciences of the GDR, Berlin-Adlershof
[2] J. Heyrovský Institute of Physical Chemistry and Electrochemistry, Czechoslovak Academy of Sciences, Prague

## ABSTRACT

Bismuth molybdate catalysts and their interaction with $H_2O$, $NH_3$ and $C_3H_6$ were investigated by special techniques of thermal analysis. A correlation between their catalytic properties, namely selectivity, and the thermoanalytical data was found.

## INTRODUCTION

From the three main possibilities of the application of thermal analysis in catalytic research, namely phase analysis, modelling of the catalyst preparation and investigation of the catalyst-reactant interaction (1, 2), the last one is stressed in this paper. A successful attempt has been made to exploit both a classical apparatus and a specialized device working with dynamic atmosphere and to correlate their results with some chosen catalytic data.

## EXPERIMENTAL TECHNIQUES AND RESULTS

The investigated samples were four catalysts of the type Bi-Mo-(P)-O on silica support, presented in Table I together with some data about their catalytic activity in a fluidized bed reactor at 703 K through which a mixture of $C_3H_6$ + $NH_3$ (1:1) with air was passed at 49.6 mmole/hr. The DTA of the samples was performed with Derivatograph (MOM, Hungary). The samples, dried and pretreated by heating at 873 K for 2 hours in air, were heated in air at 10 K/min up to 1 273 K. The typical effects, all observed

TABLE I - The catalysts investigated

| No | Analytical formula | Corr. BiMoO phase | Catalytic reaction rates (mmole/hr) | | | |
|---|---|---|---|---|---|---|
| | | | $NH_3$ conv. | $C_3H_6$ conv. | ACN form. | $CO+CO_2$ form. |
| 1 | b·2 m·8.6 s | β | 12.40 | 8.93 | 2.13 | 2.63 |
| 2 | b·2 m·8.6 s·0.05 p | β | 12.40 | 11.41 | 4.76 | 0.64 |
| 3 | b·1.08 m·6.6 s·0.04 p | γ | 26.78 | 12.40 | 7.74 | 0.64 |
| 4 | b·3 m·10.8 s·0.1 p | α | 15.87 | 10.91 | 4.56 | 0.60 |

b = $Bi_2O_3$, m = $MoO_3$, s = $SiO_2$, p = $P_2O_5$, ACN = acrylonitrile

above 900 K, were in principle the same as on the reoxidation curves (Fig. 1).

The adsorption of $NH_3$ at room temperature (RTA) and the subsequent temperature programmed desorption (TPD) was investigated with 900 mg samples in the same apparatus. Dried $NH_3$ was first led at 4-5 l/hr through the exhaustion tube of the equipment for thermogasotitrimetry (TGT) (3) in 6-8 pulses lasting 3 minutes each with 3 minutes intervals into the sample space, through which a constant stream of Ar was passed at 20-25 l/hr. The TPD was then performed at 5 K/min with TG, with samples 2 and 4 also with TGT. The RTA and TPD results are given in Table II. The lower amount of $NH_3$ found by TGT is caused by its low quantities and a certain extent of oxidation.

TABLE II - Adsorption, desorption and reduction data

| No | $NH_3$ (mg) | | TPD by TGT | | TPD by TG | | $NH_3$ reduction by TG | |
|---|---|---|---|---|---|---|---|---|
| | a | b | temp. range (K) | des. (mg) | temp. range (K) | des. (mg) | reduction start temp. (K) | total deficit (mg) |
| 1 | 4 | 2.5 | | | 323-553 | 2.5 | 653 | 57 |
| 2 | 7 | 2.5 | 353-573 | 1.4 | 348-523 | 4.5 | 643 | 57 |
| 3 | 6 | 3 | | | 403-603 | 3 | 673 | 53 |
| 4 | 5 | 3 | 333-543 | 1.3 | 323-423 | 2 | 653 | 52 |

a - adsorbed at room temperature, b - desorbed at room temperature

The Derivatograph was further used for the reduction of the samples with $NH_3$ in pulses in the way described above. Here, however, the samples were heated at 10 K/min up to ~703 K, kept ~1 hour at this temperature and then cooled in Ar down to ~313 K. A temperature programmed reoxidation, different from the thermogasomanometric TPR (4), was recorded by both the DTA (Fig. 1) and TG (Table III) during heating of the samples at 10 K/min in air.

TABLE III - Thermogravimetric reoxidation data

| No | Start of reoxid. (K) | Oxygen deficit (mg) at | | | End of reoxid. (K) | Calculated red. to $Mo^{5+}$ at 703 K (%) |
|----|------|-------|------|-------|-----|-----|
|    |      | start | m.p. of Bi | 703 K |  |  |
| 1 | 413 | 47 | 38 | 10 | 763 | 57 |
| 2 | 443 | 42 | 34.5 | 4.5 | 773 | 40 |
| 3 | 378 | 53 | 36 | 3 | 758 | 39 |
| 4 | 413 | 48 | 35 | 5 | 758 | 36 |

All the other measurements were performed with an adapted DTA device manufactured by Netzsch (FRG) (2, 5). For 450 mg granulated samples (0.2-0.6 mm) the special catalyst measuring head (6) was used. A flow of $N_2$ was passed through the sample at 60 ml/min interrupted in intervals of 5 minutes by 1 minute lasting pulses of $N_2$ containing 2.5 vol % $H_2O$, 80 vol % $NH_3$ or 20 vol % $C_3H_6$, resp. Each sample was heated (2 K/min) subsequently with pulses of $H_2O$, $NH_3$ and $C_3H_6$ up to 348 K or 523 K (for $H_2O$) or 773 K (for $NH_3$ and $C_3H_6$), the interaction with $NH_3$ being followed by a reoxidation run at 5 K/min in air up to 773 K. As the processes taking place in the system are subsequently the sorption (both ad- and desorption) of the reactive gas and its reaction with the sample, this technique was in agreement with the designation of similar ones used in catalytic research called preliminarily the temperature programmed sorption and reaction (TPSR).

The shape of the TPSR curve for a single pulse of $N_2$ + $H_2O$ is shown in Fig. 2 a. The heights of both the exo- and endothermal peaks diminish with rising temperature and sink below a perceptible limit at temperatures above 443, 373, 493 and 443 K for the samples 1-4, respectively.

The curves obtained with $NH_3$ have the same shape and behaviour up to 525-545 K, where their shape changes into that shown in Fig. 2 b; the intensity of the signal then rises with temperature, at least up to about 660 K.

The interaction of the samples with $C_3H_6$ led to an overlapping and superposition of the endo- and exothermal effects at higher temperatures, so that these could not be discerned. Fig. 3 shows the TPRS curves obtained with single pulses of $NH_3$ and $C_3H_6$ at 703 K.

## DISCUSSION

The TPSR curves obtained with $H_2O$ show only an exothermal adsorption followed by an endothermal desorption of the reactant. No correlation with the catalytic properties at 703 K was found.

The TPD of $NH_3$ preadsorbed at room temperature exhibits its distinctly highest end temperature with sample 3 (Table II). The strongest bonding of the reactant on this catalyst coincides with the highest conversion of $NH_3$ (Table I).

The TPSR curves obtained with $NH_3$ show that above 538-558 K the reactant inlet pulse is accompanied by an endothermal process, like e.g. the dissociation of $NH_3$ in the surface. After the gas stream has been switched back to $N_2$ (elution interval), in some cases an exothermal process becomes apparent, probably the oxidation of the hydrogen just formed with the lattice oxygen. This lasts still for some time during the elution interval until the reactant in the surface is consumed.

The maximum reduction (Table II) of the samples 1 and 2 corresponded to a 100 % reduction of $Mo^{6+}$ to $Mo^{4+}$ and $Bi^{3+}$ to $Bi^0$ (7), that of the samples 3 and 4 to 92-93 %.

The fine structure of the DTA reoxidation curves corresponds probably to the interaction of $O_2$ with active sites of various kinds as well as to the phase changes in the catalyst such as the melting (544 K) of metallic Bi formed by the reduction (peaks a). The effects above ~900 K indicate the phase composition of the reoxidized samples. The endothermal peaks b at 938 K with samples 1, 2 and a at 915 K (with a small shoulder b´ at 935 K) with sample 4 are near the reported melting points of the phases $\beta$ and $\alpha$, resp. (8, 9). Neither the endothermal peak c nor the exothermal one d with sample 3 could be interpreted by the phase transitions observed in this region (9) even in the presence of P (10). X-ray studies of the catalysts are therefore in progress. The distinctly highest oxygen deficit at 703 K observed with sample 1 (Table III) corresponds to the highest oxygen lability and therefore to the highest formation rate for $CO + CO_2$ (Table I).

A good correlation of the TPSR measurements with the catalytic properties of the samples was obtained, based on two findings: 1. the exothermal character of the $C_3H_6$ curves corresponds to a high total oxidation (11) - sample 1; 2. the similarity between the $C_3H_6$ and $NH_3$ curves corresponds to a high selectivity for acrylonitrile formation (12) - sample 3. Further investigations are in progress.

## REFERENCES

(1) K. Habersberger, J.Thermal Analysis 12 (1977) 55
(2) K. Habersberger, Proc. Termanal´79, Vys.Tatry 1979, 129
(3) J. Paulik and F. Paulik, Proc. 4th Int.Conf.Therm. Analysis, Budapest 1974, 789
(4) H. Miura, Y. Morikawa and T. Shirasaki, J.Catal. 39 (1975) 22
(5) K. Habersberger, to be published
(6) G. Keil, Chemie-Ing.-Techn. 35 (1963) 666
(7) K. Aykan, J.Catal. 12 (1968) 281
(8) T. Chen and G.S.Smith, J.Solid State Chem.13 (1975) 288
(9) M. Egashira, K. Matsuo, S. Kagawa amd T. Seiyama, J.Catal. 58 (1979) 409
(10) L. Ya. Margolis, Okislenie uglevodorodov na geterogennych katalizatorach, Chimiya, Moscow 1977, 20
(11) M. Křivánek, P. Jírů and J. Strnad, J.Catal. 23 (1971) 259

(12) J.E. Germain and R. Perez, Bull.Soc.Chim. France (1975) 735

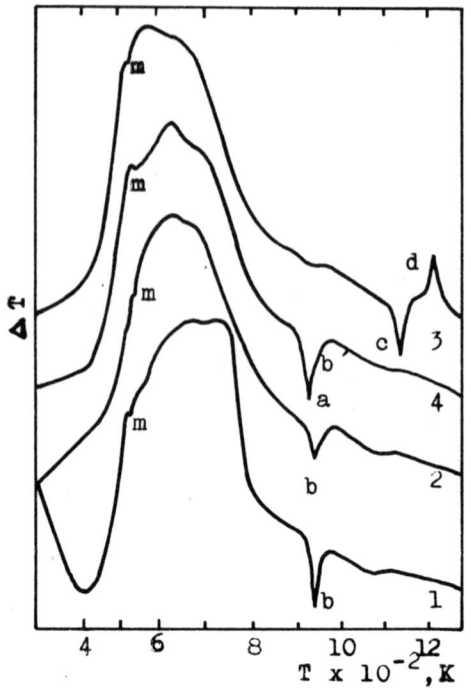

Fig.1. DTA reoxidation curves

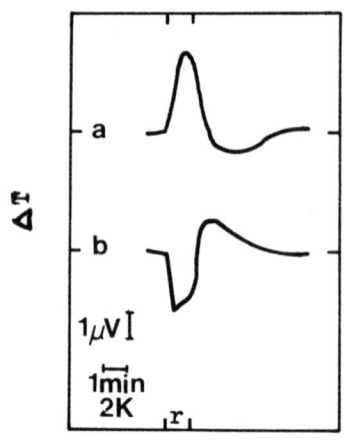

Fig.2. Shape of the TPSR curves
r - reactant inlet

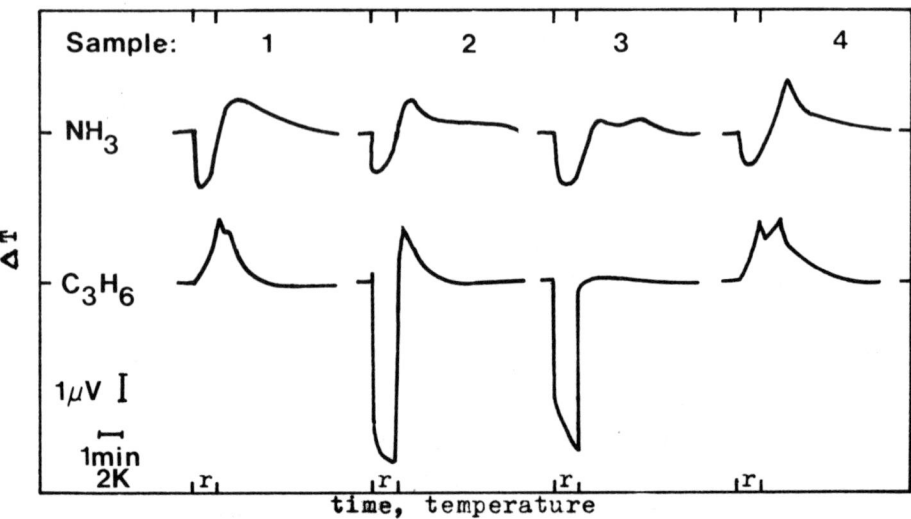

Fig.3. TPSR curves with $NH_3$ and $C_3H_6$ at 703 K

THERMAL DECOMPOSITION OF AQUEOUS MANGANESE NITRATE SOLUTIONS

T.J.W. de Bruijn, W.A. de Jong and P.J. van den Berg
Delft University of Technology, Julianalaan 136, 2628 BL Delft
The Netherlands.

SUMMARY

The mechanism and kinetics of the thermal decomposition of aqueous manganese nitrate solutions into water, nitrogen oxides and $MnO_2$ as the main solid product was investigated by thermogravimetry combined with mass spectrometry, X-ray diffraction and several other analytical methods. The decomposition was found to proceed in three steps, viz.
- partial evaporation of water to a composition containing equimolar amounts of water and manganese nitrate
- a decomposition step in which most of the residual quantity of water evolved along with some nitrogen oxides
- a second decomposition step yielding $MnO_2$, nitrogen oxides and some oxygen.

Further analysis indicates that the product of the first decomposition step is a mixture of $MnO_2$ and $Mn(NO_3)_2$. The presence of manganese (III) oxynitrate suggested in the literature is unlikely.
The main kinetic parameters of the two chemical decomposition steps were established.

INTRODUCTION

Synthetic manganese dioxide for use as depolarizer in batteries is usually made by electrolysis of manganese sulfate solutions,(1) a rather sophisticated and energy intensive process. The work reported here is part of a study dealing with an alternative process, i.e. the preparation of battery grade manganese dioxide from manganese nitrate solutions. Preliminary results of a study on the mechanism and kinetics of the thermal decomposition of aqueous manganese nitrate solutions are described here; more detailed studies are in progress.

EXPERIMENTAL

Material. An aqueous solution of reagent-grade manganese nitrate obtained from Baker Chemicals, was used in all experiments. The composition of the solution, 61.5%w $Mn(NO_3)_2$, 2.4%w$HNO_3$ and 36.1%w$H_2O$ was calculated from results of analyses for $Mn^{2+}$ and $NO_3^-$. The concentration of other components was below 0.01%w.

Theoretical weight losses up to certain compositions are calculated by assuming that nitric acid evaporated completely, in accordance with results obtained by Zdanovskii and Zhelnina (2).

Equipment. A Stanton Redcroft and a Cahn RG (TGS-1) thermobalance were employed. Air dried over molecular sieves flowed through the balance at a rate of 100 or 200 ml/min. The sample weight was normally in the range 1-5 mg , but in one series of experiments it was varied from 0.5 to 23 mg. Isothermal as well as non-isothermal experiments were performed. Water vapour could be added to the gas flowing to the Cahn balance by saturating part of the gas flow.

The off-gass of some thermobalance experiments was analysed for water vapour by a Keidel cel from Consolidated Electrodynamics Corporation, type 26-303. This instrument can measure the volume percentage water vapour in an off-gas flow of 100 ml/min. in the range 1 to 1000 ppm. Two experiments were made in which the off-gas was analysed by a Varian MAT 311 A mass spectrometer, equiped with an on-line computer for data collection.

Almost all final products and some intermediate materials obtained in the thermobalance experiments were examined by X-ray diffraction.

Results and Discussion

Thermogravimetry. When heating a sample very carefully and slowly, first water evaporates and the sample becomes more and more viscous. At our conditions evaporation continues until the overall composition is $Mn(NO_3)_2 \cdot H_2O$. The liquid sample has then turned pink and is very viscous and hygroscopic. Upon further heating decomposition sets in: small grey or black dots appear, the sample swells, bubbles escape and gradually the liquid is converted to solid. The intermediate product is grey, also hygroscopic and much more voluminous than the initial liquid sample. Finally during the second decomposition the product becomes black; the final product looks swollen and has large holes and pores. The rate of the first decomposition step normally becomes noticable at 130 - 140°C and that of the second step at 180 - 190°C. A stable weight loss corresponding to anhydrous $Mn(NO_3)_2$ was never found. A representative weight loss curve measured with the Cahn balance is shown in fig. 1, where the percentage weight loss of the initial weight is plotted against reaction time.

The sample is heated in steps; it loses water at $120°C$ up to $Mn(NO_3)_2 \cdot H_2O$ and decomposes on further heating to an intermediate which does not correspond to $Mn(NO_3)_2$ or $MnONO_3$ (3) according to the weight loss. Finally $MnO_2$ is obtained.

The representation "$Mn(NO_3)_2 \cdot H_2O$" does not imply that crystals of this hydrate are present; it only indicates an overall composition with equimolar amounts of water and $Mn(NO_3)_2$.

To find out during which decomposition step the residual water evolves from $Mn(NO_3)_2 \cdot H_2O$, the off-gass of the deconposition was analysed with the Keidel cell. Figure 1 shows clearly that all the water comes off during the first decomposition step: the amount of water vapour detected per mole of $Mn(NO_3)_2$ was approximately one mole, in accordance with the overall composition.

To establish the effect of water vapour on the decomposition, it was added to the gas phase in a known concentration. Water was found to influence both the decomposition and the water loss of the solution. When at low temperatures water vapour was added to the gas phase the weight increases, but when the temperature is raised the sample loses weight again. As expected, the resulting overall composition is a function of temperature and water vapour concentration. Water also has a very marked influence on the reaction rate of the second decomposition step and on the temperature at which it starts. Without addition of water the intermediate product does not decompose at all at $140°C$, but when the gas phase contains 2.9% water vapour the second decomposition step starts at this very low temperature and even goes to completion.

In another experiment a sample was cooled after the first decomposition at $150°C$ and then heated to $90°C$ after adding a drop of water. The stable weight obtained was higher than the weight after the first decomposition step. Reheating to $140°-150°C$, the usual temperature at which the first decomposition step occurs, again resulted in a decomposition step, but the sample weight now became stable at a lower level than after the previous decomposition step. Repeating this procedure of cooling, adding a drop of water, heating to $90°C$ and then to $140-150°C$, gave similar results: the weight obtained after the decomposition at $140-150°C$ was lower than the weight after the previous decomposition. Further heating resulted in decomposition to $MnO_2$. Thus, each addition of a drop of water causes a new first decomposition step.

It is concluded from the above experiments that water probably plays a vital role in the decomposition. This was confirmed by experiments with anhydrous $Mn(NO_3)_2$ prepared by applying vacuum according to the method of Weigel et al.(4). The $Mn(NO_3)_2$ anhydrate, when decomposed at the usual condition of 103 kPa. with 100 ml/min. air flow, showed a quite different decomposition pattern. The first decomposition step was not observed and the second step became much larger. Thus anhydrous $Mn(NO_3)_2$ decomposes in a single step whereas the decomposition of $Mn(NO_3)_2$ with some water occurs in two steps.

On-line mass spectrometry was used to confirm the above results and to find what nitrogen oxides are formed. The volatile products evolved during the first decomposition step consisted almost entirely of water vapour, except for some NO (m/e=30). The small NO peak observed may have resulted from evolution of NO and/or $NO_2$, because pure $NO_2$ gives its most important peak at m/e=30 and not at m/e=46 (intensity ratio 3 : 1).

The picture changes in the second step. Now the largest peaks are NO (m/e=30) and $NO_2$ (m/e=46). Since the ratio between the two intensities is close to 5, most of the peak at m/e=30 comes from $NO_2$. The observed peak at m/e=44 ($CO_2$ or $N_2O$) could be attributed to only $CO_2$ by operating the mass spectrometer at very high resolution. Furthermore, some oxygen is produced during the second decomposition.

Under vacuum only a single decomposition step occurs with evolution of $NO_2$ and $O_2$. The intensity ratio of the peaks at m/e=30 and m/e=46 is approximately three, indicating that the amount of NO produced is small. This agrees very well with the other data reported here, which all point to water vapour as the cause for the first decomposition step. No evidence was found for the occurrence of oxynitrate, $MnONO_3$, as an intermediate product. Rather, the intermediate product probably consists of a mixture of undecomposed $Mn(NO_3)_2$ and $MnO_2$. This conclusion is based on results of the following experiments.

First, the valence of Mn in the intermediate product was established. To that end, it was dissolved in a solution of 2 N KOH to which 5% tartaric acid had been added. This acid is known to give a stable complex with $Mn^{3+}$ ions. Polarographic analysis of the solution showed $Mn^{2+}$ and a trace of $Mn^{4+}$; $Mn^{3+}$ was not found at all. Dissolution in water and analysis for $NO_3^-$ and $Mn^{2+}$ resulted in a molar ratio $NO_3^-$ / $Mn^{2+}$ of 1.94 i.e. close to the theoretical value for $Mn(NO_3)_2$, 2.0.

Further, X-ray photographs of intermediate products revealed lines of γ and/or ρ-$MnO_2$, and occasionally very faint lines of γ-MnOOH could also be seen. In summary, everything points to an intermediate product consisting of a mixture of $MnO_2$ and $Mn(NO_3)_2$, and also to the conclusion that water vapour is responsible for the first decomposition step. The mechanism of the decomposition of the solution may be that during the first step water vapour evolves, forming bubbles in the viscous liquid. In the bubbles the water vapour concentration is fairly high, causing $Mn(NO_3)_2$ to decompose at the liquid/gas interface. The reaction stops when all the water vapour has escaped.

This mechanism should result in a dependence of the first decomposition step on sample size. Therefore, a series of non-isothermal experiments was performed in the Stanton balance, the sample size being varied between 0.5 and 23 mg. using different heating rates and 100 ml/min. air     The weight loss up to the second decomposition step increases with sample weight, irrespective of the heating rate used ( 3 - 20°C/min.). This is a logical consequence of the proposed mechanism because it is more difficult and it takes more time for the water vapour to emerge from the "particle", so it has more time to attack $Mn(NO_3)_2$ and to promote its decomposition. If the band of weight losses for the different sample sizes      is extrapolated to zero sample weight, it is found that the effect of the escaping water vapour on the $Mn(NO_3)_2$ decomposition disappears with infinitely small samples. The weight loss then is in between 36 and 41%, which agrees well with the theoretical weight loss to $Mn(NO_3)_2$ of 38.5%.

X-ray results. All X-ray photographs of final and intermediate products closely resembled each other. Compared to A.S.T.M. index data (5) the observed lines correspond well with lines from electrolytically prepared γ-$MnO_2$ (ASTM index 14-644) as well as with the five strongest lines of ρ-$MnO_2$ (ASTM index 12-714). In all products $MnO_2$ was the only or main component. Sometimes, however, faint lines of γ-MnOOH (ASTM index 18-805) or the strongest line of $Mn_2O_3$ (ASTM index 10-69) were found.

Kinetics. The kinetic parameters for each decomposition step were established by means of non-linear regression analysis. Out of 23 tested equations, the equation $[-\ln(1-\alpha)]^{\frac{1}{2}} = k.t$ described the first decomposition step best, while for the second decomposition step the equation $1-(1-\alpha)^{1/3} = k.t$ resulted.

Activation energies were calculated from the relevant Arrhenius plots, resulting in 121 kJ/mole for the first step and 143 kJ/mole for the second step. The pre-exponential factors were $8.9 \; 10^{11} s^{-1}$ and $1.6 \; 10^{12} s^{-1}$ respectively. For further details see (6).

Conclusion. It is concluded that the thermal decomposition of manganese nitrate solutions is a complex process, in which many factors play a part especially water vapour. Further research is carried out to define the product spectrum more closely and particularly to elucidate the effect of water vapour on the reaction kinetics (6).

Literature.

(1) A. Kozawa in Batteries vol.I, Manganese Dioxide (1974) ed. K.V. Kordesch

(2) A.B. Zdanovskii, G.E. Zhelnina Zh. Prikl. Khim. (Leningrad) 48 (2) (1975) 427

(3) P.K. Gallagher, F. Schrey, B. Prescott Thermochim. Acta 2 (1971) 405

(4) D. Weigel, B. Imelik, M. Prettre Bull. Soc. Chim. Fr. (1964) 836

(5) X-ray Powder Data File by American Society for Testing and Material or by Joint Committee on Powder Diffraction Standards

(6) T.J.W. de Bruijn, PhD Thesis, Delft University of Technology, to be published (1980).

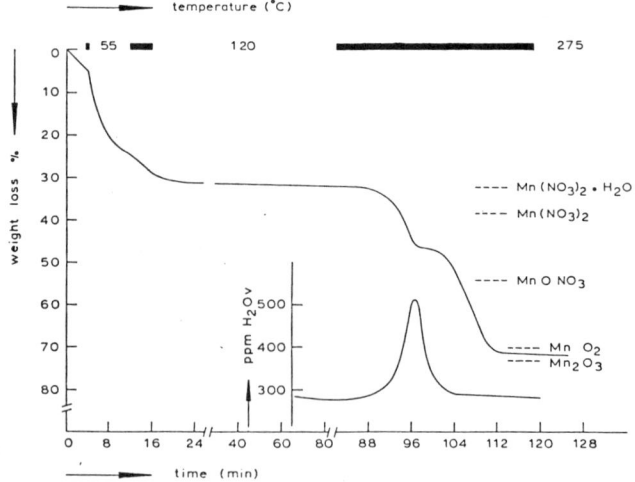

fig. 1 thermal decomposition of aqueous manganese nitrate solution.

# NITRIC ACID AND AMMONIA CONCENTRATIONS AS FACTORS CONTROLLING THE THERMAL DECOMPOSITION OF AMMONIUM NITRATE

A. Kołaczkowski and A. Biskupski
Institute of Inorganic Technology and Mineral Fertilizers
Technical University in Wrocław, Poland

## ABSTRACT

The effect of nitric acid concentration in the condensed phase on the mutual relationship between endothermal and exothermal reactions in the decomposition process of molten ammonium nitrate in the 442–548 K range under standard pressure was investigated by means of DTA and TG. The decomposition of ammonium nitrate was found to consist of consecutive reactions. The first stage of decomposition is the endothermal reversible dissociation reaction of ammonium nitrate into amonia and nitric acid. The overall thermal effect of further consecutive reactions depending on sufficient nitric acid concentration is exothermal. An increase in ammonia concentration in liquid ammonium nitrate results in a decrease in nitric acid concentration. When the ammonia concentration will reach the value limiting the dissociation reaction rate of ammonium nitrate, it will also limit the overall decomposition rate and will limit the contribution of exothermal reactions.

## INTRODUCTION

Thermal decomposition of ammonium nitrate is determined by its two specific properties: ability for endothermal dissociation into ammonia and nitric acid and ability for exothermal decomposition resulting from the presence of two nitrogen atoms, at different oxidation states, in one molecule. The contribution of the endothermal dissociation reaction

$$NH_4NO_3 = NH_3 + HNO_3; \quad \Delta H^o_{298} = + 145.2 \text{ kJ} \cdot \text{mol}^{-1} \qquad 1$$

and exothermal redox reactions in the decomposition of ammonium nitrate is determined by three basic factors: mass and heat exchange with the environment and effect of additives. They have a direct influence on the concentration of decomposition products in the system, on its temperature and on the decomposition rate and its thermal effect. The effect of ammonia partial pressure in the gaseous phase on the thermal effect of decomposition was determined in the investigation. The purpose of the investigation was to determine the factors controlling the decomposition rate and the contribution of endothermal and exothermal reactions to the process.

## EXPERIMENTAL

The effect of ammonia on the decomposition of ammonium nitrate was infered from the results of DTA and TG analyses of ammonium nitrate samples heated at constant temperature under a forced flow of the argon-ammonia mixture in the furance.
DTA and TG measurements were performed on 250 mg ammonium nitrate samples in glass crucibles. Roasted alumina was the reference system. A detailed description of the equipment and of analitical methods was provided in an earlier paper (1). Additional information on the

effect of ammonia was supplied by pH measurements of 3.2% aqueous solutions of residues left after decomposition of the samples (pHWS). Investigation was carried out in to ammonium nitrate analar grade containing:

$Cl^-$ below 0.002% by weight
$SO_4^{2-}$ below 0.01% by weight
$H_2O$ below 0.2% by weight
$NO_2^-$ below 0.0005% by weight

The mixture of gases was prepared from compressed argon containing 99.9% by vol. Ar and from compressed ammonia containing 98.9% by vol. $NH_3$ and 1.1% by vol. $N_2$.

## RESULTS AND DISCUSSION

The results of DTA and TG measurements of ammonium nitrate samples in the argon--ammonia mixture 0 to 99% by vol. $NH_3$ are summarized in Table 1. These results have confirmed the inhibition of ammonium nitrate decomposition under a gas mixture containing more than about 2% by vol. $NH_3$. This conclusion may be drawn from the themperature above which the loss of weight occurs in the sample investigated ($T_{em}$) as well as from the temperature above which the decomposition process becomes exothermic ($T_{peg}$). Worthy of special notice is the endothermic effect which was foound only in a narrow range of ammonia concentrations.

Table 1

The results of DTA and TG measurements of ammonium nitrate under a mixture of ammonia and argon flow rate 2,3 $cm^3 s^{-1}$, temperature increase rate 2,5 deg · $min^{-1}$

| $NH_3$ conc. % by volume | $T_{em}$ | | $T_{pen}$ | | $T_{peg}$ | |
|---|---|---|---|---|---|---|
| | K | °C | K | °C | K | °C |
| 0 | 456 | 183 | — | — | 469 | 196 |
| 0,5 | 455 | 182 | — | — | 471 | 198 |
| 2 | 471 | 198 | 471 | 198 | 486 | 213 |
| 5 | 481 | 208 | 481 | 208 | 495 | 222 |
| 10 | 485 | 212 | 485 | 212 | 501 | 228 |
| 16 | 492 | 219 | 492 | 219 | 511 | 238 |
| 20 | 495 | 222 | 495 | 222 | 503 | 230 |
| 33 | 500 | 227 | — | — | 507 | 234 |
| 49 | 504 | 231 | — | — | 509 | 236 |
| 66 | 508 | 235 | — | — | 515 | 242 |
| 83 | 505 | 232 | — | — | 511 | 238 |
| 99 | 502 | 229 | — | — | 517 | 244 |

Notes: $T_{pen}$, $T_{peg}$ — temperature at which the endothermic or exothermic effect begins
$T_{em}$ — temperature of mass effect on the TG curve

The results of pHWS measurements (Table 2) indicate that an increase in ammonia concentration in the gaseous phase brings about an increase in its concentration in the condensed phase. The results of studies suggest that the decomposition of ammonium nitrate is a system of consecutive reactions. Its first stage is the endothermic dissociation of ammonium nitrate into nitric acid and ammonia. Both dissociates undergo exothermic changes in subseouent consecutive reactions. A considerable decomposition pressure of ammonia and nitric acid upon ammonium nitrate indicates that reaction 1 participates in the decomposition process of ammonium nitrate (2,3). Another argument which supports the importance of the reaction 1 in the decomposition of ammonium nitrate is an increase in $T_{em}$ and the endothermic nature of the initial decomposition stage of ammonium nitrate under ammonia (Table 1). An increase in $T_{em}$ in the presence of ammonia may be explained in the following simplified manner. The decomposition of ammonium nitrate is a consequence of the system tending to reach the equilibrium of reaction 1 determined by the equation

$$c_{NH_3} \cdot c_{NHO_3} = K_c \cdot c_{NH_4NO_3} = K_c^*$$

$c_{NH_3}$, $c_{HNO_3}$, $c_{NH_4NO_3}$ — equilibrium concentration of ammonia, nitric acid or ammoniumin nitrate in the condensed phase, $K_c$ — chemical equilibrium constant. Under air, at increased temperature ($T_{em} = T_o$), having acouired the necessary activation energy, ammonium nitrate decomposes into ammonia and nitric acid and the concentration product of both components increases. Further transformation of both products and, above all, their diffusion to the gaseous phase is the reason why the equilibrium state is not reached on performing DTA and TG of ammonium nitrate under air. Therefore, the weight of sample decreases continuously. In the ammonia containing atmosphere its concentration in ammonium nitrate increases, which may be concluded from an increase in pHWS. The effect observed may results from restrictions on the diffusion of the reaction 1 products from the condensed phase to the gaseous phase, or from dissolution of gaseous ammonia in molten ammonium nitrate. An increase in ammonia concentration causes the concentration product of ammonia and nitric acid in the condensed phase to reach the value equal to constant $K_c^*$ at $T_o$, which prevents decomposition process to occur at higher temperatures ($T_1 > T_o$), at a given constant partial pressure of ammonia, since

Table 2
pHWS of samples after decomposition under $NH_3$ containing atmosphere (decomposition temperature: 464 K = 191°C, decomposition time — 14,4 ks, gaseous phase — argon with $NH_3$ admixture, concentration of aoueous solution — 3,2 g of sample in 100 g solution)

| $NH_3$ conc. in gaseous phase % by vol. | pHWS of samples with initial weights: | | |
|---|---|---|---|
| | 200 mg | 500 mg | 1000 mg |
| 0.0 | 3.80 | 3.76 | 3.75 |
| 1.0 | 4.29 | 4.22 | 4.10 |
| 2.0 | 5.36 | 5.09 | 4.87 |
| 5.0 | 5.80 | 5.78 | 5.61 |
| 16.7 | 6.10 | 5.90 | 5.73 |

$$\frac{dK_c^*}{dT} > \frac{d(c_{NH_3} \cdot c_{HNO_3})}{dT}$$

where T — temperature, $c_{NH_3}$, $c_{HNO_3}$ — concentration of ammonia or nitric acid in the condensed phase.

The increased partial pressure of ammonia in the gaseous phase causes that $c_{NH_3} \gg c_{HNO_3}$ at temperature $T_1$. The endothermic effect on the DTA curve of ammonium nitrate in the presence of ammonia indicates that nitric acid concentration under such conditions is too low to ensure a sufficient rate of the secondary exothermic reactions which is necessary to compensate for the endothermic effect of reaction 1. The dependence of exothermic reactions on nitric acid concentration indicates that these are secondary transformations with respect to reaction 1. The endothermal effect could not be attributed to possible evaporation of ammonium nitrate since this would be inconsistent with the results reported by Feicke and Hainer (2,3).

## REFERENCES

(1) Biskupski A., Kołaczkowski A., Schroeder J., Application of DTA nad TG to the estimation of thermal stability of fertilizers containing ammonium nitrate, Proceeding of the Fourth ICTA, Akademiai Kiado, Budapest 1975, vol. 3, p. 577–89.
(2) Feick G., The Dissociation Pressure and Free Energy of Formation of Ammonium Nitrate, J. Am. Chem. Soc., vol. 76, 1954, p. 5858–60.
(3) Feick G., Hainer R.M., On the Thermal Decomposition of Ammonium Nitrate and Reaction Rate, J. Am. Chem. Soc., vol. 76, 1954, p. 5860–63.

THERMAL BEHAVIOUR OF $Ge_{.25}Te_{.60}Se_{.15}$ INVESTIGATED BY EMANATION THERMAL ANALISIS AND DTA.

SANTIAGO BORDAS and MONTSERRAT GELI
Dept. de Termología, Universidad Autonoma de Barcelona,Bellaterra.Spain.

VLADIMIR BALEK and MIROSLAV VOBOŘIL
Nuclear Research Institute, 250 68 Rez,Czechoslovakia.

ABSTRACT

The ternary sample of semiconductor glass $Ge_{.25}Te_{.60}Se_{15}$ labeled by trace amount of $^{228}Th$, has been investigated by emanation thermal analisis (ETA) and DTA. The ETA results confirm the difference in the processes of structure formation of glassy or crystallin solids for thermal methods at various cooling rate. The Netzsch apparatus for ETA-DTA enables to us study the thermal behaviour of the samples. The possibility of the ETA is shown to distinguish various disorder stages.

INTRODUCTION

Semiconductor glasses have been studied for a long time as with particulary interesting properties and applications.
The aim of this work is to show the potentialities of the ETA in the study of the thermal behaviour of the ternary sample ( crystalline and glass). The ETA results are compared with the results of DTA.

EXPERIMENTAL PROCEDURE

ETA is based on the measurementof the release of radioactive inert gas from the solid previously labeled by the inert gas (1).The sample labeling was made by impregnation of the sample with an acetone solution containing $^{228}Th$ and $^{224}Ra$ isotopes. The radioactivity of the sample achieved was $5.10^5$ Bg/100 mg of the sample. The superficial distribution of the inert gas was obtained by this way. The volume distribution of the inert gas in the sample was obtained by the sample melting.
The Netzsch apparatus for ETA and DTA has been applied for the measurements. Thes analysis was performed on 100 mg sample with atmosphere of argon (flow rate: 45 ml/min).

## RESULTS AND DISCUSSION

### A) Comparison of ETA and DTA measurements

The results of ETA and DTA of the superficially labeled glassy sample, measured during heating at the rate 5°C/min are given in Fig. 1 and are listed in table I.

The temperature interval of glass transition as determined by ETA lies by 10-15°C lower than determined by DTA. The ETA of the superficially labeled sample enables us determine the first stages of the process, namely the loosening of the glass matrix which precceds to the proper glass transition. The beginning of the loosening process is indicated by ETA curve at the 115°C. Recrystallization of the powdered superficially labeled sample is indicated by the decrease of the emanation release rate E. The changes of the surface, preceding the recrystallization of the glassy sample, begin at 240°C, 30°C before the onset of the process as determined by DTA. The melting, demostrated by an endothermic effect on DTA curve, is indicated on the ETA curve by the decrease of the emanation release rate E (sample is superficially labeled).

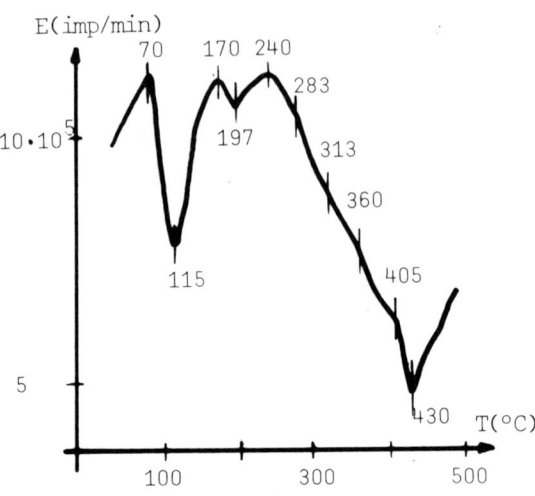
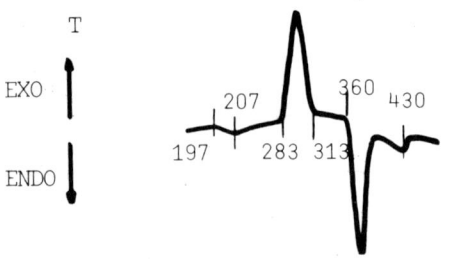

Fig. 1: ETA and DTA curves of superficial sample labeling (heating rate: 5°C/min)

According to DTA the melting starts at 360°C and finishes at 430°C. The ETA data agree fairly well with the DTA, giving more information about the melting process: the onset of the total lattice distortion is determined at 405°C. The increase of the emanation release rate between 430 and 500°C is due to the thermal diffusion of radon in the melt. Homoge-

Table I: Temperature interval of processes in $Ge_{.25}Te_{.60}Se_{.15}$ determined by ETA and DTA measurements.

| Solid state process | | Temperature °C | |
|---|---|---|---|
| | | ETA | DTA |
| Glass transition: | onset of the solid matrix loosening | 115 | – |
| | onset of the transition | 170 | 182 |
| | end of the transition | 197 | 207 |
| Recrystallization: | onset of the surface changes | 240 | – |
| | onset of the volume process | 283 | 283 |
| | maximum rate of the process | 290 | 290 |
| | end of the volume process | 313 | 313 |
| Melting: | onset of the process | 360 | 360 |
| | onset of total lattice distortion | 405 | |
| | end of the process | 430 | 430 |

nization of the radioisotopes $^{228}Th$ and $^{224}Ra$ of the superficially labeled sample takes place.

Fig. 2 shows the ETA curve of a $Ge_{.25}Te_{.60}Se_{.15}$ sample previously molten at 500°C and quenched. In this bulk sample the volume distribution of the radioactive label can be supposed. Due to the fact that the molten sample possess very small internal surface and the diffusion of radon in the glassy material at the temperatures about 200°C is negligible, the effect which should be ascribed to the glass transition of the sample is not observable on the ETA curve in Fig. 2 but the DTA curve is the same of in Fig. 1 (it is not drawing in Fig. 2). In the temperature interval 283-313°C the increase of E is controlled by the recrystallization, in the interval 315-360°C the increase of E is due to the radon diffusion in the recrystallized solid, the diffusion being enhanced by pre-melting phenomena.

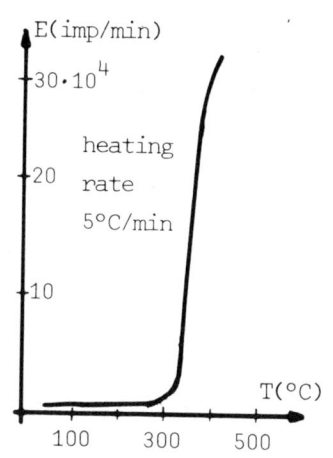

Fig. 2: ETA curve of volume sample labeling

## B) Thermal behaviour of sample at different cooling rates

The ETA enables us to investigate the behaviour of $Ge_{.25}Te_{.60}Se_{.15}$ sample during its cooling under various cooling rates. Fig. 3 shows the ETA curves measured during cooling of the sample (volume labeling) at the cooling rate 0.5°C/min (curve a) and 5°C/min (curve b). During the cooling at 0.5°C/min a crystalline state of the solid shall be established, whereas during the cooling at 5°C/min a partially glassy state of the solid is formed. During monotonous decrease of the emanation release rate two points can be determined on the ETA curve a (375 and 327°C) where changes in the material have to be expected of DTA results, and corresponds a liquidus and solidus temperatures undercooled. On the curve b, the monotonous decrease of E from 500 to 364°C is observed. An increase of E in the interval 364--345°C indicated partially crystallitation, not observed for DTA, because the critical cooling rate is $\sim$ 5°C/min. (2).

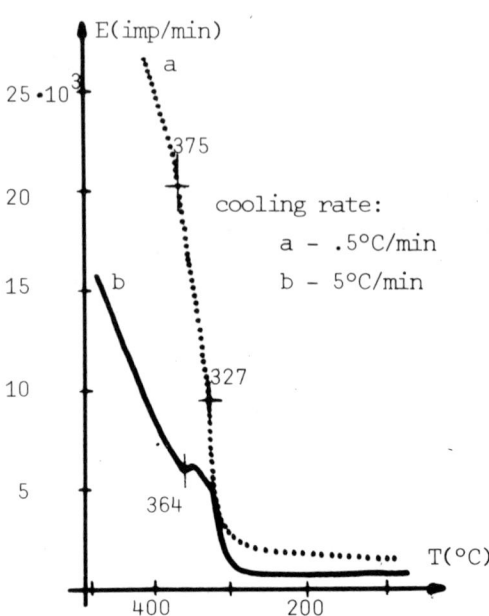

Fig. 3: ETA curves of different cooling rate.

It is interesting to describe the features of degre of order in the solids (3). We have make an intent whit ETA: a) on the base of the values $E_{20}$ of emanation release rate measured at room temperature and b) on the base of the course of ETA curve measured during heating at a constant rate.

The value $E_{20}$ characterizes generally the texture of materials (1), and normaly higher $E_{20}$ value corresponds the higher degree order, in this case the $E_{20}$ of sample 1 is more higher than sample 2.

But in glasses obtained for very fast quenching it is possible macroscopic defects. That provoke $E_{20}$ higher than the one for the same crystalline sample, and then it is necessary ETA curve measured during heating.

Fig. 4 shows heating ETA curve of two bulk samples possessing different

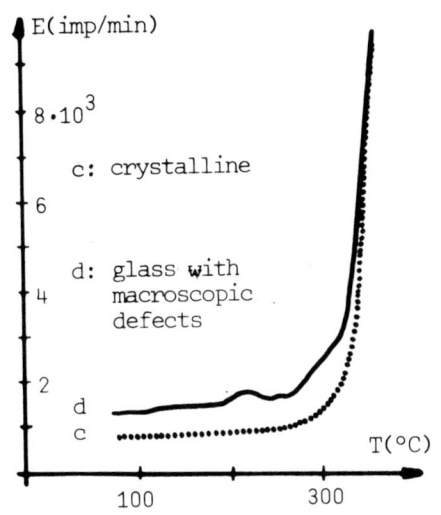

Fig. 4: ETA curves (heating rate: 5°C/min)

degrees of order. Curve c corresponds to the sample 1 (cooling rate of 0.5°C/min). Curve d corresponds to the sample 3 prepared by cooling the sample from the melt at the cooling rate 2000°C/min. Sample 1 is crystalline sample and sample 3 is gasse sample, and have structure characterized by the absence of long range order and macroscopic defects.

Both ETA curves c and d enables us to the differences in the thermal behaviour of the solid sample with different degree of order. An effect on the ETA curve d in the temperature region 190-230°C is observed due to the annealing of the macroscopic defects in the glassy structure. In the temperature range 265-318°C a linear increase of E with temperature is observed, obviously due to the crystallization. In the temperature interval 330-360°C the intense increase of E is due to the radon diffusion in the crystalline solid, which is being enhaced be pre-melting phenomena. Obviously ETA curve c no effects of the glass transition and crystallization are observed. The increase of emanation release rate E can be therefore ascribed to the radon diffusion in the crystalline phase of $Ge_{.25}Te_{.60}Se_{.15}$ sample. From the dependence $\log E_D = f(1/T)$ it is possible to evaluate the activation energy of radon diffusion in the crystalline phase (1), is 80.8 Kj/mol. As in the temperature interval from 320 to 360°C no stright line can be made of $\log E_D$ values versus $1/T$, it is proved that in this interval the radon release from the sample is controled not only by the diffusion mechanism, but it is enhaced by the pre-melting phenomena.

CONCLUSION

The ETA has been applied to the investigation of thermal behaviour of $Ge_{.25}Te_{.60}Se_{.15}$ of different thermal history. The results of ETA are compared with those of DTA.

The ETA enables us to receive a complementary information about the nature of the processes namely changes in transport properties accompanying the processes in the solids of the given physical state.

The activation energy of radon diffusion in the crystalline phase has been determined (Q=80.8 Kj/mol) as a characteristic parameter of the solid state.

The ETA can also be used for characterizing of different order degrees.

REFERENCES

(1) V. BALEK, Thermochim. Acta 22 (1977), 1
(2) S. BORDAS, Thesis, UPB, Barcelona 1977
(3) J.M. STEVELS, J.Non Cryst Solids, 6 (1971) 307

# DIFFERENTIAL THERMAL ANALYSIS OF PHOSPHATE-STABILIZED $Ca_2SiO_4$

R. Halle and V. Carin

Research Department of JUCEMA, Zagreb, Yugoslavia

## ABSTRACT

The lowering of the peak temperature for phase transformation of $\alpha'_H\text{-}Ca_2SiO_4$ into $\alpha\text{-}Ca_2SiO_4$ modification was studied by differential thermal analysis on samples of dicalcium silicate doped with $Ca_5(PO_4)_3OH$ at various levels.

The phase transition temperature as a function of $P_2O_5$ percentage incorporated in dicalcium silicate is presented in a diagram.

## INTRODUCTION

There are five polymorphic modifications of pure dicalcium silicate in the temperature range between $25°C$ and $1450°C$. Among them only the $\gamma$-form is stable at room temperature, although in certain cases the $\beta$-form can also remain at that temperature. The transformation scheme of $Ca_2SiO_4$ polymorphs and the corresponding DTA curves (1) are presented in Fig.1. When samples are heated from room temperature upward, all phase transformations are endothermal exhibiting endothermic peaks on thermal curves.

The highest and most expressed peak on the differential thermal curve is brought about by the $\alpha'_H \rightarrow \alpha\text{-}Ca_2SiO_4$ transformation with a relatively high inversion enthalpy and a narrow reaction interval.

However, by incorporating various additions it is possible to change the temperature range of stability of $Ca_2SiO_4$ modifications so that in certain cases each of its polymorphs can be stabilized at room temperature. This temperature range depends on the type and amount of the stabilizer incorporated in $Ca_2SiO_4$.

Numerous authors have dealt with the problem of phase transformation resulting from the incorporation of foreign atoms.

Pritts and Daugherty (2) observed the temperature lowering of the $\beta \rightarrow \alpha'\text{-}Ca_2SiO_4$ phase transition and on the basis of DTA results found the solubility limit of $Cr_2O_3$ into $\beta\text{-}Ca_2SiO_4$.

W.Eysel and T.Hahn (3) used DTA results for constructing the phase diagram of the $Ca_2SiO_4$-$Ca_2GeO_4$ system.

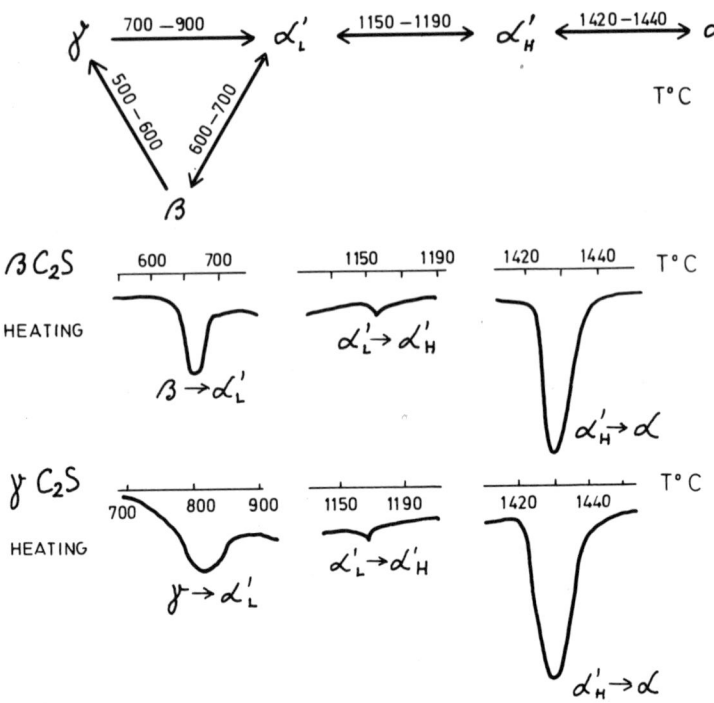

Fig.1. Transformation scheme and corresponding DTA curves

N.A.Toropov (4) studied polymorphic transformations of $2CaO \cdot SiO_2$-$2BaO$-$SiO_2$ solid solution applying DTA.
E.S.Newman and L.S.Wells (5) employed DTA to investigate the solubility of several additions in $Ca_2SiO_4$, $P_2O_5$ being one of them.
In this work the temperature lowering of the $\alpha'_H \rightarrow \alpha$-$Ca_2SiO_4$ phase transformation is followed as a function of the amount of $P_2O_5$ added.

### EXPERIMENTAL PROCEDURE

Chemicals for the preparation of doped $Ca_2SiO_4$ were as follows:
$CaCO_3$, grade p.a., Alkaloid, Skopje, Yugoslavia
$SiO_2$, grade p.a., Kemika, Zagreb, Yugoslavia
$Ca_5(PO_4)_3OH$, grade pure, Kemika, Zagreb, Yugoslavia

Initial mixes of finely powdered $CaCO_3$, $SiO_2$ and $Ca_5(PO_4)_3OH$ were intimately mixed by hand and then homogenized in a rotating homogenizer for three hours, thereafter calcined at $1000°C$, pressed into pallets and sintered in a platinum crucible in an electric furnace at $1450°C$ for 90 minutes. The products were air-quenched, ground, palleted and resintered, again at $1450°C$ for 90 minutes. The air-quenching was repeated and the samples were ground to the specific surface of approx. 3600 $cm^2/g$ determined by the Blaine permeability method.

The phase composition of the resintered samples was examined by optical microscopy and X-ray diffraction method. DTA results were obtained from Thermoanalyzer TA-1 (Mettler, Switzerland). To obtain differential thermal curves the samples were heated at a rate of $10°/min$ up to $1450°C$ in a platinum sample holder with Pt-PtRh 10% differential thermocouple and the endothermic peak of the $\alpha'_H \to \alpha$-$Ca_2SiO_4$ phase transformation was observed.

## RESULTS AND DISCUSSION

Table 1. shows the amount of $P_2O_5$ added to the samples, as well as the phase composition and average value of peak temperatures of the $\alpha'_H \to \alpha$-$Ca_2SiO_4$ phase transformation.

In Fig.2. differential thermal heating curves are presented for several samples with various amounts of $P_2O_5$ added.

Fig.3. gives peak temperatures of the endothermic peak of the $\alpha'_H \to \alpha$-$Ca_2SiO_4$ phase transformation plotted against the amount of $P_2O_5$. The diagram also includes generalized results obtained by phase analysis indicating in which concentration ranges of $P_2O_5$ addition particular $Ca_2SiO_4$ modifications are predominant at room temperature.

Proceeding from the experimentally obtained points, the phase boundary between the $\alpha'_H$- and the $\alpha$-$Ca_2SiO_4$ was found.

The slope of the limiting line between the stability field of the $\alpha'_H$ and the $\alpha$ polymorph is not constant; its change occurs at a concentration of about 5% added $P_2O_5$.

As indicated in the diagram $\beta$-$Ca_2SiO_4$ is stable at room temperature up to 5% $P_2O_5$ addition and $\alpha'$-$Ca_2SiO_4$ at higher levels of $P_2O_5$.

On the basis of the above facts it can be concluded that the slope of the limiting line changes with the change in the field of stability of dicalcium silicate polymorphs at room temperature.

Table 1. Phase composition and peak temperatures of samples with $P_2O_5$ added

| $P_2O_5$ added % wt. | Opt. Microscopy and XD | | | | Average value of $\alpha'_H \to \alpha\text{-}C_2S$ transition peak temperatures °C |
|---|---|---|---|---|---|
| | $\alpha$ | $\alpha'$ | $\beta$ | $\gamma$ | |
| 0,00  |      |      |      | ++++ | 1445 |
| 0,50  |      |      | ++++ |      | 1406 |
| 1,47  |      |      | +    | +++  | 1367 |
| 1,92  |      |      | ++   | ++   | 1336 |
| 2,42  |      |      | ++   | ++   | 1300 |
| 4,70  | (+)  |      | +++  | (+)  | 1114 |
| 5,55  | (+)  |      | +++  | (+)  | 1101 |
| 6,84  | +    |      | +++  |      | 1063 |
| 8,05  | +    |      | +++  |      | 1004 |
| 9,61  |      |      |      |      | 977  |
| 11,10 | +++  | +    |      |      | 948  |
| 12,53 | ++++ | (+)  |      |      | -    |

(+) small quantity

The difference in the amount of $P_2O_5$ solved in $C_2S$ causes a greater change of the $\alpha'_H \to \alpha\text{-}C_2S$ inversion temperature if solid solutions contain up to 5% $P_2O_5$ additions than in case of higher $P_2O_5$ percentages. It is thus evident that the difference of $P_2O_5$ solved in the $\alpha'$ modification which forms during heating from the $\beta$ modification has a greater influence on the change of the $\alpha'_A \to \alpha\text{-}C_2S$ peak temperature than such a difference in the $\alpha'$ modification which is stable within the whole temperature range from room temperature up to the inversion temperature into the $\alpha$ modification.

The sample including 12.5% $P_2O_5$ exhibited no phase transformation effect on the differential thermal curve. It has been established by phase analysis that it contains predominantly $\alpha\text{-}Ca_2SiO_4$ at room temperature.

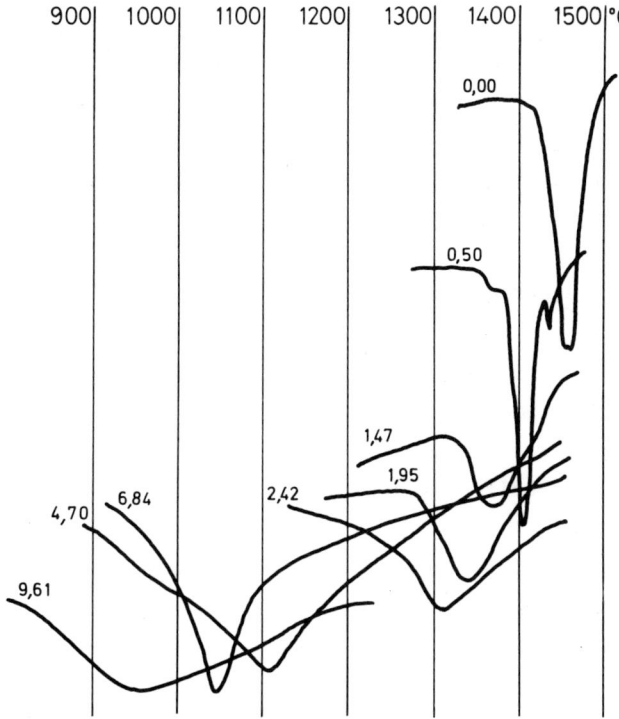

Fig. 2. DTA heating curves for $Ca_2SiO_4$ solid solutions with $P_2O_5$

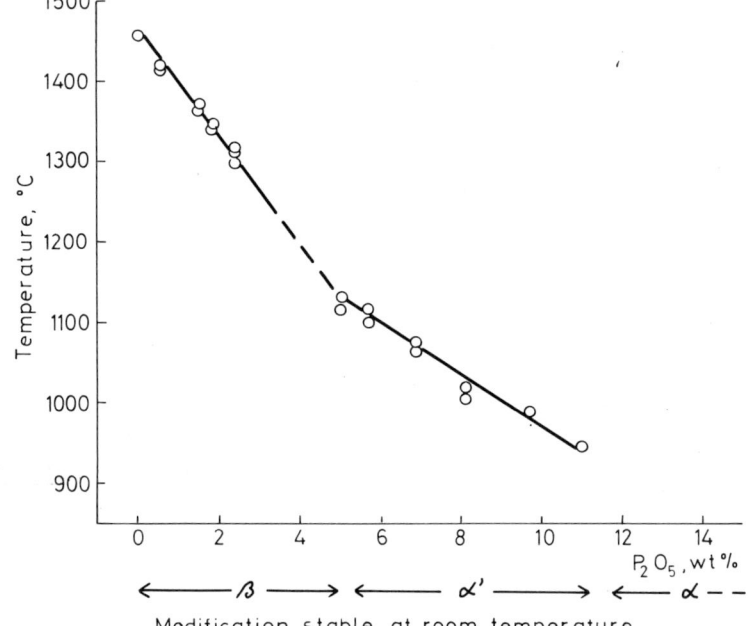

Fig. 3. Plot of peak temperatures for $\alpha'_H \to \alpha\text{-}Ca_2SiO_4$ transition

Acknowledgment

The investigation was partly supported by U.S. Department of Transportation and by the Selfmanaging Community for Scientific Research of Croatia through funds made available to the U.S. - Yugoslav Joint Board for Scientific and Technological Co-operation.

## REFERENCES

(1) M.Regourd and A.Guinier, Proc. 6th Int.Congress Chem.Cem., Moscow, 1(1974)25 (Stroiizdat, Moscow, 1976, Russ.ed.)
(2) I.M.Pritts and K.E.Daugherty, Cem.Concr.Res. 6(1976)783
(3) W.Eysel and T.Hahn, Zeitschrift für Kristallographie, Bd. 131(1970)322
(4) N.A.Toropov, Proc. 4th Int.Symp.Chem.Cem., Washington, 1(1960)113
(5) E.S.Newman and L.S.Wells, Journal of the National Bureau of Standards, 36(1946)137

PHASE STUDY OF THE PRASEODYMIUM-OXYGEN SYSTEM BY
THE MULTIPLE THERMAL ANALYSIS

Yasutoshi Saito, Seihiro Sasaki and Toshio Maruyama
Research Laboratory of Engineering Materials
Tokyo Institute of Technology
4259 Nagatsuta-cho, Midori-ku, Yokohama 227, Japan

ABSTRACT

A multiple technique coupled with thermogravimetry (TG) and evolved gas analysis (EGA) was applied to examine the phase relation of praseodymium-oxygen system under the controlled oxygen potentials. The oxygen pressure was monitored in situ in the vicinity of the sample. The evolution and uptake of oxygen were sensitively detected by EGA during the transformations in praseodymium oxides. The result of EGA was consistent with that of TG. In EGA, a oxygen sensor (OS) which consisted of a calcia stabilized zirconia (CSZ) electrolyte was found to be more effective than a thermal conductivity detector (TCD) at low oxygen potentials.

INTRODUCTION

A multiple technique coupled with TG, EGA and differential thermal analysis (DTA) is a very powerful tool for the investigation of the solid-gas reaction, and gives detailed informations under the controlled atmosphere and the programed heating and cooling schedule. Although accurately controlled temperature program is now feasible by a commercially available equipment, controlling atmosphere is not so easy. The oxygen potential is usually controlled with gas mixtures such as $CO-CO_2$, $H_2-H_2O$ and inert gas-$O_2$. The use of $CO-CO_2$ and $H_2-H_2O$ mixtures is restricted by the desired oxygen potential and temperature in spite of the excellent buffer action. The mixture of inert gas and $O_2$ is convenient to produce relatively high oxygen potential (1∼ $10^{-6}$ atm) independent of temperature. However, this system has no ability as a buffer. Due to the lack of the buffer

action, the oxygen potential is influenced even by a small
amount of oxygen leaked or evolved and up-taken during the
oxidation-reduction process. Therefore, the monitoring of
oxygen potential is necessary not only at the gas inlet or
outlet but also in the vicinity of the sample.

The praseodymium-oxygen system is one of the more complicated
and interesting metal-oxygen systems which form several
intermediate phases. In spite of a considerable number of
studies on this system (1),(2),(3),(4),(5), the phase
relations have not fully established due to the appearance
of the pseudo-phases and hysteresis (6),(7),(8). Further
investigation is still required under the exactly monitored
oxygen potentials and the accurately controlled thermal
cyclic conditions. In the present study, the phase relation
was examined by TG coupled with EGA under various controlled
temperatures and oxygen potentials. The specified oxygen
potential was obtained by mixing oxygen with helium. The
monitoring of oxygen potential was accomplished in situ in
the vicinity of the sample by the solid electrolyte cell of
a CSZ in which the sample was placed. Simultaniously with
TG, the evolution and up-take of oxygen during the transformations in praseodymium oxides were detected by the TCD
and OS.

## EXPERIMENTAL PROCEDURE

Fig. 1 shows the gas flow system employed in the present
study. Helium and oxygen were mixed in a desired ratio by
a gas flow rate controller
(FRC). The mixture was
devided into three portions
of (i)∿(iii) after eliminating
water vapor by $Mg(ClO_4)_2$.
The portion of (i) was
introduced into TCD as a
reference gas through a oxygen
sensor (OS-1) in which the
oxygen potential of the

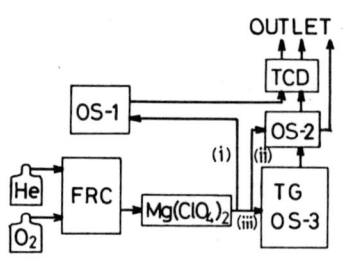

Fig. 1. Schematic diagram
of gas flow system.

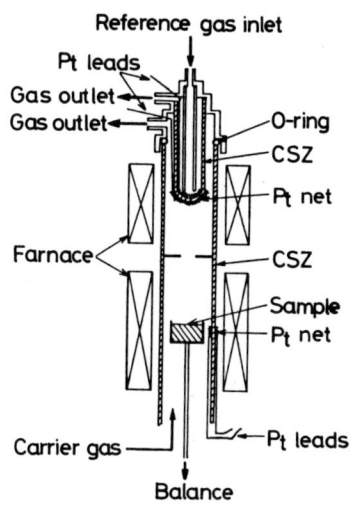

Fig. 2. TG apparatus connected with oxygen sensors.

mixture was monitored. The portion of (ii) was employed as a reference of the OS-2. The portion of (iii) was introduced into the thermobalance (TG) fitted with the OS-3, and led into the OS-2 and TCD as a measuring gas. All oxygen sensors were made of CSZ. EGA was accomplished by OS-2 and TCD. Fig. 2 shows a construction of TG connected with OS-2 and TCD in detail. The OS-3 was a tube of CSZ, and electrodes were placed near the sample holder to monitor the exact oxygen potential in the vicinity of the sample. The thermobalance was Rigaku Thermoflex, and the temperature was controlled by Rigaku Programmable Controller with a microcomputer (PTC-10A).

Praseodymium oxide having a nominal purity of 99.99% was purchased from Rare Metallic Co., Tokyo, Japan. The oxide was dissolved in hot concentrated nitric acid, and this solution was diluted after the evaporation of excess nitric acid. The oxalic acid was added dropwise to the dilute solution in order to form the precipitate of oxalate. The oxalate was decomposed to the oxide at 900°C in air. The oxide was reduced to A-$Pr_2O_3$ (hexagonal) at 900°C in the atmosphere of helium containing 5 vol% of hydrogen. About 500 mg of A-$Pr_2O_3$ was used for the experiments as a starting material.

The experiments were conducted at oxygen pressures ranging $Po_2 = 1 \sim 10^{-5}$ atm and temperatures up to 1200°C. The gas flow rate was 50 ml/min.

## RESULTS AND DISCUSSION

Fig. 3 shows a representative isobar at $Po_2=1.39\times10^{-2}$ atm obtained with a heating and cooling rate of 1°C/min. The evolution and up-take of oxygen during transformations were sensitively detected by both TCD and OS, and the curves were in excellent agreement with the result of TG. The height on the EGA peak corresponded to the rate of oxidation and reduction. The TG curves indicated that the transformations between $PrO_x$ succesively occurred accompanying hystereses with the following sequences: $A(x=1.500) \rightarrow \epsilon(x=1.800) \rightarrow \zeta(x=1.778) \rightarrow \iota(x=1.714) \rightarrow \sigma(x=1.7\sim1.6) \rightarrow \theta(x\approx1.5)$ on heating, and $\theta \rightarrow \iota \rightarrow \zeta \rightarrow \epsilon \rightarrow \beta(x=1.833)$ on cooling.

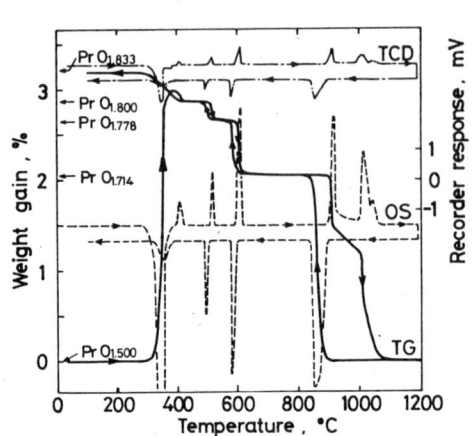

Fig. 3. Isobaric TG and EGA (TCD and OS) curves in the $PrO_x-O_2$ system at $Po_2=1.39\times10^{-2}$ atm.

This result is consistent with that of Hyde et al. (3) obtained under the same oxygen potential in an evacuated and oxygen-admitted system. Several phases such as $Pr_7O_{12}$ (x=1.714), $Pr_5O_9$ (x=1.800) and $Pr_6O_{11}$ (x=1.833) have been reported in the praseodymium-oxygen system, and the transformations between phases exhibited the hystereses in narrow ranges of temperature and oxygen potential. In the present study, the EGA using TCD and OS-2 successfully detected the oxygen evolution indicating the formation of nonstoichiometric θ phase though it was not so clear in TG.
Comparing the usefulness of TCD and OS, OS is more sensitive than TCD at $Po_2=1.39\times10^{-2}$ atm as shown in Fig. 3. Further experiments under the wider range of the oxygen potential showed that OS was more sensitive than TCD at lower oxygen potentials and vice versa at higher oxygen potentials.

Isobars were affected by heating and cooling conditions especially in hysteresis regions.

## REFERENCES

(1) R.E. Ferguson, E.D. Guth and L. Eyring, J. Amer. Chem. Soc. 76 (1954) 3890
(2) J.M. Honig, A.F. Clifford and P.A. Faeth, Inorg. Chem. 2 (1963) 791
(3) B.G. Hyde, D.J.M. Bevan and L. Eyring, Phil. Trans. Roy. Soc., London A259 (1966) 583
(4) J. Kordis and L. Eyring, J. Phys. Chem. 72 (1968) 2044
(5) L. Till, K.H. Radeke and H. Peters, J. Thermal Anal. 1 (1969) 465
(6) P.A. Faeth and A.F. Clifford, J. Phys. Chem. 67 (1963) 1453
(7) D.R. Knittel, S.P. Pack, S.H. Lin and L. Eyring, J. Chem. Phys. 67 (1977) 134.
(8) R.P. Turcotte, M.S. Jenkins and L. Eyring, J. Solid State Chem. 7 (1973) 454

INVESTIGATION OF THE DECOMPOSITION OF THE SOLID SOLUTION OF
A COMMERCIAL Al-Zn-Mg ALLOY (7015), BY MEANS OF DTA, HARDNESS
AND CONDUCTIVITY MEASUREMENTS

C.G. Cordovilla and E. Louis
Departamento de Metalurgia Física, I+D Productos
Endasa, Alicante, Spain

ABSTRACT

The decomposition of the solid solution of a commercial
Al-Zn-Mg alloy (7015) has been followed by means of DTA. The
alloy was studied in the as-quenched and naturally aged conditions. Although the sensitivity of the apparatus precluded
the study of G.P. zones formation, the precipitation of the
$\eta'$ and/or $\eta$ phases, and their redissolution processes were
identified. By varying the heating rate a value for the activation energy of the precipitation process was estimated. The
DTA peaks were interpreted by measuring the changes in the
conductivity and Vickers hardness during a linear heating similar to that provided by the DTA equipment.

INTRODUCTION

A great effort is being currently drawn to the understanding
of the processes taking place in the decomposition of the so
lid solution of Al-Zn-Mg alloys (1-8). Although depending
upon composition and other factors, various phases can be
formed (2,4,5), a simple phase sequence is generally found,
namely, G.P. → semicoherent $\eta'$ phase ($\sim Mg\ Zn_2$) → incoherent
$\eta$ phase ($Mg\ Zn_2$). Recently a metastable phase diagram for
this sequence has been inferred from different experimental
data (2). Thermal analyses are increasingly becoming a very
important tool in obtaining valuable information on these
processes (2-5). Nonetheless thermal data are hardly interpreted without the aid of other experimental measurements.
In this paper we present DTA studies of a commercial Al-Zn-Mg
alloy (7015), complemented with hardness and conductivity
measurements.

## EXPERIMENTAL PROCEDURES

The material used in this work was commercially fabricated, namely, d.c. casted, homogenized and hot rolled up to a thickness of 7.5 mm. Its composition being, in weight per cent, 4.92 Zn, 1.85 Mg, .2 Fe, .14 Si, .15 Zr, .11 Cr and .06 Mn. Two kinds of specimens were prepared, i) cylindrical specimens having a diameter of 4 mm. and a height of 7 mm., for DTA and ii) 40x70x7 mm. specimens for hardness and conductivity measurements. The samples were then solution heat treated for 60 min. at 465 ± 5ºC and quenched in water at room temperature (R.T.). Part of the specimens were afterwards studied in this condition. The natural ageing was performed during 4 days at 25 ± 0.5ºC in a thermostatted aqueous bath. The DTA measurements were carried out in a SETARAM equipment under dynamic nitrogen atmosphere. The measuring range was from R.T. up to 450ºC. Different heating rates were used although most of the experiments were performed at 2 and 4ºC/min. Conductivity ($\sigma$) and Vickers hardness (VH) were measured at R.T. on specimens previously polished and linearly heated at 2 and 4ºC/min. (in a sand fluidized furnace) up to different temperatures and then quenched in water at R.T. $\sigma$ was measured using a Sigmatest-T instrument type 2.067, and the load used to measure VH was 1.000 gr. The DTA measurements and the linear heatings to measure $\sigma$ and VH were performed at least twice; good reproducibility was found.

## RESULTS AND DISCUSSION

### i) as-quenched material (S condition)

The DTA, $\sigma$ and VH results for a heating rate of 4ºC/min. are shown in Figs. 1 and 2a. In the DTA curve we first notice an exothermic reaction between 180 and 250ºC peaked at ~220ºC (Table 1), followed by an endothermic reaction between 250 and 360ºC. The characteristics of the apparatus precluded any trustable study of reactions taking place below 130ºC, such as G.P. zone formation (2,5). The VH results show a weak maximum at 125ºC which indicates G.P. zone formation. A peak follows between 180 and 360ºC, its maximum being around 220ºC;

the end of this peak clearly coincides with the redissolution of all phases and its maximum is very closed to the maximum of the exothermic DTA peak. The $\sigma$ shows a very weak minimum between 110-130°C which is probably related with G.P. zones. Then $\sigma$ raises indicating G.P. zone redissolution and/or further precipitation; the very slight maximum around 180°C could be related with the end of G.P. zone redissolution and the beginning of $\eta'$ phase precipitation, favoring the first interpretation. The maximum $\sigma$ ($\sim$250°C) coincides with the beginning of the endothermic DTA peak, and then decreases down to its minimum value (initial value) which occurs at the same temperature than the end of that peak.

The $\sigma$ and VH results clearly support the interpretation of the exothermic and endothermic DTA peaks as precipitation and redissolution processes respectively. Moreover the maximum of the exothermic peak could be related with the change from coherent ($\eta'$ phase) to incoherent ($\eta$ phase) precipitation, as it nearly coincides with the maximum VH. Further support for this interpretation is provided by the conductivity results: this magnitude is increasing very rapidly at the maximum of the DTA peak indicating that precipitation is still occurring (growth of precipitates could also increase $\sigma$ although in a much less amount).

By varying the heating rate, the activation energy ($\Delta E$) of the exothermic reaction has been estimated by means of a modified Kissinger method (9,10). The temperature for the maximum of the DTA peak for different heating rates is given in Table 1. The resulting value for $\Delta E$ is .44 eV/molecule which is lower than those reported elsewhere (7,8).

ii) <u>naturally aged condition (NA)</u>

Results for a heating rate of 4°C/min. are shown in Figs. 1 and 2b. The exothermic peak lies, in this case, between 160 and 240°C, its maximum being at 215°C, and the endothermic reaction occurs between 260 and 370°C. It is worth noticing the plateau of $\sim$20°C between both peaks. The initial value of $\sigma$ is much lower than in the S condition; this is certainly due to G.P. zone formation (1,2,8). Then it raises up to 160°C where a weak maximum (or shoulder) is present (again it could be related to G.P. zones redissolution). The maximum $\sigma$ coincides with the end of the exothermic peak, then it remains nearly constant up to 280ºC and finally decreases up to the end of the redissolution processes. The VH starts from a value much higher than in the S condition, and then it decreases up to 160ºC where it begins to raise; the maximum VH (220ºC) again nearly coincides with the maximum of the exothermic DTA peak. The VH then decreases down to its minimum value (end of the redissolution process).

We first notice that the initial values of $\sigma$ and VH indicate that G.P. zones have been formed during the natural ageing (8). It can also be concluded that the end of G.P. redissolution occurs at 160ºC. Again the maximum of the exothermic DTA peak coincides with the maximum VH, indicating the change form coherent to incoherent precipitation. Finally we notice that the $\sigma$ and VH results suggest that along the plateau range of temperature, globulization and/or growth of particles is occurring.

In comparing our results with other experimental analyses we notice that the present thermograms are very similar to other reported elsewhere (5). Our results for G.P. zones redissolution, precipitation of $\eta'$ and/or $\eta$ phases, and redissolution of all phases are also consistent with other data (2-5). It is worth to remark that the relation between the maximum of the exothermic DTA peak and the transition from coherent to incoherent precipitation found in this work has not been previously reported.

Finally we notice that the maximum of the exothermic peak is very closed to the strongest variation of $\sigma$. This result could support the assumption made by Kissinger (9) to derive kinetic data from non-isothermal studies, namely, that the maximum of the DTA peak coincides with the maximum rate of reaction.

## REFERENCES

1. L. Mondolfo, "Aluminium Alloys, Structure and Properties"

Butterworth & C. Ltd., (1976) and references therein.
2. T. Ungar, J. Lendvai and I. Kovács, Aluminium 55,663(1979).
3. J. Lendvai, G. Honyek, I. Kovacs, Scripta Metallurgica 13 593 (1979).
4. K.H. Sackewitz et al, Kristall und Technik, 14,457 (1979).
5. A. Zahra et al, Z. Metallkde 70, 172 (1979).
6. N. Ryum, Z. Metallkde, 66 338 (1975).
7. R. Graf and L. Pelletier, Recherche aérospatiale 94,15(1963)
8. J.T. Staley, Met. Trans., 5, 929 (1974).
9. H.E. Kissinger, Anal. Chem., 29, 1702 (1957).
10. J.A. Augis and J.E. Bennett, J.Thermal Anal.13,283 (1978).

TABLE 1.- Peak temperature ($T_M$) for different heating rates ($\alpha$) for the precipitation of $\eta'$ and/or $\eta$ phases (S condition)

| $\alpha$ (ºC/min) | 2 | 4 | 8 | 10 | 12 | 20 |
|---|---|---|---|---|---|---|
| $T_M$ (ºC) | 200 | 223 | 253 | 264 | 275 | 295 |

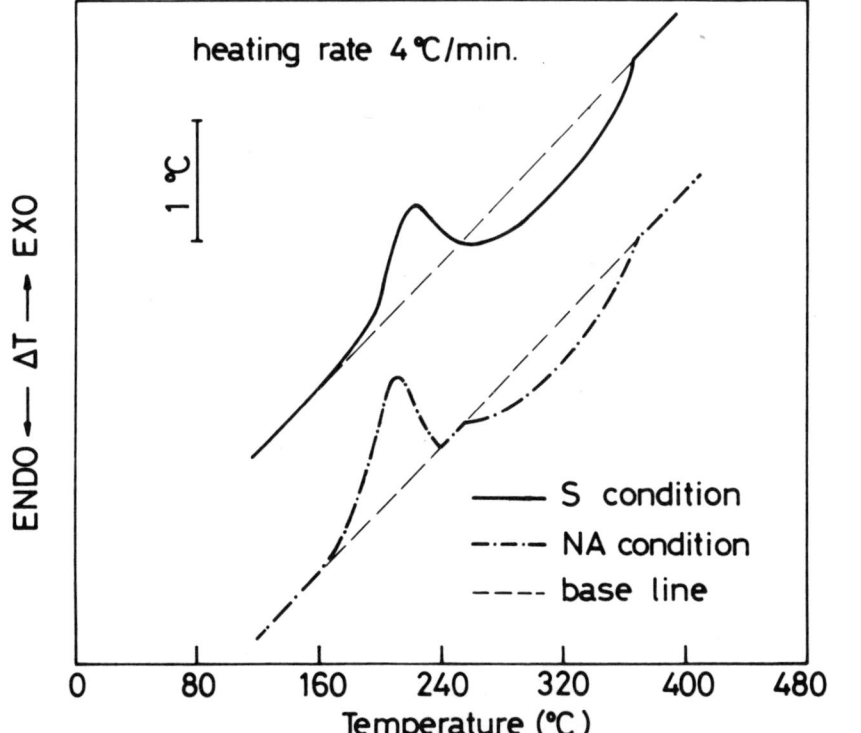

FIG. 1. DTA thermograms for the alloy studied in this work.

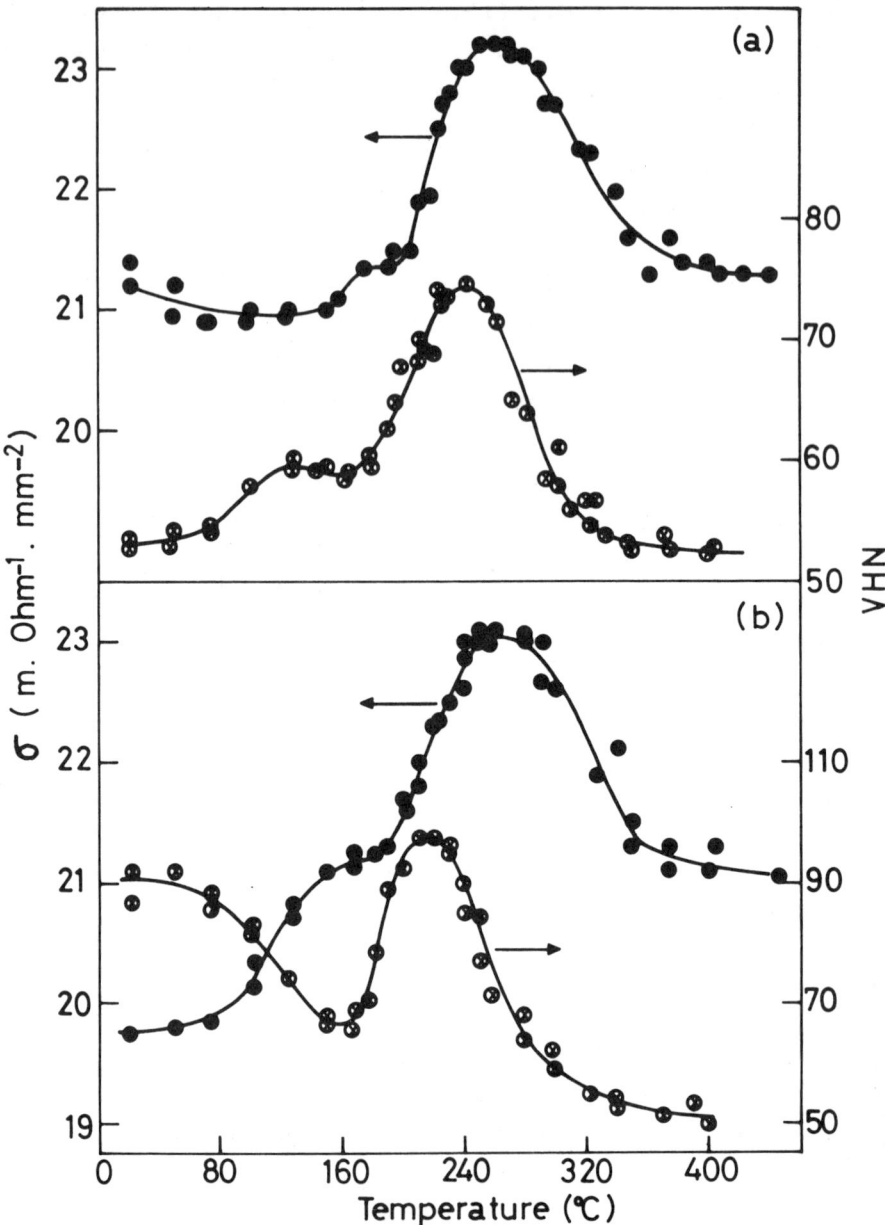

FIG. 2. Change in conductivity (●) and Vickers hardness (⊗), both measured at R.T., during a linear heating at 4ºC/min. a) as-quenched condition b) naturally aged condition. Each experimental point is the average of at least five measurements.

# THERMAL INVESTIGATION OF POLYETHYLEN USED IN POWER CABLES

G.Liptay[x], L.Ligethy[xx] and E.Brandt-Petrik[x]
[x]Department of Inorganic Chemistry, Technical University, Budapest, H-1521 Budapest, Hungary
[xx]Hungarian Cable Works, H-1117 Budapest, Hungary

## ABSTRACT

The polyethylen - used as cable insulation - changes its properties by processing and utilisation. For recording of this properties-changeing a suitable method is the isothermal-thermogravimetry carried out on a propriete temperature.

## INTRODUCTION

Polyethylene - due to its outstanding insulation property - is generally used for the core-insulation of high-voltage cables.
As a result of the specific morphology of polyethylene, the standard methods of material testing are not suitable for the qualification and classification of the materials for electrical and thermal endurance with regard to the processing and application aspects in the cable industry.
The cable core insulation operating in the electrical mains, is exposed to the simultaneous effect of the electric field intensity and varying ambient temperature. In case of emergence operating, temperature of the conductor of power cable may reach - for a short time - even $150^\circ C$, and this may occur repeatedly within the life of the cable.
The preparation of such test methods has become necessary, by which, the influence of the electrical and thermal stresses can be determined, or in case of large variety of materials, a relative stability range may be set up among the polymers (polyolefines) to be used as insulating material of high-voltage cables.

## EXPERIMENTAL PROCEDURE

The dynamic derivatographic method is suitable to follow the change in the thermal properties.

The dynamic measurements were carried out with MOM (Hungarian Optical Works) OD-2 type derivatograph is atmospheric condition at 3°C/min heating rate. The 500 mg quantity of the material cut up to prisms of 1 x 1 x 2 mm, was placed in platinum crucible and put into the furnace chamber. Experiments were carried out also with static isotherm thermogravimetric methods, which subsequently proved to be a method supplying more information. Weight variation of the samples as a function of time was recorded at a constant temperature in the isotherm measurements. First the empty furnace of the thermo-balance was heated up to the required temperature, ensuring the constancy of temperature to $\pm 0.2$ °C accuracy with a Programik (Instrument-Industrial Research Institute, Budapest) temperature control instrument. Thereafter the quartz crucible containing the 1.0 g sample was placed into the furnace chamber on the thermoelement connected to the balance, and weight change of the material was measured in atmospheric condition.

The investigations were carried out on:

A./ polyethylene with a very high molecular weight, and mol-weight distribution, containing a processing stabilisator,

B./ the same as under A./ but containing some 0.1 phr p-phenylene-diamine as voltage stabilizer,

C./ a peroxid-compound as cureing agent containing polyethylene

## RESULTS AND DISCUSSION

1./ <u>Dynamic measurements</u>. Derivatogramms of A./, B./ and C./ materials are shown on figure 1. Some differences can be seen on their thermoanalytical curves which characterises the materials, these methods cannot be developed further, because of the very small response.

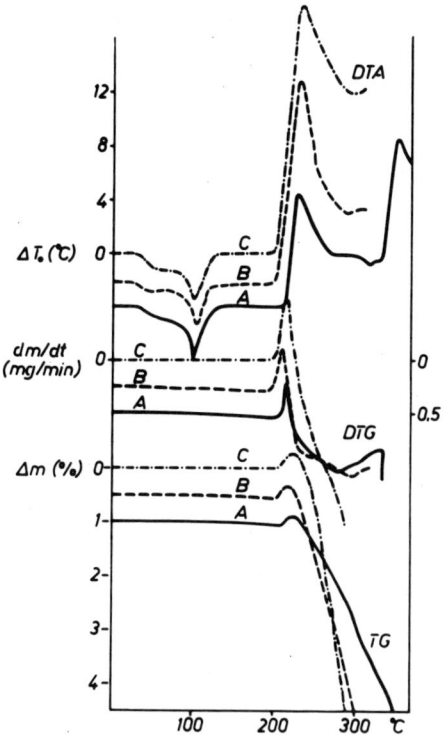

Fig.1.

2./ <u>Isothermal measurement methods.</u> The isothermal thermogravimetric investigations of polyehtylene - on a well choosen temperature, show curves with characteristic form as seen on figure 2.

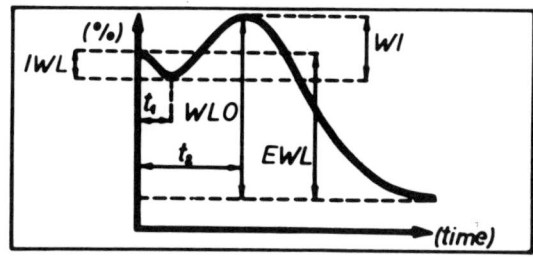

Fig.2.

Legend to Fig.2.
IWL = Initial weight loss
WI = Weight increase
EWL = Extent of weight loss during measurement
WLO = Weight loss from oxidation maximum

The general run of the curves at the isotherm measurements of the polyethylene is shown in the following schematic diagram.

Time value of IWL and WI marked with $t_1$ and $t_2$ respectively, can be determined from the diagrams, the comparison of which will give useful information.

The thermo-oxidative decomposition of polyethylenes is a consecutive process, in the course of which parallel and opposing reactions are taking place. In order to evaluate and compare the thermal test results, it is necessary to divide the curve obtained as a resultant of the processes into oxidation and degradation sections.

The changes in rate of reactions are characteristic for the pre-life (thermal and electrical) of polyethylene sample. Comparison of the isothermal curves (a) of electrically aged polyethylene (b) on fig.3 leads to the following conclusion:

1./ The slope of the decomposition curves of the non-aged and electrically aged samples are identical.

2./ Character of the decomposition of samples subjected to electric stress does not change in relation to the non-aged sample, only the time of the minimum or maximum of the isotherm will be reduced, and the pertinent weight values will be modified.

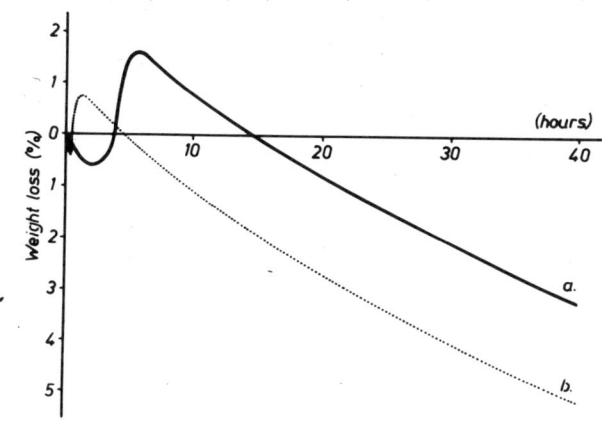

Fig. 3.

By the processing of polyethylene is also thermal stressed. On figure 4 shows the change in effects which are caused by thermal stress during processing.

The isothermal curve "a" measured on 150°C represents the original material. The curve "b" characterizes the materials one time moulded and the "c" shows a double moulded sample. Values are also given in Table I.

The result demonstrate well, that the processes taking place in the samples during the technological heat-treetment, reduced the thermal stability (WLO and EWL), and increased the number of places capable for oxidation resulting in weight increase (WT).

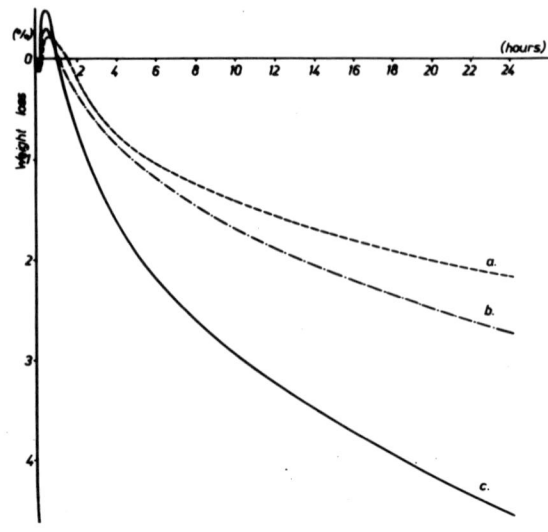

Fig.4.

Table I.

| Sample | Initial weight loss % IWL | $t_1$ hour | Weight increase % WI | $t_2$ hour | Weight loss %/24 h EWL | Weight loss the oxidation maximum %/24 h WLO |
|---|---|---|---|---|---|---|
| Granulate | 0.16 | 0.1 | 0.36 | 0.75 | 2.14 | 2.34 |
| Moulded plate | 0.14 | 0.1 | 0.39 | 0.5 | 2.73 | 2.98 |
| Double moulded disc | 0.11 | 0.1 | 0.55 | 0.45 | 4.55 | 4.99 |

Examples also showed, the isothermal thermogravimetry can be used succesfully to investigate the thermal and electric ageing of polyethylenes.

## REFERENCES

G.Liptay, L.Ligethy, E.Brandt-Petrik: Wire Industry (1977) 553-556

# LOW TEMPERATURE REGENERATION OF ACTIVATED CARBON.
# KINETIC ANALYSIS OF THERMODESORPTION OF PHENOL.

V. Amicarelli, G. Baldassarre, L. Liberti
Istituto di Chimica Applicata - Facoltà di Ingegneria-Bari
Via Re David 200 (Italy)

## INTRODUCTION

Use of granular activated carbon (GAC) widely employed for water and wastewater treatments, where removal of potentially mutagenic substances is frequently required (1), is limited by the cost of thermal regeneration. In order to visualize possible economies in thermal regeneration procedures, a systematical investigation on thermo-release mechanisms of several common pollutants from GAC has been undertaken.

A cyclic low-temperature regeneration procedure, in particular, showed promising indications for a cheaper thermal regeneration of GAC (2). According to this procedure, exhausted carbon is routinely regenerated (partially) by heating in inert atmosphere up to the peak temperature for each adsorbate, with a drastic thermal treatment occurring every 15-20 cycles.

Operational performances of the accordingly regenerated carbon have been evaluated, with losses of adsorption capacity comparable, or even lower, then those occurring with the standard, more expensive procedure.

A previous evaluation of gravimetric curves (TG, DTG) indicated that various steps are consecutively occurring during thermal desorption, in inert atmosphere, of phenol, aniline and their nitro-derivatives (3-5). In every case, the first step, which allows the largest part (70-90%) of adsorbate to be released, occurs below 500 °C.

In this paper a kinetic analysis of thermodesorption of phenol from GAC is presented.

## EXPERIMENTAL

About 100 mg of 16-25 (US) mesh dry GAC (Filtrasorb 400 from

Calgon Corp., Pittsburg, Pa.) were equilibrated with 100 cc
of acqueous solution containing 80 ppm of phenol. The adsorbed amount was determined spectrophotometrically by analysis of the acqueous phase at the wavelength of maximum absorption (i.e., 270 nm).
The samples were vacuum filtered and then heated upto 500 °C in a dynamic nitrogen atmosphere, with heating rate of 2,4,6, and 10 K°/min in a TA1 Mettler thermoanalyzer.
Mathematics were solved through an HP 9845 A computer.

## RESULTS AND DISCUSSION

Thermogravimetric desorption of phenol from GAC, conducted in inert atmosphere at 10 °K/min, has been shown to occur with 4 consecutive steps within the thermal range 150-950 °C (3). The first step which closes before 400 °C, is responsible for as much as about 90% of over-all thermo-desorption. In order to obtain a quantitative evaluation of the nature of physico-chemical processes involved in this step, experimental thermogravimetric data (TG) betwen 150 °C and 400 °C have been submitted to kinetic analysis.
Among the various integral models of thermogravimetric kinetic analysis available in the literature, the method of Ozawa (6), indicated among the affordable ones in a foundamental review by Flynn and Wall (7), appears particularly suitable for thermal processes dealing with several, not well distinct steps (i.e., without clearly recognizable initial and final plateaux), as in our case. By evaluation of thermodesorption curves carried out at different heating rates, $\underline{a}$, Ozawa's method allows for the various steps involved to be discerned and, for each step, the so-called kinetic parameters, activation energy, $\underline{E}$, reaction order, $\underline{n}$, and preexponential factor, $\underline{Z}$, to be calculated. Accordingly, thermogravimetric analysis of step between 150-400 °C has been performed at $\underline{a}$=2, 4, 6, 10 °K/min., Fig. 1.

In Fig. 2 the reciprocal absolute temperatures, at a given thermodesorption degree ($\alpha$), corresponding to different heating rates are plotted: as expected, straight lines are obtained, each corresponding to eq.:

$$\log a = -0.4567 \, E/RT + \text{const.} \qquad (A)$$

By the constancy of slopes within the thermal ranges 201-299, 206-324, 214-330 and 222-351 °C for the above mentioned heating rates respectively, Ozawa's plot confirm that in these ranges the first step can be definitely delimited. From such figures, the E value for this step is calculated according to equation (A). As resumed on Fig. 2, results the average value E=22.0 Kcal/mol.

Follwing Ozawa modified method (8), in order to evaluate the reaction mechanism, suitable functions of α have been plotted versus the so-called reduced time θ, defined according to eq:

$$\theta = E/aR \cdot p(E/RT) \qquad (B)$$

Within the thermal ranges and heating rates investigated only for 1st-order reaction a good linearization as been obtained. Consequently a value around $1.3 \times 10^6$ is obtained for the pre-exponential factor Z.

Finally, with such values of E, n and Z the thermogravimetric curves for first step of phenol release within the ranges investigated have been calculated. As shown by Fig. 3, a good agreement is obtained by curve-fitting the experimental to calculated values, which confirms the affordability of Ozawa kinetic model for our case.

By use of this model, the major thermal release of phenol from GAC under the conditions investigated appears to occur by a 1st-order kinetic reaction, which can be associated to pyrolitic fragmentation of such strongly bound, relatively highly boiling adsorbate.

As far as its major release is concerned, thus, phenol thermodesorption can be included in the class of organic compounds undergoing 1st-order, thermal craking desorption processes for which, as described by Suzuki (9) and by Seewald (10), an over-all E value around 20-25 Kcal/mol can be expected.

## CONCLUSIONS

Ozawa model appears to provide an affordable method for kinetic analysis of phenol thermodesorption from granular activated carbon, for which a 1st-order, thermal craking mechanism has been demonstrated. A quantitative analysis of thermodesorption of other pollutants, already investigated, will be performed by a similar procedure, eventually modified to reduce the exstensive experimentation (i.e., use of several heating rates) required by this model.

The evaluation of the mechanism of the reaction involved, and the delimitation of the thermal ranges where the major releases of the adsorbate occur, well below the temperature around 950 °C usually employed for such processes, is expected to provide useful suggestions for a cheaper thermal regeneration of GAC.

## REFERENCES

(1) J.A. Cotruvo, C.Wu , JAWWA Nov. (1978) 590
(2) V. Amicarelli, G. Baldassarre and L.Liberti, J. Thermal Anal. 1 (1980)
(3) V. Amicarelli, G. Baldassarre and L.Liberti, Thermochim. Acta 30 (1979) 247
(4) V. Amicarelli, G. Baldassarre and L.Liberti , ibid. 30 (1979) 255.
(5) V.Amicarelli, G.Baldassarre, L.Liberti and V.Balice , ibid. 30 (1979) 259.
(6) T. Ozawa, Bull. Chem. Soc. Japan, 38 (1965) 1881
(7) G.H.Flynn, L.A.Wall, J. Res.Nat.Bureau Standards-A 70, 6 (1966) 487.
(8) T. Ozawa, J. Thermal Anal. , 9 (1976) 369.
(9) M. Suzuki, D. M. Misic,O.Koyama, K.Kawazoe, Chem. Eng. Sci. 33 (1978) 271.
(10) H.Seewald, Untersuchungen zur Sorption organischer Stoffe an Aktiv kohlen. Bergbau-Forschung GmbH (1974).

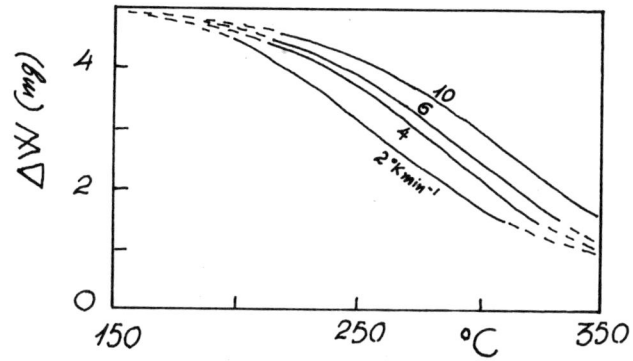

Fig.1. EXPERIMENTAL THERMOGRAVIMETRIC CURVES

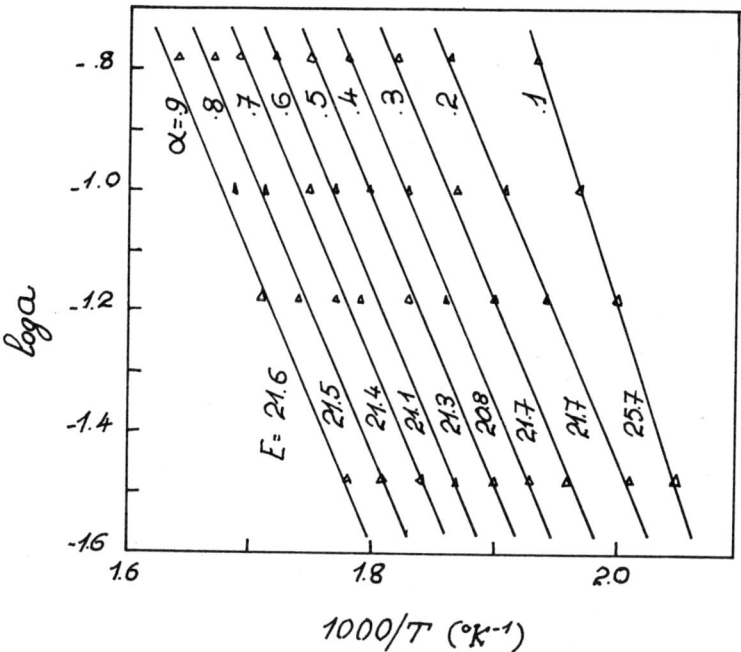

Fig.2. OZAWA PLOT OF THE INVESTIGATED STEP

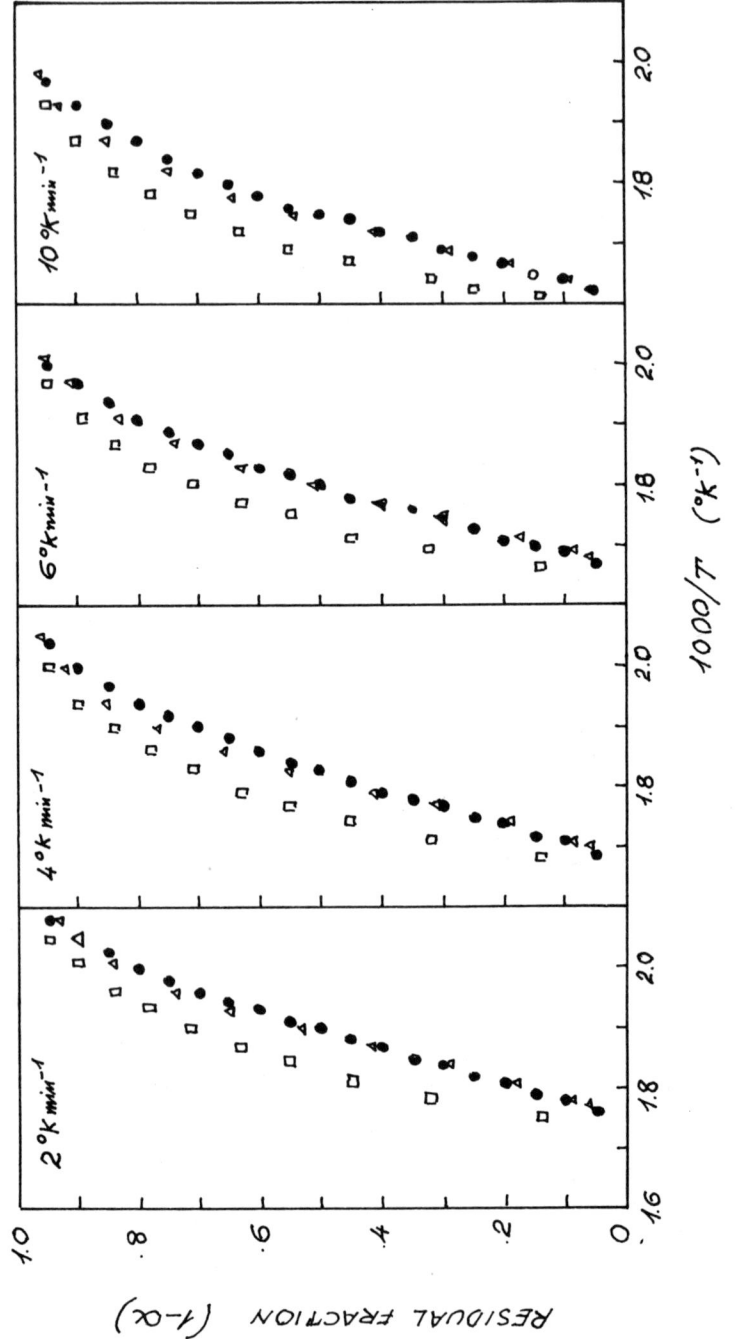

Fig. 3. EXPERIMENTAL (●) AND CALCULATED THERMOGRAVIMETRIC CURVES (1st-order reaction △; diffusion □)

# DEVELOPMENT OF TEMPERATURE DISTRIBUTION INSIDE THE REACTION ZONE OF A BURNING CIGARETTE

Richard R. Baker
Group Research and Development Centre,
British-American Tobacco Co. Ltd.,
Southampton, GB

## ABSTRACT

The development of the temperature distribution as a cigarette is smoked has been determined in order to provide a framework for rationalising the combustion processes occurring.

## INTRODUCTION

Products in cigarette smoke such as carbon monoxide and carbon dioxide are formed by two distinct mechanisms: combustion of tobacco and thermal decomposition of tobacco (1-4). These processes occur over specific temperature regions (5-7) and consequently the products are released within specific localities of the combustion coal (i.e. burning zone) of the cigarette (4,8,9). The development of the temperature distribution during the smoking cycle is therefore of prime importance in determining the yield of products formed inside the cigarette.

## EXPERIMENTAL PROCEDURE

There are two temperature distributions within the burning zone which must be considered - that of the solid matrix of the zone, and that of the gases within the zone (10,11). The solid phase temperatures have been measured by the infra-red emission from the burning zone, which was transmitted along a fibre optic probe placed inside the cigarette, to an infra-red detecting system (Vanzetti 1074 thermal probe system). The burning zone of a cigarette is virtually a blackbody radiator (12). A platinum/platinum - 13% rhodium thermocouple inside the burning zone effectively measures the temperature of the gas phase within the burning zone (10).

The internal temperature distributions during the third puff of the smoking regime, and subsequent smoulder period, have been monitored.

The cigarettes were smoked singly on a smoking machine, taking a 35 cm³ puff of 2 seconds duration, once per minute.

The electrical outputs from either the infra-red system or the thermocouple were monitored on a multichannel data logging system. The results were processed on an off-line computer. By combining together the results obtained with the probes initially placed into different positions inside the burning cigarette, contour distributions of the gas and solid phase temperatures at different times in the smoking cycle have been obtained.

Further details of the experimental and calculation procedures are given in (11).

## RESULTS

### Internal Temperature Distribution

Non-filter cigarettes have been used, containing flue-cured tobacco, 70 mm long and 8 mm in diameter, and were wrapped in paper of permea--bility 17 cm min$^{-1}$ (10 cm water)$^{-1}$ (1 cm water = 98 N m$^{-2}$).

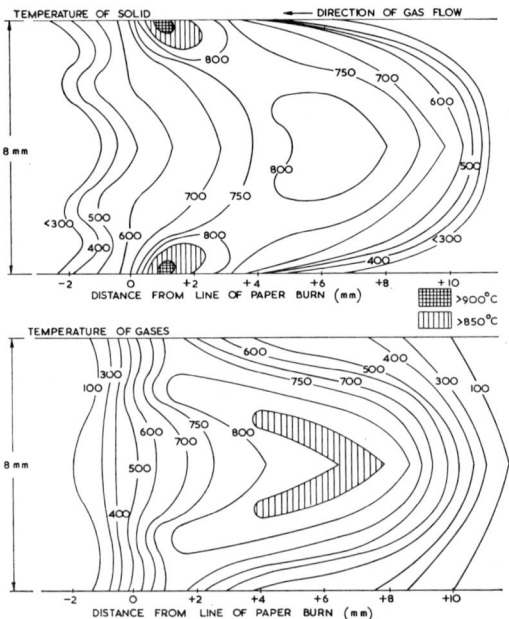

Fig 1. Temperature (°C) distribution, 1.0 second after the start of a puff (reproduced by permission of Academic Press).

At the end of the smoulder period, prior to the puff being taken, the temperature distributions of both the solid and gas phases are similar. Both phases have their highest temperatures (780-790°C) occurring in the centre of the coal, and cooling to about 500°C at the periphery in the region of the paper burn line. Thermal equilibrium between the two phases in the coal is close to being attained. However, during the puff the two phases have very different temperature distributions near the surface of the coal, although they are similar (at about 800-850°C in the central regions (e.g. Figure 1). The most dramatic change that occurs during the puff is that the solid phase temperatures at the periphery of the coal, just in front of the paper burn line, have increased from 500°C to 910°C. This is the region where the maximum amount of combustion is occurring, where the main air flow is entering the coal.

The highest gas phase temperature on the surface of the coal occurs at the same place as the highest solid phase temperature, but its value is much lower (770°C). The maximum gas phase temperature (860°C) occurs in the centre of the coal, 7 mm in front of the paper burn line.

When the puff ends, the solid phase temperature on the periphery of the coal cools from over 900°C to 600°C in one second. Within about four seconds the two phases have attained quasi-thermal equilibrium throughout the coal.

## Variation Of Maximum Surface Temperature Of The Coal With Volume Of Air Drawn Through The Coal

The volume of air drawn through the coal during the puff may be varied in several ways: varying the air flow drawn from the cigarette, varying the permeability of the cigarette paper, or varying the puff number. As the paper permeability is increased, a higher proportion of the air drawn into the cigarette by passes the coal and enters via the paper. Similarly, as the cigarette is smoked the length of the unburnt tobacco rod continuously decreases, so a larger proportion of the puff volume enters the cigarette through the coal.

The maximum surface temperature of the coal has been measured in three series of experiments:

1. the third puff (2 seconds duration) in the smoking regime, with puff volumes from the drawing end of the cigarette of 25, 35 and 45 cm$^3$ (cigarette paper permeability = 17 cm min$^{-1}$ (10 cm water)$^{-1}$,

2. the third puff (35 cm$^3$, 2 seconds), for cigarettes wrapped in paper of permeabilities 9.5, 17, 37, 80, 93 and 138 cm min$^{-1}$ (10 cm water)$^{-1}$,

3. consecutive puffs as the cigarette is smoked, ignoring the first (lighting) puff (each puff 35 cm$^3$ and 2 seconds long, cigarette paper permeability 17 cm min$^{-1}$ (10 cm water)$^{-1}$).

The 95% confidence limit of each quoted mean maximum surface temperature (mean of ten replicates) is about $\pm$ 25°C.

For each set of experimental conditions, the volume of air drawn through the coal $V_c$ (cm$^3$) into the unburnt tobacco rod can be calculated using equations published elsewhere for the gas flow through cigarettes (13-15). In order to use these equations, values for the impedance of the tobacco rod behind the coal, and the pressure drop across the burning cigarette were required. These were determined as described elsewhere (13,16). The quoted values of $V_c$ are probably too high by 20% due to net formation of gases in the coal (17). However, the quoted values should be correct relative to one another.

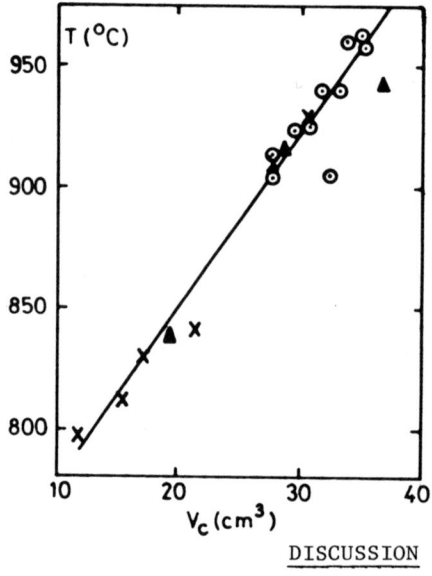

Fig 2. Maximum surface temperature (T, °C) v. volume of air entering coal ($V_c$, cm$^3$).

▲ Drawing flow varied
x Paper permeability varied
⊙ Puff number varied.

## DISCUSSION

From an examination of the development of the temperature distribution during the smoking cycle, many of the physical processes occurring during cigarette combustion can be rationalised, as follows.

In a cigarette, the rate of combustion of the tobacco is controlled by the rate of oxygen transport to the tobacco surface (7,18). Because of increases in the viscosity and velocity of the gases in the combustion coal with temperature, the pressure drop across the coal is high (16). Consequently, during a puff air tends to enter the cigarette at the base of the coal, near the paper burn line, where the draw resistance is smallest (11,19,20). During the puff heat release at the base of the coal from surface oxidation is greater than heat losses, and the peripheral temperatures of the solid phase increase (Figure 1). Consequently, during a puff it is mainly the periphery of the coal that advances (19). During about the first half of the puff the pressure drop across the coal increases rapidly (16), due to the increased viscosity and velocity of the gases caused by the rising gas temperatures.

Thus, as the puff progresses the volume of the cigarette consumed tends towards a constant value, and an increasing proportion of the mainstream smoke consists of air which has by passed the coal and entered via the paper (19). Since the permeability of paper to air increases sharply at about $300°C$ (21), there is a large influx of air just behind the paper burn line (20). Therefore, since much of the incoming air during the puff by passes the central regions of the coal, the gas and solid phase temperatures in the central regions increase by a smaller amount than the temperatures at the periphery.

When the puff ends, there is a greatly reduced transport of oxygen to the surface of the coal, and therefore a greatly reduced amount of exothermic surface oxidation. The periphery of the coal cools rapidly, since it radiates heat rapidly to the surroundings; its main source of heat is now the inner core of the coal. Oxygen diffuses into the back of the coal via the cigarette paper. The central region of the coal advances to re-establish a flat region at the back of the coal. Thus, immediately after a puff there is often a time delay of up to fifteen seconds before the paper burn line visibly moves.

## REFERENCES

(1) J. E. Baxter and M. E. Hobbs, Tobacco Sci. __11__ (1967) 65.
(2) W. R. Johnson, D. H. Powell, R. W. Hale and R. A. Kornfield, Chem. Ind. (1975) 521.
(3) R. A. Kornfield, R. H. Newman, L. E. Brown, Jr. and W. R. Johnson,

Jr., Chem. Ind. (1979) 664.
(4) R. R. Baker, Combust. Flame 30 (1977) 21.
(5) R. R. Baker, Beitr. Tabakforsch. 8 (1975) 16.
(6) H. R. Burton, Beitr. Tabakforsch. 8 (1975) 78.
(7) R. R. Baker, Thermochim. Acta 17 (1976) 29.
(8) R. R. Baker and K. D. Kilburn, Beitr. Tabakforsch. 7 (1973) 79.
(9) H. V. Lanzillotti and A. R. Wayte, Beitr. Tabakforsch. 8 (1975) 219.
(10) R. R. Baker, Nature 247 (1974) 405.
(11) R. R. Baker, High Temp. Sci. 7 (1975) 236.
(12) A. T. Lendvay and T. S. Laszlo, Beitr. Tabakforsch. 8 (1976) 283.
(13) R. T. Jarman, Paper presented at the Tobacco Chemists' Research Conference, Winston-Salem, N.C., USA, October 1960.
(14) K. M. Meyer-Abich, Beitr. Tabakforsch. 3 (1966) 307.
(15) P. Somasunderan and K. J. Mysels, Paper presented at the CORESTA/Tobacco Chemists' Joint Conference, Williamsburg, Va., USA, October 1972.
(16) R. R. Baker, Beitr. Tabakforsch 8 (1975) 124.
(17) P. Somasundaran, Tobacco Sci. 14 (1970) 176.
(18) K. Gugan, Combust. Flame 10 (1966) 161.
(19) Sir A. Egerton, K. Gugan and F. J. Weinberg, Combust. Flame 7 (1963) 63.
(20) R. R. Baker, Nature 264 (1976) 167.
(21) R. R. Baker, Tappi 59 (1976) 114.

## THERMAL ANALYSIS AS AN AID TO THE CRIMINALIST

R. Halonbrenner
Wissenschaftlicher Dienst, Stadtpolizei Zürich, Switzerland

When a criminalist is faced with the task of analyzing the traces secured after an offence or a crime and to make them talk, it is his duty to use the latest methods and processes of investigation.
But the range of application of new investigation techniques in the field of criminalistics is limited from two sides. Since many traces of fact are available only in microscopic amounts, on the one hand all those analytical procedures are ruled out which require a relatively large sample. On the other hand, microscopic traces will preferably be submitted to nondestructive methods of investigation, so as to preserve the corpora delicti both as evidence and, if needed, for checking of test results. For these reasons, it is not surprising that criminalists did not pay attention to thermal analysis while relatively large amounts of sample material were required.
Thermal analysis has not yet been accepted by institutes of criminal investigation as a general method. (1), (2).
In our capacity as a laboratory consulted on criminal investigation from all over Switzerland and abroad, we have made it our business to incorporate thermal analysis into our set of standard test methods. It is our policy to apply thermal analysis in all those cases where it will provide evidence or results economically. At the same time, this is the only limitation for the application of thermal analysis in our field.
To meet requirements in the best possible way, we have acquired a Mettler Thermoanalyzer TA-1 which enables us to combine thermogravimetry and differential scanning calorimetry.
Investigations relating to cases of fire damage are in the foreground. While trying to determine the cause of a fire, it

is often to be found out whether certain materials stocked or deposited in the zone of initial fire have a tendency to self-ignition and might therefore have led to spontaneous ignition. Previously applied methods of investigation, for instance the often used Mackey-test, produced only qualitative or at best semiquantitative results. Even abstaining from a precise quantitative determination of all effects - which is usually sufficient for practical work - thermal analysis allows more eloquent results. Let me explain this by a group of substances which often play a role when determining the causes of a fire with regard to their possible self-heating or self-igniting properties, namely the so-called drying oils such as linseed oil and others. To obtain a large surface we apply such substances on kieselguhr.

In interpreting such a TG/DTA curve, we always speak of "self-heating, possibly with ensuing spontaneous ignition" since firstly the quantitative proportions are not determined by the diagram and, secondly, specific conditions of the fire site also have to be taken into consideration. Among these other factors, there is the distribution of a self-inflammable substance at the fire site as, for instance, a shutter impregnated with linseed oil cannot ignite spontaneously whereas a ball of cottonwaste soaked with linseed oil can.

At any rate, thermal analysis indicates whether spontaneous ignition will have to be considered as a possible cause of fire or not.

The cause of a fire in a hair-dresser's laboratory had to be determined. He prepared his shampoos and emulsions himself, and among other ingredients he also used a colourless oil of the guttiferae family. Testing the various substances thermoanalytically, we only found an exothermic heat of reaction at about 100 °C for the colourless oil, accompanied by a slight weight increase. In order to determine whether this self-heating trend could possibly have led to spontaneous ignition of the oil under the circumstances prevailing at the site of the fire we also determined the oil's iodine number.

It was 82 which places this oil among the quasi-nondrying oils. Normally, these will not ignite spontaneously. Partly based on these conclusions, the hair-dresser admitted in the end that he had set fire to his own shop.

This example shows that spontaneous ignition can only be excluded in those cases with no exothermic heat of reaction. However, as soon as there is an exothermic heat of reaction, it is necessary to know the exact conditions at the site of the fire.

As a further example, let me refer to a fire in a factory manufacturing insulation cork slabs. We localized there the zone of initial fire in the stock of raw cork. On the eve of the fire, freshly produced slabs had been deposited there. They were manufactured by heating bitumen with an oil burner to, allegedly, slightly less than 200 °C and mixing it with roasted cork meal. But our measurements showed that temperatures of 220 °C could easily be reached not causing an alteration of the end product. Whereas dry cork is not known to self-ignite, freshly processed insulation cork materials in blocks is counted among the substances with a tendency to autoxidation. We therefore had to establish the temperature limit above which exothermic reactions occur in freshly processed material. Thermal analysis showed the critical temperature to be 190 °C. This was the first marginal value for the experiments to be carried out in our fire test lab. When heating a block of cork to 200 °C, we found the dissipation of heat to be sufficient to avoid further heating of the block in spite of the exothermic reactions taking place. If, however, the block of cork was heated beyond 200 °C, the exothermic reactions led to ignition even hours after the heating had been turned off. With that the case was closed.

Let me give you another example:
In a new building vertical plastic ventilation ducts were installed. They were insulated by a layer of rock wool coated by aluminum foil; the rock wool was held together by a plastified wire-mesh. Shrunk-on sleeved placed at intervals served to fix the rock wool to the pipes. The sleeves were heated

by a gas burner. As a workman was proceeding with this work, the flame of the gas burner ignited the plastified parts of the ventilation ducts above, causing a fire in the new building which resulted in a property damage of several million Swiss francs. Trying to find out the cause of the fire, the ventilation manufacturer made incendiary tests which led to the conclusion that the ventilation duct system could not have made a contribution in spreading the fire, but the reason should be found elsewhere. Our service was entrusted with a superior expert evidence and had to give an answer on this question, too.

We started our tests with the thermal analysis of all materials used in the ventilation ducts. All of them showed an endothermic heat of reaction when heated with the exception of the plastified material on the wire-mesh. The latter however showed an exothermic heat of reaction between 190 and 220 °C, connected with a substantial weight decrease. This and the following investigation provided the key to the cause of the fire: Mass spectroscopic analysis proved that the volatile matters set free at 200 °C were primarily combustible substances. Now all we had to do was to find the key-hole. This we found out by varying the draft in our incendiary tests. Without any artificial aeration in our "ventilation duct" - as had been the case in the tests made elsewhere - the fire did not spread upwards. But as soon as a slight draft was simulated artificially, - corresponding to the conditions at the destroyed building - the flames rose quickly to the top.

Let me now leave the field of fires and show you how thermal analysis can be applied in other areas in order to solve criminalistic problems.

After a fire in a garden pavilion, the clothes and shoes of a suspected arsonist were examined. Earth was found clinging to his shoes. The suspect declared that he had been near the grave of his son and that he had not walked on any other ground or meadows. For a comparative analysis of the soil samples we took earth from the area around the son's grave

and from the possible access roads to the fire-site. 20 mg
each of these samples were weighed into crucibles for thermal
analysis and heated at a rate of 6 °C/min to 900 °C.

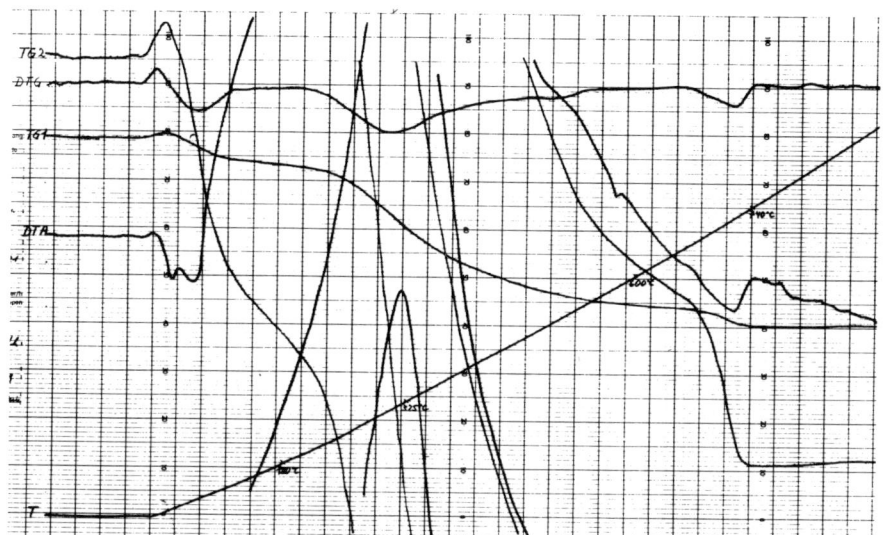

Results showed clearly that the soil taken from the shoes was
completely different from the one near the grave. Yet the
differences to the other three samples, all of which were
taken from the vicinity of the fire area, were minimal. Since
soil as such is a heterogeneous mixture, analytical tests can
generally not prove any identity. But the differences found
between the earth on the shoes and the earth from the grave
proved that the suspect had not told the truth. For this
reason he was sentenced for arson.

In the chemical industry, thermal analysis has been long
known as a method of testing safety specifications of different products and for analyzing explosives (3). Obviously, we
also apply thermal analysis for clarifying accidents in chemical plants. Let me tell about cases where thermoanalytical
tests of partly unknown substances or mixtures made a substantial contribution in finding out the truth.

The first case refers to a stick of some fairly old explosive
found outside. We had to find out if this material would
still detonate. The test in a high-pressure crucible showed

the beginning of an exothermic reaction at 150 °C. With a weight-in quantity of 50 mg, this reaction continued with a deflagration at 170 °C. This provided a positive answer to the question posed. Thin-layer chromatography showed that the sample contained the main components nitroglycol, 2,4-dinitrotoluene and 2,4,6-trinitrotoluene. Thus, it would also have been possible to find out whether the detonating capacity of the found explosive was still the same as that of a new or well-stored sample of the same material.

In another case we had to analyze some material that had been found in the possession of a prisoner. It was a small cardboard box fitted with an electric ignition device and containing red powder. We had to determine whether this device was suitable to attempt a blow-up assault. Comparative thermal analysis showed that the red powder were scrapings from the tops of matches. Consequently, the box was well suited as an incendiary device.

When choosing a test method in a criminalistic laboratory one has to note that thermal analysis is not completely nondestructive. This restricts the application of this method. For instance, chemical fibers and their mixtures which are often found in lengths of only a few millimeters will not be tested thermoanalytically but by the nondestructive microspectrophotometry. On the other hand, metal particles which may be found as welding beads in a burglar's pockets, are better suited for X-ray fluorescence analysis.

Nevertheless, thermoanalytical tests performed to date have shown that this method provides a valuable addition to the investigation techniques of the criminalist.

## References
(1) W. Schiller & E. Röhm, Kriminalistik **31**, 439 (1977)
(2) G. Hellmiss, Zeitschrift für Forschung und Technik im Brandschutz **27**, 37 (1978)
(3) G. Krien, Explosivstoffe **13**, 205 (1965)

# Industrial Applications

G. Hentze
Bayer AG
Leverkusen, FRG

R. Bauer
Brown, Boveri & Cie AG
Mannheim, FRG

THERMOANALYTICAL CHARACTERIZATION OF CURING BEHAVIOR OF CLEAR VARNISHES

H. Möhler, K.D. Gauler, A. Henig, G. Janik and M. Jäth
Fachrichtung Kunststofftechnik der FH Würzburg-Schweinfurt, FRG

ABSTRACT AND INTRODUCTION

Because of the claims for short drying durations combined with low drying temperatures and small solvent retention in order to realise high conveyor-belt velocities in serial manufacture of automobiles, the chemical composition and properties (hydroxyl and carboxyl value, average molecular weight) of high-solid varnishes and the dispersing conditions with additional variation of hardeners for powder varnishes were varied. Some dependences between thermoanalytical values as for example temperatures and specific heat of reaction and the above mentioned parameters were found.

EXPERIMENTAL PROCEDURE AND VARNISHES FORMULATION

For our thermoanalytical investigations the Heraeus-System TA 500 S 2 equipped with the differencemicrocalorimetric cell (DCS) and the thermobalance TGA 500 was used. The samples were weighted with an accuracy of $\pm$ 0,05 mg.

Powder Varnishes

As components for the formulation were used:
Epikote 1055 (Shell Chemie, Frankfurt), Uralac P2228 (Scado GmbH, Emden), Dicyandiamid-hardener (Süddeutsche Kalkstickstoff-Werke, Trostberg), Acidanhydride-hardener XB2622 (Ciba-Geigy AG, Basel).

The actual formulations of the powders are given in table 1. The blends were dispersed in two different ways. For the first way the coarsely grounded and dry-blended materials weré extruded and then for thermoanalytical measurement grounded to a particle size of less than 80 $\mu$m. The blends marked by an asterisk were only grounded to a particle size of less than 80 $\mu$m.

Polyurethane/polyester resin varnishes

Five high-solid polyesters with a quota of about 80 % were produced by varying the acid-components. The polyol-components were kept constant.

Table 1: Formulations and dispersing conditions

| Blend | proportion in weight | | | Cylinder temp [K] | nozzle temp [K] | passing period [s] |
|---|---|---|---|---|---|---|
| 1055/DCD | 90:3,2 | 90:4,0 | 90:4,8 | 353 | 393 | 50 |
| 1055/XB2622 | 90:8,8 | 90:11 | 90:13,2 | 353 | 393 | 50 |
| 1055/P2228 | – | 50:50 | – | 323 | 385 | 50 |
| " | – | 50:50 | – | 333 | 388 | 50 |
| " | 30:70 | 50:50 | 70:30 | 353 | 393 | 50 |

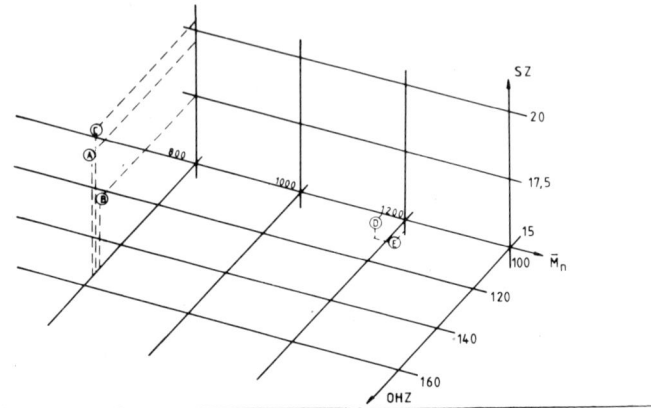

Fig. 1: Polyester resins

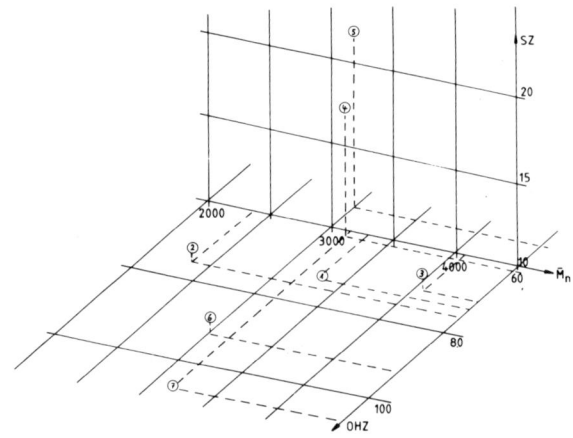

Fig. 2: Alkyd resins

Fig. 1 shows their chemical properties. As isocyanate Desmodur N (Bayer AG, Leverkusen) was chosen. Most measurements were done with a reaction index of 100, sometimes with one of 75.

## Alkyd/melamine resin varnishes

In the same way seven high-solid resins with a solid quota of about 60 % were produced only by variation of the molar concentration of the components. Fig. 2 shows their chemical properties. As melamine resin Cymel 301 (American Cyanamid, Wayne) was taken. For all Cymel/alkyd formulations a ratio of 30/70 parts in weight was chosen.

## RESULTS AND DISCUSSION

### Epoxide powder varnishes

Fig. 3 shows a typical DCS-thermogram. The first peak is no thermal effect but is due to the switching on of the temperature program of DSC. At $T_{g1}$ the glass transition range begins. In most cases the glass transition of both components $T_{g1}$ and $T_{g2}$ culd be bound. Depending on the thermal history of the powder an endothermic relaxationpeak may be superimposed to the glass-transition. $T_{of}$ is the temperature at the onset of flow indicated by the actual melting of the powder. $T_{oc}$ is the temperature at which the curing of the powder starts. The interval between $T_{of}$ and $T_{oc}$ is therefore important for a good flow of the varnishes. The curing of the powder is visible as an exothermic peak, the area under the peak

Table 2: Characteristic temperatures of EP-powder varnishes

| Blend | $T_{g1}$ [K] | $T_{g2}$ [K] | $T_{of}$ [K] | $T_{oc}$ [K] | $T_m$ [K] | $T_e$ [K] |
|---|---|---|---|---|---|---|
| 1055/DCD | | | | | | |
| 90:3,2 | 332 323 | 354 349 | 373 373 | 442 429 | 483 468 | 538 515 |
| 90:4,0 | 329 327 | 352 353 | 375 – | 439 425 | 479 470 | 511 515 |
| 90:4,8 | 331 326 | 351 352 | 374 376 | 442 434 | 482 473 | 521 505 |
| 1055/XB2622 | | | | | | |
| 90:8,8 | 330 328 | 353 352 | 376 – | 395 394 | 454 443 | 526 518 |
| 90:11,0 | 329 326 | 350 349 | 377 – | 397 393 | 451 438 | 530 508 |
| 90:13,2 | 332 327 | 353 352 | 389 – | 402 394 | 450 438 | 528 491 |
| 1055/P2228 | | | | | | |
| 30:70 | 330 336 | 347 – | 379 381 | 412 385 | 450 433 | 489 473 |
| 50:50 | 335 335 | 349 355 | – 370 | 409 389 | 451 444 | 491 478 |
| 70:30 | 331 335 | 345 – | 384 371 | 406 398 | 450 444 | 481 468 |
| 1055/P2228 | | | | | | |
| 323K | 334 | 350 | 372 | 384 | 444 | 476 |
| 333K | 335 | 357 | 370 | 380 | 442 | 474 |
| 353K | 335 | 355 | 370 | 389 | 444 | 478 |

being a measure for the specific heat of reaction $\Delta h$. At $T_m$ the maximum curing velocity is reached and at $T_e$ curing is "over". The sharp endothermic peak at 484K was found only for the handmixed blends with DCD as hardener and is in good agreement with the melting point of DCD.

Table 2 shows the characteristic temperatures of the blends from table 1. The reproducibility is $\pm$ 4 K. The first temperature values of each column belong to the handmixed blends. So, the glass transition and the onset of flow point show no assured dependence on the different parameters, but the curing range. Acid anhydride shifts the onset of cure and the maximum of curing velocity of about 20K towards lower temperatures (1) and the values of the extruded blends are lower than those of the handmixed ones. Only the "final temperatures" of the acidanhydride blends after the dispersion by extrusion show a dependence on the concentration. A dependence of the characteristic temperatures on the extrusion parameters was not found, so that no pre-curing took place during the extrusion. For the evaluation of $\Delta h$ and the reaction time t from dynamic and isothermic DSC-measurements the formerly (2) described procedure was used. For dynamic as well as for isothermic measurements only the part of reaction was used which precedes the extreme rise in viscosity

Table 3 shows the reaction values. The first values of each column belong to the handmixed blends. On the basis of the reaction time t it is shown more clearly than by the characteristic temperatures that the cheaper handmixing is no acceptable alternative to extrusion. The extrusion conditions don't show any influence on reaction time. The reaction times obtained from isothermic measurements are shorter by a factor of two than those obtained from dynamic ones and very well correspond to practice values. So, one can say that the values from isothermic measurements seem to be more realistic than those from dynamic ones, but there is no change in the order of the systems. Therefore the dynamic ones which are easier to do are preferred.

Polyurethane/polyester resin varnishes

In contrast to solvent-free powder varnishes the switch-on peak is followed by a strong endothermic slope of the heat flux curve, which belongs to a spontaneous solvent evaporation, as TGA-measurements show. After curing is finished often an abrupt evaporation of the remaining solvent may be observed (Fig. 4). In many cases this evaporation takes place more or less continously during the exothermic curing reaction. This is manifested by

Table 3: Reaction values

| Blend | Δh [J/g] | | t [min] at 433K, α=0,95 | |
|---|---|---|---|---|
| 1055/DCD | | | | |
| 90:3,2 | 99,0 | 104,2 | $3,4 \cdot 10^4$ | $1,3 \cdot 10^3$ |
| 90:4,0 | 96,7 | 99,2 | $1,3 \cdot 10^4$ | $2,4 \cdot 10^3$ |
| 90:4,8 | 100,4 | 99,2 | $2,5 \cdot 10^4$ | $1,4 \cdot 10^4$ |
| 1055/XB2622 | | | | |
| 90:8,8 | 60,2 | 64,0 | 140 | 66 |
| 90:11,0 | 73,3 | 69,1 | 215 | 68 |
| 90:13,2 | 73,3 | 68,3 | 130 | 57 |
| 1055/P2228 | | | | |
| 30:70 | 16,4 | 19,5 | 27 | 11 |
| 50:50 | 20,9 | 29,6 | 11 | 16 |
| 70:30 | 17,0 | 18,5 | 21 | 16 |
| 1055/P2228 extruded at | | | | |
| 323K | | 28,1 | | 12 |
| 333K | | 27,4 | | 12 |
| 353K | | 26,8 | | 16 |
| 1055/P2228 extrudet at (isothermic DSC) | | | | |
| 323K | | 26,4 | | 6,4 |
| 333K | | 28,4 | | 7,3 |
| 353K | | 26,4 | | 6,6 |

a notched curve ($Y_1$-curve). With higher heating rates the characteristic temperatures are shifted to higher values (Fig.5), but they proved to be independent of the reaction index. If one takes the $T_{oc}$-points extrapolated to heating rate OK/min as a measure for an early start of the reaction, then two groups exist, first the polyesters A and B at 315K and second the polyesters C, D and E at 332K. In A and B the adipinic acid part was kept constant, whilest the phthalic acid was replaced by isophthalic acid. If both are replaced by hexahydrophthalic acid (C) or if only phthalic acid is chosen (E), then the beginning of reaction to the same degree

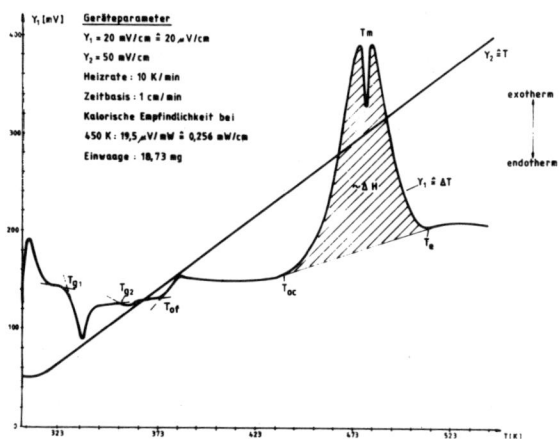

Fig. 3: Typical dynamic DSC-thermogram of EP-varnishes

as if without any changing of the components in resin A only the hydroxyl value is lowered (D). A self-curing of the varnish-components could not

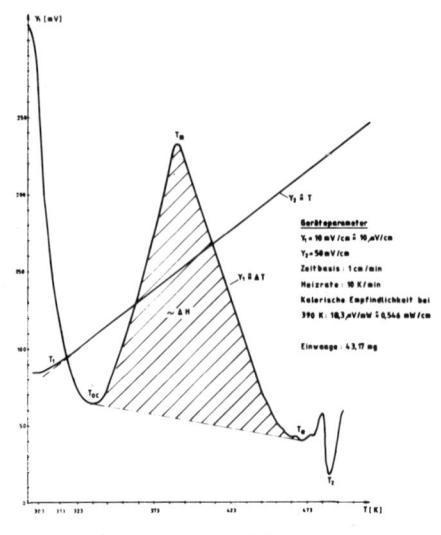

Fig. 4: Dynamic DSC-thermogram of PU/polyester varnishes

be observed within the reaction range of the varnishes. Regarding the reproducibility, no differences in the reaction time could be found. Only Δh grows with higher hydroxyl or carboxyl values and lower average molecular.

Alkyd/melamine resin varnishes

Here the curing reaction is superimposed by the endothermic solvent evaporation, as typical dynamic DSC-thermograms with and without pre-drying indicate (Fig. 6 and 7). The glass transition in Fig. 7

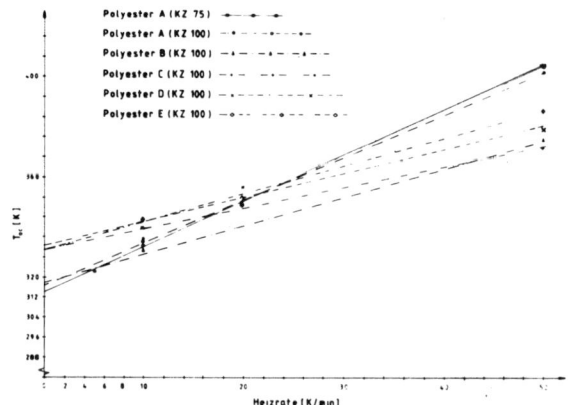

Fig. 5: Characteristic temperatures as a function of heating rate

belongs to a varnish only physically dried. When dried above 393K the $T_g$ value is shifted to higher temperatures, which indicates some curing (3). Also endothermic as well as exothermic reactions of the pure resins are observed within the reaction range. So, a border area for the application of thermoanalytical method is reached. Dynamic TGA-measurements show a spontaneous solvent evaporation with a maximum at 371K (Fig. 8). Beginning with 420K the weight again decreases. This is due the evaporation of methanol which is formed during the polycondensation.

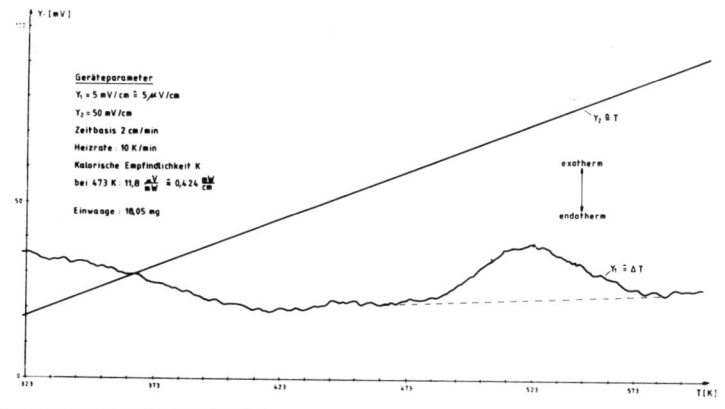

Fig. 6: Dynamic DSC-thermogram alkyd/melamine without pre-dry.

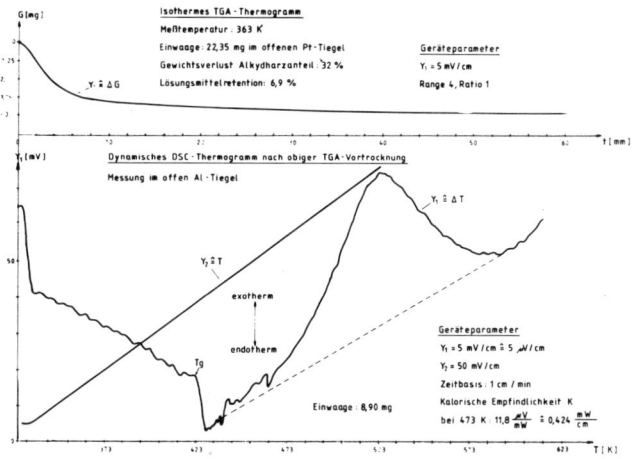

Fig. 7: Dynamic DSC-thermogram of alkyd/melamine with pre-dry

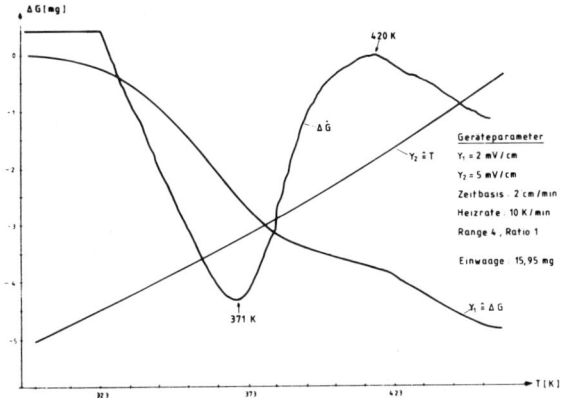

Fig. 8: Dynamic TGA of alkyd/melamine varnish without pre-dry.

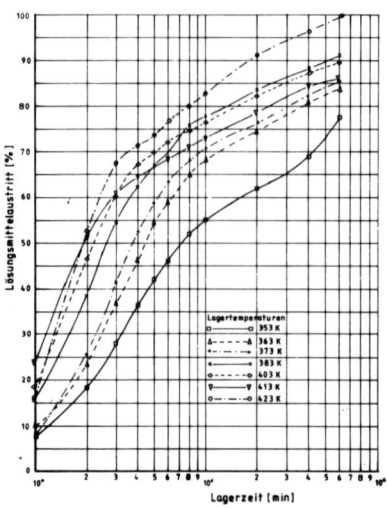

Fig. 9: Solvent evaporation of alkyd/malamine varnish (TGA)

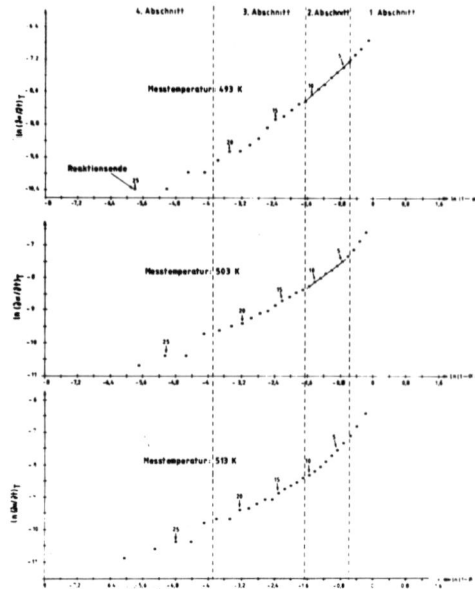

Fig. 10: Reaction-curves from isothermic DSC for alkyd/melamine

So, to reduce the effects, which partly oppose each other, the varnishes were dried for 60 min at 353K before being cured. As shown in Fig. 9 the solvent is then evaporated down to 5 %. For kinetic evaluations of the dynamic measurements an exact base-line at the end of the reaction is necessary. This is reached at temperatures above 600K (Fig. 7), where already the decomposition process takes place. So, only isothermic measurements were possible. Because of the manyfold possibilities of reaction the reaction-lines especially at higher temperature are more or less continously bended (Fig. 10). So no reliable reaction values could be found. Only $\Delta h$ showed the same dependence on the hydroxyl and carboxyl value and the average molecular weight as with isocyanate/polyester resin varnishes.

## REFERENCES

(1) S. Gabriel, J.Oil Chem. Assoc. <u>58</u> (1975) 52

(2) H. Möhler, farbe und lack, 1980, Heft 3, S. 211 - 215

(3) W.P.Brennan, Analysentechnische Berichte, Heft 49 (1977), PE Corp.

# CALORIMETRIC ANALYSIS OF POLYMER FABRICATION

H. Weber and U. Guggisberg
Processanalyses, Turgi, Swizzerland

## ABSTRACT

By use of the energy- equation for a typical polymerisation process, the reaction enthalpy and its rate during the process is evaluated with appreciable accuracy from a measurement of only a few temperatures and the mass flow of the monomer. Through the known enthalpy of polymerisation the course of the free monomer during the process can be determined and it is shown that these data are correlated with the viscosity of the final product as a function of the shear- rate. It is hoped that this correlation enables a control of the process towards a fixed viscosity shear- rate relation.

## 1. MAJOR PROBLEMS OF POLYMERISATION

Polymerisation processes have very complicated reaction-mechanisms and the course of the reaction is very often incalculable even under welldefined conditions. The amount of polymer formation during the reaction is not easy to determine. This leads to large stochastic deviations in the quality of the product. The factors that influence the course of the polymerisation are not well understood and as a cosequence, no optimisation of the process with respect to energy, time or quality is possible. The polymerisation is commonly run by an empirically fixed prescription and no variations are risked as soon as the quality of the product is acceptable. A better result can be obtained as soon as it is possible to monitor and influence the course of the polymerisation.

## 2. CALORIMETRIC PROCEDURE TO CONTROL THE POLYMERISATION

Accounting is the instrument for the quantitative measurement of complex commercial processes and is used for their control and optimisation. In complete analogy, the measurement and evaluation of the energy- balance of physical, chemical or biological processes could be used for their control and optimisation. The optimal process is the process with the best economy. In order to measure economy, quantitative criteria for an economical process must be found. There are several of them, e. g. time, energy, quality and safety. Optimisation of a process can only be done if the dependence of these criteria on a set of free process- parameters is known. This dependence should be verified experimentally.

Fig. 1 shows a simplified scetch of a polymerisation-plant for emulsion- polymerisation of vinylacetate[1]. The process is carried out as follows. Two tons of emulgator is prepared in the system with all the catalysts at a temperature of $60^0C$. Two tons of monomer is added linearly in time during 90 minutes and the system is held at a temperature of $90^0C$ for 4 hours by use of a jacket-

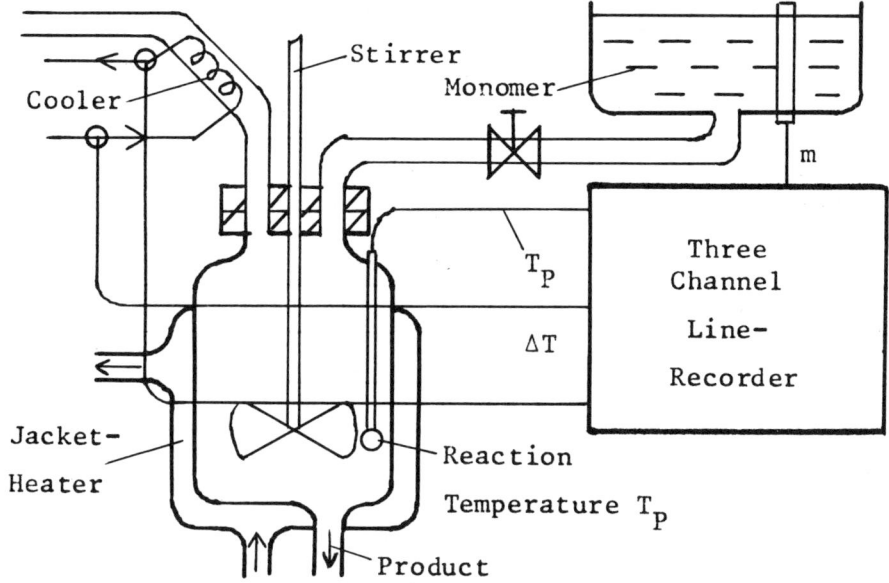

Fig. 1: Polymerisation- Plant

heater. Fig. 2 shows the temperature of the system, the temperature- difference over the condensation- cooler and the course of the monomer input as a function of time.

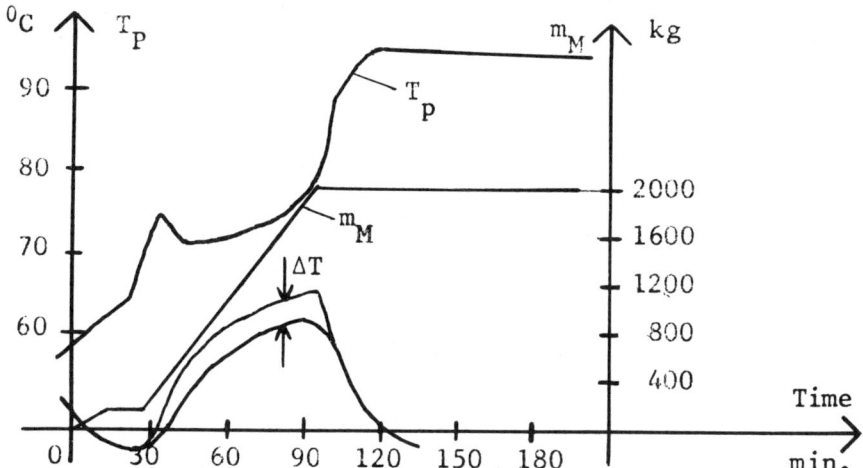

Fig. 2: Recorded Data of a Polymerisation

The energy- and heat- balance of the system can be represented as a thermal circuit with resistence R, capacity C and power- currents i (fig. 3). The energy- balance-

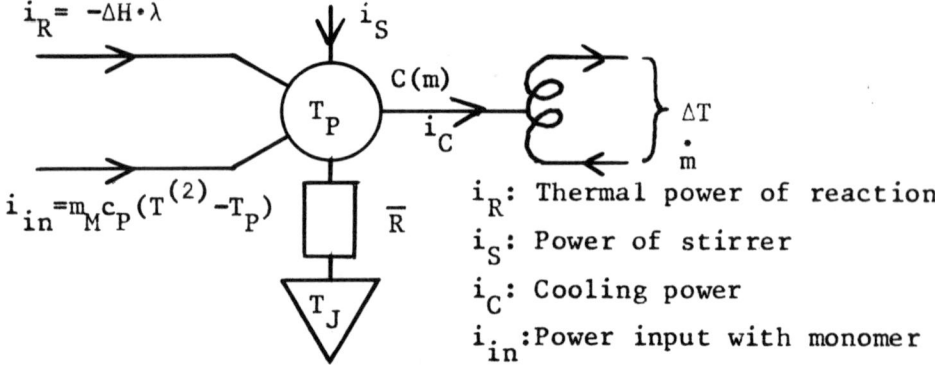

Fig. 3: Thermal Circuit for Reacting System

equation can now easily be written down as a power equation:

$$C_P \cdot \frac{dT_P}{dt} = i_R + i_S - i_C - i_j + i_{in} \qquad (1)$$

$i_R$ is the thermal power of the reaction, $i_S$ the power of the stirrer, $i_C$ the power leaving the system through the condensation- cooler, $i_j$ the power leaving through the jacket and $i_{in}$ the power- input to the system through the monomer.

## 3. EVALUATION OF THE AMOUNT OF POLYMERISATION

The thermal power of the reaction $i_R$ is given by:

$$i_R = -\Delta H \cdot \dot{\lambda} \qquad (2)$$

where $\Delta H$ is the enthalpy of polymerisation per mole of monomer[2] and $\dot{\lambda}$ is the time- derivative of the reaction-

parameter. If the power of the stirrer $i_S$ is assumed constant during the process, the two terms $i_S - i_j$ can be evaluated as:
$$i_S - i_j = \frac{T_\infty - T_P}{\overline{R}}. \qquad (3)$$

$T_\infty$ is the convergence temperature in the absence of any reaction and of the cooler- current and can be evaluated from the temperature- time- curve (fig. 2). $\overline{R}$ is the total thermal resistance of the jacket and is evaluated from the temperature- rate prior to the start of the reaction. The cooling- current is:
$$i_C = \dot{m} \cdot c_P \cdot \Delta T, \qquad (4)$$
where $\dot{m}$ is the constant mass- flow in the cooler and $\Delta T$ the recorded temperature difference between its exit and its entrance. $c_P$ is the specific heat of the water. The input- current $i_{in}$ finally is found as:
$$i_{in} = \dot{m}_M \cdot c_P^{(2)} \cdot (T^{(2)} - T_P), (5)$$
where $\dot{m}_M$ is the mass- flow of the monomer, $T^{(2)}$ its temperature and $c_P^{(2)}$ its specific heat. The power- equation (1) can now be written in the form:
$$C_P \cdot \frac{dT_P}{dt} = -\Delta H \cdot \dot{\lambda} + \frac{T_\infty - T_P}{\overline{R}} - \dot{m} c_P \Delta T + \dot{m}_M c_P^{(2)} (T^{(2)} - T_P). (6)$$

This equation can be solved for $\dot{\lambda}$ and $\lambda(t)$ by numerical integration. Fig. 4 shows the result for the process of fig. 2. From the reaction- parameter $\lambda(t)$ and the monomer input as a function of time, the free monomer in the reaction- system can be evaluated. Fig. 5 shows the results for three different polymerisation- processes together

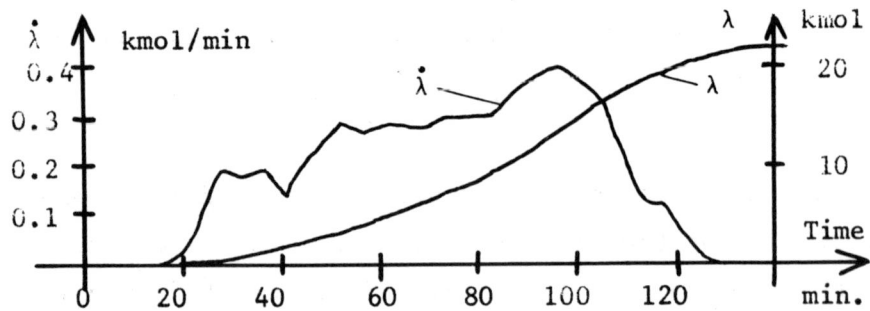

Fig. 4: Plots of $\dot\lambda(t)$ and $\lambda(t)$.

with the viscosity- shear- rate- functions. The two set of curves seem to correlate.

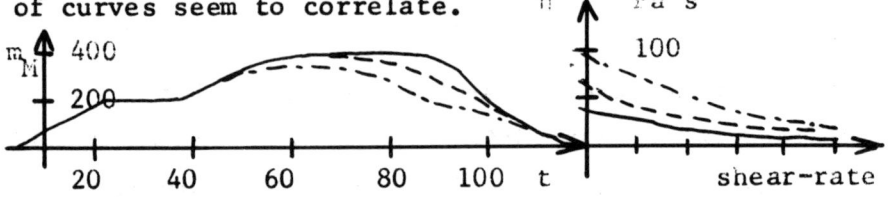

Fig. 5: Free- Monomer- viscosity- correlation.

## 4. CONTROL OF THE POLYMERISATION- PROCESS

A number of further off- line analyses are planned in order to investigate more closely the nature of this correlation. A knowledge of this correlation enables an automatic control of the process for a selected viscosity- shear- rate- function of the product.

### REFERENCES

(1) The process under discussion is run at the swiss firm Ed. Geistlich Söhne AG in Schlieren, Zurich.

(2) Landolt Börnstein, Vol. II, part 4, p. 29.

THERMO-ANALYTICAL TECHNIQUES IN THE ASSESSMENT OF
MECHANICAL PROPERTIES AT ELEVATED TEMPERATURES OF
POLYESTER-GLASS COMPOSITES

P. K. Datta and T. R. Manley,
Newcastle upon Tyne Polytechnic,
Ellison Place,
Newcastle upon Tyne,
NE1 8ST.

## ABSTRACT

Thermoanalytical techniques (DTA, TG, TMA) have been used on glass fibre reinforced polyester resins (GRP). Mechanical tests (tensile, flexural and buckling) were performed on GRP at various temperatures. The thermal methods give a rapid and simple indication of the temperature ranges where critical changes in mechanical properties occur.

## INTRODUCTION

With increasing knowledge of the mechanical properties of glass reinforced polyester resins (GRP) structural applications are now consuming increasing amounts of the material. An important parameter in a structural material is the resistance to fire. This has two aspects, the first one being that of flammability for which the critical oxygen index was used to study the material [1]. In many instances however the effect of the resultant heat on the mechanical properties of the material can be a more serious factor than flammability alone. When GRP is used in buildings, aircraft, and ships, it is important that the material retains its structural integrity. Mechanical testing is very expensive and time-consuming and this deters manufacturers from undertaking it. The object of this work was to see whether the much more rapid and convenient thermo-analytical techniques could be used to reduce the amount of mechanical testing used.

## EXPERIMENTAL

A Du Pont 900 instrument was used in the DSC mode. Commercial resins supplied by Messrs. Scott Bader Limited were "A" a preaccelerated orthophthalic maleic propylene glycol polyester. Chlorinated paraffin, antimony trioxide and an inert filler are incorporated to give flame retardance to the BS 476 Part 7 surface spread of flame class 2. "B" an ethylene glycol polyester containing 2% of a cobalt accelerator and approved by the Lloyds Register of Shipping for the construction of boat hulls. Resin C is an isophthalic based propylene diethylene glycol polyester. The catalyst is a 50% solution of methyl ethyl ketone in dibutyl phthalate and the accelerator is cobalt octoate in styrene (0.42% metallic cobalt). In all cases the formulation was 10 parts by weight of resin, 2 of catalyst and 2 of accelerator, with styrene as the cross linking solvent. Chopped strand mat "E" glass (Marglass), 2 oz./sq.f. (0.61 kg/m$^2$), was used to make samples of one two and four layers of glass. These contained 20.4, 22.8 and 27.4% of glass respectively. The laminates were postcured at 64°C for 36 hours.

Tensile strength, flexural strength, buckling strength and fracture toughness of the laminates were measured as a function of temperature at a constant cross-head speed of 50 mm/minute in an environmental chamber fitted to an Instron machine. Each specimen was allowed to equilibrate for 12 minutes at the test temperature which was maintained constant with ± 2°C. Three to four specimens were tested at each temperature. Tensile tests were carried out in accordance with BS 2782 Part 3. Flexural specimens were 101 mm x 13 mm x 7 mm. For buckling tests specimens of width 50 mm and of varying length (65 mm to 115 mm) and thickness (1.79 mm to 2.25 mm) were used. The fracture toughness was determined using the linear elastic fracture mechanics (LEFM) technique on specimens of one and two mat laminates with resin A. Single edge notched (SEN) specimens were fractured in tension and the work of fracture $\gamma_i$ pertaining to the initiation of fracture was determined. The fracture toughness was obtained for four mat laminates with resin A using the work of fracture technique. The work of fracture ($\gamma_F$) relating to the whole fracture process was determined using notched bend tests as described by Tattersall and Tapin [2].

The maximum load required to break the SEN specimens was used to calculate an apparent stress intensity $K_{IC}$ using the analytical formula given by Srawley and Brown [3]. The fracture energy $\gamma_i$ to initiate the crack was then obtained from $K_{IC}$ using

$$\gamma_i = \frac{(1 - \nu^2)K_{IC}^2}{2E}$$

## RESULTS

The DTA results are shown in Figure 1. It can be seen that there are two transitions in the material, the first in the region of 50°C and the other around 120°C. The first transitions were confirmed using a thermomechanical instrument [Stanton Redcroft penetrometer 678. The values obtained were as usual slightly above those indicated by DTA viz 68°C for Resin A, 65°C for B and 60°C for C. Similar results were obtained with the glass reinforced materials. This indicates that the transitions are associated with the resin. Comparison with the mechanical tests in Figures 2 to 6 show that there is a good correlation between these transitions and the marked changes in mechanical properties, tensile strength, flexural strength, buckling strength and modulus of elasticity. In Fig. 2 the 50 deg. transition is associated with a marked decrease in tensile strength whilst the 120 deg. transition marks a more gradual reduction. In Fig. 3 the modulus shows a similar trend; the low values for resin C at 180° are indicative of incipient degradation. In Fig. 4 the two transitions are clearly seen; the differing amounts of glass do not seem to have a great effect on the relationship between flexural strength and temperature. Fig. 5 shows the two transitions in the buckling strength.

DTA traces showed exotherms beginning at 180°C and thermogravimetry also shows a departure from baseline around this point. In the thermal balance when the sample is held at 200°C prolonged decomposition takes place. This has been previously observed in this temperature region about 180°C for ester linkages of epoxy compounds [4] and it is quite clear that this is the terminal decomposition of the material. Above this temperature no mechanical strength whatever would remain and the material would have completely lost its structural value.

Figure 6 shows as expected that $\gamma_i$ did not vary with temperature, but the apparent fracture toughness $K_{IC}$ showed strong dependence on temperature varying by $\sim 8$ MNm$^{-3/2}$ at $20°C$ to $\sim 4$ MNm$^{-3/2}$ at $130°C$. Within the experimental scatter, the one and two mat laminates did not show any significant difference in $K_{IC}$ and $\gamma_i$ values. The bending tests showed that the fracture occurred in a controlled manner indicating that all the stored elastic energy in the specimen-machine system was absorbed in forming the fracture surface. The fracture energy $\gamma_F$ obtained from the area under the load deflection curve divided by twice the specimen cross-section area represented the average energy for the whole fracture process. Unlike $\gamma_i$, the fracture energy $\gamma_F$ for a/w = 0.167 displayed considerable variation with temperature. $\gamma_F$ decreased from $\sim 16$ KJm$^{-2}$ to 5 KJm$^{-2}$ for a change of temperature from $20°C$ to $130°C$. Using the maximum load required to break the bend specimens, a few values of $\gamma_i$ in the work of fracture method were calculated. These values of $\gamma_i$ compared reasonably well with those obtained from the SEN specimens.

## DISCUSSION

The transition temperatures that are obtained vary according to the rate of heating of the specimens (Fig. 1). This effect may easily be allowed for by standardising on one given heating rate usually $10°C$ per minute. The speed of testing has a marked effect on the mechanical properties of a material and indeed the relationship between the transition temperature and speed of testing is known.

In all the mechanical and thermoanalytical tests two transitions are clearly seen, and although there are minor variations in the precise position of this temperature, it is clear that the thermoanalytical techniques are indicating transitions that have a critical effect on the mechanical properties of the composites. Either DTA or the penetrometer may be used, care however must be taken with the former when very highly loaded samples are obtained in order that a homogeneous specimen be obtained. Thermogravimetry indicates the chemical degradation of the material which gives a temperature limit above which mechanical testing is of little practical value. The highest temperature used in the tensile tests was $180°$, above this as expected the degradation was apparent in colour changes of the material and obviously no mechanical testing would be done above this point. TG therefore does give an indication of the temperature at which all mechanical strength is lost whilst the DTA and penetrometer indicate the regions where transitions may be obtained and which must be covered by mechanical testing if a true indication of the mechanical properties of the material in service are to be found.

Both the $K_{IC}$ and $\gamma_F$ are temperature dependent as can be seen from Fig. 6. $\gamma_i$ however is temperature independent.

## CONCLUSION

Although further work is required in order to ascertain more exactly the temperatures of the transitions that are occurring, and to elucidate the structural changes associated with them, it is clear that mechanical tests at room temperature alone will not give an accurate picture of the behaviour of this composite material. On the other hand the full series of mechanical tests is extremely time-consuming and it is not practical to do tests of this nature on completed structures. Having established the relationship between the mechanical tests and the thermoanalytical technique the latter are obviously useful for quality control work since very small samples are required.

## REFERENCES

[1] Manley, T., Datta, P. and Hedley, J. A. in preparation.
[2] Tattersall, H. G., Tappin, G., J. Matls. Sci. $\underline{1}$, 296 (1966).
[3] Srawley, J. F. and Brown, W. F., Fracture Toughness Testing and application ASTM 1964 and NASA TM X52030, 1964.
[4] Manley, T. R. in "High Voltage Applications of Epoxy Resins", Ed. M. J. Billings, University of Manchester, April 1967.

## FIGURES

### Figure 1
The effect of variations in heating rate on transitions in Resin B containing glass fibre, (i) 25 deg/min, (ii) 20 deg/min, (iii) 15 deg/min.

Figure 2
Tensile strength $MNm^{-2}$ of single mat GRP laminates at various temperatures.
+ Resin A; Δ Resin B; X Resin C.

Figure 3
Modulus of elasticity $GNm^{-2}$ of GRP at various temperatures
• Resin A -- -- ; x Resin B ——— ; o Resin C --------

Figure 4

Flexural strength $MNm^{-2} \left[ \dfrac{1 - SPL}{Wd^2} \right]$ of GRP at various temperatures.

- Single mat; x 2 mat; o 3 mat; • 4 mat laminate

Figure 5
Fracture stress $MNm^{-2}$ under buckling of GRP at various temperatures and aspect ratios.

Figure 6
Fracture toughness parameters of GRP at various temperatires.
ΔΔ $\gamma_F$ from bend specimens for $a/w = 0.167$
•• $K_{IC}$ from SEN specimens for $a/w = 0.3$
xx $\gamma_i$ from SEN specimens for $a/w = 0.3$
x x $\gamma_i$ from bend specimens for $a/w = 0.167$

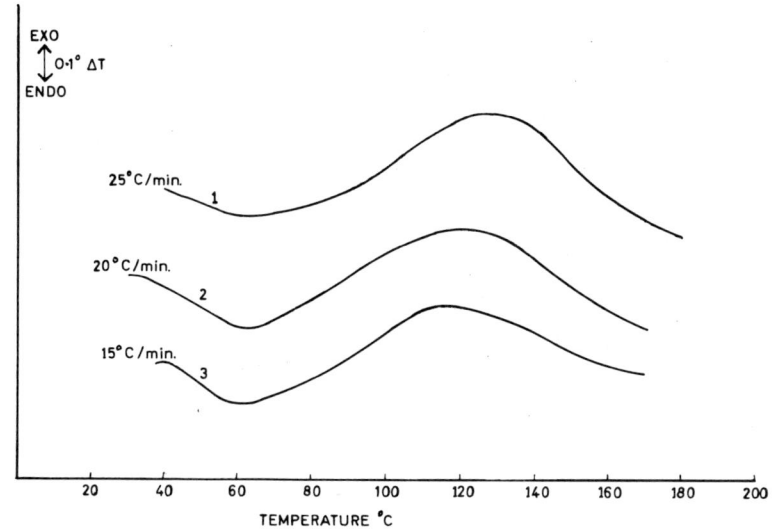

FIG. 1. DTA OF RESIN B + GLASSFIBRE

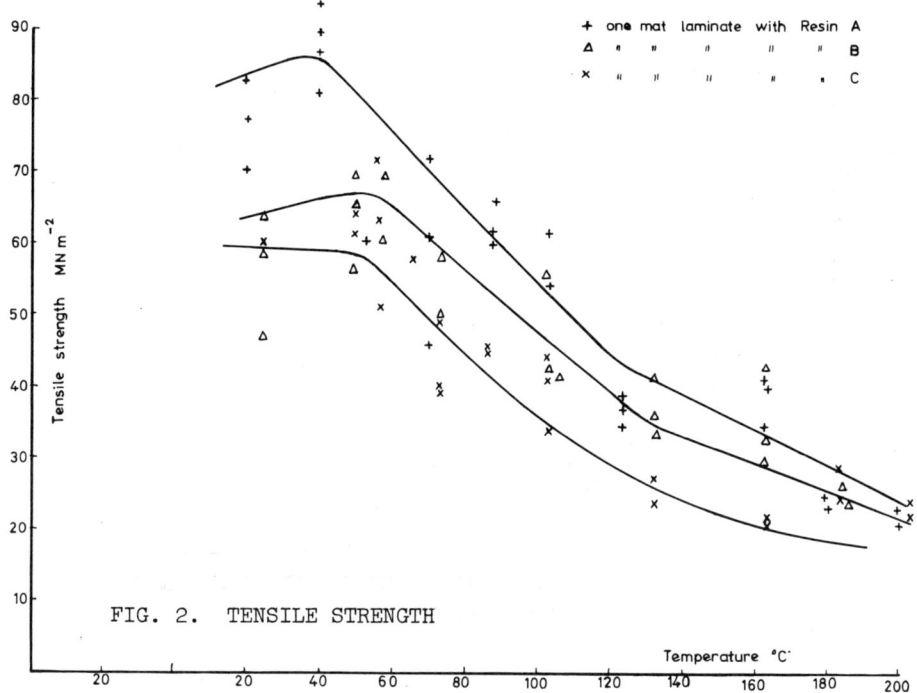

FIG. 2. TENSILE STRENGTH

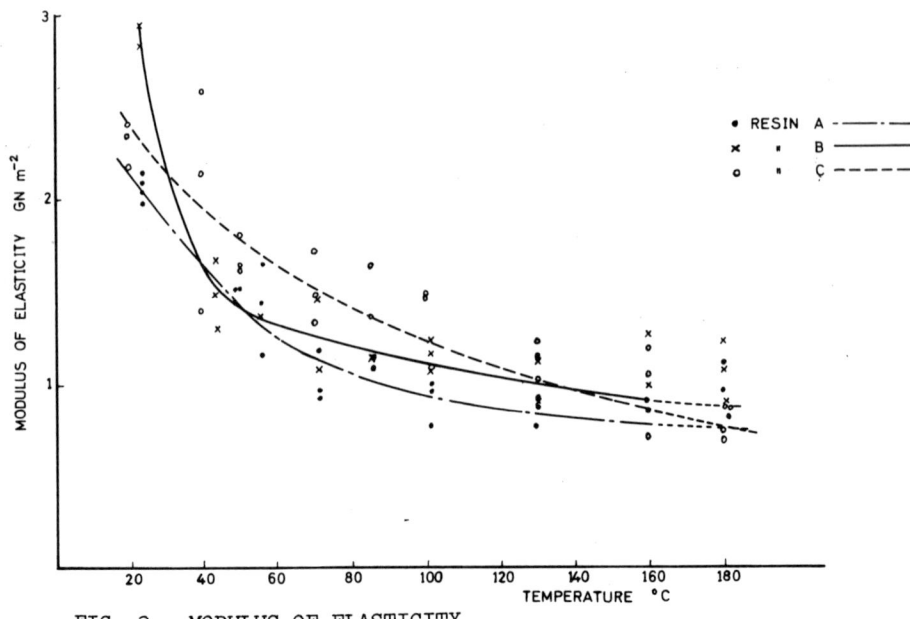

FIG. 3.  MODULUS OF ELASTICITY

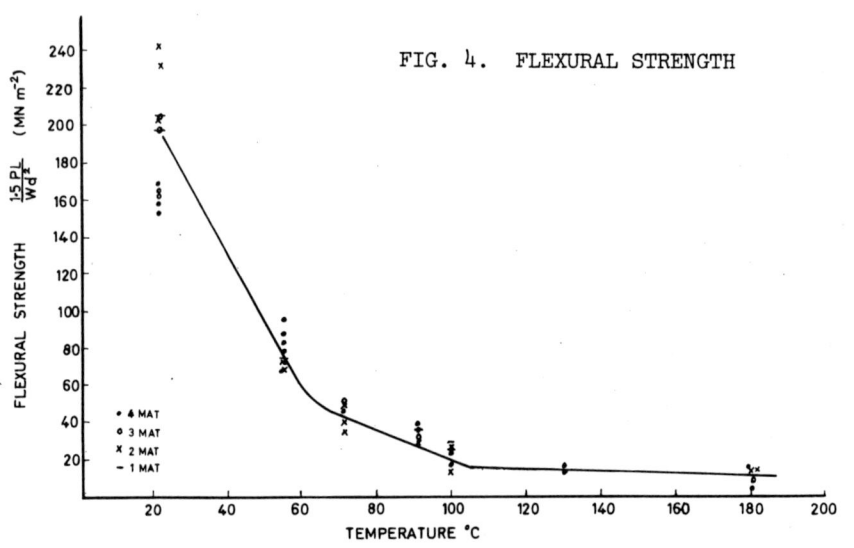

FIG. 4.  FLEXURAL STRENGTH

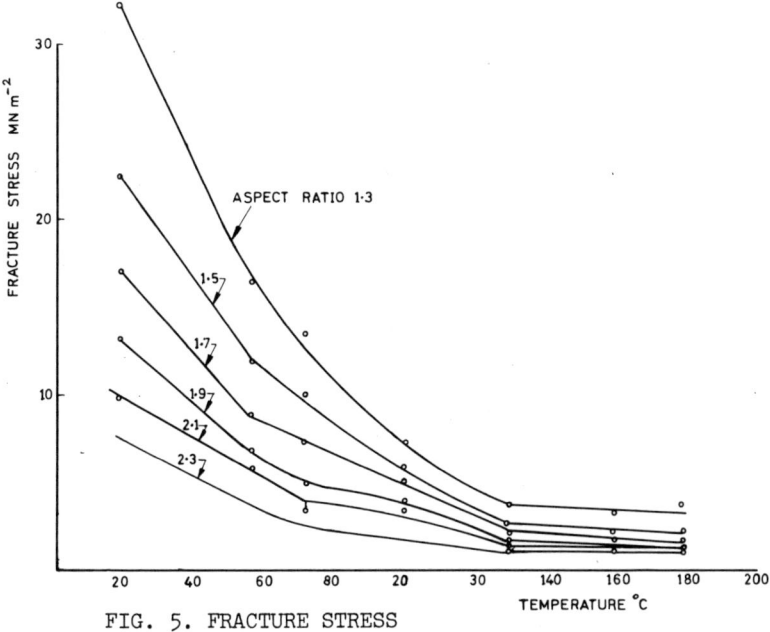
FIG. 5. FRACTURE STRESS

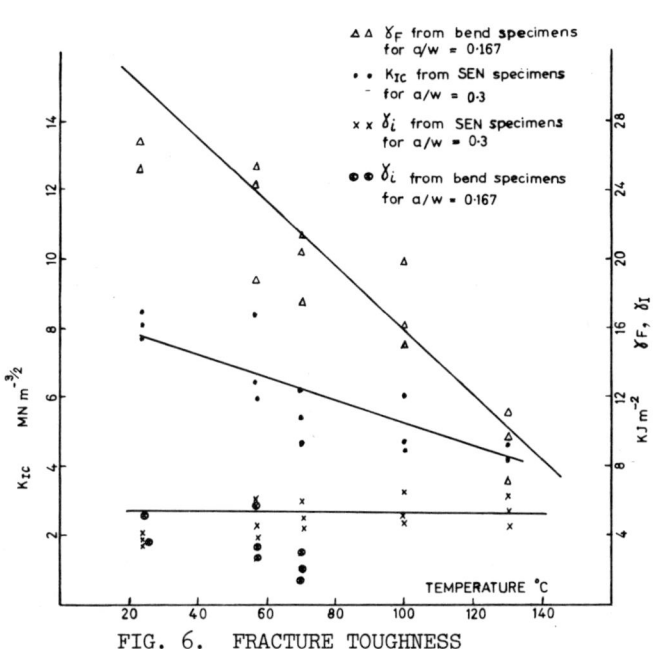
FIG. 6. FRACTURE TOUGHNESS

# THERMAL ANALYSIS OF RTV SILICONE RUBBERS USEN IN HIGH VOLTAGE CABLE ACCESSORIES

G.Liptay[*], L.Ligethy[**] and J.Nagy[*]
[*]Department of Inorganic Chemistry, Technical University, Budapest, 1521-Hungary
[**]Hungarian Cable Works, Budapest, 1117-Hungary

## ABSTRACTS
Processing parameters of silicone rubbers of high voltage cable termination influences the service life. The curing process were studied by thermal analysis.

## INTRODUCTION
Room temperature vulcanizing silicone rubbers are applied in making terminations for high voltage polyethylene insulated cables. The extraordinary advantegous mechanical properties of this materials enables to use as prefabricated terminations that is slipped on cables in ready made form. This composite material system is exposed to a complex stress (electrical, thermal and environmental).
As earlier we found, the methods of thermal analysis can be applied to study the thermal stability, also the effect of thermal stress caused by processing and utility can be investigated.

## EXPERIMENTAL PROCEDURE
The investigations are carried out partly dinamic simoultaneous methods by Derivatograph made by Hungarian Optical Works, Budapest type OD-2 in atmospheric with a heating rate 5 °C/minute. Samples 200 g were weighted in a platinum crucible. Bulk-surface ratio was helt constantly during the investigation. Every occasion the same quantity of materials were weightened in, so it succeded to avoid any differences in bulk-surface ratio.
Some investigations were also carried out by isothermal thermogravimetry, where the weightloss of samples were

recorded in function of time at constant temperature. The oven of thermo-balance was heated up alone. The temperature was controlled ±0.2 °C accuracy by PROGRAMIC (Research Institut for Instrument Industrial, Budapest) temperature controll device. Afterwards 1.0 g of the sample in a quarz crucible was put into the oven on the thermocouple, and the weightloss was measured in atmospheric conditions.

The samples were pre-cured in little crucible belonging to the equipment and were studied that vulcanised silicone rubbers.

The samples investigated were two types of electrical semicondutive and other two types of insulating silicone rubbers made by Wacker Chemie GmbH München by which the thermination are made for industrial purposes with special applications technique.

## RESULTS AND DISCUSSION

First the thermal properties of semiconductive rubbers were studied, which is applied as stress control element in

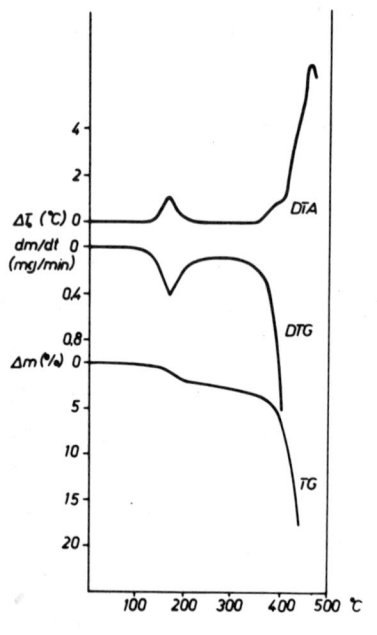
Fig.1.

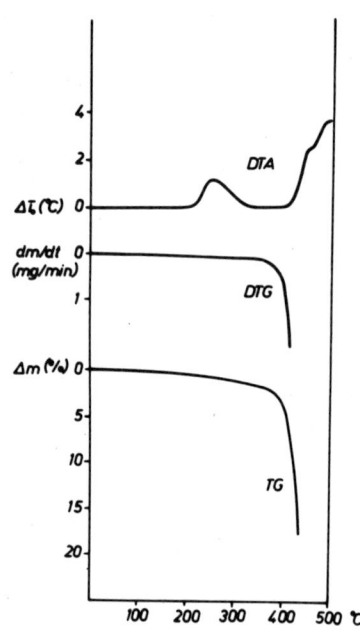
Fig.2.

cable terminations, and which are embedden voidfree in the
insulating silicone rubber. Figure 1 shows the derivatogram
of R 570/70 $C_1$ semiconductive silicone rubber which contains dicumilperoxyde as cureing agent. On thermal curves
can be seen the cureing at about 150°C as an exotherm process, which shows an characteristic weightloss. The weihgt
loss of the semiconductive rubber part in the compact termination embedded can cause apparence of voids which des-
troys the device.

Figure 2 shows a derivatogram of silicon rubber type
VPR-3275 cured with addition-catalyst system. The exotherm
cureing process of this material can be obtained at 250°C
with constant weight practically.

This effect ensures the void free termination product. The
pot-life of that material has a great influence on the properties of ready made products. Fig.3 shows a derivatogram
of a sample compounded and stored for one week in refrigerator, where a weight loss (10-15 %) is obtained being
great enough to damage the ready made cable termination
product. When the components are stored in refrigerator
separately no wieght loss causing processes can be obtained
as shown in figures 4. and 5.

This experiment shows that the prolonged pot-life can effect the properties of products and this effect can be
followed by thermal methods.

The second components of a cable termination are the insulating silicone rubber produced for high voltage purposes
especially (RTV-ME 622 and 628). Their cureing conditions
does not effect the service life of product substantially
either investigated without cure (a) or cured for 24 hours
in room temperature (b) or cured for 1/2 hour on 120 °C
degree (c).

Figur 6 shows the the isothermal curves of RTV-ME 622
measured on 180°C degree. No substantial differences in
obtained depending on curing conditions.

The one component semiconductive silicone rubber (R 570/70
$C_1$) was studied with changeing bulk-surface rate (the

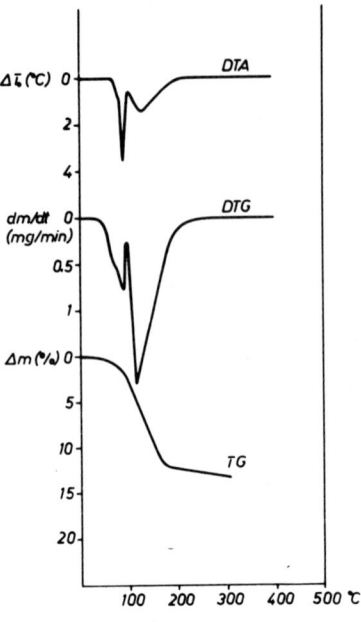

Fig. 3

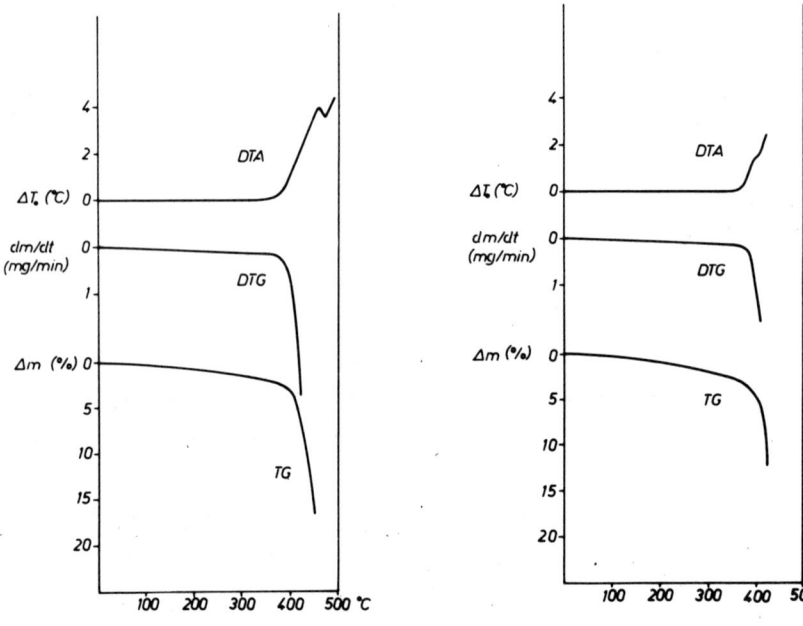

Fig. 4

Fig. 5

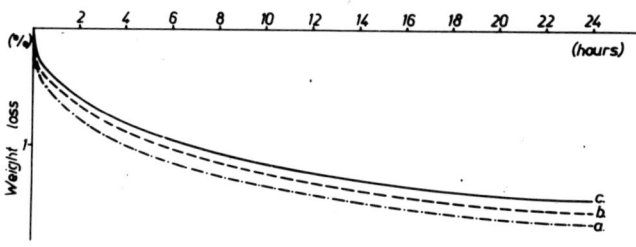

Fig.6

quantity of sample was changed in the crucible having same surface.) The increasing proportion of sample quantity causes decreasing in absolute weight loss and the time consumption would be longer. The weight losses measured in the same time are plotted in the function of sample weight and analysing the obtained curves numerically stipulating on exponential degradation rate we get the limit weight loss of null quantity of the material.(Fig.7)

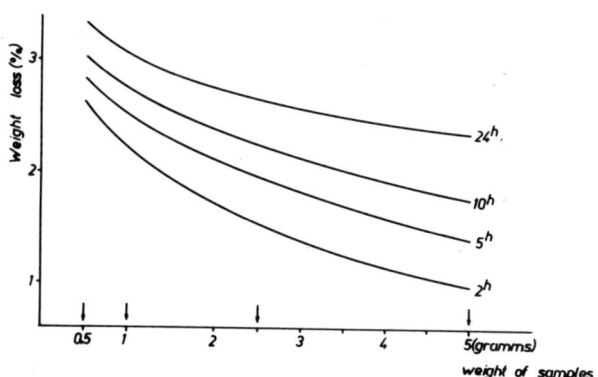

Fig.7

## REFERENCE
I.Porubszky, L.Ligethy, G.Liptay: Wire Industry (1970) 1009

# THE NEW METHOD OF THE POWER CURRENT ELECTROTECHNOLOGY - THE METHOD OF THERMAL ANALYSIS

Zbyněk Kraus and Václav Mentlík
Department of Electrotechnology of the Division of Electrical Machinery
College of Mechanical and Electrical Engineering - 306 14 Plzeň, ČSSR

All the time increasing demands on large energetic installations become evident in rising claims to one of their operationally most important parts - i.e. high-voltage insulation systems. These facts contribute conceivably to the tendency of raising statement ability of the testing methods for insulation property checking. Hitherto used methods determine from the major part both original properties and changes which arise in materials due to operational influences by means of transferred - communicated data, as e.g. loss coefficient, permittivity etc. In this way, of course, it cannot be obtained entirely single-valued information of changes in the structure of high-voltage insulations and conclusions for further practice need not be quite exact, too.

The decisive part for the operational degradation in the high-voltage insulation systems is formed from the technological point of view by necessary bonds of organic origin mainly on the basis of epoxide resins. The effect of the whole complex of operational factors - namely of temperature - results in total small changes of their initial properties caused by commonly complex chemical properties at which it is possible to assume that they preferably cause the degradation and influence the reactions with oxygen and the process determining in the major part the insulant degradation in the range of operational temperatures is the thermooxidational ageing.

It is known that the concentration of the initial substance changes during reactions - the concentration of the substances entering into reaction falls while the concentration of the arising ones grows and the reactive speed is proportional to the product of concentrations of the reacting initial substances. For the appearance of these reactions the level of energetic barrier is decisive which is given by the value of the activated energy of the material. To the change of molecules of the initial substances two presumptions must be satisfied, i.e. sufficient energy and the possibility of their collision. It can be seen then that in the case of insulation degradation the activated energy influences its speed - the greater the speed the more slowly the material changes. Owing to the fact that the concentration of the active molecules is closely connected with the change of the physical properties of materials it is possible to record the time change of the studied physical property of the insulation P respecting the Arrhein´s law:

$$\frac{dP}{dt} = - A \cdot e^{-E/RT} \cdot f(P) \qquad (1)$$

where A is the preexponential factor, E - the activated energy, R - the universal gas constant, T - the absolute temperature and f/P/ is the function expressing the order of passing reactions and respecting the continuity between the concentration of reacting parts and outer manifestation of the changes of the properties of P.

The time interval has a great importance for the electrotechnological practice in which the material satisfies reliably its function, i.e. the time of its life during which the studied property from the initial value $P_o$ changes to the limit /critical/ one $P_{crit}$. It is the limit after which the material lost its functional ability.

From the equation /1/ by integration the expression for the time and for the period of life can be obtained. It holds

$$t_{\ell} = - e^{E/RT} \cdot \frac{g(P_{crit}) - g(P_o)}{A} \qquad (2)$$

where g/P/ respects the change of the function f/P/ by integration. It is possible to express graphically the given connections by means of the l i n e s  o f  s e r v c e  l i f e which can be obtained under presumption that

$g/P_o/-g/P_{crit}/ > 0$ takes the logarithm /2/ at the base of ten

$$\log t_z = \frac{E}{2{,}303\,RT} + \log\left[\frac{g(P_o)-g(P_{crit})}{A}\right] = \frac{k_1}{T} + k_2 \qquad (3)$$

From the equation /3/ is evident how the gradient of this characteristic /in semilogarithmic coordinates $\log t_z$ and $1/T$ represented by the straight line/ expressing the speed of the gradation procedure depends on the activated energy of the material while its shift is then influenced by the choice of the ageing criterion.

It is clear from the above mentioned that just the activated energy is the factor characterizing the state of the material and from the point of view of the objective expression of its development for the estimation of the insulation behaviour with organic bonds it is necessary to use such methods that include in their principles and directly proceed from the study of this characteristic.

If e.g. the materiality of the material G which is the function of temperature and time is considered as the studied property in the equation /1/, then for its change with the temperature at the substance heating with a certain speed of the temperature growth we may write

$$\frac{dG}{dt} = -\frac{A}{\beta} \cdot e^{-E/RT} \cdot f(G) \qquad (4)$$

Modifying and introducing the new variables $x = E/RT$ and considering in harmony with the relation /3/ the changes of the materiality from the starting $G_o$ to the critical $G_{crit}$ we receive

$$\frac{1}{A}\left[g(G_{crit}) - g(G_o)\right] = \frac{E}{\beta R} \cdot \left[\mathcal{F}(x_{crit}) - \mathcal{F}(x_o)\right] \qquad (5)$$

where the function $g/G/$ expresses again the integration $f/G/$ and the elements $F/x_{crit}/$ and $F/x_o/$ represent the degradation values of the studied insulant and namely $F/x_o/$ at the temperature growth from the absolute zero to the ambient temperature $/x = \infty$ and $x = 0/$ what enables in contradiction to the element $F/x_{crit}/$ expressing the degradation from the starting to the critical value its neglect and we can write

$$\frac{1}{A}\left[g(G_{crit})-g(G_0)\right]= \frac{E}{\partial R}\cdot \mathcal{F}(x_{crit}) \qquad (6)$$

what is in substance the equation characterizing the TG analysis where /as we may see/ it is the important member of the activated energy of the studied insultant.

Similarly we may find in the basic relations for DTA the close connections with our problems especially in the fact that here the activated energy of the studied reactions is the important characteristic element.

Rightly we can then say that while the classical diagnostic methods express the development of the insulation properties indirectly - by the reflection of the outer manifestation of substances - the thermal analyses have the direct relation to the study of the insulation changes with organic components because they enable to follow the important characteristic factors for the state of substances - the development of their activated energy. Valid information for the life of insulation study with the bonds of organic character can be obtained by the study and analyses of the thermal analysis results because these methods have full theoretical presumptions for the application in power current electrotechnology.

In our Department of Electrotechnology of the Division of Electrical Machinery of the Faculty of Electrical Engineering of the College of Mechanical and Electrical Engineering in Plzeň we have studied the possibilities of the application of the thermal analyses in power current electrotechnology especially in two spheres and namely partly at the direct study of the changes of materials during their expedition and partly with the aim of effective gaining of the life lines.

In the first case the attained results at the minimal variation coefficients guarantee statistically substantial estimations of the life of material at which the direct connection of the results of thermal analyses was proved and in power current electrotechnology the most often used tests, i.e. the measuring electrical strength.

From the results of the second group gained up to now may be seen the great time energetic and material saving which is possible in this sphere thanks to thermal analyses because the present experience advises the tenfold shortening of necessary work at gaining statistically more efficient results than the classical methods afford.

The given survey cannot, of course, comprehensibly cover the whole range of this complicated and for application in the industry important problems. It gives only the main ideas of the research carried in our department.

# OBJECTIVE DETERMINATION OF THE STATE OF INSULATIONS BY THE DIFFERENCE THERMAL ANALYSIS

Václav Mentlík and Zbyněk Kraus
Department of Electrotechnology of the Division of Electrical Machinery
College 0f Mechanical and Electrical Engineering - 306 14
Plzeň, ČSSR

Modern high-voltage insulation systems contain organic bonds on the basis of epoxide resins as an essential technological component. This part of insulation determines the momentary state of material by its character because due to operation effects as temperature, oxygen and vibration in the range of operating temperatures of these insulations the material is very easily subjected to a number of reactions resulting in the deterioration in its complex and with it in the shortening of the length of its life.

As generally known the degradation reactions have a very complicated mechanism and in the course of their ageing they change their speed and intensity. The change of the thermal discolouration in dependence on the temperature occurs too what is closely connected and has a contact with the insulant temperature resistance. As it is known too just the temperature is one of the decisive degradation effects. We may say that this dynamical development of the considered reactions reflects in the development and changes of the activated energy of the substance. In accordance with the widespread valid knowledge of the course of chemical reactions just this quantity is the direct and immediate image reflecting above mentioned agents of reactions without distortion and at the same time the changes occuring in the material.

We can see then that one of the possibilities which occur in the power current electrotechnology for the objectivation of the measuring methods used up to now is the introduction of such a method which would enable to follow the development and changes of the activated energy of the material and also to give direct information of the immediate state of the substance. The method satisfying the above mentioned requirement is the method DTA which in its principle expresses well the development of the activated energy of the insulations - the image of the changes of the properties with their stress, especially by the most active degradation agent - the temperature.

After having analysed the thermograms of the DTA insulations after their completing - in the original state - we may identify the reactions depreciating the materials resulting in the successive loss of the mechanical properties by embrittlement and finally the loss of electroisolating properties caused by the great growth of the polar products of degradation - what results in the insultant loss of its functional capacity. At the same time we may convince of the decisive influence of the bond on the life of the material from the DTA records of the studied insultation components - glass fabric and bonds on the basis of epoxide resins. And while the course of the first component almost agrees with the zero line of the thermogram a great peak corresponding to the vigorous thermooxidizing reactions appears on the record.

When studying the changes of the insultant properties by their stress, especially by temperature, we shall consider their state from the DTA record analysis, namely from the exothermal peak development corresponding to the passing reactions. If namely either during operation or in laboratory test the insultant is subjected to the action of temperature it is more or less intensively effected by the temperature heigth of the above mentioned degradation epoch for a certain period of time. DTA records of such exposed samples show the changes because they express the remaining part of the degradation processes - i.e. these reactions which were not able to take place sooner during the exposure either during operation or in the laboratory. Essentially we record the dissounding of the participating reactions corresponding to the growing degradation drop of the material enthalpy.

Experimental verification of the above mentioned part can be demonstrated e.g. on the results gained in our department by means of the equipment of the firm STANTON REDCROFT - London, Great Britain - DTA Module 671B together with the recorder Servoscribe 2S on the classical high-voltage insulant Sklotextit ARV /non-alkaline glass texture and dian epoxide resin - contents 25 - 30% - manufacturer Kablo, Bratislava, ČSSR/.

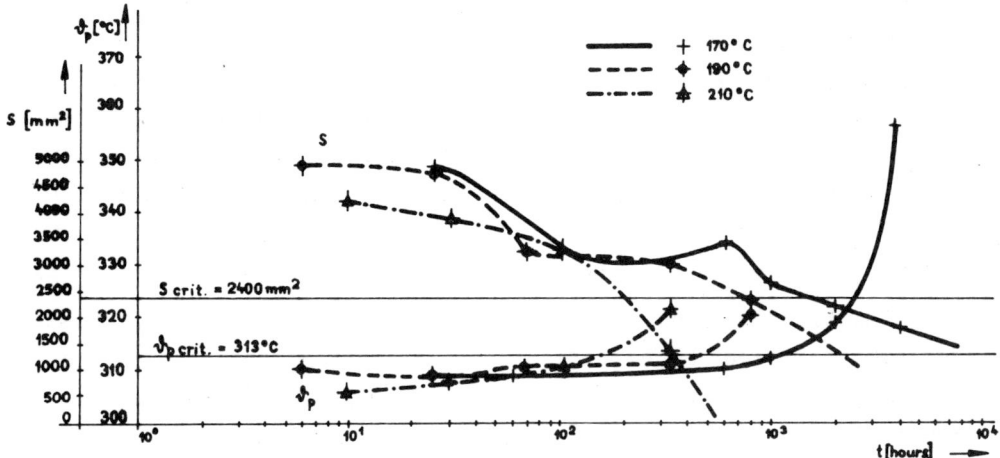

In the attached figure we may very well follow the development of two characteristic parameters DTA the peak surface of the reactions S /mm$^2$/ and temperatures corresponding to the maximum of these peaks $\vartheta_p$ /°C/ on the period of the material stress t/h./ at three exposed temperatures 170, 190 and 210°C. The difference of the courses S for 170°C and of the two remaining exposed temperatures responds to the dissimilar mechanism of ageing when the growth of the products of the reactions disposed to participate in the further reactions at the lowest temperature for a certain period of time - till 600 hours of exposure - occurs while the other courses illustrate the practical dissounding of the reactions and the persisting drop of the material enthalpy. On the courses for $\vartheta_p$ we follow the growth $\vartheta_p$ after the initial well-balanced course - i.e. the reaction peak displacement with the exposure time to the sphere of higher temperatures.

For the application of materials in the electrotechnics it is necessary to know the so-called life of material, i.e. the limit of the studied property under what the insulation loses its functional ability.The figure contains also the limits necessary for our studied parameters of DTA records gained by the introduction into mutual connection with the DTA results and classical methods for the study of life - of electrical strength and mass defect by the statistic method whose principle is the estimation of the probability by means of the relative frequencies on the basis of the law of great numbers.

The method DTA plus its principle guarantees statistically more substantial results and by it more reliable estimation at determining the life of materials in contrast to classical methods.

The described method being quite new in the study of the life of materials enables the objective estimation of the state of materials partly by comparison of records for the studied exposed sample with the album records for different samples exposed by the known manner and partly enabling to judge the remaining period of life of the material from the development of $S$ and $\vartheta_p$. Indisputable advantage of the method is the minimal want of experimental material at statistically substantial results.

# A DSC STUDY OF GLASSES OBTAINED FROM ORGANOMETALLIC GELS

V.Gottardi, G.Scarinci and G.Carturan
Istituto di Chimica Industriale - University of Padova - Italy
and
A. Marchetti and V. Frosini
Istituto di Chimica Industriale ed Applicata - University of Pisa-Italy

## ABSTRACT

A comparison has been made between the thermal behaviour of inorganic glasses obtained from organometallic gels and from oxides. Measurements of heat capacity and absorption of thermal energy at Tg (enthalpic relaxation) have been carried out. Although the composition of the two glasses is the same, the glass obtained by organometallic gel shows a greater change in heat capacity at Tg and a higher activation energy associated with a sharper glass-liquid transition.

## INTRODUCTION

There is a considerable interest in the achievement of the glass state with non-tradictional procedures and, in particular, starting from gels obtained by hydrolisis of organometallic precursors (1-4). Owing to the different preparation methods, the glasses obtained from gels might display a different structure and a different response to the thermal treatment with respect to the materials prepared by oxides melting even if with identical chemical composition. These differences should be difficult to evidenciate and we tested this hypothesis using the DSC analysis which has been demonstrated a suitable method to appreciate even small structural differences in the case of the polymers (5-8). We explored the glass $B_2O_3$ 32%, $P_2O_5$ 34% and $Na_2O$ 34% since it was well described (9) and requires simple preparation procedures either by melting the oxides, either by the gel method.

## EXPERIMENTAL

The gel was prepared on refluxing an appropriate $H_3BO_3$, $H_3PO_4$ and $NaOCH_3$ solution in ethanol during 3 hours. The solution was concentrated to small volume by distillation and on adding some wather it collapses to the gel. This by heating up to 700°C, gives a **trasparent glass** (Glass I) with chemical composition identical with that obtained from oxides melting (Glass II). Both materials have the same density and result to be completely amorphous to X-ray. Calorimetric measurements have been carried out on a Perkin-Elmer DSC-1B and DSC-2 apparatus.

## RESULTS AND DISCUSSION

Plots of $C_p$ versus T for samples of Glass I and Glass II are shown in fig.1. For both samples the specific heat increases linearly with temperature up to 675°K. In the range 700-750°K the step rise in the specific heat demontrates the existence of the glass transition. $T_g$ is taken as the temperature at which the specific heat reaches a value midway between those corresponding to the glassy and the liquid states. The experimental data of fig.1 show that Glass I and Glass II have the same $T_g$, however appreciable differences exist in the shape of the $C_p$ vs. T curves of the two samples in the glass transition range. The fact that for Glass II the transition extends over a much larger temperature interval than in Glass I is a clear evidence of the much more diffuse nature of the transition in the former than in the latter. On the other hand, the heat capacity jump at $T_g$, $\Delta C_p = C_{pl} - C_{pg}$, is somewhat higher for Glass I than for Glass II. Besides, a more evident endothermic peak superposed

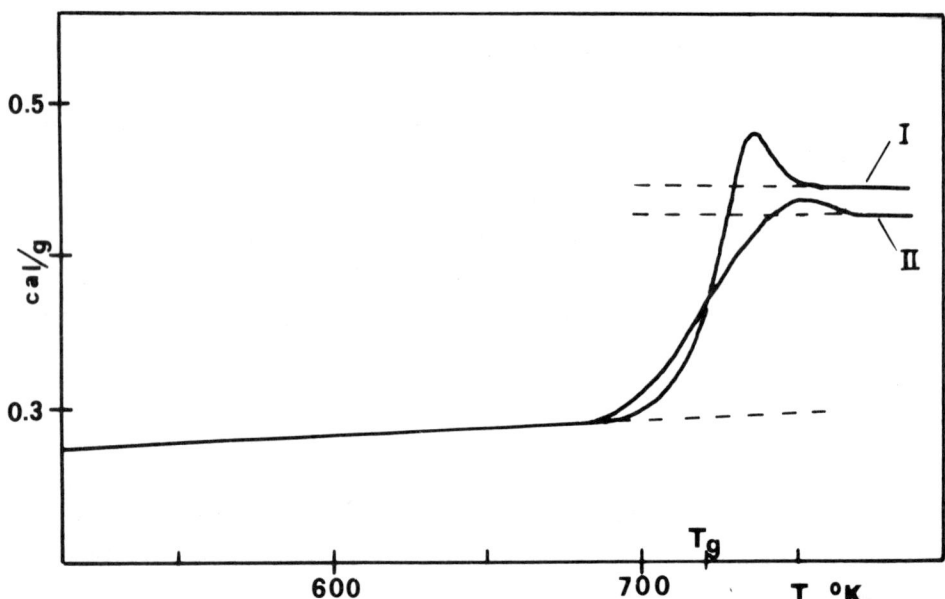

Fig.1. Cp vs. temperature plot for Glass I and Glass II.

on the baseline shift associated with the glass transition is observed in the $C_p$-T curve of sample I. Because the glass transition phenomenon is dominated by a molecular relaxational mechanism, the kinetic aspects associated to the glass transition rest on relaxational processes occurring in the material in the vicinity of $T_g$. These rate processes are of paramount importance in determining the thermal relaxation phenomena of the amorphous material in the transition range. In fact, for such a material, on slow cooling from the melt, a relatively small fraction of free volume is frozen in at $T_g$. On subsequent fast heating the equilibrium free volume fraction is not immediately achieved, as the fast heating rate does not allow time for the necessary structural rearrangements to occur. Once through the glass transition, molecular relaxation times decrease very rapidly and the glass is able to achieve its equilibrium free volume fraction, with a parallel absorption of energy so that an endothermic peak is produced in the thermogram. This thermal relaxation is easily observed in the dynamic measurements performed with DSC apparatus. The results obtained for Glass I and Glass II are reported in fig.2. In Table I are reported the amounts of the thermal energy absorbed in the

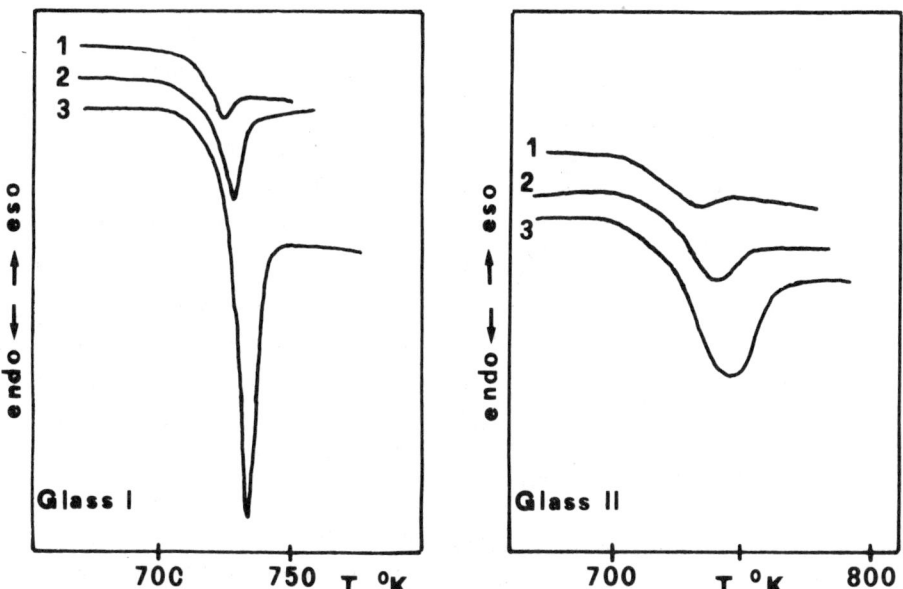

Fig.2. DSC curves for Glass I and Glass II previously cooled at 0.5°/min as a function of heating rate. Heating rates: 1) 8°/min, 2) 16°//min, 3) 32°/min.

TABLE I

| q °/min | ΔH Glass I cal/g | ΔH Glass II cal/g |
|---|---|---|
| 8 | 0.92 | - |
| 16 | 1.60 | 0.75 |
| 20 | 1.95 | 0.85 |
| 32 | 3.34 | 2.52 |
| 40 | 3.71 | 3.02 |

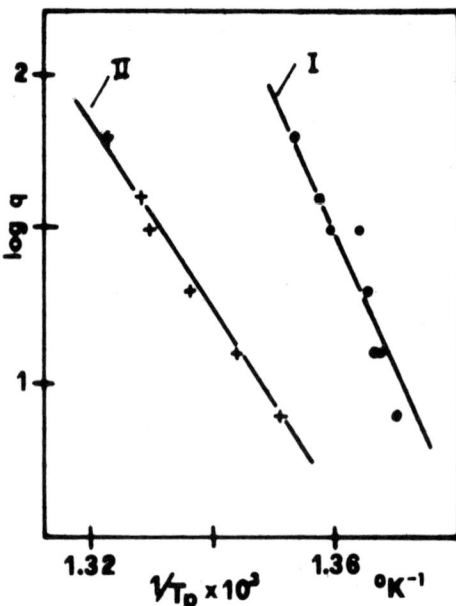

Fig.3. Plots of logarithm of heating rate vs. reciprocal of peak absolute temperature for Glass I and Glass II.

two glasses. In both cases the amount of the energy absorbed at $T_g$, as measured by the area of the endothermic peaks, and the temperature of the peaks increase with increasing heating rate q. In any case much sharper endothermic peaks are obtained for Glass I than for Glass II. This substantiates the idea that in the former the glass transition process is characterized by a narrower spectrum of relaxation times. We may assume that the hole theory (5) which proooved to be valid to describe the dependence of $C_p$ at $T_g$ on heating or cooling rate for most amorphous polymers, is also applicable to this case. According to this theory the peak temperature $T_p$ can be expressed over four decades of heating rates by $\lg q = A - B/T_p$, where q is the heating rate, A is an approximate constant and B is related to the activation energy for hole formation $\varepsilon_j$ by the expression: $B = 0.434 \varepsilon_j/R$, where R is the gas constant. Thus plotting lgq as a function of the reciprocal absolute temperature of the endothermic peaks for Glass I and Glass II, two straight lines are obtained from the slope of which $\varepsilon_j$ can be calculated. The values for the activation energy of a "mean hole" so obtained are of the order of 133 Kcal/mole for Glass II and 200 Kcal/mole for Glass I. It can be admitted that the observed difference in

$\varepsilon_j$ for the two samples reflects different structural arrangements of atoms or ions in the glasses. In fact, the preparation of glasses from alkoxides by mixing the components in solution, followed by gelling and dessiccation, unlike the method based on the melting of oxides, ensures a much higher degree of homogeneity of the starting mixture. Therefore, it should be expected that the two different preparation methods lead to glasses with different network structures, to which the observed differences in thermal behavior of Glass I and Glass II, can be related.

## CONCLUSION

This is a preliminary study on the thermal behavior of glasses obtained by organometallic gels, only one borophosphate glass of this type being examined. However the experimental results reported in this work allow to draw the following conclusions:

1) D.S.C. analysis demonstrated to be a very suitable method to evidence details in the thermal behaviour associated to the structural relaxation of such glasses at $T_g$.
2) Two glasses of the same chemical composition, and density, may exhibit different thermal behaviours depending on the particular method adopted for their preparation. Thus, the glass obtained by organometallic gels though displayng the same $T_g$ as the glass obtained by oxides melting, is characterized by a sharper glass-liquid transition with a higher activation energy.

## REFERENCES

(1) R.Roy, J.Amer.Ceram.Soc. 39 (1956) 145
(2) H.Dislich, Angew.Chem. 83 (1971) 428
(3) G.Carturan, V.Gottardi, M.Graziani, J.of Non-Crystal.Sol. 29 (1978) 41
(4) B.E.Yoldas, J.Mater.Sci. 12 (1972) 1203
(5) B.Wunderlich, D.M.Bodily, M.H.Kaplan, J.Appl.Phys. 35 (1964) 95
(6) S.E.B.Petrie, J.Polymer Sci. A2 10 (1972) 1255
(7) V.Frosini, A.Marchetti, E.Butta, La Chimica e l'Industria 59 (1977) 415
(8) V.Frosini, G.Levita, A.Marchetti, Proceedings of "Journées de Calorimetric et d'Analise Thermique", Torino, 28-30 June 1978 - Vol.IX-A B5-35
(9) K.Takahashi, "Advances in Glass Technology" Technical Papers of the VIth I.C.G., 366 (1962).

THERMAL ANALYSIS STUDIES OF THE DECOMPOSITION OF AMMONIUM URANYL
CARBONATE (AUC) UNDER SIMULATED INDUSTRIAL CONDITIONS

Lars Hälldahl, ASEA-ATOM, Västerås, Sweden
O. Toft Sörensen, Risö National Laboratory, Roskilde, Denmark

ABSTRACT

In a previous study (1) the decomposition and subsequent reduction of
AUC were examined by conventional thermal analysis (TG and DTA) and
the nature of the intermediate products formed during these processes
in different atmospheres was established. A new series of measurements
has been carried out in a modified thermobalance system in which the
rapid heating of the AUC powders, characteristic of the industrial
process, can be simulated. The equipment constructed for these measurements and the results obtained by this method in an actual industrial
atmosphere are discussed in this paper.

INTRODUCTION

Ammonium uranyl carbonate (AUC) has become an important intermediate
product in the conversion of $UF_6$ to $UO_2$ powder, used for fabricating
fuel elements for nuclear reactors. AUC is the precipitate from a wet
chemical process and has to be reduced in a hydrogen atmosphere to $UO_2$
powder, which is pressed and sintered to $UO_2$-pellets used in the fuel
elements.

In an earlier paper (ref 1), we studied the intermediate products formed
by AUC in different atmospheres by thermal analysis (TG and DTA). The
purpose of the present work is to simulate the industrial process in a
modified thermobalance in which the material is heated very rapidly in
an atmosphere of steam and hydrogen. In this equipment the change in
composition has been determined as a function of temperature and type
of atmosphere, and from the data obtained the kinetics of the reactions
involved in the decomposition of AUC and the subsequent reduction has
been evaluated.

## THEORY OF KINETIC EVALUATION

The reaction rate can generally be described by the equation:

$$d\alpha/dt = Z \exp(-E/RT) f(\alpha) \tag{1}$$

where $\alpha$ is the fraction reacted at time t, Z is the frequency factor, T the absolute temperature, R the molar gas constant, and E the activation energy. $f(\alpha)$ is a function of $\alpha$ describing the reaction mechanism (see ref (2)). By separating the variables, and introducing the heating rate, $q = \frac{dT}{dt}$, the equation can be written

$$d\alpha/f(\alpha) = (Z/q) \exp(-E/RT) dT \tag{2}$$

The integrated left side of the equation is the $g(\alpha)$-function.

Sestak (ref (3)) showed that for the correct $g(\alpha)$-function, the plot of logarithm $g(\alpha)$ versus $\frac{1}{T}$ gives a straight line.

To calculate the activation energy, the logarithm of Eq (2) is used

$$\ln(d\alpha/f(\alpha)) = \ln(ZdT/q) - E/RT \tag{3}$$

and from the plot of $\ln(d\alpha/f(\alpha))$ versus $\frac{1}{T}$, the activation energy E can be calculated from the slope.

From the above it is also clear that $d\alpha/f(\alpha) = g'(\alpha)$ and when the correct function describing the reaction mechanism is found, the activation energy can thus be obtained directly by plotting the derivative of this function versus $\frac{1}{T}$.

## EXPERIMENTAL

### Equipment

The TG equipment set-up used in these experiments is schematically shown in Fig. (1). (Abbreviation of Fig. later on). It is built around a Netsch STA 429, which has been complemented with a steam generator (1), a condenser for the escaping gases (2), a movable furnace (3) and a special furnace tube arrangement. Fig. 2 shows the functioning of this arrangement. It consists of three concentric tubes surrounding the sample crucible placed on the balance. In the inner tube a flow of helium comes from below, which protects the balance from contact with steam. The reaction gas comes from above, a mixture of steam and hydrogen. Before the analysis is started, valve 1 is opened and valve 2 closed. The protection gas forces the reaction gas to escape through valve 1 without coming in contact with the sample. The analysis is started by closing valve 1 and opening valve 2; this causes the reaction gas to come into contact with the sample on its way toward escaping through valve 2. The sample is always heated to just above $100°C$ before the valves are switched, to prevent water condensation on the sample.

The furnace is heated to a desired temperature in its upper position, and then moved downwards over the furnace tube to produce rapid heating of the sample. This is a fairly good simulation of the industrial process, in which the AUC is fed into a fluidized-bed reactor, at a specified temperature and a steam-hydrogen atmosphere.

## Materials

All analyses were performed on AUC-powder taken directly from a production line in ASEA-ATOM's fuel factory. The uranium content was 45.82%. The only detectable impurity was fluoride at the level of 50 ppm.

## Experimental conditions

Thermogravimetric measurements were obtained in different steam-hydrogen mixtures, the hydrogen content ranging from 0% to 20%. The total gasflow was in all cases 10 l/min. Buoyancy corrections were negligible. The starting temperature of the furnace in its upper position ranged from $380°C$ to $650°C$ to produce different heating rates and to evaluate the equilibrium composition at different temperatures. The samples were cooled to room-temperature in helium before they were taken out for X-ray analysis.

## RESULTS AND DISCUSSION

As mentioned previously, the logarithm $g(\alpha)$ plot versus $\frac{1}{T}$ gives straight line for the correct reaction mechanism. This theory is based on the assumption that only one reaction is involved and that this is controlled by one mechanism throughout the whole range of $\alpha$. From the previous study (ref 1) we know that this is not the case when AUC is reduced in a hydrogen atmosphere at these temperatures, but that the first step is:

$$(NH_4)_4UO_2(CO_3)_3 \longrightarrow UO_3(H_2O)_x + 4NH_3 + 3CO_2 + (2-x)H_2O \qquad (4)$$

Where x at maximum can be equal to 2.

Then a stepwise dehydration follows, where $UO_3$ is formed:

$$UO_3(H_2O)_x \longrightarrow UO_3 + XH_2O \qquad (5)$$

and if the temperature is high enough formation of $U_3O_8$ takes place: (6)

$$3UO_3 \longrightarrow U_3O_8 + \frac{1}{2}O_2 \quad (T \quad 400°C)$$

Finally, if the atmosphere is sufficiently reducing, the last step is

$$U_3O_8 + 2H_2 \longrightarrow 3UO_2 + 2H_2O \qquad (7)$$

To resolve these reactions from each other, we assumed that $UO_3(H_2O)_2$ was to be formed, and then defined $\alpha_I$ equal to 1 for the weight loss corresponding to this composition. A plot of logarithm $g(\alpha_I)$ versus $\frac{1}{T}$ (see Fig.) gives a straight line as long as reaction 1 is the rate controlling step, and if $g(\alpha_I) = (1 - (1-\alpha)^{1/3})^2$ correspond to three-dimensional, spherically symmetric diffusion - the Jander equation. From Fig. 3 it is also clear that reaction 2 starts to influence the mechanism when the temperature has reached a certain value. This means that in a transition region, two reactions are competing about the reaction rate control. The point at which the first straight part of the plot starts to deviate from the straight line is then taken as a starting point for the second step. The endpoint is assumed to be the weight loss corresponding to the composition $UO_3$. In Fig. 3 these plots are shown for a typical analysis. After the transition region, part II becomes straight where reaction 2 controls the reaction rate.

The derivative of $g(\alpha)$ is: $g'(\alpha) = \frac{2}{3}((1-\alpha)^{-2/3} - (1-\alpha)^{-1/3}) = d\alpha/f(\alpha) \qquad (8)$

and according to Eq. (3), the activation energy can be determined from the slope of the straight lines obtained when $\ln g'(\alpha)$ was plotted versus $\frac{1}{T}$ (see Fig. 4). The results from the analysis are presented in Table 1. The mean values are: $E_I = 79.5 \pm 6.9$ kJ/mole and for $E_{II} = 48.5 \pm 4.9$ kJ/mole. These values are independent of the hydrogen concentration, heating rate and final temperature. In the literature, no data are available on the decomposition of AUC. A lot of work has been done, however, on ADU (see ref 5) and one paper (ref 6) calculated the activation energy for the dehydration of $UO_3(H_2O)_2$ to $\beta$-$UO_3$, to 75 kJ/mole. The mechanism is a phase-boundary process of spherical symmetry.

## TABLE 1

| No | %$H_2$ in $H_2O$ | Max temp °C | q °C/min | X-ray | $E_I$ kJ/mole | $E_{II}$ kJ/mole |
|----|------|-----|-----|------|-------|------|
| 1  | 0    | 650 | 130 | -    | 70.2  | 53.6 |
| 2  | 0    | 650 | 133 | -    | 74.0  | 49.7 |
| 3  | 3.3  | 650 | 109 | $UO_2$ | 71.8 | 50.1 |
| 4  | 7    | 650 | 128 | $UO_2$ | 76.8 | 35.2 |
| 5  | 10.1 | 650 | 120 | $UO_2$ | 83.7 | 47.5 |
| 6  | 13.8 | 650 | 118 | $UO_2$ | 77.5 | 54.5 |
| 7  | 13.8 | 650 | 136 | $UO_2$ | 74.7 | 49.2 |
| 8  | 13.8 | 650 | 115 | -    | 102.7 | 47.2 |
| 9  | 19.4 | 650 | 118 | $UO_2$ | 87.7 | 47.7 |
| 10 | 13.8 | 630 | 214 | $UO_2$ | 74.2 | 47.1 |
| 11 | 13.8 | 600 | 128 | -    | 80.2  | 43.9 |
| 12 | 13.8 | 550 | 72  | $UO_2$ | 81.9 | 54.3 |
| 13 | 13.8 | 480 | 95  | $U_3O_8$ | 92.7 | 51.7 |
| 14 | 13.8 | 380 | 53  | $\alpha\text{-}UO_3$ | 87.7 | 47.9 |

The mean value for $E_I = 79.5 \pm 6.9$ kJ/mole, $E_{II} = 48.5 \pm 4.9$ kJ/mole.

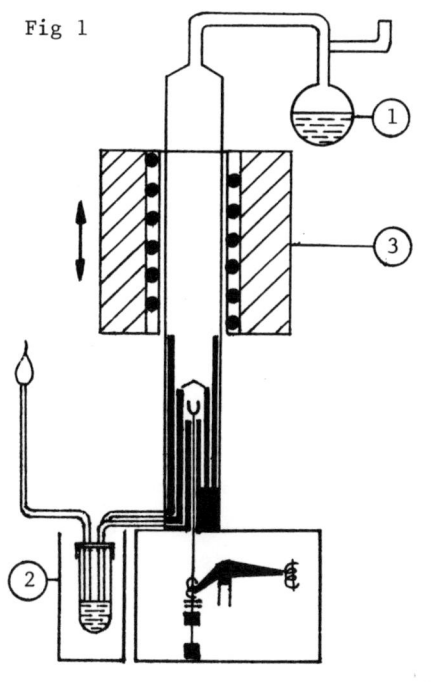

Fig 1

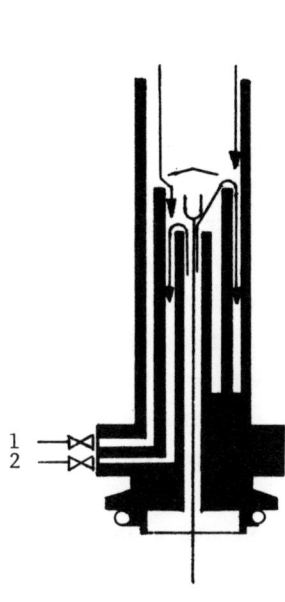

Fig 2

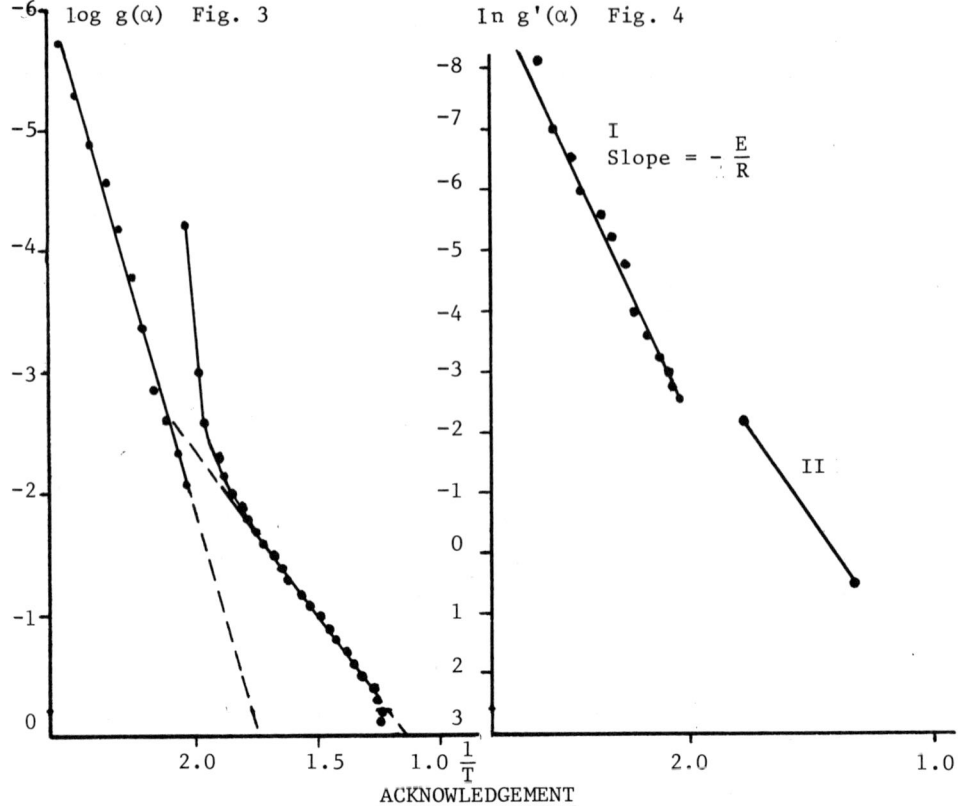

## ACKNOWLEDGEMENT

The authors wishes to thank H. Jensen, K. Larsen and C. Klittholm for building the special furnace and running the experiments.

## REFERENCES

1. L. Hälldahl, O. Toft Sörensen "Thermal Analysis of the Decomposition of AUC in Different Atmospheres" Thermochimica Acta 29 (1979) 253-259.
2. V. Sátava, F. Skvara "Mechanism and Kinetics of the Decomposition of Solids by a Thermogravimetric Method" Journal of The American Ceramic Society vol. 52, No 11 591-595.
3. V. Sátava "Mechanism and Kinetics from Non-isothermal TG-traces" Thermochimica Acta 2 (1971) 423-28.
4. J. Sestak. Note on the applicability of the p(x)-function to the determination of reaction kinetics under non-isothermal conditions". Thermochimica Acta 3 (1971) 150-54.
5. Gmelin, Handbuch des Anorganischen Chemie, Band 55, Uran. Ergänsungsband. Teil C3, Verbindungen, pp 54-69.
6. M.C. Ball, C.R.G. Birkett, D.S. Brown, M.J. Jaycock "The Thermal Decomposition of Ammonium Diuranate" J. inorg.nucl. Chem., 1974, Vol. 36, pp 1527-1529.

# THERMAL STUDIES OF CONTRACTING GAS BUBBLES IN MOLTEN GLASS

C. Parton and D. Dollimore
Pilkington Brothers Limited
Research and Development Laboratories
Lathom, Ormskirk, Lancs L40 5UF, England
and
Department of Chemistry & Applied Chemistry
University of Salford, Salford, England

## ABSTRACT

The behaviour of oxygen gas bubbles in a barium-aluminium alkali-silicate glass is described. The contracting sphere equation was found to describe the kinetics of bubble contraction in a series of isothermal experiments performed in the temperature range 850-1400°C. The experiment involved direct photographic recording of bubble size at recorded time intervals of a bubble rising freely in a bath of molten glass. Use was made of reduced time plots to indicate that the process was isokinetic over this range of temperature and that the contracting sphere equation gave the closest fit with the experimental data. On the basis of existing theoretical treatments this would indicate a phase boundary controlled process rather than a diffusion controlled kinetic process. The activation energy for the process is 142 kJ$^{-1}$ and the pre-exponential term is 0.925 cm s$^{-1}$ if the experiments are recorded in terms of change in diameter (cm) with time (s).

## INTRODUCTION

The behaviour of freely moving contracting bubbles of oxygen in a molten soda-lime glass melt has been described in a previous paper (1). The data were first tested to indicate that the process was isokinetic over the temperature range studied. The contracting sphere equation was shown to apply. Similar results were obtained for helium and neon bubbles in molten soda-lime glass (2).
In this study the contraction of oxygen bubbles in a barium-aluminium-alkali-silicate glass is treated similarly and the activation energy for the process established.

## EXPERIMENTAL PROCEDURE

Bubbles of oxygen (supplied from a cylinder) were introduced into the molten glass using an apparatus and experimental technique described previously (1). Single bubbles of oxygen were blown and detached from a platinum/rhodium capillary tube. Photographs were taken at suitable intervals of time as the bubble rose through the glass melt. Negatives on 35 mm film were projected onto a screen, bubble diameters measured and bubble diameter against time curves plotted. Each experimental run was carried out at constant temperature within the range $800°-1400°C$.

## RESULTS

Typical plots of bubble diameter against time are shown in Figure 1. The curves show that the diameters decreased with time until a point was reached when the rate of absorption became very much slower or the bubble ceased to change size. This residual volume, representing less than 5% of the original volume, contained one or more foreign gases which are more insoluble in the glass than oxygen. Such additional gases may have originated from dissolved gases in the glass.

### Influence of Temperature on the Rate of Contraction

It is clear from Figure 1 that the rate of contraction in the molten glass melt increased as the temperature increased. The gradients of each curve were measured and the rate of contraction S calculated, the units being $cm\ s^{-1}$. Values of $\log_{10}S$ were plotted against the reciprocal absolute temperature and a straight line graph obtained. This is shown in Figure 2. The equation of the line is

$$\log_{10}S = -0.742 \cdot \frac{10^4}{T} - 0.034$$

An estimate of the energy of activation E associated with the absorption of oxygen into the glass gives a value of $142\ kJ\ mol^{-1}$.

### Use of Reduced Time

The data were replotted on a reduced time scale in Figure 3. In this diagram the ordinate is represented by the fractional volume contracted $\alpha$ and the abscissa by $t/t_{0.5}$ where t was recorded and $t_{0.5}$ the time when $\alpha = 0.5$. The value of $\alpha$ is given by $(V_i - V_t) / (V_i - V_f)$ where $V_i$ is the initial volume of the bubble, $V_t$ is the volume at time t and $V_f$ is the final or residual volume.

It can be seen that a common curve can be drawn for the results indicating that the behaviour is isokinetic in the range 850-1400°C. The theoretical line drawn on this graph is that expected for a contracting sphere, viz

$$1 - (1-\alpha)^{1/3} = kt$$

where $k = \dfrac{v}{r}$

and v = velocity of advance of interface (assumed constant)
    r = initial radius of spherical bubble

The radius is used here and not the diameter because most quoted users of the contracting sphere equation employ a radius term.

## DISCUSSION

The analytical treatment utilised here has been applied in the past to solid state decompositions which are largely governed by the geometry of the reaction interface and by the diffusion of material up to and away from the interface. It is considered that a similar treatment may be made regarding the changing dimensions of a bubble in molten glass. The task is to recognise the appropriate equation to use and this is done here by utilising the method of Sharp et al. (3) which uses the concept of reduced time.

The general equation $F_{(\alpha)} = kt$, where $\alpha$ is the fraction of the process that has taken place and has a limited value when completed of unity and t is the time, may be used. The concept of reduced time may now be introduced when $F_{(\alpha)} = A(t/t_{0.5})$ where $t_{0.5}$ is the time for 50% completion of the process and A is a calculable constant depending on the form of $F_{(\alpha)}$. The reduced time $t/t_{0.5}$ is dimensionless. The contracting sphere mechanism is then expected for a changing interface on the assumption that the rate process is phase boundary controlled and the shape of the bubble is spherical. We then have

$$R_{(\alpha)} = 1 - (1-\alpha)^{1/3}$$
$$= \left(\dfrac{v}{r}\right) t$$
$$= 0.2063 \, t/t_{0.5}$$

where the radius of the bubble was initially r and the constant velocity of contraction was v. Using this method the first order decay law becomes

$$F = \ln(1-\alpha)$$
$$= kt$$

$$= -0.693 \, (t/t_{0.5})$$

If diffusion played a part in controlling the rate of bubble contraction then the resultant laws based on diffusion processes could be expressed in a similar form. The Jander equation, for example, recognises the combined presence of diffusion and the changing interface for contracting spherical surfaces and takes the form

$$D_{(\alpha)} = \left[1 - (1-\alpha)^{1/3}\right]^2$$

$$= \left(\frac{k}{r^2}\right) t$$

$$= 0.0426 \, t/t_{0.5}$$

One must first establish that the kinetics are isokinetic. This is done by inspection of Figure 3 where it is seen that all the experimental points lie close to a common curve. Theoretical data can also be represented on the same graph and it is found that the experimental points in Figure 3 cluster round the line representing the contracting sphere. The fact that a contracting sphere model is obeyed implies that the rate of contraction is determined by phase boundary conditions and not by a diffusion process. The activation energy for this process for oxygen bubbles in molten barium-aluminium-alkali-silicate glass is 142 kJ mol$^{-1}$ and the pre-exponential term is 0.925 cm s$^{-1}$. This compares with a value of 236 kJ mol$^{-1}$ and 24390 cm s$^{-1}$ for the activation energy and the pre-exponential term respectively for the behaviour of contracting oxygen bubbles in a soda-lime-silica glass (1). It should be noted that helium and neon bubbles in molten glass also behave according to a contracting sphere equation (2). Helium in a soda-lime-silica glass had an activation energy of 83 kJ mol$^{-1}$ and for neon 89 kJ mol$^{-1}$ with corresponding pre-exponential terms of 0.0553 cm s$^{-1}$ and 0.00482 cm s$^{-1}$ respectively.

## ACKNOWLEDGEMENT

This paper is published with the permission of the Directors of Pilkington Brothers Limited and Mr. A.S. Robinson, Director of Group Research and Development.

## REFERENCES

(1) C. Parton and D. Dollimore, Thermochim. Acta. **19** (1977) 25.
(2) C. Parton and D. Dollimore, Glass Technol. **18** (1977) 181.

(3) J.H. Sharp, G.W. Brindley and B.N. Narahari Achar, J. Amer. Ceram. Soc. __49__ (1966) 379.

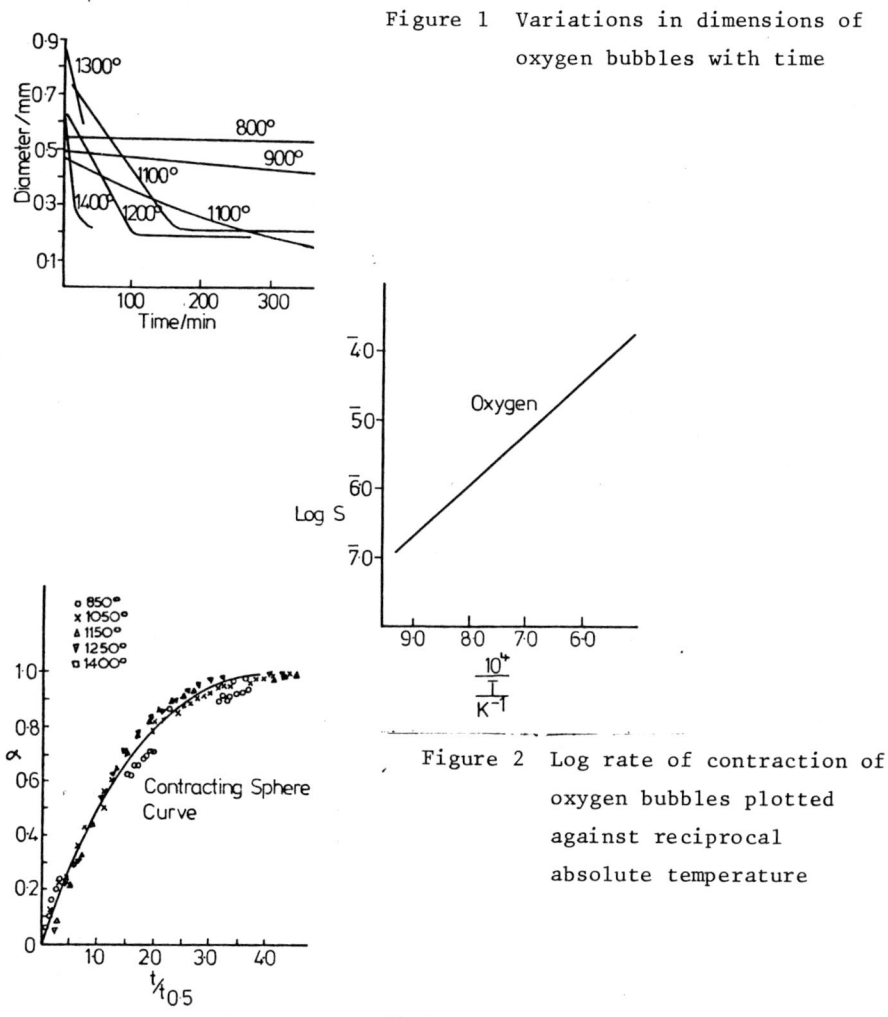

Figure 1  Variations in dimensions of oxygen bubbles with time

Figure 2  Log rate of contraction of oxygen bubbles plotted against reciprocal absolute temperature

Figure 3  Values of α plotted against $t/t_{0.5}$

QUANTITATIVE PHASE ANALYSIS IN THE SYSTEM
$CaSO_4 \cdot xH_2O$ BY TG-DTA METHODS

Winfried Hädrich and Erwin Kaisersberger
NETZSCH-Gerätebau, 8672 Selb, FRG

ABSTRACT

The quantitative determination of different components in the system $CaSO_4 \cdot xH_2O$ is necessary for the production and application of technical plaster qualities. Applying the common method this analysis needs four divergently prepared samples and a period of about 200 hours. Using a simultaneous TG-DTA apparatus with water vapour atmosphere on the sample, the result on one sample only can be obtained in a four hours' run. Advantages of this method are: easy preparation of samples, considerable time reduction and reasonable precision.

INTRODUCTION

During thermal dehydration in the system $CaSO_4 \cdot xH_2O$ (x = 0, 0.5, 2) seven different components arise, where the α- and β-formes of the half-hydrate (HH in the following) and the anhydrite III (hereinafter A III) differ only energetically. Depending on the fact, whether one works in dry or humide atmosphere, dehydration of the dihydrate (DiH) proceeds divergently up to the anhydrite II ( A II). The following scheme makes this more evident.

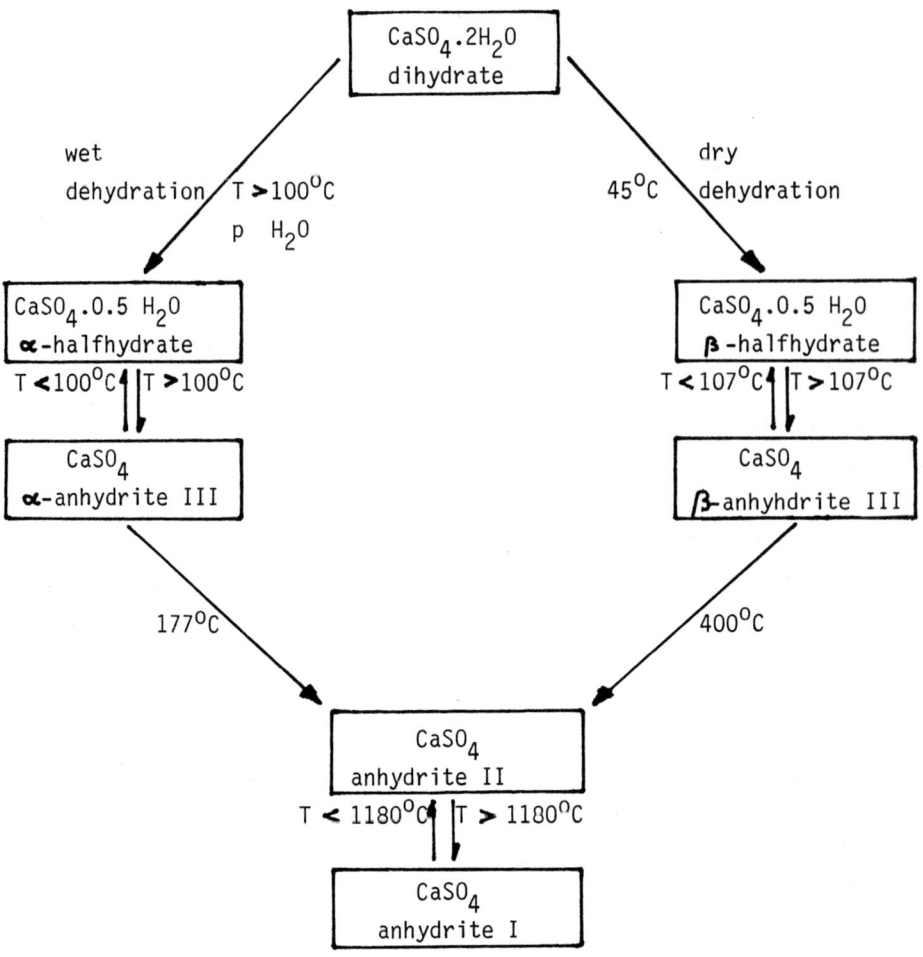

Fig. 1

The reversible polymorphic transistion of A II to anhydrite I (A I) is only of theoretical interest.

Technical plaster may be composed of a mixture of DiH, the two HH and A III formes and A II, as well as pollutions. By means of a combined TG-DTA measurement carried out under an increased water vapour partial pressure as well as by the traditional chemical analysis of the content of $Ca^{2+}$ and $SO_4^{2-}$ all seven components can be detected quantitatively. Fig. 2 shows the schematized diagram of a TG-DTA measurement.

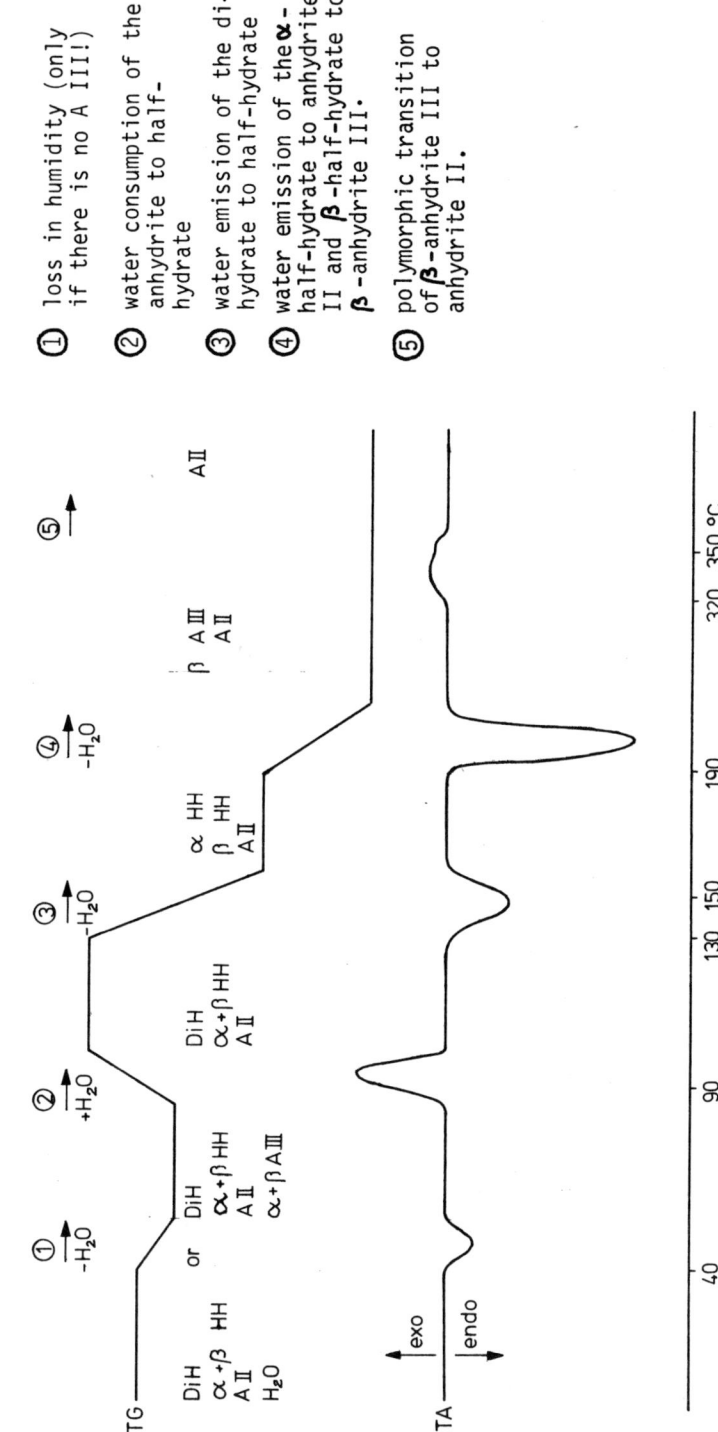

Fig. 2

## EXPERIMENTAL PROCEDURE

Plaster samples of ~100 mg were studied on the NETZSCH STA 409, provided with a furnace for vapour gas atmosphere. Evaluation is made as follows by means of the scheme in fig. 2:

The material to be tested consists of
$$\text{plaster } X \text{ mg} = (a \text{ DiH} + b \text{ HH} + c \text{ A III} + d \text{ A II} + e \text{ V})$$
(a, b, c, d = contents in mg of the corresponding components, e V = e mg pollutions).

The weight increase on ② shows the absolute content of A III according to the equation:
$$\text{A III} + 0.5 \text{ H}_2\text{O} \longrightarrow \text{HH}$$
increase in weight (in mg) x 15.113 = c (mg A III).

The decrease in weight on ③ shows the absolute content of DiH:
$$\text{DiH} \longrightarrow \text{HH} + 1.5 \text{ H}_2\text{O}$$
decrease in weight (in mg) x 6.371 = a (mg DiH).

The decrease in weight on ④ shows the absolute content of HH:
$$\text{HH} \longrightarrow \text{A} + 0.5 \text{ H}_2\text{O}$$
[(decrease in weight on ④ - increase in weight on ②) x 16.113] - 0.843 x a = b (mg HH).

Provided that for pollutions (V) portions of $Ca^{2+}$ or $SO_4^{2-}$ are admitted, the calculation of the A II portion and the pollutions (V) is based on the following: the content of $Ca^{2+}$ and $SO_4^{2-}$ in X mg plaster is determined chemically. If no pollutions containing calcium or sulfate are existant, the weight quotient of $Ca^{2+} : SO_4^{2-}$ must be 1:2.398. Deviating cases must be reduced to this proportion and a content of f mg $CaSO_4$ is obtained. Now, it must be respected, that this f mg is composed of the $CaSO_4$ content in DiH, in HH, in A III, and A II. By means of the thermogravimetric results the portion d (in mg) in A II can be calculated:

$$d \text{ (mg A II)} = f - (0.791 \times a) - (0.938 \times b) - c$$

At least the portion of pollutions V is determined from all received values:
$$e \text{ (mg V)} = X - a - b - c - d$$

Finally, the determination of α-A III and β-A III portions can be effected in the following manner:
A known quantity of pure β-A III (or β-HH) is heated in the same measuring arrangement. A DTA peak is obtained at ∼350°C for the transition of β-A III → A II, the peak area is linear proportional to the quantity of β - A III. This allows to calculate correspondingly the β- A III portion from the plaster measurement and even the portion of α - A III, α-HH and β - HH by means of the qualitative and quantitative knowledge of the course of reaction.

## RESULTS AND DISCUSSION

Fig. 3 shows a section of a measurement with natural plaster (sample weight 100 mg).

Fig. 3

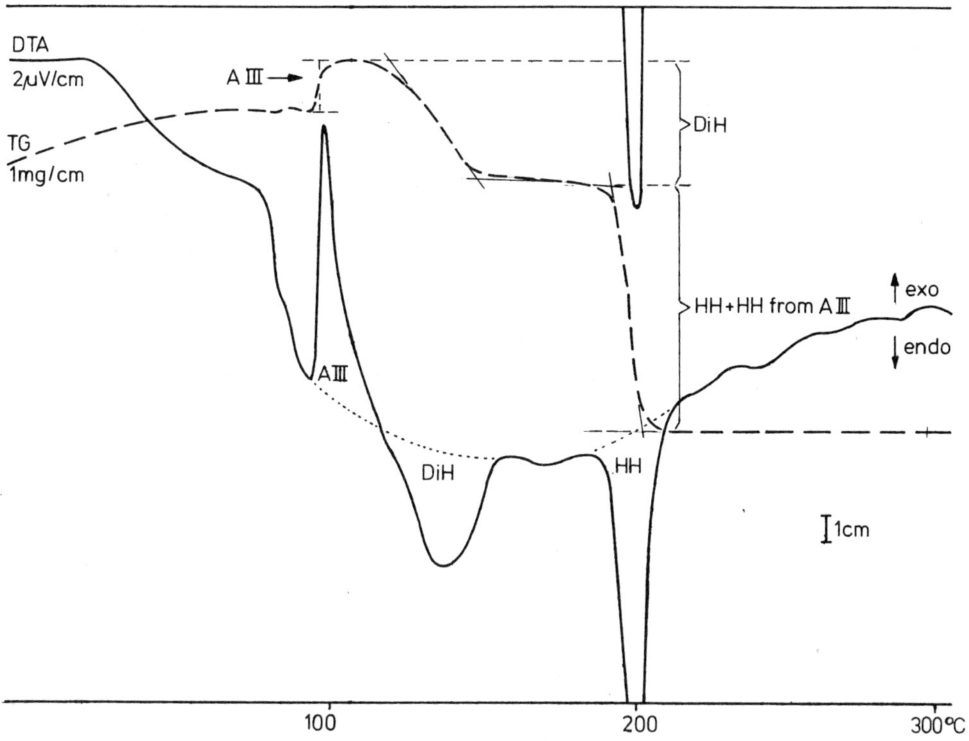

The content of A III is 12.69 mg, of DiH 13.06 mg, the HH 41,85 mg. Analysis has shown a contents of $CaSO_4$ of 78.18 %, accordingly the contents of A II is 15.92 mg, the contents of pollutions amounts to 16.4 mg.

Having a reading accuracy of 0.5 mm ≙0.05 mg for the TG curve, an accuracy of better ±1.5 % will result for the measured portions of the sample (DiH, HH, and A III), referred to the total quantity. Hence it follows the calculation of the portions of A II and V with an accuracy of better ±4.5 %. A further reduction of errors for the determination of the relative contents of the different components can be easily achieved by increasing the sample weight to double and a half as much.

The discussed method allows a rapid, complete and quantitative analysis of all $CaSO_4 \cdot xH_2O$ species in natural and artifical mixtures including the summary determination of pollutions. The DSC method principally cannot realize this with the presence of A III, because the exothermal effect of A III hydration and the endothermal effect of DiH dehydration are partially superposed. If thereby the encapsulated sample is hermetically sealed, as usually done, the HH value is falsified additionally by the water consumption of the anhydrite III to half-hydrate, because the total concentration of HH is increased.

The method demonstrated is suited for an automatic evaluation in on-line and off-line operation. Provision is, that existant pollutions do not adsorb or evolve water in the temperature range between 100 and 400°C in water vapour atmosphere, will not decompose by releasing gases, and, if a separation of the $\alpha$-and $\beta$-modifications of the half-hydrate and anhydrite is required, do not show solid state reactions.

# REACTIVITY OF LIME AND RELATED MATERIALS WITH SULPHUR DIOXIDE

D.R. Glasson and P. O'Neill

John Graymore Chemistry Laboratories, Plymouth Polytechnic, England, U.K.

## ABSTRACT

Dynamic TG studies were made on the uptake of sulphur dioxide by lime and related materials used as absorbents for the desulphurisation of industrial flue gases. Samples of quicklime, hydrated lime, limestone, magnesite and dolomite of widely different surface areas and particle sizes were calcined on a mass-flow balance in atmospheres containing various amounts of sulphur dioxide. Their relative effectiveness in reacting with sulphur dioxide could be seen rapidly by comparing the reaction rates with temperature and gas composition for each solid reactant. This indicated that only calcium oxide or calcium hydroxide could be reasonably considered as industrial desulphurising agents for use below 500°C. The greater reactivity of the hydrated lime samples at temperatures lower than those required for the other absorbents was related to the simultaneous formation of activated quicklime, especially at temperatures above 400°C.

## INTRODUCTION

In the desulphurisation of industrial flue gases the reactivity of sulphur dioxide absorbents, such as lime and related materials, depends mainly on their alkalinity and surface activity (1) in relation to their thermochemical properties (2). Suitable absorbents must react rapidly with sulphur dioxide in the temperature range 150-1750°C, but preferably between 150°C (flue gas temperature) and 400°C (upper temperature at which gas leaves economiser); the latter condition would allow the absorption system to be incorporated with the minimum upset to the heat extraction system. In the present research, samples of quicklime, hydrated lime, limestone, magnesite and dolomite of widely-different surface areas (1) and particle sizes are calcined on a mass-flow balance in atmosphere containing various amounts of sulphur dioxide. Thus relative reaction rates can be compared at different temperatures on the same sample. The technique also indicates the reactions occurring when cold particles of desulphuriser are raised in temperature to that of the hot gas into which they are injected industrially.

## EXPERIMENTAL PROCEDURE

Absorbent samples were calcined on a Stanton-Redcroft mass-flow balance in atmospheres containing (a) 2vol-% $SO_2$ in $N_2$, (b) 0.8 vol-% $SO_2$ in $N_2$, (c) 0.8 vol-% $SO_2$ + 8.2 vol-% $O_2$ in $N_2$ and (d) $N_2$ alone, with heating rates of 5°C $min^{-1}$ and flow rates of 1 $lmin^{-1}$ (3).

## RESULTS AND DISCUSSION

Thermochemical data relevant for interpretation of the DTG studies are presented in Fig. 1, (extension of reference 2). The data for the DTG studies are given in Fig. 2 and 3; the fully-lined curves represent the uptake of sulphur dioxide (and oxygen, where applicable), after allowance has been made for any decomposition of the absorbents indicated by the broken-lined curves.

Quicklimes. For the more active quicklime ($\underline{S}$ = 5.0 $m^2g^{-1}$), Fig.2(a), the small amount of hydrated lime (ca. 5%) mainly on the surface, decomposes between 350--400°C to very active CaO (4) which reacts rapidly with some of the $SO_2$, enhancing initial $SO_2$ uptakes. The remaining CaO reacts rapidly with $SO_2$ between 500--700°C, after which the reaction becomes slower, with no further weight gains above 945°C (2vol-% $SO_2$) and 888°C (0.8vol-% $SO_2$), i.e., at temperatures about 100°C higher than the theoretical dissociation temperatures of 840°C and 785°C (Fig.1). Thus measurable dissociation of calcium sulphite, $CaSO_3 \rightarrow CaO + SO_2$, begins after time for heat distribution and crystal nucleation has elapsed; subsequent weight losses occur, with rising temperature as dissociation predominates over any small weight gains from the secondary reaction $4CaSO_3 + 2SO_2 \rightarrow 4CaSO_4 + S_2$. Some disproportionation of calcium sulphite, $4CaSO_3 \rightarrow 3CaSO_4$ + CaS, gives more thermally stable products.

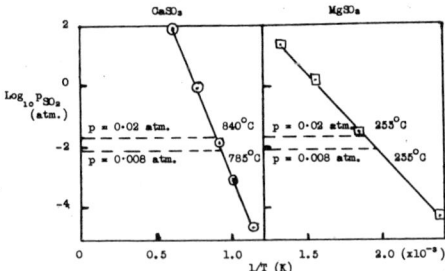

Figure 1(a). Dissociation pressures of calcium and magnesium sulphites at different temperatures.

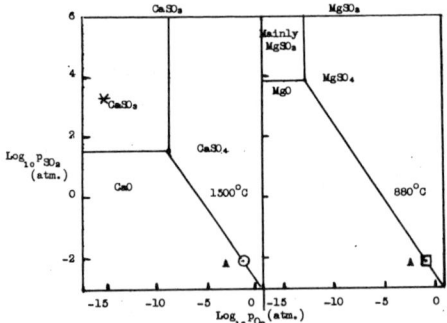

Figure 1(b). Predominance area diagrams for calcium and magnesium sulphites.

*$CaSO_3$ with some disproportionation:- $4CaSO_3 \rightarrow 3CaSO_4$ + CaS.

Figure 2(a). Reaction of CaO with $SO_2$

Figure 2(b). Reaction of calcium hydroxide with $SO_2$

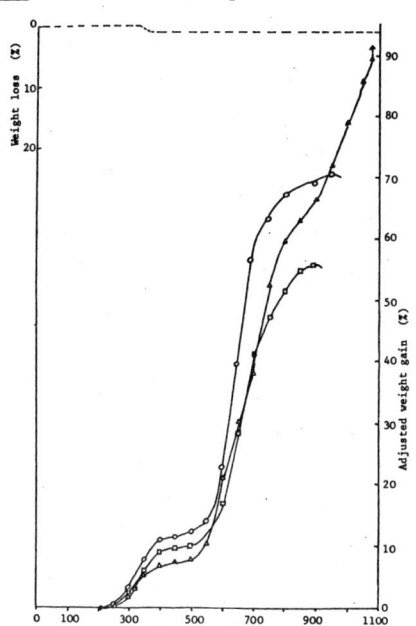

Figure 2(c). Reaction of $CaCO_3$ with $SO_2$

Figure 2(d). Reaction of magnesium carbonate with $SO_2$

Figure 2(e). Reaction of dolomite with $SO_2$

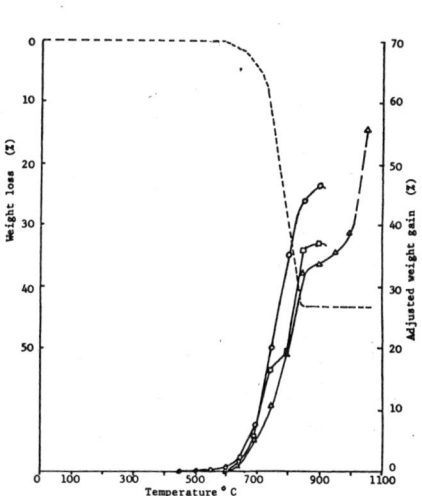

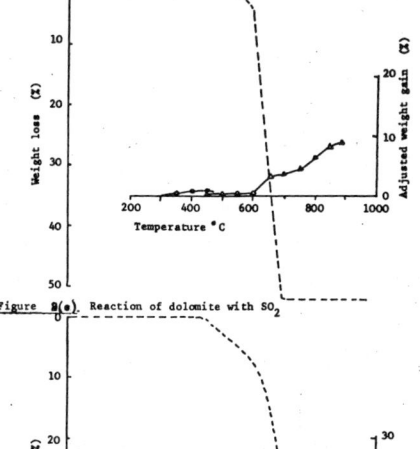

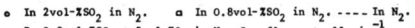

○ In 2vol-%$SO_2$ in $N_2$.   □ In 0.8vol-%$SO_2$ in $N_2$.   ---- In $N_2$.
▲ In 0.8vol-%$SO_2$ + 8vol-%$O_2$ in $N_2$. Gas flow rate 1l $min^{-1}$.

The maximum adjusted weight gains correspond to 62% and 49% conversion of CaO to $CaSO_3$, most of which (50% and 37% respectively) takes place at temperatures up to 700°C. The decrease in rate of uptake of $SO_2$ above 700°C is due also to impedance caused by the surface area and porosity of the lime being reduced by the reaction products. It has been found (5,6) that the CaO used was highly porous (apparent density 1.5--1.6 $gcm^{-3}$, cf. 3.3 $gcm^{-3}$ when completely fused). Nevertheless, the voidage between the lime particles (55% of the total volume) was only sufficient to accommodate the increased volume for 61% conversion of CaO to $CaSO_3$ without external expansion being required; thus considerable loss of porosity occurred (1). Disproportionation would decrease the volume of the products, permitting $SO_2$ uptakes of up to 83% theoretical without external expansion being required, but the comparatively low-melting CaS formed (m.p. 722°C) would promote sintering.

In the presence of oxygen, some of the $CaSO_3$ is oxidised directly to $CaSO_4$, while some disproportionates to $CaSO_4$ and CaS, before part of the CaS is converted either to $CaSO_4$, viz., $CaS + 2O_2 \to CaSO_4$ or to lime, $2CaS + 3O_2 \to 2CaO + 2SO_2$, depending on temperature and partial pressures of $SO_2$ and $O_2$. The thermochemical data for these reactions have been discussed recently (2), but in practice the ractions are impeded by protective layers of sulpate. Therefore, comparing the actual weight increase with the theoretical increase due to sulphite formation would be likely to give too high a value for the absorbent utilisation factor, because of the effects of oxidation; conversely, a comparison with the theoretical increase due to sulphate formation would give too low a value for the utilisation factor, if all of the sulphite and sulphide had not been oxidised. However, most of the $CaSO_3$ formed at the lower temperatures oxidises to sulphate or disproportionates as the temperature rises, indicated by an indefinite inflexion at about 850°C, just above the dissociation temperature, 785°C, of $CaSO_3$ [Fig. 1(a)] (cf. the maximum given at 880°C in the experiment using 0.8vol-% $SO_2$ without $O_2$). Similarly, the CaS formed by disproportionation will oxidise to $CaSO_4$ between 500--1000°C, so that $CaSO_4$ is the exclusive product at higher temperatures (1000--1100°C). Hence at 1000°C the adjusted weight gain represents 55% conversion of the lime to sulphate. Both CaO and $CaSO_4$ sinter appreciably above 900°C (7) when crystal lattice diffusion becomes more prominent, enabling the chemical reaction to proceed by ionic

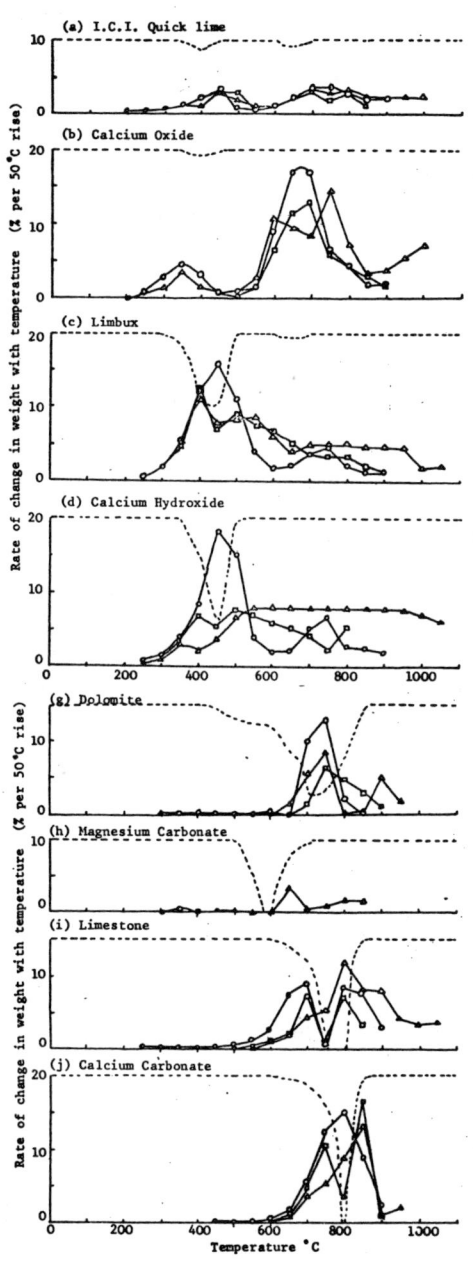

Figure 5. Rates of uptake of $SO_2$ by various alkaline earth metal compounds

● In 2volume-%$SO_2$ in nitrogen
■ In 0.8volume-%$SO_2$ in nitrogen
▲ In 0.8volume-%$SO_2$ + 8volume-%$O_2$ in nitrogen
----- In nitrogen

(a) I.C.I. Quick lime
(b) Calcium Oxide
(c) Limbux
(d) Calcium Hydroxide
(g) Dolomite
(h) Magnesium Carbonate
(i) Limestone
(j) Calcium Carbonate

diffusion. Thus the amount of sulphation increases somewhat more rapidly with rising temperature and at the highest temperature, 1086°C, weight gains of 89--104% were recorded representing 62--73% sulphation of the lime.

A less active commercial CaO ($\underline{S}$ = 0.8 $m^2g^{-1}$), containing a small quantity of $Ca(OH)_2$ and $CaCO_3$, reacted more slowly with $SO_2$, although there was some increase in rate when the $Ca(OH)_2$ and $CaCO_3$ decomposed to form some more active CaO.

Hydrated and carbonated limes

The maximum uptake of $SO_2$ in all cases corresponds to the decomposition of $Ca(OH)_2$ or $CaCO_3$ forming active CaO, Fig. 2(b), (c) and 3, consistent with earlier research (4,5).

Magnesite and dolomite

There is very little reaction between magnesite ($\underline{S}$ = 0.6 $m^2g^{-1}$) and $SO_2$ in $N_2$, Fig. 2(d), as expected from dissociation temperature values for $MgSO_3$, Fig. 1(a), (b), although small amounts of dolomitic impurities do react. In the presence of $O_2$, some uptake of $SO_2$ occurs as the magnesite decomposes to MgO, but subsequent decomposition of $MgSO_4$ [Fig. 1(b)] prevents further uptake. Similarly, dolomite initially shows poor uptake of $SO_2$, but the formation of dolomitic lime (8) corresponds to a period of higher reaction rates, Fig. 2(e) and 3.

## CONCLUSIONS

The relative effectiveness of the sulphur dioxide absorbents can be seen rapidly by comparing the variation in reaction rates with temperature and gas composition for each solid reactant. This indicates that only quicklime or hydrated lime could be reasonably considered as industrial desulphurising agents for use below 500°C. The greater reactivity of the hydrated lime samples at temperatures lower than those required for the other absorbents is related to the simultaneous formation of activated quicklime, especially at temperatures above 400°C.

## ACKNOWLEDGMENTS

The authors thank Mr. L.C. Anderson for his interest and encouragement in this research; also Imperial Chemical Industries Ltd. (Mond Division) and the Science Research Council, U.K., for sponsorship.

## REFERENCES

(1) D.R. Glasson and P. O'Neill, "Characterisation of Porous Solids" (ed. S.J. Gregg), pp. 351-7, Society of Chemical Industry, London (1979).

(2) D.R. Glasson and P. O'Neill, "Industrial Use of Thermochemical Data" (ed. T.I. Barry), in press, N.P.L. - Chemical Society, London (1980).

(3) Stanton-Redcroft Ltd., Copper Mill Lane, London; see also P.O'Neill Ph.D. Thesis, Plymouth Polytechnic (C.N.A.A.), 1978.

(4) D.R. Glasson, J. Chem. Soc., 1956, pp. 1506-10.

(5) D.R. Glasson, J. Appl. Chem., Lond., $\underline{8}$, (1958), 793-7; ibid., $\underline{11}$, (1961), 201-6.

(6) L.C. Anderson and J. Vernon, J. Iron Steel Inst., $\underline{208}$, (1970), 329-35.

(7) D.R. Glasson, J. Appl. Chem., Lond., $\underline{17}$, (1967), 91-6.

(8) D.R. Glasson, J. Appl. Chem., Lond., $\underline{14}$, (1964), 121-5.

# DETERMINATION OF NITRIC ACID IN NITRATING MIXTURES
F. Oehme

Polymetron AG, Hombrechtikon, CH-8634

## ABSTRACT

Using the principle of injection enthalpiemetry the determination of $HNO_3$ in nitrating mixtures can be carried out. The analytical reaction is the nitration of benzene which reacts rapidly and - depending on the ratio of sample and benzene - selectively with $HNO_3$. The repeatability of the method within a concentration range of 20 to 60% wt./wt, of $HNO_3$ is about $\pm 0,4\%$ $HNO_3$. The new method compares favorable with other analytical method used until now.

## INTRODUCTION

In the production of aromatic nitro compounds and/or of explosives nitrating mixtures are beeing used on a very large scale. Only a small fraction of the $HNO_3$ of the initially present concentration will be used-up in such processes. It is common practice to replenish such mixtures by adding strong $HNO_3$ again. For this reason a fast, simple and reliable method to determine the concentration of $HNO_3$ should be available. Table 1 (see Appendix) summarises the more important of the until now used methods. None of these methods suits too well to the task, however. This has been the reason that we started an investigation based on the use of the heat of nitration as an analytical tool. Interesting enough a similar principle has been used earlier to determine the concentration of benzene and/or toluene in the presence of cyclohexane (1), (2). We looked for the other side of the reacting partners, however.

The concentration of nitric acid in freshly made nitrating mixtures and/or that of the used ones is specified in Tab.1 and Tab 2.

Typical concentration of nitrating mixtures:

Tab. 1: Producing TNT. Values in bracketts = after use
% (wt./wt.) $HNO_3$ 22,4±0,4 (19,5), $H_2SO_4$ 69,5±0,6 (71,0)

Tab.2: Producing nitro glycerol
% (wt./wt.) $HNO_3$ 55,0±0,5 (26,5), $H_2SO_4$ 43,3±0,5 (64,4)
The difference to 100% in each case is made up by the water content of the mixture. The values given indicate the need for rather precise analytical methods which until now have been cumbersome and time consuming.

## EXPERIMENTAL PROCEDURE

The methode applied is known as injection enthalpiemetry(4). The sample of interest is brought together with a suited reagent which is beeing applied in an excess. The heat of reaction expressed by an increase of temperature will be measured. It is a direct mean for the concentration.

The experimental set-up can be seen in Fig.1 (Appendix). The sample injection can be carried out with a syringe in the usual way provided that the addition is a fast one. Fig. 2 illustrates a typical transient time function in which the temperature increase $\Delta T$ is directly proportional to the concentration. We have found, however, that a much better reproducibility will be obtained when the resulting temperature $\Delta T$ after a defined time from sample injection is measured (see Fig. 2, time interval t(c)).

Instead of recording of the temperature against time function modern Mikroprocessor-based methods of measurement could be used of course indicating the $HNO_3$ concentration of the sample immediately.

Another point of importance is the selection of the model substance to be nitrated by $HNO_3$ of the sample.

Fig. 3 illustrates the reactivity of a number of aromatic compounds. Keeping the high $HNO_3$ concentrations of Table 1 and Table 2 under consideration benzene is the best matching substance. It is interesting to know that the introduction of the first nitro group slows down the reactivity of the formed nitro benzene considerably. The poorly defined formation of polynitro benzenes is prevented by this resulting in a better analytical precision. nitrobenzene compares with monochlorobenzene so far (Fig. 3, (3)).

Handling samples of $HNO_3$ concentrations lower than 10% results in a low speed of reaction. Under such conditions benzene can be replaced by toluene which is much more reactiv (Fig. 3).

Fig. 4 shows a number of typical temperature against time functions. diagrams of such a kind give valuable information of the reaction under consideration. The analytical conditions should be fixed in such a way that a record in accordance with Fig. 2 results.

## RESULTS AND DISCUSSION

A typical calibration graph plotted with the readings obtained by an experimental set-up shown in Fig. 1 can be seen in Fig. 5. The repeatability of a single measurement is added as an error band in temperature increase. This band is coming up with an error in $HNO_3$ concentration of ±0,4% which is within the specifications to be found in Tab. 1 and/or Tab. 2. It certainly can be assumed that an improvement of the experimental equipment will narrow down the error band quite a bit.

The enthalpiemetric determination of may be an inte ting alternative to the existing methods to analyse nitrating mixtures (see Tab. 1).

## REFERENCES

(1) B.B.Corson and L.J.Brady, Ind.Engng.Chemistry, Anal.Ed. 14(1942),531

(2) R.L.Bishop and E.L.Wallace, l.c.(1), 15(1943),563

(3) W.Hückel, Theoret.Grundlagen d. organ. Chemie, Vol.2, page 583, Leipzig 1954

(4) L.D.Hansen, R.M.Izatt and J.J.Christensen, Applications of thermometric titrimetry to analytical chemistry, Treatise in titrimetry, Vol.2, J.Jordan, New York 1974

## Table 1

Known methods to analyse $HNO_3$ in nitrating mixtures.

1. <u>Differential titration</u> (Standard Method
   a) titrate the total of the acids ($HNO_3$ + $H_2SO_4$)
   b) evaporate $HNO_3$ (+$H_2O$) out of a weighed sample,
   c) titrate the remaining $H_2SO_4$

2. <u>Non-aqueous titration</u> (Kucharsky + Šafařik)
   Titrate the sample in acetone with tetrabutylammonium-hydroxide, record pH.
   Separate pH jumps can be observed for each of the acids.

3. <u>Redox reactions</u> (Pélouze, improved by Leithe)
   Boil the sample with ferrous chloride and hydrochloric acid:
   $$HNO_3 + 3FeCl_2 + 3HCl \longrightarrow NO + 3FeCl_3 + 2H_2O$$
   Titrate the formed 3-valent iron with $TiCl_3$

4. <u>Gas-volumetric methods</u>
   a) (Tiemann + Schulze), proceed as under 3), collect the formed NO in a gas measuring tube
   b) (Lunge). Use a Nitrometer to collect NO:
   $$2HNO_3 + 6Hg + 3H_2SO_4 \longrightarrow 2NO + 3Hg_2SO_4 + 4H_2O$$

5. <u>UV-Photometry</u> (Oehme + Koebel)
   Dilute sample 1:10 with water, use a cuvette with a length of 0,1...0,5 cm, read the optical density at a wavelength of 298 nm.

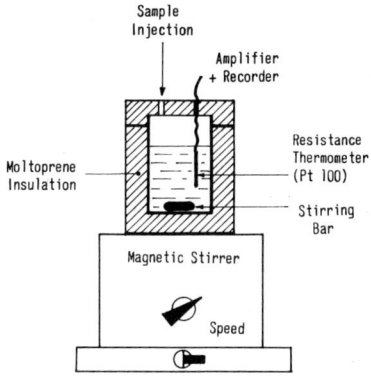

Fig.1 Experimental set-up of the principle of injection enthalpiemetry

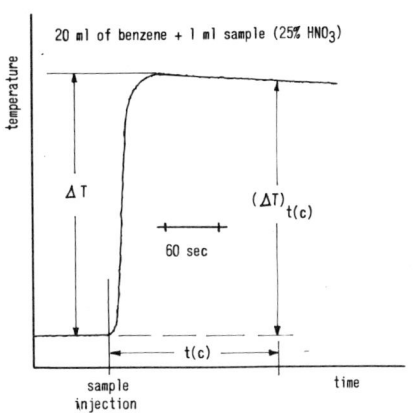

Fig.2 Transient time function in nitrating of benzene

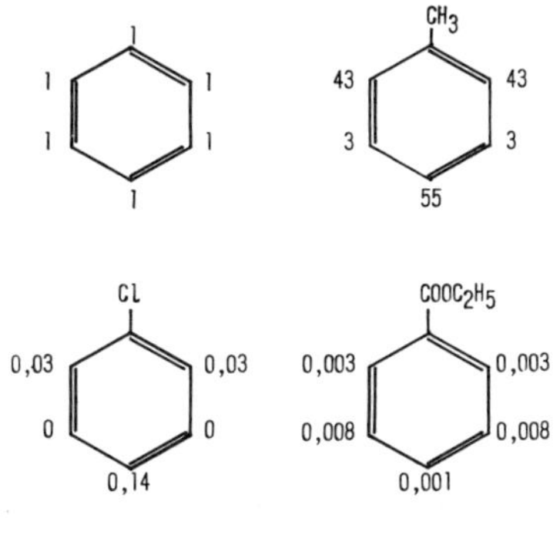

Fig.3 Relative reactivity of aromatic compounds

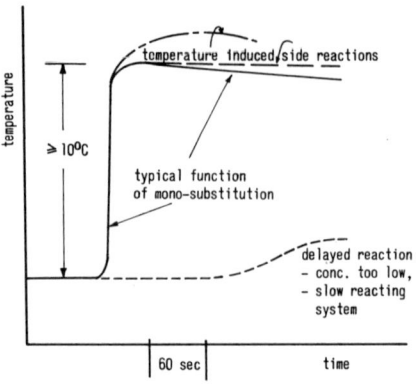

Fig.4 Transient time functions of different substituting reactions

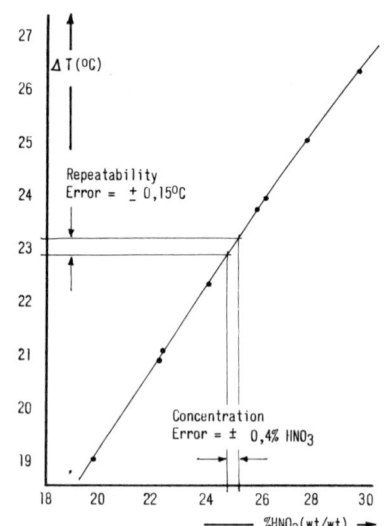

Fig.5 Calibration graph for the $HNO_3$ determination

# WATER DETERMINATION BY DIE METHOD

P. MARIK-KORDA
Institute for General and Analytical Chemistry, Technical University, Hungary

## ABSTRACT

Using a solution calorimeter /Firma: Hungarian Optical Works/ a direct thermometric method /DIE/ was worked out for the water-content determination in several organic solvents and solid samples. The measurements are based on that fact that the heat of reaction between Karl Fischer reagent and water is fairly great and under appropriate conditions the temperature change is proportional to the water content of the samples.

## INTRODUCTION

It is a very important and frequent task in the testing of industrial row materials, intermediates and products to determine their water content, because it effects their stability applicability directly or indirectly.
Water content determination is a conventional procedure. Karl Fischer titration is widely used for the determination of water in organic solvents, solid substances and gases /1/.
The procedure is based the oxidation of sulphur dioxide by iodine in the presence of water.

$$I_2 + SO_2 + 2H_2O \rightleftharpoons 2HI + H_2SO_4$$

Iodine and sulphur-dioxide are solved in piridine and methanol. So the water determination is a redox reaction, taking place in non-aqueous medium.
The end-point of the titration can be determined by visual, photometric and dead-stop method.
The titre of the reagent must be controlled frequently because of its fast change. In the methods mentioned heat evolved in the solution during titration interferes. It has been observed by several authors that the reaction is high-

ly exothermic −16.1 kcal/mole of water /67.4 kjoule/ /2/.
Wasilewsky and Miller used this − otherwise interfering −
phenomenon for the water determination in methanol /3/.
According to our investigations it can be made a simple and
fast water determination with thermometric devices produced
in Hungary /4/, which operate on the basis of the principle
of direct enthalpimetry /DIE/ /5/. As in this method a great
excess of the reagent is added to the sample solution in
one portion, work with the K. Fischer method would be expensive and troublesome. The DIE technique therefore been modified /6/ in a way that samples of small volume were added
successively to a large volume of the highly concentrated
reagent being in the adiabatic measuring unit, without opening it, or changing the reagent.
This method had been used for the determination of water
in some crystal hydrates /7/.

### EXPERIMENTAL PROCEDURE

The measurements were carried out by means of a Hungarian
DIE apparatus /Hungarian Optical Works/. This is actually
solution calorimeter, in which the temperature change is
sensed by a thermistor switched in a Wheatstone bridge.

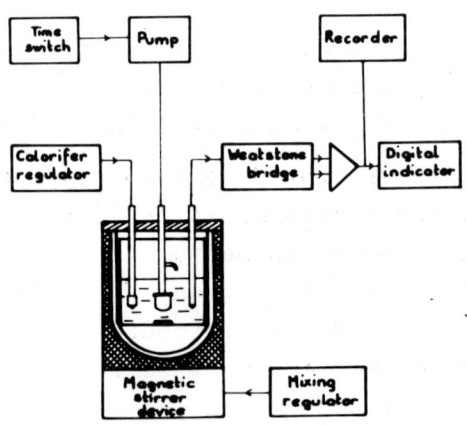

Fig.1. Schematic diagram of DIE apparatus

A compensograph type Radelkis OH-814/1 was used to record the results. The signal proportional to the temperature change was read off the recording.

The measurements were carried out at room temperature. 200 mls of Karl Fischer reagent /R/ were introduced into the measuring vessel.

Sample introduction was effected through a boring with a diameter of 10 mm, using a funnel for solid /Fig. 2a/ and a syringe for liquid samples /Fig. 2b/.

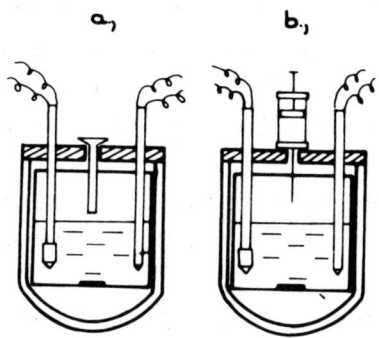

Fig. 2. Adiabatic measuring cell

Calibration: "Thermometric titre" of the reagent

a., for solutions

    2-3 drops /10-15 mg/ of distilled water were added to the R from a hypodermic syringe and the amount added was determined by weighing. Samples can be introduced one in a minute.

b., for solid samples

    analytical grade $Al_2O_3$ powder with known water content was added to the R in a way described above, in every 2-3 minutes.

The scale deflection is proportional to the water weighed in. The thermometric factor corresponding to 1 mg water:

$$f_t = \frac{\text{scale deflection}}{\text{mg water}}$$

The result of 11-11 parallel measurements /n/:

a.,        b.,

The mean:
$\bar{X} = 6.49$        $\bar{X} = 6.52$

The standard deviation

$$S = \sqrt{\frac{(X - \bar{X})^2}{n - 1}} = \pm 0.29 \,;\, \pm 0.35$$

Investigation of side-reaction effects.

The heats of dilution - dissolution - solvation also appear in the temperature change, or one of the components of the R itself might react with other components of the sample, in addition to water. These all should be substracted from the total heat produced to obtain that corresponding to the reaction of water alone.

We have tried to solve this problem empirically from the other side: the sample was added to just exhausted - also iodinefree - reagent.

If there is no side-reaction, we can use the factor, $f_t$ to calculate the results. If a side-reaction occurs a calibration curve must be taken with models with known water content.

In the following section three examples of the application of the method are described:

1. Determination of water in acetone
   Three times about 200 mg of acetone /min. 1%, max. 8% water content/ is added from a syringe. The measurements follow each other every minute, but after that we can use the reagent, only after the acetone-iodine reaction has been completed.

2. Determination of water in milk powder
   Three times about 100 mg powder are weighed from a hermetically closed plastic bag into the R. The measurements follow each other in every 2-3 minute.

3. Determination of water in vitamin-C tablet
   Whole tablets with presumably the same weight /480 mg/ can be dropped into the R in required parallels. The measurements should be made from powder, too. Ascorbic

acid was determined both thermometrically and using dead-stop titration.

## RESULTS AND DISCUSSION

According to our observations the rate of the reaction of water with the Karl Fischer reagent always higher than that of the others. The determination of water in aceton shows this, where we get a well evaluable curve with a break-point at the end of the reaction of water and the beginning of the slow reaction of acetone. So on the basis of kinetic difference the determination can be carried out. The allowed water content of milk powder within the guaranteed time is specified by standards. As a control water content was also determined by taking DTA and TG curves, and a fairly good agreement was found.

The water in vitamine-C tablets can not be determined by Fischer titration with dead-stop detection, as the reagent consumed by ascorbic acid is commensurable with that consumed by the water content. The molar heat of reaction of ascorbic acid is about 4% of that of water, thus it can be corrected for in the thermometric determination.

In general, if a multicomponent sample contains components other than water which react with the Fischer reagent /the excipient or vehiculum is inert/, but the matrix is the same, whereas the water content may be different, the letter can still be determined by thermometric method.

If the dead-stop titration yield results in good agreement with the theoretical water content, it means that the water present is completely released in the non-aqueous medium. An agreement of the results of the thermometric determination with those of the dead-stop titration is indicative of the fact that the energy changes accompanying the dissolution /dilution, solvation, structural changes/ are negligable compared with the heat of reaction, $\Delta H$ used for the determination.

The reagent can be used until complete exhaustion, so the absolute factor of the reagent need not be known. It means that the specific heat and weight as well the ionic strength remain practically constant, and the subsequent samples do not change the original great heat capacity significantly. The effect of the moisture in the ambient air only causes a base line shift and is constant.

The economicalness appears in fastness, too: samples may follow each other at a rate depending on the reaction time, and different substances can be measured in succession, if the samples do not react each other. The measuring unit need not be opened between measurements.

The modified method is suitable to compare whole tablets in respect of homogeneity, and might be very useful in following the water uptake.

## REFERENCES

/1/ E. Eberius, Wasserbestimmung mit Karl Fischer Lösung, Verlag Chemie GmbH, Weinheim, 1958

/2/ L.S. Bark and S.M. Bark, Thermometric Titrimetry, Pergamon Press, 1969

/3/ J.C. Wasilewsky and C.D. Miller, J. Anal. Chem. $\underline{38}$ /1966/ 1751

/4/ L. Erdey and P. Marik-Korda, Magy. Kém. Lapja $\underline{11}$ /1970/ 584

/5/ I. Sajó, MTA Kémiai Közl. $\underline{26}$ /1966/ 119

/6/ P. Marik-Korda, J. Therm. Anal. $\underline{13}$ /1978/ 357

/7/ P. Marik-Korda, Proc. 1st Czechoslovak Conference on Calorimetry, Liblice 1977, C12-1

# INVESTIGATION OF VEGETABLE OILS WITH DERIVATOGRAPH

I. Buzás, S. Gál[+] and J. Simon[+]
Department of Natural Sciences I., Hungarian Academy of Sciences, Budapest, [+]Institute of General and Analytical Chemistry, Technical University of Budapest, Hungary

## ABSTRACT

Thermal and oxidative stability of sunflower and rapeseed oils were investigated by a complex thermoanalytical method using a derivatograph both in dynamic and static operating systems. A new heating unit facilitated a highly reproducible isothermal heating program. Thermal curves obtained under dynamic conditions were characteristic for the composition and pretreatment of the samples. Results gained under static conditions gave information on storability and oxidative changes occuring in the course of regular storage.

## INTRODUCTION

In the oil industry thermal and oxidative stability of edible fats and oils is one of their most important properties concerning manufacturing, storing and consumption. This stability can be estimated also by thermal analysis. DSC in isothermal operation at normal and elevated pressures of oxygen, as well as micro TG both in static and dynamic operating systems using oxygen flow have been suggested for this purpose, and both methods showed good agreement with the standard tests. The advantage of these procedures over standard methods lies in their fastness, continuous measurement and better reproducibility /1-4/.

Our aim was to study the application of the derivatograph for the complex thermoanalytical examination of the thermal and oxidative behaviour of edible oils.

## EXPERIMENTAL PROCEDURE

Fresh and aged sunflower and rapeseed oils were used as models. Due to its fatty acid composition, fresh rapeseed oil is more resistant to oxidation than fresh sunflower oil. Aged samples were obtained from commercial fresh oils either by storing at room temperature, or by aeration at 100°C, and in a few cases also by frying potatoes several times at 180°C, modeling the household process. Decomposition of the oils was characterized by peroxide and acid values /POV, AV/ determined according to the standard tests /5/.

Samples were dispersed as a thin film on a thermostable ceramic block of considerably large constant specific surface /1 m$^2$/g measured by argon adsorption/, fitting in the regular platinum crucible or plate. A similar block was the reference material for DTA measurements. Sample weight was about 100 mg.

The simultaneous TG, DTG and DTA curves were registered using the derivatograph /MOM, Hungary/ as a function of temperature /dynamic program/ and also as a function of time under isothermal conditions /static program/. Heating of the furnace was regulated using the prototype of a linear temperature programmer /CHINOIN, Hungary/. This tiristor-regulated instrument allows highly reproducible heating, cooling and isothermal programs with the accuracy of 1 %, by controlling the heating current based on comparison of the temperature of the heating wire and the inner airspace of the furnace with the programmed value. In the isothermal program, temperature was raised rapidly /in approx. 10 minutes/ to the reaction temperature, which was kept constant /$\pm$ 0.5°C/.

## RESULTS AND DISCUSSION

As can be observed in Figure 1., the decomposition of different oils proceeded very similarly under identical dynamic experimental conditions. In oxygen atmosphere three steps of decomposition took place due to oxidative and thermal effects.

The overall enthalpy change was exotherm. The curves obtained in nitrogen flow proved that decomposition up to 250°C is caused by oxidation, and sample volatilization does not effect the measurements. In agreement with literature data the first step of oxidative decomposition is decisive in the study of oxidative changes. Initiation temperature of weight gain, temperature of TG and DTG maxima, the total weight gain and loss depend on the composition /stability/ of the samples.

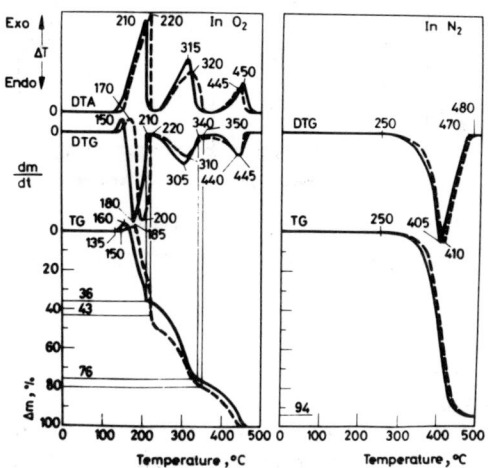

Fig.1. Decomposition of fresh sunflower and rapeseed oils in oxygen and nitrogen flow /20 l/h/; heating rate 5°C/min; _____ sunflower oil /POV=1.0, AV=0.1/; ------ rapeseed oil /POV=2.0, AV=0.1/

At lower heating rate, differences between sunflower and rapeseed oils could be better observed in air flow, mainly due to the fact that in oxygen atmosphere temperature of the samples deviated from the heating program during the exothermal process /Fig.2./. When thermal and oxidative stability of the samples was lowered by frying /strong degradation/, significant differences could be detected also between fresh and fried samples in air flow /Fig.3./.

No substantial difference could be detected, however, between the decomposition of fresh and mildly oxidized /stored, aerated/ oils under dynamic conditions.

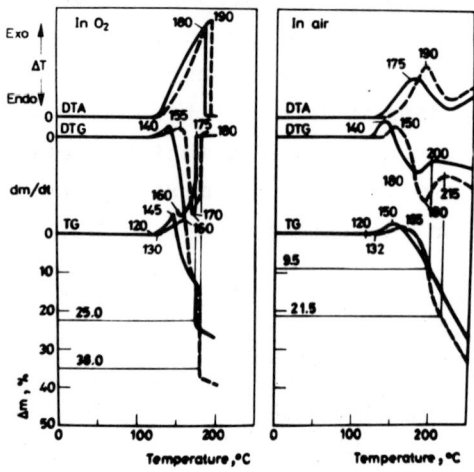

Fig.2. Decomposition of fresh sunflower and rapeseed oils in oxygen and air flow; heating rate 2°C/min; for further legend see Fig.1..

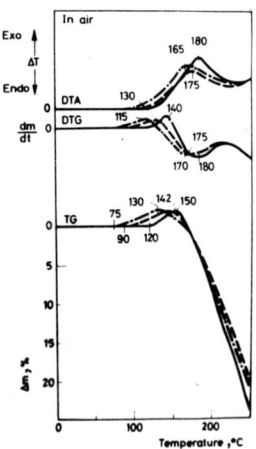

Fig.3. Decomposition of fresh and thermally treated /fried/ sunflower oils in air flow /20 l/h/; heating rate 2°C/min; ——— fresh /POV=1.0; AV=0.1/; ----- fried: used 10 times /POV=5.0; AV=0.6/; —·—·— fried: used 20 times /POV=5.0; AV=0.9/

Isothermal /static/ conditions proved to be suitable for detecting milder autoxidative changes in the samples, i.e., for the evaluation of minor differences in storability, generally needed in practice. Oxidative changes measured at different temperatures under air flow, showed 90°C to be suitable for the comparison of different sunflower and rapeseed oils /Fig.4./. Evaluation of the TG curves was based on classical methods /6,7/. Oxidative stability of the oils could be characterized by the length of induction periods /IP, min/ and TG, DTG and DTA maxima as a function of time. Total weight gain /$\Delta m$,%/, depended practically on the experimental conditions.

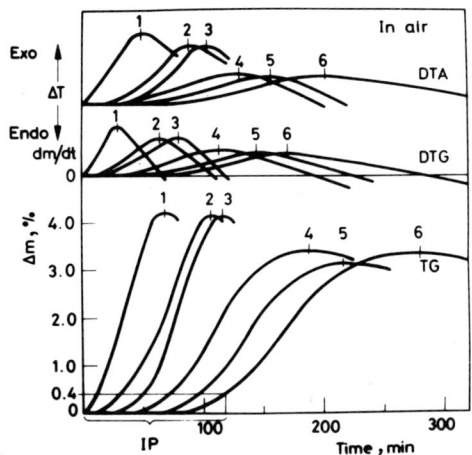

Fig.4. Thermoanalytical curves of fresh, stored and aerated sunflower and rapeseed oils at 90°C in air flow /20 l/h/;
1: sunflower oil oxidized at 100°C /POV=120.0; AV=0.3/;
2: stored sunflower oil /POV=12.0; AV=0.1/; 3: fresh sunflower oil /POV=1.0; AV=0.1/; 4: rapeseed oil oxidized at 100°C /POV=90.0; AV=0.3/; 5: stored rapeseed oil /POV=15.0; AV=0.1/; 6: fresh rapeseed oil /POV=2.0; AV=0.1/

Good reproducibility of the measurements was ensured by the high accuracy of weighing.

## REFERENCES

/1/ C.K. Cross, J.Amer.Oil Chem.Soc. **47** /1970/ 229
/2/ H.J. Nieschlag et al., Anal.Chem. **46** /1974/ 2215
/3/ R.L. Hassel, J.Amer. Oil Chem.Soc. **53** /1976/ 179
/4/ J.W. Hagemann and J.A. Rothfus, ibid. **56** /1979/ 629
/5/ L.V. Cocks and C. van Reede, "Laboratory Handbook for Oil and Fat Analysts", Acad. Press, 1966
/6/ H.S.Olcott and E.Einset, J.Amer.Oil.Chem.Soc.**35** /1958/159
/7/ id. ibid., 161

DIFFERENTIAL SCANNING CALORIMETRY - SCOPE AND LIMITATIONS
OF ITS USE AS A TOOL FOR ESTIMATING THE REACTION DYNAMICS
OF POTENTIALLY HAZARDOUS CHEMICAL REACTIONS

Ruedi Gygax, Max W. Meyer, Franz Brogli
Central Safety Research Laboratory,
CIBA-GEIGY Ltd., Basle, Switzerland

ABSTRACT

In addition to the use as a screening instrument for the appraisal of thermal hazards of chemical reactions, differential scanning calorimetry (DSC) can give information also on reaction dynamics. This can be achieved, for example, by selecting samples at strategic process stages, and by comparing their DSC-curves, the heat produced during the selected time increment can be monitored. Thus it is possible to get an idea of the heat production rate due to the desired as well as to the follow-up reactions under given conditions. Furthermore, the dynamics of a reaction can be inspected by the use of DSC in the isothermal mode. For estimating the thermal hazard of the reactivity of chemical compounds, information on reaction dynamics is especially important when materials are involved which decompose only after an induction time at a given temperature. The isothermal technique is also valuable for detecting consecutive reactions and catalytic effects since it is more sensitive than the scanning technique in this respect. The merits and limitations of DSC in this context will be discussed.

Introduction

Differential scanning calorimetry (DSC) can be used as a screening tool for the appraisal of thermal hazards of chemical reactions. As discussed in the previous paper (1) the information drawn from a single DSC-experiment in many cases is sufficient to determine the approximate heat of reaction, calculate the temperature attainable if no cooling is provided and estimate the consequences of secondary processes that may be initiated at this temperature. If, however, the desired and undesired reactions are not well separated on the temperature scale, further insight is necessary to judge the safety of a certain process. With the aid of some practical examples we like to present a few considerations and techniques that are currently used in our thermal hazards testing laboratory when approaching such cases.

Investigation of the intended reaction

Two typical chemical processes are the quaternisation of pyridine by a halide in a solvent (case A) and the amination of a chloronitroaromatic compound in an autoclave (case B). DSC-curves (Fig. 1) of the respective reactant mixtures exhibit a contingent series of exothermal signals (2). Starting from the proposed process temperature ($T_p$) (arrows in Fig. 1), the total energy observed is by far sufficient potentially to cause the temperature and pressure to exceed values tolerable in the plant. It is necessary, therefore, to gain information also on the dynamics of the exothermal processes taking place.

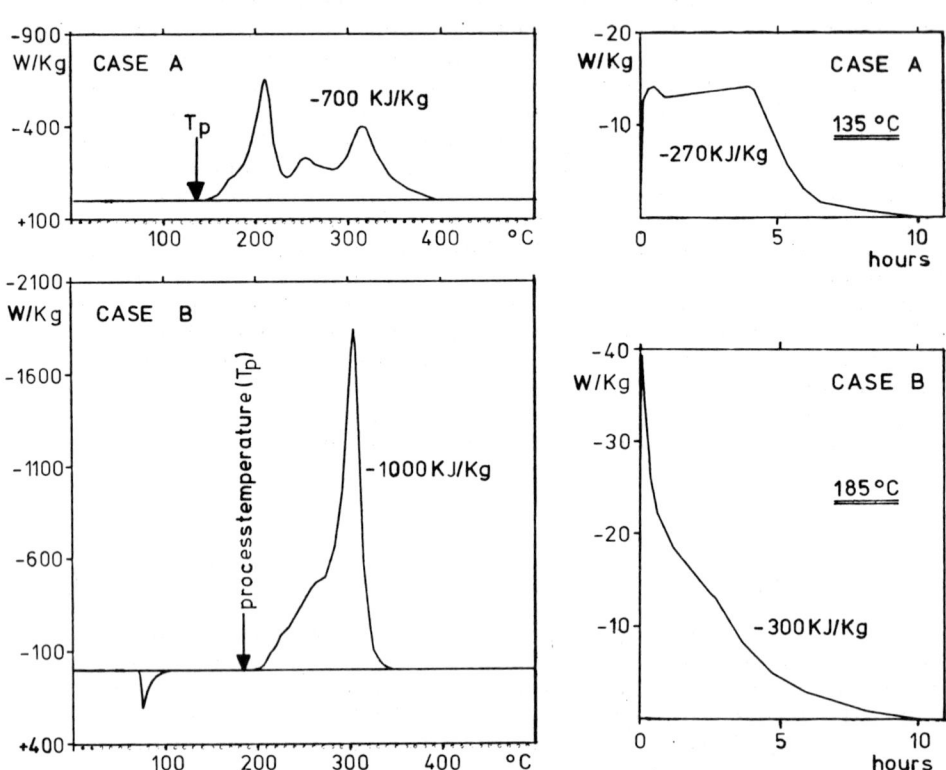

Fig. 1: Temperature programmed DSC-curves of reactant mixtures.

Fig. 2: DSC-curves of isothermal treatment of reactant mixtures A and B at process temp.

In order to separate the heat due to the desired reactions, we shall proceed by thermally treating the reactant mixtures according to the desired procedure. In particular, this can be done in the DSC pressure tight capsules while the heat production rate is monitored. Fig. 2 displays the signals of the two reaction masses familiar from Fig. 1. These experiments give indications on the heat of reaction and approximative information on the time until completion of the process, which usually agrees with the process duration in the pilot plant and in the plant procedure. For example, the reaction times of the above reactions in the plant are 10 and 8 hours, respectively. For case B the reaction proceeds with steadily decaying heat evolution while the shape of the isothermal curve in case A suggests a more complex mechanism. Closer inspection revealed that the formation of two phases during the course of the reaction was a possible explanation for this.

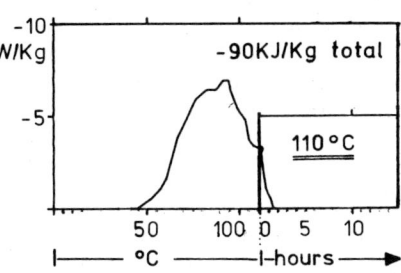

Fig. 3: Case C
Heating the reactant mixture at 0.2 K/min from 50 to 110 °C then isothermal treatment at 110 °C.

In another example, the sulfonation of a nitroaromatic compound (Case C), a slow steady temperature increase from 60 °C to 110 °C followed by an isothermal treatment at 110 °C as applied in the plant was simulated on the microscale (Fig. 3) of a DSC experiment. The dynamics of the heat evolution as well as the heat of reaction was in excellent agreement to a bench scale calorimeter experiment (3,4).

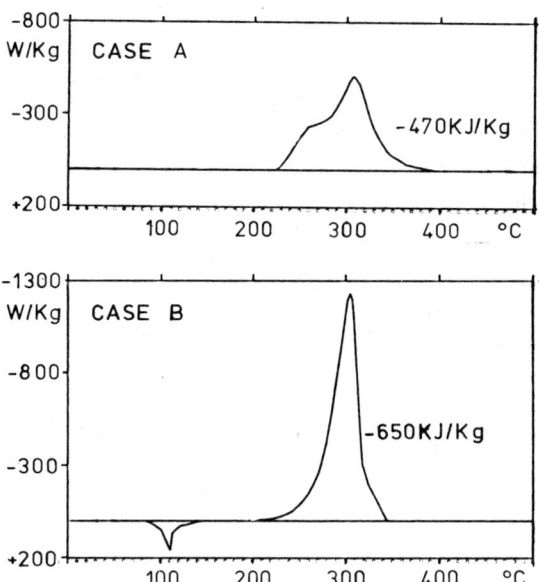

Fig. 4: Temperature programmed DSC-curves of thermally treated samples.

Returning to our two first examples the next step may be to
run a temperature programmed DSC experiment of the samples
after the isothermal treatment (Fig. 4). Comparing Fig. 4
and Fig. 1, the difference of the integral heat evolved
with a thermally treated and a freshly mixed reactant
sample gives yet another estimate of the heat of reaction.
Product samples originating from an authentic run according
to process instructions are usually checked against the
reactant mixtures treated in the DSC crucibles to make sure
the smallness of the sample and the absence of stirring
have no drastic effect on the result. Once the heat of reaction is thus discriminated, the potential temperature
rise due to the heat of the desired reaction can now be
evaluated and checked against external conditions (1).

Investigation of secondary reactions

The maximum attainable temperature thus established, an
assessment of the risk due to follow-up reactions can be
discussed asking questions of the following type:

- What approximate heat production rate due to secondary
  reactions is to be expected at the specified temperature?
- If the cooling breaks down what rate of temperature rise
  must be expected? How much time will be available to take
  measures against the temperature rise, before the reactions become uncontrollable?
- Many decompositions proceed with a self-accelerating
  mechanism (autocatalysis). The stability of a reaction
  mass is then a function of time as well as temperature.
  It is important to know the dynamic behaviour of the decompositions and their approximate induction time.

Clearly, such questions again call for an experimental
method which is time based rather than temperature programmed. Isothermal experiments are now performed at temperatures at which decomposition reactions take place. For
example, Fig. 5a displays such an experiment run at 200 $^o$C
of a sample drawn at the end of the procedure of case A.

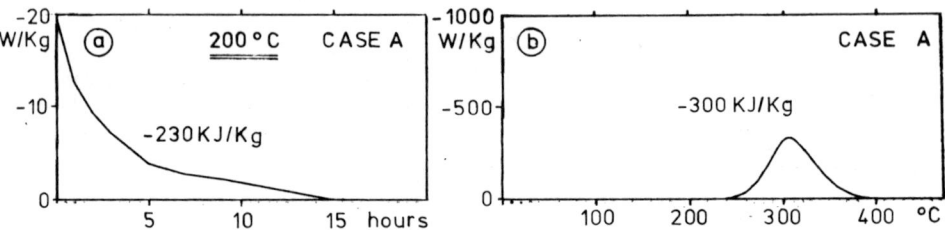

Fig. 5:  a) isothermal run on final reaction mass at 200 $^o$C
        b) temperature programmed DSC-run on sample from
           Fig. 5a.

No accelerating mechanism is observed and the heat production rate due to decomposition at 200 °C is similar to that observed for the desired reaction at 135 °C. Since in this case the decomposition proceeds rather moderately and the refluxing solvent provides an intrinsic cooling capacity, the process is considered safe if constant stirring is guaranteed.

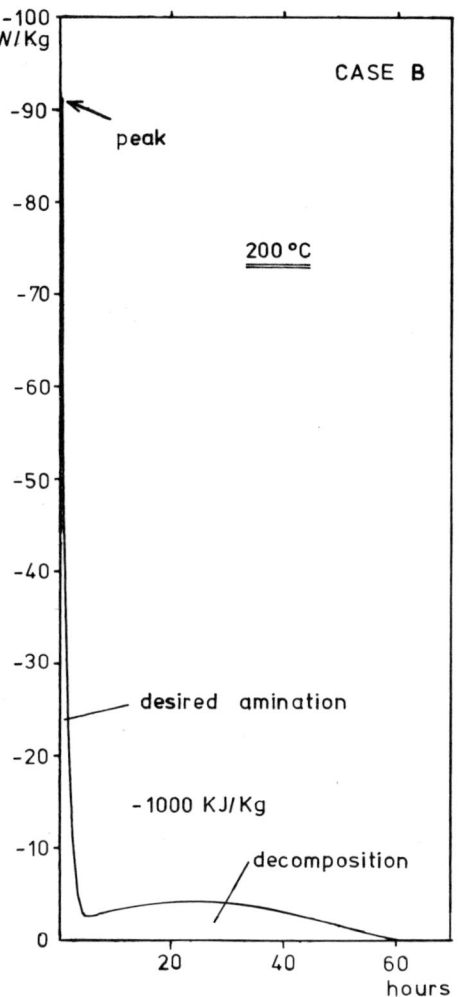

Fig. 6: *Isothermal experiment on reactant mixture B run at 200 °C.*

Concerning the amination reaction, case B, an isothermal experiment with the reactant mixture at 200 °C (Fig. 6) indicates that the decomposition reaction partially overlaps the desired reaction and produces all of its total exotherm of - 1000 kJ/kg within 65 hours even if the temperature limit of 200 °C is not exceeded. The huge amounts of heat involved originating both from the amination and the decomposition of the nitroaromatic compound necessitate the existence of very effective cooling which is fairly easily realized in the miniature scale of the DSC experiment but involves a major investment on the plant level. Safe performance of the process is therefore possible but relies heavily on technical means allowing to dissipate the heat under any circumstances. It is clear from our investigations, however, that the consequences of a runaway situation in this case will be severe.

## Kinetic Aspects of Decomposition Reactions

Most nitroaromatic compounds dissolved in anhydrous sulfuric acid, a reaction mass very typical for the dyestuff industry, decompose in the characteristic way exemplified by the series of isothermal DSC traces reproduced in Fig. 7. The mixture decomposes according to a self-accelerating mechanism formally identical to autocatalysis. Experimentally, the time to maximum heat evolution is observed to increase with decreasing temperature. Using a formal kinetic model, this regularity may be used to interpolate the results and, by mathematical simulation, to apply them to conditions other than isothermal; e.g. runaway situations may be modelled (3).

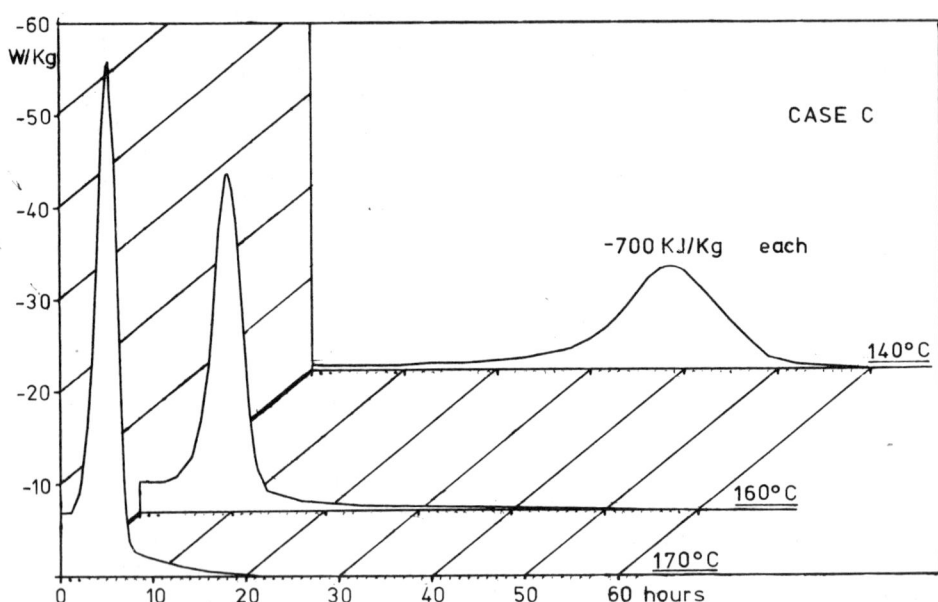

*Fig. 7:* Three-dimensional representation of isothermal runs at different temperatures (samples drawn after the completion of the desired reaction).

## Comparative Thermal Stability Tests

When comparing different compounds, say, the reactants and the products of a process, the time to maximum heat evolution may serve as a measure of their relative stabilities and thus give indications of what are the most critical phases of the process with respect to decomposition.

Isothermal experiments have also proven to be fairly sensitive to influences of catalysts on the dynamic stabilities of a given compound. Specific steel particles or other contaminants are sometimes added to the samples to test for possible influences of reactor construction materials and impurities present in the reaction mass.

Discussion and Conclusions

By using the DSC method in the way described we try to gain a general idea of the thermal behaviour of the materials we are dealing with and to establish a rough but consistent scenario of the thermal course of a reaction process, including possible undesired occurences. This is achieved by selecting reaction masses at strategic steps of a planned procedure, using the DSC technique to get a handle on the potential heat content left in the sample, and by performing isothermal experiments to gain insight in the magnitude and the time dependence of an ongoing heat evolution. A statement concerning stability - which is not regarded as an absolute quantity - depends not only on the properties of the test equipment but also on the plant situation. Knowlegde gained from DSC experiments in combination with facts known about the plant circumstances in most cases allows to draw practical conclusions in collaboration with the chemist responsible for the plant. Typically, a plant procedure is not run very closely to critical conditions which would necessitate much more accurate thermal investigations.

The microthermoanalytical approach using DSC is in many cases an efficient alternative to an attempt to design experiments which duplicate plant conditions as closely as possible. These are often very difficult to achieve. On the other hand, we are well aware of the problems stemming from possibly poor representativeness of the small samples used in DSC and the fact that unstirred samples may react differently from stirred ones. Problems arising from heterogeneous mixtures can in most cases be solved by refined sample preparation techniques.

In addition to relying on microthermoanalytical results, our thermal hazards testing laboratory takes advantage of heat flow calorimetry (4), whenever the specific way in which the desired reaction is conducted is important. Heat accumulation tests (5) are performed when a judgement of long term stabilities is essential, adiabatic experiments (4,6,7) give information on the runaway dynamics of chemical reactions and thermomanometric tests (8,9) serve to investigate hazards originating from vapour pressure changes and gas evolutions.

REFERENCES

(1)  F. Brogli, R. Gygax, M.W. Meyer, this conference
(2)  Experimental procedure see ref. (1)
(3)  R. Gygax, unpublished results
(4)  W. Regenass, this conference
(5)  Jahresberichte 1968 und 1972, Bundesanstalt für Materialprüfung, Berlin
(6)  D.J. Townsend, Chem. Eng. Progr. $\underline{73}$ (1977) 80
(7)  L. Hub, Dissertation Nr. 5577, ETH, Zürich (1975); Proceedings 3rd International Symposium on the Prevention of Occupational Risks in the Chemical Industry, Frankfurt/M (1976) p. 141
(8)  R. Janin, ibid, p. 89
(9)  A.V. Zatka, Thermochimica Acta $\underline{28}$ (1979) 7

DIFFERENTIAL SCANNING CALORIMETRY - A POWERFUL SCREENING
METHOD FOR THE ESTIMATION OF THE HAZARDS INHERENT IN INDUS-
TRIAL CHEMICAL REACTIONS

Franz Brogli, Ruedi Gygax, Max W. Meyer
Central Safety Research Laboratory,
CIBA-GEIGY Ltd., Basle, Switzerland


ABSTRACT

The systematic assessment of risks inherent in chemical reactions is essential for the safe design and operation of
chemical plant processes. It is advantageous to determine
the basic thermal data for the estimation of the two components of risk - probability and severity of an undesired
event - in a stepwise procedure.
In the first step, the screening phase, scanning microthermoanalytical methods like DSC and quantitative DTA have proved
to be powerful tools. Provided the necessary pressure tight
sample vessels are available, these methods of investigation
represent in most cases a quick and cheap way to get primary
information like heat of reaction and decomposition as well
as their respective temperature ranges, specific heat and
information on the kinetic behaviour of the chemical system
under investigation.
Selected case studies illustrate the use of these data in
combination with a priori knowledge such as the chemistry
and the physical properties of the reaction mixture, either
to estimate the risk inherent in a given chemical process
or as a guide to the design of further experiments.


Introduction

The main hazards connected with chemical production stem
from the reactivity of compounds and reaction mixtures being
processed. Under inappropriate operating conditions the process may get out of control resulting in a thermal explosion
(1) and, as a consequence, release large quantities of energy and gases within a very short time. Secondary events
of high severity may result, if the material is confined or
toxic, or if its dust or vapour are inflammable.

These hazards may originate from (i) the intended process
itself, e.g. a chemical reaction, or from (ii) unintended
secondary reactions such as rearrangement, polymerisation
and decomposition as well as reactions involving the surrounding media like solvent or air (2,3).

## Relevant Data, Safety Criteria

The aim of the thermal safety investigation is to provide reliable information on the risks inherent in a given chemical process i.e. data which allow to estimate qualitatively probability and severity of a possible event.

An essential quantity for the estimation of the severity of a given chemical process will be the maximum temperature increase to be expected under adiabatic conditions, $\Delta T_{ad}$. It allows to calculate the maximum final temperature in case of a runaway situation. Comparing this final temperature to the temperature range of secondary reactions and to the physical properties of the reaction mixture under investigation such as melting point, vapour pressure and boiling point, yields direct information about the consequences of a runaway reaction. The picture may be completed by additional elementary knowledge of the chemistry and physics of the system in question.

The difference between the range of the process temperature, taking into account possible temperature increases caused by accumulation of heat due to the intended process, and the temperature at which secondary reactions are observed in e.g. the DSC-experiment, give a first indication of the probability of an undesired event. Of course, the probability depends essentially on the way the process is conducted and on the actual plant conditions.

In most cases a quantitative knowledge of risk is not necessary, a sound idea about the nature of a possible incident is often sufficient.

## Test Procedure

For the judgement of simple physical operations (e.g. drying) usually a few characteristic material properties are adequate (4) whereas in the case of storage of bulk materials or of complex chemical reactions a detailed knowlege of the thermal and kinetic data of the process including secondary reactions may be essential.

In this situation, especially in the investigation of industrial chemical reactions, Differential Scanning Calorimetry (DSC) has proved to be a valuable screening method. With little effort and expense it provides the relevant data which, together with a priori knowledge of chemical and physical properties, allow either to assess safe process conditions immediately or to design appropriate further experiments (3).

The application of the outlined concept will be illustrated by four typical examples.

Experimental

The experiments were performed on either a METTLER DSC TA 2000 or a PERKIN-ELMER DSC-1B using gold plated sealed steel capsules with a volume of 50 µl (5). The capsules are tight to a pressure of 200 bar. The usual sample size varies between 5 to 20 mg. A standard heating rate of 4 K/min was applied throughout. The detection threshold of the experimental procedure is between 5 to 30 W per kg of the test material.

Diamino-anthraquinone

Case A: 2,6-Diamino-anthraquinone is produced from its 2,6-disubstituted precursor using a 25 % aqueous ammonia solution and a catalyst. The mixture is held at 190 °C for 18 hours in an autoclave at 40 bar. Figure 1 shows the DSC-curve of the starting materials mixed at room temperature. The only signal up to 350 °C is observed between 200 and 320 °C and corresponds to the intended amination reaction since the final reaction mixture does not show an exotherm in this range. Thus the main hazard results from the heat of amination which amounts to - 400 kJ per kg of the reaction mixture and which, in case of loss of control, could heat the mass to a final temperature of approximately 300 °C assuming a specific heat of 4 kJ/kg/K. At this temperature the vapour pressure of the mixture reaches 80 bar which would likely cause the autoclave to burst.

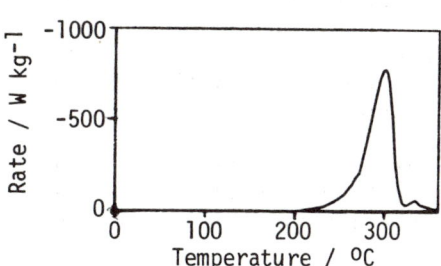

Figure 1. *DSC-curve of an amination reaction yielding 2,6-diamino-anthraquinone.*

It can be concluded that the consequences of a runaway reaction are severe and that such a situation will be initiated whenever cooling or stirring fails, unless appropriate technical measures are provided.

## Diazotization of Aromatic Nitrocompounds

Case B: A nitroaromatic amine is diazotized in an organic solvent. At a temperature of 5 to 10 °C sodium nitrite is added over a period of one hour. The DSC-curve of figure 2 was obtained by preparing the starting materials in the test capsules at -10 °C at which temperature the DSC-measurement was started. The diazotization of the amine can not be observed under these conditions indicating that the intended reaction is fast and therefore the rate is feed controlled at the proposed plant operating temperature. The small signal (- 80 kJ per kg of the mixture) between 40 and 80 °C could cause an adiabatic temperature rise of about 40 to 50 °C (estimated specific heat 2 kJ/kg/K). By experience and from computer simulation it is known that runaway situations which are induced by self-heating do not proceed at fast rates if the total temperature increase is in the range of some 50 °C (6). If the process is run below 10 °C the decomposition of the diazonium salt solution which, of course, may take place slowly even at this temperature might self-heat under unfavourable conditions to 50 to 60 °C over a long period of time. At this temperature at which the rate constant would become important the product is almost entirely consumed.

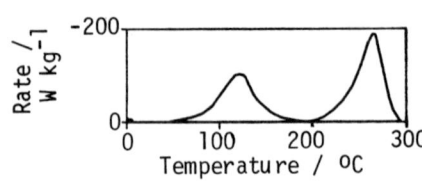

Figure 2. *DSC-curve of decomposition reactions of a diazonium salt solution in an organic solvent.*

To conclude, the diazotization reaction is feed controlled at the process temperature. The decomposition reaction which becomes relevant only some 20 to 40 °C above the process temperature is likely to be induced but does not take a violent course. The worst thing to happen would be foaming of the reaction mixture and frothing over, eventually.

Case C: In contrast to case B a different diazotization reaction (figure 3) which is run at 25 °C in concentrated sulfuric acid with a total reaction period of 40 minutes shows a first signal between 60 and 150 °C (- 390 kJ/kg) and a second one between 150 °C and 225 °C (- 460 kJ/kg). The final reaction mixture shows an identical DSC-curve. The adiabatic temperature increase due to the two exotherms amounts to 250 and 300 °C, respectively, the heat liberated by the first reaction being able to trigger the second one.

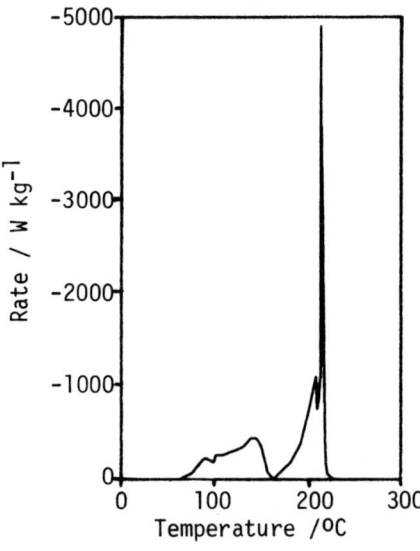

Figure 3. *DSC-curve of decomposition reactions of a diazonium salt solution in concentrated sulfuric acid.*

The diazotization reaction is, as in the previous example, feed controlled. The secondary reactions, becoming relevant 20 to 40 °C above the process temperature, would proceed in two stages under runaway conditions, liberating large quantities of gas within a short time. Nitrogen is produced, the sulfuric acid is evaporated and possibly gaseous decomposition products are produced as well. Facing these consequences appropriate technical measures have to be provided in order to prevent the temperature from exceeding 25 to 30 °C.

## Sulfonation of Aromatic Nitrocompounds

Case D: It is common practice to produce 2-chloro-5-nitro-benzene-sulfonic acid by mixing chloro-4-nitrobenzene and fuming sulfuric acid at 50 °C and initiating the sulfonation reaction by heating the mixture to 110 to 120 °C. The sulfonation is completed after a holding period of up to 8 hours at this temperature. The DSC-curve (figure 4) of the initial reaction mixture shows that the sulfonation takes place at a reasonable rate above 100 °C (- 85 kJ/kg). Above 200 °C a violent decomposition reaction (- 845 kJ/kg) is observed under the experimental conditions. The adiabatic temperature increases due to sulfonation and decomposition are 50 and 470 °C, respectively. The shape of the decomposition signal in figure 4 is typical of an autocatalytic reaction (7) indicating a large apparent activation energy in the temperature programmed experiment. In the case of autocatalysis decomposition might become relevant some 50 to 100 °C below the onset temperature observed in the temperature programmed DSC-curve after a temperature dependent induction period.

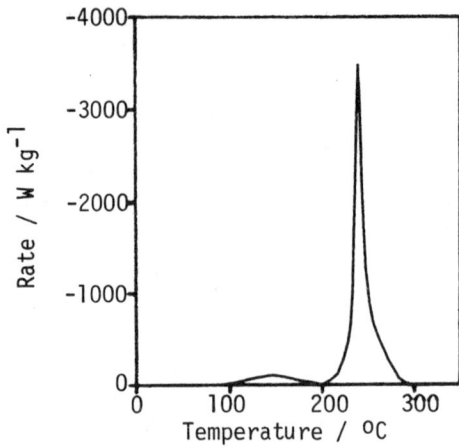

Figure 4. *DSC-curve of the sulfonation of chloro-4-nitrobenzene and of the decomposition of the final reaction mixture.*

It is likely that under unfavorable conditions (e.g. energy failure) the heat liberated by the sulfonation reaction triggers the decomposition reaction. Once the decomposition has started it will accelerate rapidly and finally produce large amounts of gaseous products (decomposition compounds, evaporated sulfuric acid). It is obvious that this is a case where the process conditions have to be carefully optimised using more sophisticated methods (8) and taking into account dangerous decomposition reactions as well (3,9).

References

(1) P. Gray and P.R. Lee, Oxidation and Combustion Reviews, Vol. 2, p. 1, Elsevier Publ. Comp., Amsterdam (1967)

(2) K. Eigenmann, Proc. 2nd Int. Symp. on Loss Prevention and Safety Promotion in the Process Industry, p. III - 124, Heidelberg (1977)

(3) F. Brogli and K. Eigenmann, Proc. "Colloque sur la Sécurité dans l'Industrie Chimique", p. 54, Mulhouse (1978)

(4) J. Lütolf, Staub, Reinh. Luft 31 (1973) 93

(5) K. Eigenmann, P. Jordi and V. Stocks, to be published

(6) F. Brogli and K. Dixon-Jackson, unpublished results

(7) F. Brogli, P. Grimm, M.W. Meyer and H. Zubler, Proc. 3rd Int. Symp. Loss Prevention and Safety Promotion in the Process Industry, Basle (1980)

(8) W. Regenass, these proceedings

(9) R. Gygax, W.M. Meyer and F. Brogli, these proceedings

# NON-ISOTHERMAL KINETICS – A GENERALIZED APPROACH

Hem Shanker Ray
Department of Metallurgical Engineering
Indian Institute of Technology, Kanpur 208016, India

## ABSTRACT

The article describes a generalized approach on non-isothermal kinetic investigations which use large samples and nonlinear heating programme. It proposes a moving boat experiment in which a large volume of sample is spread thinly in a long boat and the boat introduced into a furnace hot zone at a uniform speed. When the boat is withdrawn, it gives a series of successive volume elements which have been subjected to an identical temperature-time programme but for different periods provided heat conduction along the bed is small.

The heating programme, which is variable, would depend primarily on the temperature profile of the furnace, the boat speed and the heat transfer coefficient. The temperature-time plots can be obtained by having thermocouples positioned in the moving boat.

Derivation of nonisothermal kinetic equations for some nonlinear programming is indicated

## INTRODUCTION

Kinetic data obtained under rising temperature conditions are analyzed by using a combination of the following basic equations.

(a) The kinetic law
(b) The Arrhenius equation for rate constant
(c) Equation describing variation of temperature with time

If heating rate is constant, as is the case in thermogravimetric (TG) experiments then the basic equation for analysis of TG data is written as

$$g(\alpha) = \int_0^\alpha \frac{d\alpha}{f(\alpha)} = \frac{A}{B} \cdot \int_0^T \exp(-E/RT) \, dT \qquad (1)$$

where $g(\alpha)$ is a function as defined. If the form of $f(\alpha)$ is known then $g(\alpha)$ is easily determined. It is, however, difficult to evaluate the R.H.S. integral and only approximate solutions can be considered.

Traditionally rising temperature experiments have been mostly restricted to very small samples and linear rates of heating only. The importance of a more general approach need not be overemphasized. The present article outlines an experimental approach which allows rising temperature experiments with large samples and also indicates some methods of analyzing data when the heating programme is not linear but complex.

## EXPERIMENTAL PROCEDURE

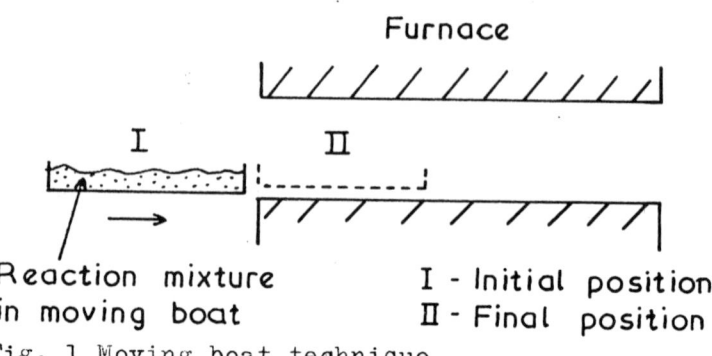

Fig. 1 Moving boat technique

Fig. 1 outlines an arrangement which can be used to impose a well defined heating programme on a large sample. The set-up employs a furnace at a constant temperature and a moving boat in which the reaction mixture (sample) is spread as a thin layer. The leading edge of the boat initially rests at the furnace mouth. It is then introduced into the furnace at a constant speed using a special movement control unit. As soon as the entire boat goes into the furnace it is quickly withdrawn and the reaction mass quenched by an appropriate technique.

In this technique the reaction time of a volume element is known from the total length of the boat, the total boat travel time and the location of the volume element. Analysis of the reaction mixture, from various locations, therefore, yields kinetic data for simultaneous reaction of a large number of volume elements at different temperatures. The heating programme can be changed widely by changing boat speed. It can be measured by employing a thermocouple. The use of the concept of moving boat experiments in studying high temperature heterogeneous reactions in general has been discussed in detail elsewhere[1].

## THEORY

Suppose that temperature changes exponentially.

$$T = T_f - (T_f - T_o) \exp(-qt) \qquad (2)$$

The nonisothermal kinetic equation for this exponential temperature rise is obtained as

$$\int \frac{d\alpha}{f(\alpha)} = \frac{AE}{Rq} \int \frac{e^{-E/RT}}{(T_f - T)} \cdot dT$$

$$= -\frac{AE}{Rq}\left\{\left[\log\left|\frac{1}{T}\right| - \frac{E}{RT} + \frac{1}{2 \cdot 2!}\left(\frac{E}{RT}\right)^2 - \frac{1}{3 \cdot 3!}\left(\frac{E}{RT}\right)^3 + \ldots\right]\right.$$

$$- e^{-E/RT_f}\left\{\log\left(\frac{T_f - T}{T}\right) - \frac{E}{RT_f}\left(\frac{T_f - T}{T}\right)\right.$$

$$\left.\left.+ \left(\frac{E}{RT_f}\right)^2 \cdot \left(\frac{T - T_f}{T}\right)^2 \cdot \frac{1}{2 \cdot 2!} - \left(\frac{E}{RT_f}\right)^3 \cdot \left(\frac{T_f - T}{T}\right)^3 \cdot \frac{1}{3 \cdot 3!} + \cdot\right\}\right] \quad (3)$$

**A more general heating programme: The S-shaped variation of temperature with time**

Suppose now that a volume element enters a furnace hot zone which is not isothermal. If the furnace temperature profile is exponential like then the furnace temperature $T_f$ at a given distance l within the hot zone is given by

$$T_f = T_M(1 - e^{-jl}) \quad (4)$$

where $T_M$ is the maximum temperature and j is a constant. If the speed of the volume element be v then l equals vt. Variation of temperature with time is now expressed approximately as

$$T = \frac{T_M}{q - jv}\left[q(1-e^{jvt}) - jv(1-e^{-qt})\right] \quad (5)$$

This indicates a S shaped curve which is a more general heating programme. Unfortunately it is inconvenient to incorporate this equation into the kinetic equation.

Polynomial expression for time-temperature relationship

Assume that the S shaped temperature variation is expressed approximately by a polynomial

$$t = -A' + B'T - C'T^2 + D'T^3 \quad (6)$$

where $A'$, $B'$, $C'$ and $D'$ are constants (positive numbers). In this case we have

$$\int \frac{d\alpha}{f(\alpha)} = \frac{AE}{R(m-n)}\left[e^{-\frac{E}{Rm}}\left\{\log\left|1 - \frac{m}{T}\right| - \frac{E}{RT}\left(\frac{T}{m} - 1\right) + \frac{E^2}{R^2 T^2}\left(\frac{T}{m} - 1\right)^2\right.\right.$$

$$\left.\frac{1}{2 \cdot 2!} + \ldots\right\} - e^{-\frac{E}{Rn}}\left\{\log\left|1-\frac{n}{T}\right| - \frac{E}{RT}\left(\frac{T}{n}-1\right) + \frac{E^2}{R^2T^2}\left(\frac{T}{n}-1\right)^2\right.$$
$$\left.\frac{1}{2 \cdot 2!} + \ldots\right\}\bigg] \tag{7}$$

This equation is similar to eq. (3) where $\frac{T-T_f}{T}$ is replaced by $[T-m \text{ (or } n)]/m \text{ (or } n)$.

Eq. (7), is valid only if the roots m and n are real and positive. If they are not real then a numerical integration procedure will be required.

Series summation

Eq. (3) and (7) both involve series summations of the type

$$\sum = w + \frac{w^2}{2 \cdot 2!} + \frac{w^3}{3 \cdot 3!} + \frac{w^4}{4 \cdot 4!} + \ldots \tag{8}$$

where w is a variable (generally negative).

This series has been evaluated for some values of w using a computer. It has been found that for negative values of w the summation approaches a finite value which is rather small if w is less than 40 beyond which there is a rapid increase. It has also been found that $\Sigma$ approaches a finite value even for positive values of w provided it is not too large.

The graphical approach

Preceeding discussions show that the evaluation of exponential integral in the nonisothermal kinetic expression becomes rather difficult when the heating rate is nonlinear. A simple graphical procedure can be employed to evaluate the integral for any heating programme provided the time-temperature relationship is accurately known.

If it is assumed that the activation energy remains unchanged throughout then one can calculate the values of exp.(-E/RT) at various values of t by considering the instantaneous values of temperature. The total shaded area of a graph between exp(-E/RT) against t then gives

$$\int_0^t \exp.(-E/RT).dt$$

To obtain the value of E from nonisothermal kinetic data one should first evaluate $g(\alpha)$ from the known value of for a given value of t. The value of A must be known. Then $g(\alpha)/A$ gives the value of the integral. It will now be necessary to obtain the shaded area for a set of E values then choose, by using a calibration procedure, the E value which gives the correct area equal to the integral. This is the activation energy provided it has remained constant through out the course of heating.

Symbols

$A$ = Preexponential constant in Arrhenius equation
$B$ = Heating rate (constant)
$B'$ = A constant
$C'$ = A constant
$D'$ = A constant
$E$ = Activation energy (constant)
$f$ = A function
$g$ = A function
$j$ = A constant
$k$ = Rate constant
$l$ = Distance inside furnace hot zone
$n,m$ = Real and positive roots of eq. $B'-2C'T+BD'T^2 = 0$
$q$ = A constant
$R$ = The gas constant
$t$ = Time
$T$ = Temperature
$T_o$ = Initial temperature
$T_f$ = Final (furnace) temperature
$T_M$ = Maximum temperature
$v$ = Velocity of boat in moving boat experiment
$w$ = A variable
$\alpha$ = Fraction reacted

References

1.  H.S. Ray & P. Sewell, 'Kinetic studies under nonisothermal conditions', Proc. Int. Conf. on Advances in Process Metallurgy (ICMS-79), Bhabha Atomic Research Centre, Trombay, Bombay 400 085, India, Jan. (1979).

# INDUSTRIAL EXPERIENCE WITH HEAT FLOW CALORIMETRY

W. Regenass

CIBA-GEIGY, Basle, Switzerland

## ABSTRACT

A "bench scale heat flow calorimeter" designed for hazard analysis and process design of industrial organic reactions has been in use at Ciba-Geigy for more than ten years. The experience in the application of this instrument for organic synthesis, kinetics, hazard assessment and design data is reviewed.

## INTRODUCTION

The potential of thermokinetic methods for process research and development is widely recognized (1), (2), but their application in industry is still sparse and often confined to safety investigations. This is mainly due to the lack of appropriate instrumentation which fits the needs of the industrial investigator.

An instrument which fills this gap has been described some years ago (3), (4), (5). This "bench scale heat flow calorimeter", which was developed at Ciba-Geigy in the late 60's has now found a wide acceptance within the Ciba-Geigy group. More than 20 units are in use, mainly in process development laboratories and in hazard testing facilities. Due to its wide application and due to a number of fundamental studies (6), (7), (8), (9), (10) its virtues and also its limitations are well known by now.

A brief description of its design and a review of its applications are given here.

## INSTRUMENTATION

In the experience of the author, bench scale heat flow calorimetry is the thermal method most suited for the investigation of industrial organic reactions, because all standard operations carried out with industrial stirred tank reactors can be performed. Conventional micromethods with passive or active heat flow (DTA, DSC) are used for screening.

A discussion on accumulation methods versus heat flow methods (adiabatic calorimetry) and on micro methods versus macro methods is given elsewhere (4), (5). There, a review on operating principles and instrumentation for heat flow calorimetry is also given.

Among the generally used methods for active heat flow control (compensation heating, peltier heat transfer, adjustment of environment temperature) the latter one was chosen for the Ciba-Geigy instrument. Its basic design is outlined in Figure 1 (BSC = Bench Scale Calorimeter). The stirred tank reactor (A) is surrounded by a jacket in which a heat transfer fluid is circulated at a very high rate. A cascaded controller (B) adjusts the temperature of the circulation loop (C) so that heat transfer through the reactor wall equilibrates the heat evolution in the reactor. Injection of thermostated hot or cold fluid is used to adjust the temperature in the loop.

The rate of heat transfer q (which equals the rate of heat evolution) is related to the observed temperature difference ΔT between the jacket fluid and the reaction mixture by the relation

$$q = U \cdot A \cdot \Delta T = f_c \cdot \Delta T \quad (1)$$

where the calibration factor $f_c$ is the product of U, the overall heat transfer coefficient, and A the active (= wetted) heat transfer area. Because both A and U depend on the reactor contents and on the stirring conditions, specific calibration is required. This is done by producing a known heat input rate to the reaction mixture by means of an electric heater (D). The need of frequent calibration is of some inconvenience as compared with heat balance calorimeters. On the other hand, the method chosen permits the use of an uninsulated glass reactor and thus allows visual observation of phase changes, colour changes and mixing conditions. This is a distinct advantage for process development work.

The 1975 series of instruments is equipped with a refrigeration unit, electronic controls for temperature programming, automatic calibration and with feeding systems for gases, solids and liquids.
The specifications are as follows:
reactor temperature       : -20 to 200°C
temperature programm      : ± 1 to 200°/hour
pressure (glass reactor)  : -1 to 2 bar
volume of reactor         : 0.5/2.5 litres (exchangeable)
volume of reactant        : 0.3-2.5 litres
sensitivity               : 0.5 Watts for low viscosity reaction mixtures
heat removal capacity     : 500 Watts (for temp. >30°C)
response time (to a step change of the heat release rate):
20 seconds for 50%, 200 seconds for 99% of the full signal.

Special units for high temperature (250°C), low temperature (-60°C, ref. 7) and moderate pressure (50 bar) are also in use.

A new modular design with digital electronics is currently being implemented (Figure 2).
The basic module is the reactor unit containing a very fast thermostat with a programmable digital controller. Units for a) refrigeration, b) automatic calibration, c) reflux, d) programmable variation of stirrer speed and e) automatic data acquisition and evaluation, may be added as options.

Many instruments with similar goals have been described in the literature in the past ten years (11,12,13,14); none of them is available commercially. On the other hand, there is a distinct need for better instrumentation in reaction calorimetry, particularly for safety investigations. Therefore, the Ciba-Geigy instrument will be made available for third parties.

EXPERIMENTAL RESULTS AND THE EVALUATION OF THERMAL AND KINETIC DATA

The record of a temperature programmed decomposition of a diazonium-salt in aqueous sulfuric acid solution (Figure 3) illustrates the primary information obtained from a heat flow experiment.

The rate of heat evolution which is of importance for safety and design considerations, is obtained immediately from the observed temperature difference and calibration.

The total heat of reaction follows by integrating the area under the heat flow curve.

Heat capacities and specific heats are obtained from temperature programmed runs. When the rate of imposed temperature change, $\dot{T}$, is altered, there is a step change in heat flow (S in Fig. 3).

$$\Delta q = \Delta(\dot{T}) \cdot (w + m_r c_{p,r})$$

In this relation, w denotes the proportionate heat capacity of the calorimeter (a quantity which depends on temperature and on the volume of the calorimeter contents and has to be calibrated for a specific reactor), $m_r$ and $c_{p,r}$ are the mass and the specific heat of the mixture under investigation.

Heat transfer data can be calculated from the calibration factor $f_c = U \cdot A$, as outlined in citations (5) and (9).

Automatic data reduction (calculation of true heat release rates from recorded temperature differences, determination of heats of reaction by integration, etc.) is not trivial because of complex baseline changes. These are less severe than in micro heatflow calorimetry (DTA, DSC) but quite disturbing nonetheless. Computer programs for true heat release rates and true heats of reaction are incorporated in the new modular instrument.

From such reduced release data, it is easy to calculate the kinetic parameters (rate constants, activation energies, etc.). For statistical parameter estimation (using numerical integration of the differential equations defining the reaction model), the reduced data may be transferred to a larger computer system.

## ORGANIC SYNTHESIS

This is by far the most important application of our BSC's. The experimental worker can immediately see when a reaction starts and when it is completed, and information on the influence of temperature and of the concentration of reactants and catalysts on the reaction rate is easily obtained. Furthermore, the BSC is a very convenient automatic reactor unit. Its use considerably reduces cost and time required to optimize a procedure.

## KINETICS

The elaboration of kinetic models and their parameters is one of the primary aims of all the heat flow calorimeters described in the literature (1). The most common objections to the use of thermal methods for kinetic work are related to the facts that most reactions are complex and that heat evolution is very unspecific information. However, in the authors experience of several hundred reactions of widely different kinds, very often the main reaction thermally dominates to such an extent, that the influence of concentration and temperature on the reaction rate can be obtained by heat flow experiments alone. In most other cases, thermal data provides valuable information additional to the data obtained by classical means, a fact which considerably

speeds up kinetic work. Of course, thermal methods are not appropriate for selectivity determination.
Only few kinetic investigations by means of the BSC have been published so far:
- the Arbusov isomerisation of trimethylphosphite (4,6)
- the hydrolysis and aminolysis of some acid anhydrids and chlorides (7)
- the oxidation of aromatic methyl groups by molecular oxygen (10,16).

The latter example

$$Ar-CH_3 \xrightarrow{O_2} \text{Intermediates} \xrightarrow{O_2} Ar-COOH$$

which is related to important industrial processes (benzoic acid, terephthalic acid) is particularly interesting. This free radical chain reaction is catalyzed by cobalt and bromine salts; acetic acid is the solvent commonly used. When the ratio of catalyst to aromatic compound is applied as indicated in the patents, the reaction order with respect to the methyl-aromatic compound is - after an induction period - almost zero. When the catalyst concentration is reduced, a dramatic change of the reaction rate pattern is observed (Fig. 4): whereas the average reaction rate is low, sharp rate maxima (up to more than 50 times the average rate) are found. This very surprising reaction behaviour would not have been detected by classical means (e.g. periodic GLC analysis of the reaction mixture). This phenomenon, which is caused by chain branching has not been reported in the literature to date.

## ASSESSMENT OF THERMAL HAZARDS

Thermal explosions do occur, when the heat evolution rate of a reaction (with a high latent adiabatic temperature rise) exceeds the heat transfer capacity of the reactor. Events of this type which have happened in industry, may be divided into two classes:
1) exothermic decomposition or polymersation of thermally instable mixtures (e.g. nitro compounds, benzyl-halides, etc.)
2) "run away" of an intended reaction.

Type 1 hazards are easily assessed using micro heat flow (DSC, DTA) or heat accumulation methods (15). However, the elaboration of safe reaction conditions (which avoid type 2 hazards) is still a problem. Whenever possible, highly exothermic reactions are performed in such a way, that the reactants disappear by reaction as they enter the reactor (semibatch or continuous operation). Under these conditions, any accumulation of reactants in the reactor is potentially hazardous. It may have several causes, e.g.
21) too low temperature
22) insufficient mixing
23) wrong kinetic assumptions (a case often encountered in process development: the reaction is assumed to be fast and the heat evolved after the addition of the reactants is not noticed on laboratory scale; then at the pilot stage, there is a run away).
24) incorrect initiation

Even slightly exothermic reactions may become dangerous, when at a higher temperature an exothermic decomposition is triggered off. For this reason, the potential adiabatic temperature rise of industrial reactions and also the heat release rate as a function of temperature, concentrations, etc. are of fundamental interest. When synthetic work is done in the bench-scale-calorimeter, the required data are obtained without additional effort.

## SCALING UP AND PLANT DESIGN

The main problems in scaling up chemical reactions are mixing and heat transfer. Mixing problems arise mainly in connection with viscous media and/or very fast reactions; as a rule, their study requires experiments on an intermediate scale or even on plant scale. Heat transfer problems are generally met (particularly for reactions in stirred tanks) due to the decrease of the specific heat-transfer-area with increasing reactor size; however, heat transfer problems can be solved by bench experiments and calculation.

As mentioned above, heat release rates as well as heat transfer coefficients are obtained from BSC-experiments.

The overall heat transfer coefficient U is the inverse of the sum of the resistances of the wall and of the liquid films on either side of the reactor wall

$$\frac{1}{U} = \frac{1}{h_j} + \frac{s}{\lambda} + \frac{1}{h_r}$$

Whereas the thermal resitance of the wall and the film in the reactor jacket are usually known or easily determined experimentally, the reaction side film resistance is specific to each reaction mixture. A procedure for determining the characteristic fluid properties and for calculating film coefficients of a specific reaction mixture in any reactor under various flow (stirring) conditions is given in citations (5) and (9).

Because heat transfer coefficients can be influenced by reactor geometry and stirring speed only to a limited extent the scale up problem of matching heat release rates and heat transfer capacities, must be solved by influencing the heat release rate (or by increasing the heat transfer surface).

Industrial batch reactions are - if possible - controlled by the addition of one component. Due to the facts described above, the duration of the addition is, as a rule, much longer on plant scale than in conventional laboratory procedures. It is important that process optimization is performed for conditions which are feasible on an industrial scale, because the addition rate influences not only the heat release rate but in many cases also selectivities (formation of by-products, etc.).

In our standard scale-up procedure, we first determine heats of reaction and heat transfer capacities and calculate from this data the required duration of the dosing. In this context, the BSC is an ideal minipilot reactor, which permits a direct scale-up to plant scale, where conventional means would necessitate a pilot stage. Here again, it reduces drastically time and cost for process development.

## REFERENCES

(1) F. Becker; Chem. Ing. Techn. 40 (1968), 933
(2) F. Brogli and K. Eigenmann; Coll.sur la sécurité dans l'industrie chimique, Mulhouse (France), 27.9.1978

(3) W. Regenass, W. Gautschi, H. Martin and M. Brenner,
    Proc. 4th Int.Conf.Thermal Anal. 3, 834, Budapest 1974
(4) W. Regenass, TCA 20 (1977), 65
(5) W. Regenass, ACS Symp. Ser. 65 (1978), 37
(6) H. Martin, Ph.D.Thesis, Basel, 1973
(7) W. Kanert, Ph.D.Thesis, Basel, 1977
(8) W. Gautschi, Ph.D.Thesis, ETH Zürich, 1975
(9) M. Bürli, Ph.D.Thesis, ETH, Zürich, 1979
(10) J. Beyrich, Ph.D.Thesis, ETH, Zürich, to be published
(11) W. Köhler et al., Chem.Ing.Techn. 45 (1973), 1289
(12) L. Hub., Ph.D.Thesis, ETH, Zürich, 1975
(13) J. Schildknecht, 2nd Int.Symp. on Loss Prevention, Heidelberg, 1977
(14) B. Hentschel, Chem.Ing.Techn. MS 725/79
(15) T. Grewer, Chem.Ing.Techn. 51 (1979) 928
(16) J. Beyrich, W. Regenass and W. Richarz, Chimia 34 (1980), May-issue

FIGURES

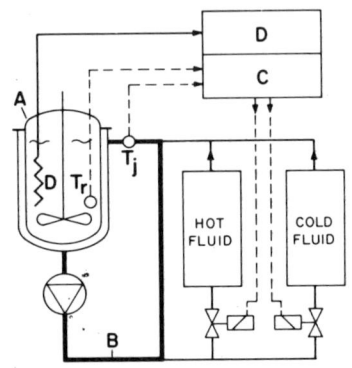

Fig. 1: BSC Operating Principle

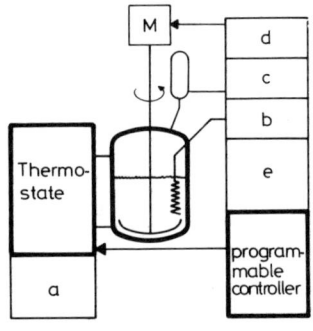

Fig. 2: BSC Modular Design

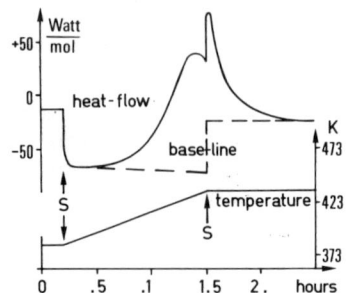

Fig. 3: Decomposition of
        Diazonium Salt

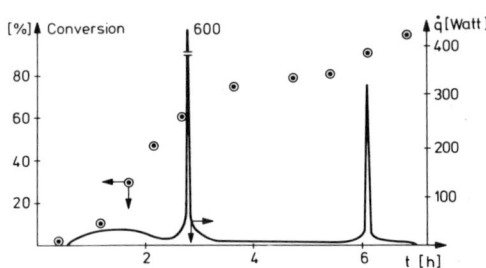

Fig. 4: Oxidation of
        4-Chloro-toluene

# THERMOANALYTICAL STUDY ON FERTILIZER COMPONENTS AND THEIR MIXTURES

Maija-Liisa Friman,[a] Markku Leskelä, and Lauri Niinistö
Department of Chemistry, Helsinki University of Technology,
SF-02150 Espoo 15, Finland
Erkki Aalto, Antero Hörkkö, and Juhani Poukari
Research Centre of the Kemira Oy, P.O.Box 14, SF-02271
Espoo 27, Finland
[a]Present address: Kemira Oy, Harjavalta Plant, P.O.Box 13, SF-29201 Harjavalta, Finland

## ABSTRACT

Thermal decomposition of several fertilizer components and their most important mixtures has been studied by simultaneous recording of the TG, DTG, and DTA curves up to 900 °C. The seven mixtures studied were $NH_4NO_3+(NH_4)_2SO_4$, $NH_4NO_3+KNO_3$, $0.7NH_4NO_3 \cdot 0.3KNO_3+2NH_4NO_3 \cdot (NH_4)_2SO_4$, $0.7NH_4NO_3 \cdot 0.3KNO_3+NH_4H_2PO_4$, and $(K,NH_4)_2(NO_3)_2SO_4$ with $2NH_4NO_3 \cdot (NH_4)_2SO_4$, $NH_4H_2PO_4$, or $0.7NH_4NO_3 \cdot 0.3KNO_3$. In addition to simple mixtures, synthetic and industrial compound fertilizers with varying N:P:K ratios were studied.

The effect of additional elements (Cu, Fe, Mg and B) and of experimental conditions, such as the amount of sample and the heating rate, was investigated for a synthetic fertilizer preparation containing $N:P_2O_5:K_2O$ in the ratio 20:10:10.

## INTRODUCTION

During the last few years considerable attention has been paid to the safety aspects involved in the manufacture, storage and transportation of fertilizers. Especially compound fertilizers with a high ammonium nitrate (AN) content have received attention because of the hazardous nature of AN. This compound may bread into spontaneous combustion or, in the worst instance, explode (1), (2). Certain additives, like copper and chloride ions, are known to promote the thermal decomposition of AN (3),(4), but the factors affecting the thermal behaviour of compound fertilizers are much less well understood.

The present investigation was initiated to obtain, through systematic study involving combined thermoanalytical methods, thermal data which might prove useful and contribute to a higher level of safety in the fertilizer industry.

## EXPERIMENTAL PROCEDURE
### Materials

Most of the single components studied were double salts prepared from analytical grade reagents in appropriate proportions by crystallization form aqueous solution. The purity of the crystalline products was checked by X-ray diffraction.

The binary and complex mixtures were prepared from these compounds. In addition, industrial compound fertilizers from the Siilinjärvi and Kokkola plants of Kemira Oy were studied.

### Methods

TG, DTG and DTA curves were recorded simultaneously in a dynamic air atmosphere (90 $cm^3$/min) with a Mettler TA-1 Thermoanalyzer. The finely ground samples were packed uniformly in standard platinum crucibles (diam. 7 mm, height 19 mm). In DTA measurements alumina was used as reference material.

Several heating rate/sample weight combinations were tested. Generelly, for the single phases and compound fertilizers two combinations, 2 $^\circ min^{-1}$/20 mg and 10 $^\circ min^{-1}$/200 mg were used. For mixtures only the intermediate combination of 6 $^\circ min^{-1}$/100 mg was employed, in order to reduce the time needed for measurements.

## RESULTS AND DISCUSSION
### Single phases and binary mixtures

Table 1 lists the fertilizer components and the binary mixtures included in the study. The TG curves are relatively simple; there is one major weight change which occurs generally at 200-300 °C. DTA curves may contain several peaks as phase changes are also registered. By way of example the curves for $(K,NH_4)_2(NO_3)_2SO_4$ are shown in Fig. 1.

Table 1. Fertilizer components and their mixtures.

| Single phases | Mixtures[a] |
|---|---|
| A. $NH_4NO_3$ | A + B |
| B. $(NH_4)_2SO_4$ | A + C |
| C. $KNO_3$ | A + E |
| D. $(NH_4)_2HPO_4$ | F + H |
| E. $NH_4H_2PO_4$ | F + E |
| F. $(K,NH_4)_2(NO_3)_2SO_4$ | G + H |
| G. $0.7 NH_4NO_3 \cdot 0.3 KNO_3$ | G + E |
| H. $2NH_4NO_3 \cdot (NH_4)_2SO_4$ | |

[a] Five proportions were studied: 10/90, 30/70, 50/50, 70/30, 90/10 (w/w).

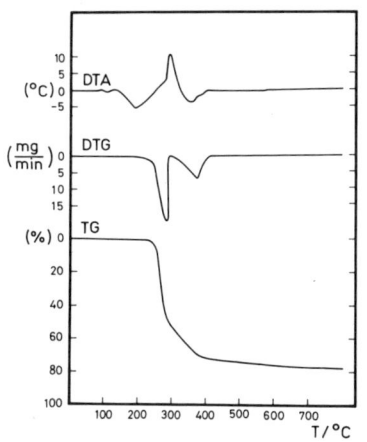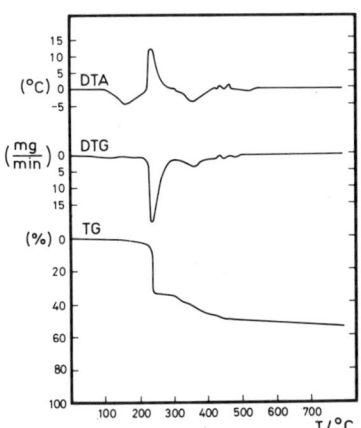

Fig. 1. Thermal decomposition of $(K,NH_4)_2(NO_3)_2SO_4$ (left) and a 15-15-15 compound fertilizer (right). Sample weights are 200 mg and heating rate is 10 °min$^{-1}$.

Compound fertilizers

Three industrial fertilizer preparations of 20-10-10, 15-15-15, and 20-10-10 composition ($N-P_2O_5-K_2O$) were studied, along with a synthetic 20-10-10 mixture. The most noteworthy feature of the TG, DTG and DTA curves of these compound fertilizers was their resemblance to the curves of $(K,NH_4)_2(NO_3)_2SO_4$; for an example see Fig. 1.

Effect of heating rate and addititives.

The heating rate had a clear but expected effect on the curves. Additives (Cu, Fe, B, Mg, oil) also tend to change their appearance. Thus, 0.1 and 1 % Cu additions lowered the DTA maximum for the 20-10-10 compound fertilizer by 10 and 20 °C, respectively. Moreover, a considerable increase in the peak height was observed. On the other hand, a 1-2 % addition of B stabilized the fertilizer slightly.

## REFERENCES

(1) G. Perbal, Proc. Fert. Soc. 1971, No. 124, 63 pp.
(2) K. Barclay, Kongr. "Chem. Polnohospod." [Pr.], 2nd 1972, 1,A7, 21 pp.
(3) A.G. Keenan, K. Notz and N.B. Franco, J. Am. Chem. Soc. 91 (1969) 3168
(4) R. Kümmel and F. Pieschel, J. Inorg. Nucl. Chem. 36 (1974) 513

# DSC MEASUREMENTS ON THE EFFECT OF ADDITIVES IN CUTTING OILS

U. Kurpjuweit, E. Wappler
W.C. Heraeus GmbH, Hanau, FRG

W. Keil
Aral, Bochum, FRG

## ABSTRACT

Cutting oils are extremely stressed by thermal and mechanical influences. To improve their efficiency, additives are added. These substances, e.g., Cl-paraffins, or compounds containing sulphur, react with the metal being worked, thus forming surface layers which are instable to shearing stress and, therefore, function like a solid film lubricant. The range of thermal stability of these reaction layers is limited. This range, the optimal concentration of additives, and the probable life of the oil can be determined by DSC measurements, which give information about the limits of application of a particular additive.

## INTRODUCTION

Thermoanalytical methods have hitherto been used mainly to investigate motor oils, gear lubricant oils, and hydraulic oils. In these cases, thermal stability, oxidation stability, and a low tendency to ageing are most important (1). Less attention has been paid to drilling, honing, lapping and cutting oils. With these oils for working metals, emphasis is laid upon a diminution of wear under a given load (2), rather than upon thermal and oxidation stability.

With both these categories of oils, the desired properties are obtained by means of additives.

It is generally assumed that the effects of additives in oils for working metals, namely a longer service life of the tools and a better quality of the product, are due to the fact that the metal to be worked forms reaction layers, instable to shearing stress, with the reactive component of the additive (3). The layers thus formed are chlorides, sulphides or phosphates, depending on the additive used. These layers are stable only in a limited temperature range; they form at a certain minimum temperature only, disintegrating again when a specific limit temperature is exceeded (4).

An additive, or mixture of additives, is especially suitable for the working of metals if it is capable of forming such reaction layers at all temperatures arising during the working process.

The methods by which the efficiency of an additive is determined are time-consuming and expensive standardized tests, e.g., the FZG test according to DIN-Standard 51354.

To limit the number of these tests, many attempts have been made to establish criteria designed to permit preselection of additives. For instance, in the case of additives containing chlorine, the release of HCl was determined at constant temperature, with the additive in contact with the material to be worked.

A rapid and informative method of characterizing additives is thermal analysis, especially differential scanning calorimetry (DSC). The reaction of the additive with the metal to be worked is an exothermic process. The characterizing values furnished by this process are the heat of reaction, the temperature limits of the reaction, and the progress of the reaction with time.

By way of example, the reaction of different additives with aluminium is to show the variation of these values.

## EXPERIMENTAL PROCEDURE

The investigation was carried out with the Heraeus TA 500 Thermal Analysis System. The measurements were made in a DSC cell in the range between 100 °C and 270 °C, with a heating rate of 10 K/min.

Samples of 15 mg of basic oil each, with different types of additives, were weighed in aluminium crucibles, which were then sealed hermetically.

The measured values were recorded by the XY-recorder incorporated in the Control and Recording Unit TA 500 S/2. The X-axis was controlled by the sample temperature measured by a Pt 100 resistance directly on the sample. The $\Delta T$ signal between the sample and an aluminium reference was recorded on the Y-axis. The resolution on the Y-axis was 0.1 mW/cm, that on the X-axis, 10 K/cm.

## RESULTS AND DISCUSSION

The shape of the curve is characterized by an increase in specific heat of the oil, leading to increased heat absorption by the sample, as compared with the reference, and, therefore, a descent of the curve (Fig. 1).

At a specific temperature, 155 °C in the present instance, the reaction of the additive with the metal, an exothermic effect, begins, producing an upward peak corresponding to the increase in temperature of the sample, as compared with the reference.

The peak area is proportional to the heat transformation. It can be indicated most conveniently with the help of the Data Processing System TA 500 DV, in J or in J/g.

There are certain additives which release large quantities of gas, causing the hermetically sealed crucible to bulge at elevated temperatures. This process manifests itself by a sharp upward drift of the curve, due to inadequate thermal contact between crucible and temperature sensor.

A number of selected examples are to show the relationship between the chemical structure of an additive and the variation of the characteristic values, viz. heat of reaction, temperature limits, and progress of reaction with time.

Figure 2 shows the reaction of two Cl-paraffins with aluminium.
One of these additives is a narrow fraction of monochlorinated n-paraffins, obtained by dewaxing mineral oil. The other is a mixture of different chlorinated n- and iso-paraffins. The chemically uniform substance leads to a narrow, high peak with a half-width value of 19 °C. The differently bound chlorine atoms of the paraffin mixture react over a wide temperature range, thus producing a peak equal in area, but broad, with a half-width value of 35 °C.

While the reaction is completed at 200 °C in both cases, the reaction of the paraffin mixture begins at 145 °C, or 18 K below the temperature at which the reaction of the uniform fraction begins.

The reaction temperatures are distinctly higher in the following example, concerning a dichlorinated aromatic compound (nitropara-dichlorobenzene) (Fig. 3).

The reaction takes place between 170 °C and 235 °C, with two maxima, at 175 °C and 203 °C respectively.

The difference in polarization of the two chlorine atoms, one of which is in an ortho-position, the other in a meta-position in respect of the $NO_2$ group, obviously accounts for the fact that the atoms react at different temperatures.

The reaction temperatures are still higher in the case of aromatic additives containing sulphur (dibenzyl disulphide) (Fig. 3). The reaction of this compound takes place in the range from 170 °C to 252 °C, with a maximum at 220 °C.

The reaction properties of various additives can be combined by mixing the additives. By mixing the aromatic compound of chlorine with the wide paraffin fraction, a very convenient reaction process is obtained, with a reaction area from 168 °C to 232 °C and a broad maximum (Fig. 4). This means that the effect of the additive is constant over a wide temperature range.

Figure 5 shows the reaction of an additive at different concentrations. What is surprising, in the first place, is the uniform heat of reaction of the additive at all concentrations between 5% and 0.6%.

The reaction is not limited by the crucible surface available. This is proved by the fact that neither the area nor the shape of the peak is changed by the addition of aluminium powder.

The heat of reaction is independent of the concentration of the additive. This invariance of the heat of reaction with the concentration of the additive is accounted for by the observation that the crucible with the highest concentration of additive is the only one to swell before the end of the temperature programme already, because it is in this crucible that the highest pressure is produced and the largest amount of gas release.

This observation agrees with the results obtained from the investigation of additives according to the classical methods. Also in this case, only a small portion of the chlorine reacts with the metal, while most of it is released in the form of gaseous HCl.

Hence it follows that a higher concentration of additives does not lead to increased reaction with the metal, but reduces the reaction temperature.

Figure 6 represents the relationship between the concentration of the additive, the temperature at the beginning of the reaction, and the maximum temperature of the reaction. The diagram shows a linear dependence of these temperatures on the concentration of the additive.

This linear relationship allows conclusions to be drawn as to the additive still contained in a used oil and, therefore, the time for which the oil may continue to be used.

The above results are to present thermal analysis as a method of characterizing additives. Within the 20 minutes required for measurement, a preliminary decision can be taken as to the use of a specific additive, and combinations of additives, assumed to be suitable, can be developed and tested for their ageing behaviour in a relatively short time.

More comprehensive investigations will follow, perhaps with wider temperature ranges, using pressure crucibles, or tubes of glass or quartz, resistant to pressure, at adjustable pressures. This allows investigations to be carried out also on additives reacting only at elevated temperatures or evolving large quantities of gases.

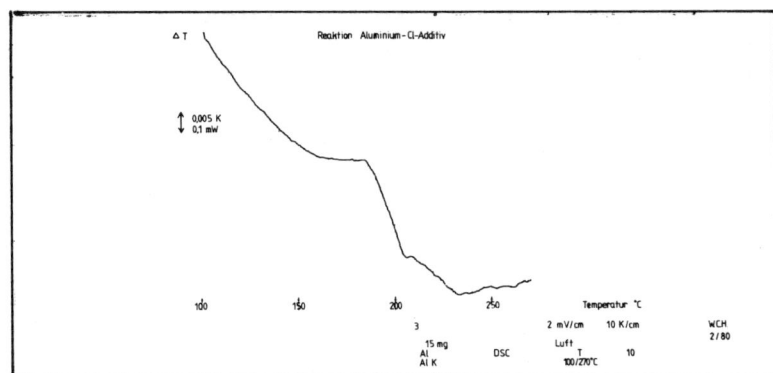

Figure 1

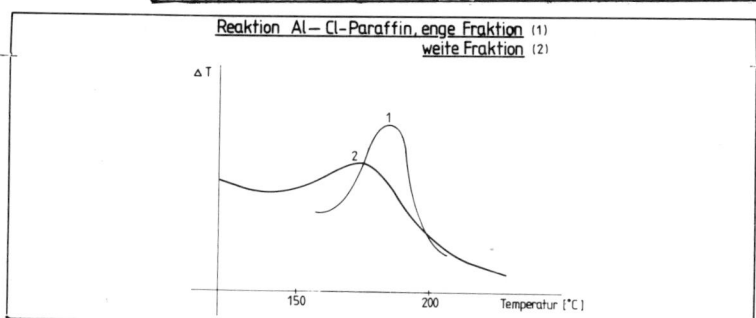

Figure 2

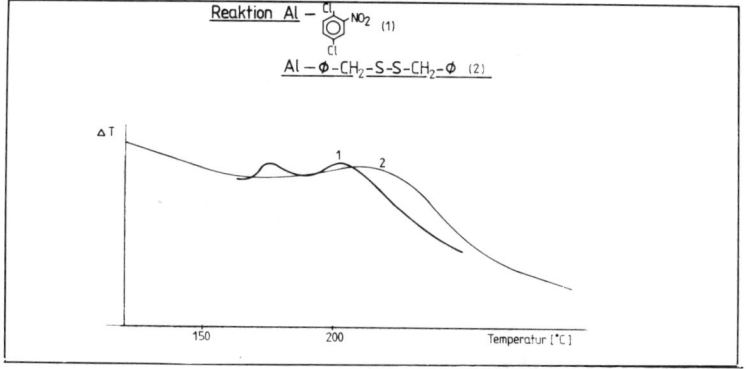

Figure 3

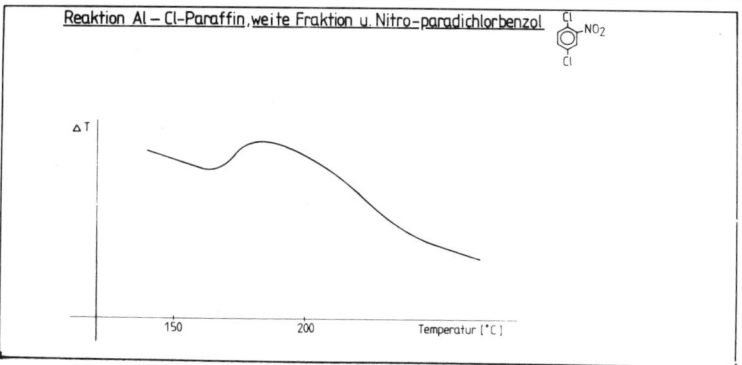

Figure 4

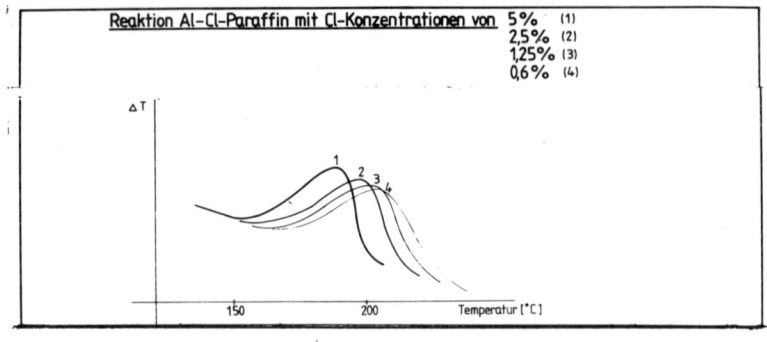

Figure 5

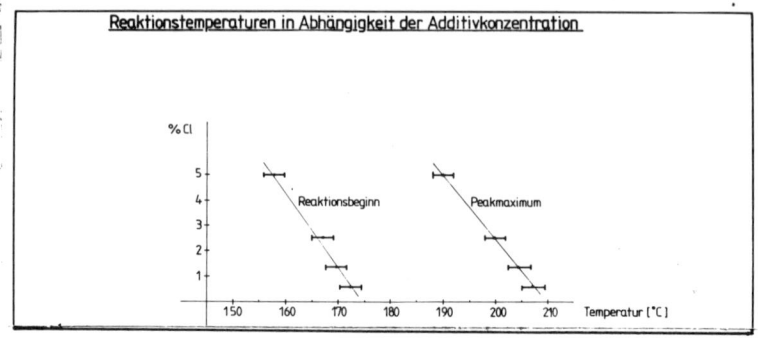

Figure 6

## REFERENCES

1 A. Commichau, Beitrag zur Bestimmung der Alterungsstabilität von Schmierstoffen durch DTA, Erdöl und Kohle 25 (6), 322-27 (1972)

2 C. Lovász, Thermische Stabilität von Schmierstoffen, Erdöl und Kohle 30 (5), 219-23 (1977)

3 C. Weber, R. Büchner, D. Hötel, J. Wagner
Zum Zusammenhang Grenzflächenreaktivität, Lasttragevermögen von EP-Additives

4 B. Essiger, H. Berndt, S. Hummel, R. Voigt
Grenzflächenchemische Untersuchungen zur Reaktivität schwefelorganischer Verbindungen gegenüber Metallen
Schmierungstechnik 10(12), 356-62 (1979)

THERMOANALYTICAL STUDY OF β-FeOOH OBTAINED BY HOMOGENEOUS
PRECIPITATION WITH UREA

Mª Emilia García-Clavel and Sara Goñi-Elizalde
Sección de Termoanálisis y Reactividad de Sólidos.Departamento de Química Analítica del C.S.I.C.Facultad de Ciencias Químicas.Ciudad Universitaria.Madrid-3.Spain.

## ABSTRACT

Akaganéite specimens were obtained from Fe(III) hydrochloric solutions by the technique "precipitation from homogeneous solution",using urea as the hydrolytic reagent.
It has been studied the influence of different precipitation conditions: concentration of the iron solution,concentration of urea,temperature and time of heating.We have established the optima conditions of akaganéite obtention: $[Fe(III)]=3.7 \times 10^{-2}$M; $[(NH_2)_2CO]=0.74$M; boiling point temperature;time of heating,30-45 minutes.If it is employed a longer time of heating,the akaganéite changes to hematite.

## INTRODUCTION

One of the most frequently used methods of "precipitation from homogeneous solution" is to increase the pH of an unsaturated solution,in a controlled way. This is done hydrolysis of a reagent that generates ammonia"in situ".Urea is an almost ideal and often used reagent in this type of process,because it is inexpensive and soluble in water(1)(2)(3).Its hydrolysis rate depends on temperature and therefore the precipitation rate of the compounds can be easily controlled.
We believe it would be very interesting to use this urea method to precipitate Fe(III) compounds under various controlled conditions,and furthermore to study composition,structure and thermal behavior of the precipitates.

## EXPERIMENTAL PROCEDURES

Reagents:Iron wire of 0.2mm of diameter.Concentrated hydrochloric acid, Merck,A.R.grade.Urea,Carlo Erba.
Hydrochloric solutions of Fe(III):Iron wire is dissolved in hydrochloric acid to boiling temperature.$H_2O_2$ is used to oxidize the possible Fe(II) to Fe(III),because it is easy to remove it from the solution by boiling. As the lowest pH at which the precipitates begin to form is 1.5,the starting pH,at room temperature,is adjusted at 0.7.The ferric solution concentrations are $3.7 \times 10^{-3}$;$1.8 \times 10^{-2}$ and $3.7 \times 10^{-2}$M.
Urea solutions:0.33;0.74;1.3;2.2;5.0;8.3M.The 5.0 and 8.3M solutions are over saturated,therefore the hydrolysis of the urea does not fulfill the requeriment of homogeneity.The 0.33M solution is too dilute,so the increase of the pH in the ferric solution is negligible.Because urea hydrolyzes very slowly in the range of temperatures below 90ºC,precipitation processes were carried out at boiling points of the solutions.
Preparation of the samples.Samples vere prepared into a two neck flask. A thermometer and a combined electrode(available for high temperatures) are immersed in the solution through the stopper of one the flask using the second neck.The conderser has to be used because the high temperature

and long time of hydrolysis would yield evaporation water losses. The apparatus is placed into a silicone bath which is heated by an electric hot plate conveniently joined to the flask.Solution were stirred at controlled speed during the process.

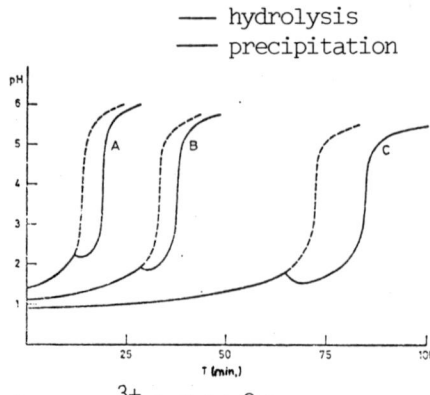

Fig.1. $Fe^{3+}=1.8x10^{-2}M$
$(NH_2)_2CO; A=2.2; B=1.3; C=0.74M$

The two Types of curves,urea hydrolisis and precipitation of Fe(III),giving the change of pH with the time, gotten at the various conditions before mentioned,have the same profiles.Fig.1.shows some of them.The second type differs from the first only because the abrupt change in the pH is delyed.The pH encreases slowly when the precipitation begins,due to the fact of two different reactions are taking place simultaneously:urea acid hydrolysis (CNO$^-$+2H$^+$+2H$_2$O ⎯⎯ ⎯⎯H$_2$CO$_3$+NH$_4^+$)and Fe(III)precipitation

The start point of the precipitation is showed by the inflexion point of the curve.
In all the cases the Fe(III) is totally precipitated.
The precipitates are collected on porcelain filtering crucibles A1(porosity 6µ)and rinsed with distilled water till the wash liquid does not contain Cl$^-$.After washing the precipitates are dryed at 100ºC during 1h.All of them are dark brown colour.
Heating time.The structure of the solids depends largely on the length of time from the moment that precipitation starts till filtering operation is carried out.
The periods of heating time studied are: 30,45,60,75 and 90 minutes.
Chemical analysis.Determination of Cl$^-$,NH$_4^+$ and Fe$^{3+}$ have been carried out on the before mentioned solids.Fe$^{3+}$ has been determinated too in the residua of the calcination of these solids at 600ºC.Both Fe$^{3+}$ determinations may establish that there are not losses of Fe as FeCl$_3$ during the ignition processes.NH$_4^+$ is in the concentration range of traces.The proportion of Cl$^-$ is:≈6% of the sample weight for the precipitates heated during 30min;≈3%(45 and 60 min.) and≈0.5%(75 and 90 min.)

Apparatus
pH-meter Metrohm AG CH 9100 Herisau,E 516,Titriskop.Combined electrode Metrohm AG.9100.Magnetic bar stirrer with thermostatic heating Rühromag.
TG apparatus.Chevenard thermobalance,model 93 from Adamel.Photographic recording.Heating rate,300ºCh$^{-1}$.Sample holder,Staalich Berlin porcelain crucible(A 4/0).DTA apparatus,Dupont 990,atm.N$_2$ or O$_2$,200ml/min.Electron microscope,AEI EM6,the samples were prepared by evaporation of the solid suspensions on a copper grid coated by a carbon film.X-ray powder diffraction,Philipps P.W.1010 generator,diffractometer P.W.1050,Geiger counter, recorder P.W. 1051,CoK$_1$ radiation, Fe Filter.

## RESULTS AND DISCUSSION

Structural and morphological study. Fig.2. shows the random powder X-ray diffraction patterns of the previously mentioned precipitates. Tabla I summarizes the main features of these patterns

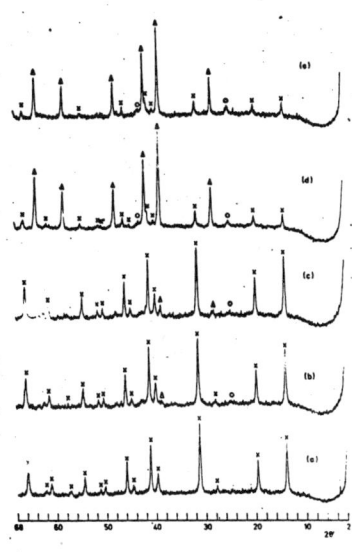

Fig.2

Table I

| Pattern | Aging time min. | Crystal species |
|---------|-----------------|-----------------|
| a) | 30 | A |
| b) | 45 | A |
| c) | 60 | A + H* |
| d) | 75 | A* + H |
| e) | 90 | A* + H |

A akaganéite; H hematite
* minoritarie phase

The pattern b) shows a very week reflection at 4.18Å, that is slight increased in the patterns c),d) and e). It may be due to the ocurrence of small amounts of goethite. The formation of these two phases in hydrochloric solution has been already pointed out by Feitknecht et al. (4). The results by the X-ray diffraction are consistent with what appears under the electron microscope (Fig.3,a (30 and 45 min.),b (60 min.),c(75 min.) and d(90 min.). In all of these samples akaganéite has the morphology that is termed as somatoids, which are assembled in two different fashions, either as star-shaped twins and or as tactoids (5)(6). The electron micrograph (Fig.3.b), shows "chains of crystallites", it suggests that a new phase is being formed. Hematite occurs

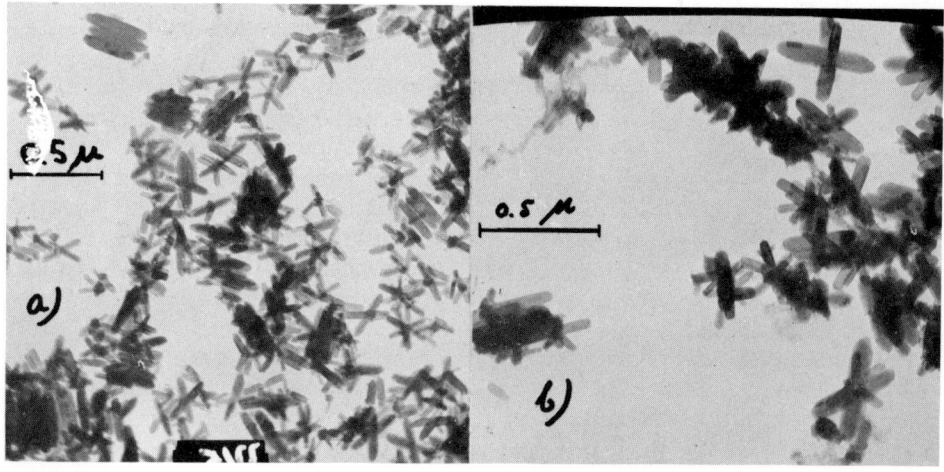

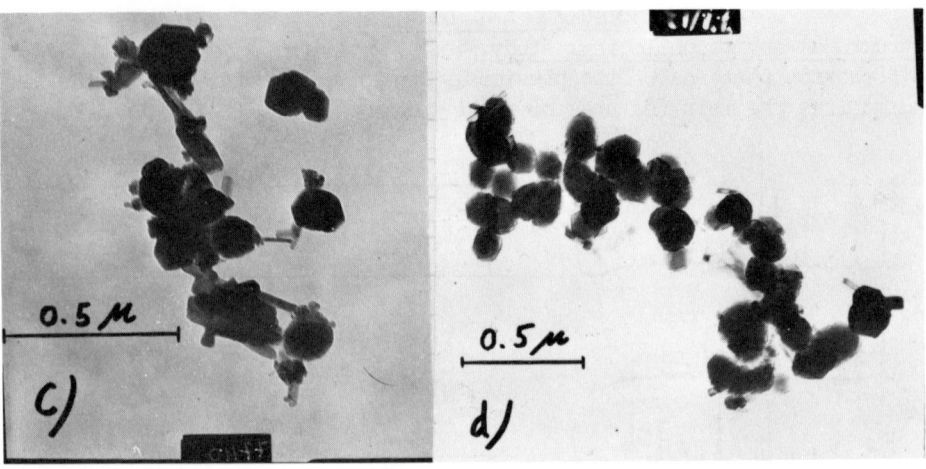

as rhombohedral crystals in samples aged 75 and 90 minutes(Fig.3,c and d ).
These micrographs can be observed that as bigger the hematite crystals
are as smaller are the akaganéite ones. This fact indicates that the
β-FeOOH,which is easily formed due to the presence of structural $Cl^-$ ions,
redisolved with a longer aging time and the hematite phase, that is formed slower but is more stable, appears.

Thermoanalytical study. The loss of $Cl^-$ is perfectly observed in the TG
curves(Fig.4)corresponding to the samples 30,45 and 60 minutes aged and
it takes place in an abruptly manner.

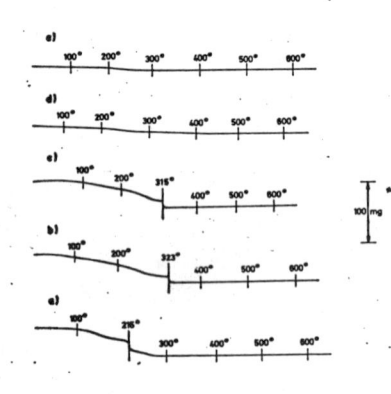

Fig.4.

In the 30 minutes aged sample,curve a),
the loss $Cl^-$ is inserted in the water
one, while in the 45 and 60 minutes
(curves b and c) the loss of $Cl^-$ is observed after dehydration is complete.
Otherwise,it is not able to distinguish
between loss of $Cl^-$ and dehydration in
the TG curves of the samples 75 and 90
minutes aged (curves d and e).Calculations made on the TG curves reveal that
the proportions of $Cl^-$ agree exactly
with the values from chemical analysis.
Furthermore,DTA curves run in $N_2$ and $O_2$
atmosphere have identical profiles.
These findings suggest that $Cl^-$ are removed as HCl.

It is not possible to stablish the difference between the loss of adsorbed water and that coming from dehydroxylation, even if the sample is heated at 150ºC/h.

Fig.5.shows DTA curves in $N_2$ atmosphere.In the curve a),the abrupt scope
of $Cl^-$,already observed by TG,is indicated by a very sharp endothermic
peak,which is not overlopped with the one due to dehydration;it does not
happen in the cases of curves b) and c).(It should be remind that in both

cases, the losses of water and Cl⁻ are not superimposed).

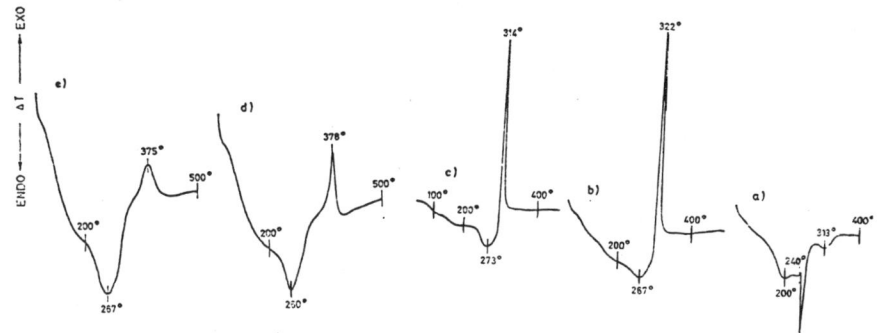

Fig.5.

DTA does not show up a distinction between dehydration and dehydroxylation.
The exothermic peak corresponds to hematite crystallization, fact that is well-stablished by X-ray difrraction.
The sample 45 minutes aged, akaganéite, has been studied, after being calcinated at different temperatures, by X-ray diffraction and electronic microscopy. Temperatures were chosen from the DTA curve.
We know that at 270ºC the akaganéite is close to be totally dehydrated.
The Cl⁻ determination has supplied the same value for the calcinated and no-calcinated samples(3.2%), which shwos that the Cl⁻ ions remain in the akaganéite crystals, which allows the morphology of the β-FeOOH persist.
In the Fig.6 we can see star-shaped and somatoids crystals, characteristic of the akaganéite, but holed as can be expected, due to dehydratation. The

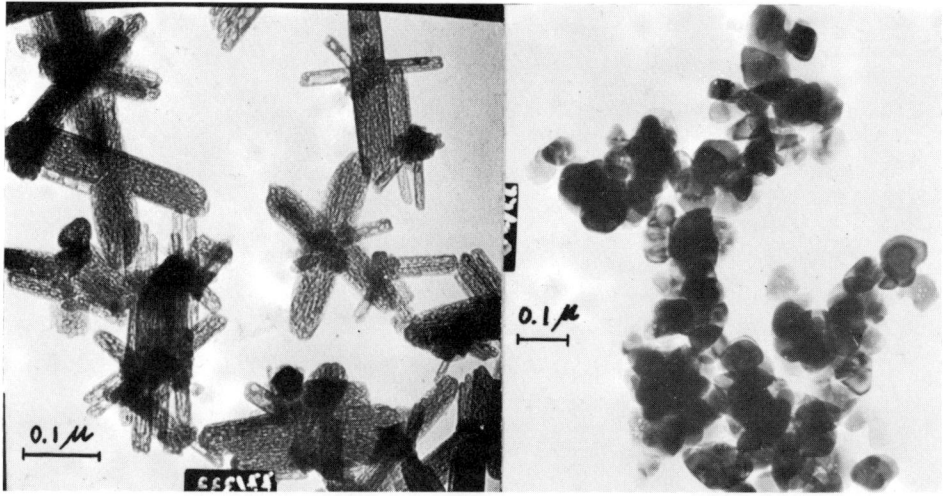

Fig.6.-Sample 45 min. aged, calcinated at 270ºC

Fig.7.-Sample 45 min. aged, calcinated at 550ºC

X-ray diffraction shows that the β-FeOOH lattice is closed to be destroyed,a new crystalline face is appearing, corresponding to hematite.
The X-ray diffraction of a sample calcinated at 325ºC,corresponds to an hematite with Cl⁻ ions as impurities (0.4%).
At 550ºC,the X-ray diffraction corresponds to a very well crystalized hematite,with rhomboedral morphology (Fig. 7).

Acknowledgement.
We thank Dr. Mª I. Tejedor-Tejedor, her support and the performance of the electronic microscopy photographs.

## REFERENCES

(1) H.H.Willard and N.K.Tang,J.Am.Chem.Soc.59 (1937) 1190.
(2) H.H.Willard and J.L.Sheldon,Anal.Chem.32 (1950) 1162
(3) L.Gordon,M.L.Salutsky and H.H.Willard,Precipitation from homogeneous solution.J.Wiley and Sons,Inc.,Publ. (1959).
(4) W.Feitknecht,R.Giovanoli,W.Michaelis und M.Müller,Z. anorg.allg.Chem. 417 (1975) 114.
(5) A.L.Mackay,Min.Mag.32 (1960) 545.
(6) J.A.Gard, ed. The Electron-Optical investigation of clays. Mineralogical society.Clay Minerals group 41 (1971) 315.

# THERMAL ANALYSIS IN EARTH SCIENCES

G. Lombardi
Istituto di Mineralogia e Petrografia
Città Universitaria, 00185 Roma

Earth sciences include mineralogy, petrography, geology, geochemistry and related subjects where the knowledge of the chemical and mineralogical composition of rocks and the estimation of reciprocal percentages of the constituents is often of major concern.

Though important mineral groups typical of high temperature consolidation , such as feldspars, olivimes, pyroxenes, are thermally inert - and therefore our techniques can offer very little to the problems of their determination - thermoanalytical methods may be very convenient tools for the estimation of many other mineral species and as auxiliary techniques to deduce their chemical composition and variation.

Several examples of well-known and less common applications of thermoanalytical methods to earth sciences problems are presented, together with an analysis of their development in the last decades, following instrumentation progress both in our and other related fields.

However, notwithstanding the wide range of possibilities offered few are the earth sciences laboratories where these methods are applied in the routine analysis of rocks and minerals. And whilst in other branches of science there has been in the last decade a marked increase in terms of use and level at which thermal analysis methods are employed, in the field of earth sciences, on average, their possibilities are not fully exploited.
This occurs even though earth scientists can claim a historical standing in the use of these methods, with the first DTA experiments of Le Châtelier carried out on clays and the substantial contribution given by them to their development in the 50s and 60s.

The possible reasons for this general underutilization of

thermoanalytical methods in earth sciences and the pathways to obtain both a larger diffusion of our methods in this field and a better characterization of minerals and rocks components are discussed. Finally, some problems of earth sciences where thermoanalytical methods have received up to now only a very limited interest are pointed out, to suggest possible future developments.

# Appendices

# Appendices

MEMORIES OF KYOTO

Hirotaro Kambe
Past President, ICTA

PROLOGUE

The possibility of invitation of ICTA Conference to Japan was first discussed among Prof. Toshio Sudo, Prof. Syuzo Seki, and myself in a room of dormitory of Holy Cross College at the 2nd ICTA in Worcester, Mass., August 1968. When I was nominated for Vice President at the 3rd ICTA in Davos, I made my mind to invite the 5th ICTA to Japan. With a support of the Society of Calorimetry and Thermal Analysis, Japan and particularly of Prof. Seki, I started to organize the Conference. I asked Prof. Seki to be the Chairman of the Organizing Committee and also the President of the Japanese Society for the period of Conference in 1977. We proposed Council in a formal procedure and at a meeting of Council at the 4th ICTA in Budapest it was decided that the 5th ICTA would be held in Kyoto in August 1-6, 1977.

SITUATIONS

Kyoto was capital of Japan from 794 to 1868 A.D. There remain many temples and shrines with classical Japanese gardens, which were not destroyed by the 2nd War. Japanese Government built a huge International Conference Hall at the hillside embracing a natural pond as a foreground. The place was considered the most suitable to our ICTA in Japan. However, it is usual that the summer in Kyoto is terribly hot. In August 1977, we had such a hot summer as expected. Inside of the Conference Hall, we had nicely cooled rooms, of course, but outside, the participants, particularly from European countries must have exhausted to overcome the heat.

PARTICIPANTS

Total number of registered participants was 199(including 141 Japanese) from 20 countries, and 9 persons accompanied active members. The number cecreased so much from previous conferences as seen in Table, but long dis-

TABLE

| Year | Place | Participants | Countries |
|---|---|---|---|
| 1965 | Aberdeen | 304 | 29 |
| 1968 | Worcester | 284 | 19 |
| 1971 | Davos | 362 | 31 |
| 1974 | Budapest | 615 | 30 |
| 1977 | Kyoto | 199 | 20 |

tance from other countries should be considered.

## SESSIONS

We used three rooms for sessions. The number of contributed papers which were scheduled in the program was 160, but actually 18 papers were withdrawn and 3 papers were added. The award lectures by Profs. P.D.Garn and H. Kambe were given after the opening session in Monday morning, and plenary lectures by Profs. Seki and Otsuka were presented on Thursday for introducing scientific achievements of Japanese in calorimetry and thermal analysis, respectively.

## PROCEEDINGS

Proceedings of the 5th ICTA were edited in prior to the conference and distributed to the participants. It was compiled of extended abstracts of 3-4 pages for each contribution, by the policy

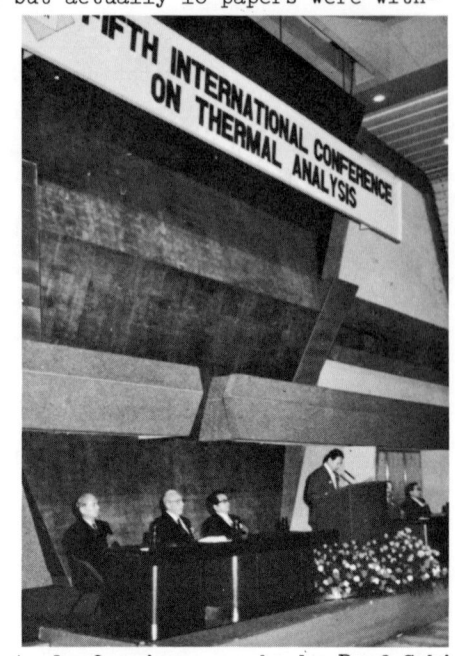

Photo 1. Opening remarks by Prof.Seki.

of Council to shorten the period for publishment. The 18 papers involved in Proceedings were not orally presented. The proceedings is still available from the Heyden & Son Ltd., London.

## SOCIAL MEETINGS

A number of foreigners must have personal chance to meet Japanese participants, even though they are not very familiar with English conversation. At several occasions as mixer, reception, excursion, and banquet, Japanese expressed their friendship with foreign guests. Ladies were not participating more than expected. In Ladies'programs, however, they must have enjoyed Japanese gardens, culture, handicrafts, and music. At excursion, many participants went to Todaiji Temple(Big Buddah) in Nara and Horyuji Temple(the most ancient Buddism temple built in 693 A.D.) The hottest day during the conference we experienced on that day.

Photo 2. Presidents' meeting at mixer.

Photo 3. Dragon dance at reception.

Photo 4. Toast by First President at banquet.

Photo 5. Geisha dance at banquet

Photo 6. Ladies' program at old palace

Photo 7. Excursion in deer park, Nara

## ICTA - DU PONT AWARD

From the 5th ICTA, it was started to give an award sponsored by E.I. du Pont de Nemours Co. at every conference. Prof. Paul D. Garn, University of Akron, Ohio received this award as the first recipient.

Photo 8. ICTA Council met a representative of du Pont

## EPILOGUE

First ICTA Conference in Asia was finished successfully. Whether it was successful or not may be a matter of personal sentiment. Financially it raised something for ICTA. As a memeory of our collaborations for ten years to ICTA, I engraved letters of "1977 ICTA" on a pair of watches as Christmas gift shared with my wife.

REPORT
of the
COMMITTEE ON STANDARDIZATION

The certification of five magnetic reference materials for temperature calibration in thermogravimetry was completed in time for this report. Issuance by the United States National Bureau of Standards prior to Sixth ICTA is expected. These materials are a metal--nickel--and four magnetic alloys. The materials were chosen partly for the ease of obtaining a well-defined change of apparent weight and partly for resistance to oxidation. They can be purchased under the catalog number GM 761 from the Office of Standard Reference Materials, U. S. National Bureau of Standards, Washington, D. C. 20234, U.S.A.

Magnet reference materials are necessary because the temperatures of other events such as decompositions were found to vary with the thermodynamic environment. The change in magnetic properties is independent of thermodynamic environment except for possible contamination. The mean apparent weight changes in these materials are ca. 259, 353, 381, 455, and 751°C. Three points on the curve are defined.

The ICTA-Certified Reference Materials now available are:

| | | |
|---|---|---|
| GM | 754 | DTA-DSC glass transition |
| GM | 757 | DTA-DSC below 350 K |
| GM | 758 | DTA-DSC 125-435°C |
| GM | 759 | DTA-DSC 295-675°C |
| GM | 760 | DTA-DSC 570-940°C |
| GM | 761 | TG temperature calibration |

GM 759 is offered as convenience for users interested in this range; the materials are duplicated in GM 758 or GM 760.

The demand for the ICTA-Certified Reference Materials has been great enough to cause re-supply problems. The ICTA, through this committee, is responsible for supply of the materials.

There is a large time span between our purchase of the materials and the return of a portion of the sale price from NBS to ICTA. In the NBS fiscal year ending 30 September 1979, the total sales were 236 units. Some orders had to be delayed waiting for our re-supply, which in turn had to wait for returns from earlier sales. Recent payments have made it possible to re-supply the materials in short supply, which are also the most expensive. Now that funds for re-supply have been coming, new delays should be rare.

One re-certification of a material has been necessary because no more of the original batch was available from the manufacturer; another re-certification is presently in progress.

This committee has the goal of providing reference materials for all purposes for which there appears to be a reasonable need. We continue to welcome suggestions concerning such needs, preferably with an estimate of the number of possible users and identification of probably-useful materials. The invitation includes the extending of temperature ranges of present sets. The areas of active interest at this time are high-temperature materials (>1000°C) and inorganic glasses.

In addition to the chairman, members of the Committee on Standardization during this term were K. Heide, H. Kambe, G. Lombardi, R. C. Mackenzie, O. Menis, H. R. Oswald, T. Ozawa, F. Paulik, J. P. Redfern, O. T. Sørenson, and H.-G. Wiedemann (vice chairman).

Paul D. Garn
Chairman

# REPORT OF THE PUBLICATIONS COMMITTEE

J. P. Redfern (Chairman, Publications Committee)
Stanton Redcroft, Copper Mill Lane, London, England

The main publications activities of ICTA are as follows:-
1. Thermal Analysis Abstracts    publ. bi-monthly  Ed. J. H. Sharp
2. Proceedings of ICTA Conferences
3. Newsletter                              Ed. C. J. Keattch
4. For Better Thermal Analysis             Ed. G. Lombardi

Thermal Analysis Abstracts  Vol. 1 (1972) - date
This abstract service with full keyword and author indexes covers all thermoanalytical methods, calorimetry, thermodynamics and the kinetics of the solid state for every type of material. Every year some 2000-2500 abstracts are listed. Dr. J. H. Sharp with his two assistant editors Drs. G. M. Clark and A. D. White, team of regional editors and abstractors have combined to produce a most valuable on-going service for all thermal analysts.

Newsletter
Dr. C. J. Keattch edits a regular ICTA Newsletter available to all ICTA members. It gives details of forthcoming meetings, keeps members up-to-date on the latest developments and contains other articles of general interest.

For Better Thermal Analysis
First published in 1977 a second edition (publ. 1980) is now available. The brochure serves as a concise introduction to ICTA and to thermal analysis.

Books and Monographs
The first list of books and monographs devoted principally to thermal analysis was published in 1971 (Proc. 3rd ICTA, 1971, Vol. 1, 615-622) updated to 1976 in the 1st edition of 'For Better Thermal Analysis' (1977, 31-36) and updated again in the 2nd edition of 'For Better Thermal Analysis'. The latest revision is reproduced at the end of this

report. Any amendments or additions would be appreciated to keep the list as accurate and up-to-date as possible.

Membership of Committee
The current membership of the Committee is:
Dr. J. P. Redfern (Chairman),   Dr. J. H. Sharp (Vice-Chairman),
Dr. H. G. McAdie (Canada),   Dr. R. C. Mackenzie (U.K.),
Prof. R. Otsuka (Japan),   Dr. E. Buzagh (Hungary),
Dr. P. C. Gravelle (France),   Mr. J. E. Kruger (South Africa),
and Dr. Ing. W. Krajewski (Germany),
Ex-officio:   Dr. T. Sorensen (Secretary),
              Prof. P. D. Garn (Standardisation).

Books and Monographs, continuation list

1971   Atlas of Thermoanalytical Curves   Vol. 1
       G. Liptay
       Akademiai Kiado, Budapest and Heyden & Son, London.


       Calorimetry, Thermometry and Thermal Analysis - 1971 Edition
       Edited by the Society of Calorimetry and Thermal Analysis.
       Kagaku Gijutsu Sha, Tokyo.


1972   Termická Analyza (Thermal Analysis).
       A. Blazek
       SNTL, Prague.


       Progress in Vacuum Microbalance Techniques. Vol. 1.
       Th. Gast and E. Robens, editors.
       Heyden & Son, London.


       Differential Thermal Analysis. Vol. 2: Applications.
       R. C. Mackenzie, editor.
       Academic Press, London.


       Calorimetry, Thermometry and Thermal Analysis - 1972 Edition,
       edited by the Society of Calorimetry and Thermal Analysis.
       Kagaku Gijutsu Sha, Tokyo.


       Analiza Termica A Mineralelor. (Thermal Analysis of minerals).
       D. N. Todor
       Editura Technica, Bucarest.


       Thermogravimetrie. Etude Critique et Theorique, Utilization,
       Principaux Usages.
       P. Vallet
       Gauthier-Villiars, Paris.

Thermal Analysis: Proceedings of Third ICTA, Davos, 1971
Vol. 1, 2 and 3.
H. G. Wiedemann, editor.
Birkhaüser Verlag, Basel and Stuttgart.

1973 Progress in Vacuum Microbalance Techniques Vol. 2.
S. C. Bevan, S. J. Gregg and N. D. Parkyns, editors.
Heyden & Son, London.

Thermal Analysis
T. Daniels
Kogan Page Ltd.

Atlas of Thermoanalytical Curves, Vol. 2.
G. Liptay
Akadémiai Kiadó, Budapest and Heyden & Son, London.

Termanal '73.
Slovenská chemická spoloncnost pri sav. Bratislava.

Calorimetry, Thermometry and Thermal Analysis - 1973 Edition,
Edited by the Society of Calorimetry and Thermal Analysis.
Kagaku Gijutsu Sha, Tokyo.

The Study of Heterogeneous Processes by Thermal Analysis.
Thermochimica Acta, Vol. 7, No. 5
Elsevier, Amsterdam.

Pyatoe Vsesoyuznoe Soveshchanie po Termografii: Tezisy Dokladov
(Novosibirsk, 3-6 Iyulya, 1973). (The Fifth All-Union Conference
on Thermal Analysis: Abstracts of Papers (Novosibirsk, 3-6 July
1973).
Izd. Nedra, Moscow.

1974 Thermal Analysis
A. Blázek
Van Norstrand Reinhold Co. Ltd., London.

Polymer Characterization by Thermal Methods of Analysis
J. Chiu, editor.
Marcel Dekker Inc., New York.

Termischeshkii Analiz Mineralov i Gornykh Porod. (Thermal Analysis of Minerals and Rocks).
V. P. Ivanova, B. K. Kasatov, T. N. Krasavina and
E. L. Rozinova
Izd. Nedra, Leningrad.

Thermal Analysis. Comparative Studies on Materials.
H. Kambe and P. D. Garn, editors.
Kodansha Ltd., Tokyo and John Wiley & Son, New York.

Atlas of Thermoanalytical Curves Vol. 3
G. Liptay
Akadémiai Kiadó, Budapest and Heyden & Son, London.

Analytical Calorimetry Vol. 3
R. S. Porter and J. F. Johnson, editors.
Plenum Press, New York.

Differential Thermal Analysis: Application and Results in Mineralogy
W. Smykatz-Kloss
Springer Verlag, Berlin.

Calorimetry, Thermometry and Thermal Analysis – 1974 Edition
edited by the Society of Calorimetry and Thermal Analysis
Kagaku Gijutsu Sha, Tokyo.

Handbook of Commercial Scientific Instruments Vol. 2:
Thermoanalytical Techniques.
W. W. Wendlandt
Marcel Dekker Inc., New York.

Thermal Methods of Analysis. Second edition.
W. W. Wendlandt
Wiley, New York.

1975 Thermal Analysis: Proceedings Fourth ICTA, Budapest, 1974.
Vols. 1, 2 and 3.
I. Buzás, editor
Heyden & Son, London and Akadémiai Kiadó, Budapest.

Progress in Vacuum Microbalance Techniques. Vol. 3
C. Eyraud and M. Escoubes, editors
Heyden & Son, London.

Netsu Bunseki
H. Kambe, editor
Kodansha Ltd., Tokyo.

An Introduction to Thermogravimetry. Second edition
C. J. Keattch and D. Dollimore
Heyden & Son, London.

Atlas of Thermoanalytical Curves. Vol. 4
G. Liptay
Akadémiai Kiadó, Budapest and Heyden & Son, London.

Acoustic Methods of Investigating Polymers.
I. I. Perepechko (translated by G. Leib)
Mir Publications, Moscow.

International Rev. Sci. Physical Chemistry. Series 2 10.
Thermochemistry and thermodynamics 5.
H. A. Skinner, editor.
Butterworths, London and Boston.

Calorimetry, Thermometry and Thermal Analysis - 1975 edition.
Edited by the Society of Calorimetry and Thermal Analysis
Kagaku Gijutsu Sha, Tokyo.

Thermometric Titrations
J. Barthel
John Wiley & Son, New York.

1976 Proceedings of the First European Symposium on Thermal Analysis
D. Dollimore, editor.
Heyden & Son, London.

Thermal Methods in Analytical Chemistry
C. Duval
Elsevier Scientific Publishing Co., Amsterdam.

Proceedings of a Roundtable Discussion on Thermoanalytical Techniques
P. D. Garn, editor
U.S. Dept. of Health, Education and Welfare

Metody Kalibrace a Standardizace v ta (Methods of Calibration and Standardization in TA).
P. Holba, editor
D.T. CSVTS, Usti, n. Luz, Czechoslovakia.

Atlas of Thermoanalytical Curves. Cumulative Index Volumes 1-5
G. Liptay
Heyden & Son, London and Akadémiai Kiadó, Budapest.

Termanal '76. Proceedings 7th Czechoslovakian Conference on Thermal Analysis.
OSTA, SVST, Bratislava.

Calorimetry, Thermometry and Thermal Analysis - 1976 edition.
Edited by the Society of Calorimetry and Thermal Analysis
Kagaku Gijutsu Sha, Tokyo.

Thermal Analysis of Minerals.
D. N. Todor
Abacus Press, Tunbridge Wells, Kent, England.

Benchmark Papers in Analytical Chemistry Vol. 2 Thermal Analysis
W. W. Wendlandt and L. W. Collins, editors.
Dowden, Ross and Hutchinson, Stroudsbourg, Pennsylvania, USA

1977 Prístrojová Technika a Standardy Termické Analyzy (Instrumentation in TA investigations).
M. Beranek, editor.
D. T. CSVTS, Pardubice, Czechoslovakia.

Shin Jikken Kagaku Koza 2, Kiso Gijutsu 1, Netsu Atsuryoku.
Chemical Society of Japan, Maruzen, Tokyo.

Thermal Analysis: Proceedings of the 5th ICTA Conference
H. Chihara, editor.
Heyden & Son, London and Sanyo Shuppan Boeki Inc., Co., Tokyo.

Summaries of the 1st Czechoslovakian Conference on Calorimetry in Liblice, CSAV - CSCHS, Prague.

Non-Isothermal Reaction Kinetics
E. Koch
Academic Press, London.

Dielectric Spectroscopy of Polymers
P. Hedvig and A. Hilger
Bristol and Akadémiai Kiadó, Budapest.

Atlas of Thermoanalytical Curves Vol. 5
G. Liptay
Akadémiai Kiadó, Budapest and Heyden & Son, London.

For Better Thermal Analysis
G. Lombardi
ICTA and Università di Roma, Rome.

Differential Thermal Analysis.
M. I. Pope and M. D. Judd
Heyden & Son, London

Calorimetry, Thermometry and Thermal Analysis - 1977 Edition
Edited by the Society of Calorimetry and Thermal Analysis
Kagaku Gijutsu Sha, Tokyo.

Kobunshi no Netsubunkai Gas Chromatography (Pyrolitic Gas Chromatography of High Polymers).
T. Takeuchi and S. Tsuge, editors
Kagaku Dojin, Kyoto.

1978 Emanation Thermal Analysis
V. Balek
Thermochimica Acta, Vol. 22, No. 1, Elsevier

Metody Analiza Veshchestvennogo Sostava Gornykh Porod i vod pri Geokhimicheskikh Issledovaniyakh. (Methods of analysis of the constitution of Rocks and Waters for Geochemical Investigations Chapter on DTA.)
V. A. Kuzuetsov and J. A. Dobrovolskaya, editors
Nauka i Tekniko, Minsk.

Metody Geokhimicheskikh Analizov Gornykh Prod i Prirodnykh vod. (Methods of Geochemical Analysis of Rocks and Natural Waters, Chapter on DTA.)
V. A. Kuzuetsov, editor
Riso an. Bssr, Minsk.

Thermodynamic Properties of Minerals and Related Substances at 298.15K and 1 bar ($10^5$ Pascals) Pressure and at Higher Temperatures
R. A. Robie, B. S. Hemingway and J. R. Fischer
Geol. Surv. Bull. 1452, Washington.

Zpusoby Vyhodnocování Krivek Termické Analyzy (Means for the Evaluation of TA Curves)
J. Sesták, editor
D.T. CSVTS, Prague

Calorimetry, Thermometry and Thermal Analysis - 1978 Edition
Edited by the Society of Calorimetry and Thermal Analysis
Kagaku Gijutsu Sha, Tokyo.

1979 Termická Analiza a Kalorimetrie Polymeru (TA and Calorimetry of Polymers)
M. Berka and J. Vanicek, editors
Z.P. CSVTS, Silon, Plana n. Luz, Csechoslovakia.

Dynamische Thermische Analysen Methoden
K. Heide
VEB Deutscher Verlag fur Grundstoffindustrie, Leipzig.

Netsu Bunseki Jikken Gijutsu Nyumon (Introduction to Experimental Techniques in Thermal Analysis)
S. Nagasaki, editor
Shinku Riko and Kagaku Gijutsu Sha, Tokyo.

Termanal '79 Zbornik VIII Celosták. Konf. Termicke Analyze
(Proc. 8th Czechoslovakian Conference on Thermal Analysis)
OSTA, SVST, Bratislava, Czechoslovakia.

Calorimetry, Thermometry and Thermal Analysis - 1979 Edition
Edited by the Society of Calorimetry and Thermal Analysis
Kagaku Gijutsu Sha, Tokyo.

Combustion Calorimetry
S. Sunner and M. Manson, editors
Pergamon Press, Oxford.

1980 For Better Thermal Analysis 2nd edition
G. Lombardi
ICTA and Universita di Roma, Rome.

# Author Index

Aalto, E. 567
Abdel-Rehim, A.M. 357
Akechi, K. 219
Alsdorf, E. 387
Amicarelli, V. 433
Andrejs, B. 225
Arnold, M. 69

Baker, R.R. 439
Baldassarre, G. 433
Balek, V. 375, 403
Barendregt, R.B. 105
Barnes, P.A. 327
Baró, M.D. 155
Bayer, G. 279
Beattie, A.J. 245
Berndt, H.J. 345
Bertrand, G. 81
Biskupski, A. 399
Bordas, S. 403
Bouster, C. 51
Brandt-Petrik, E. 427
Brennan, W.P. 265
Brogli, F. 541, 549
Buttler, F.G. 381
Buzás, I. 535

Cammenga, H.K. 149
Carin, V. 409
Charsley, E.L. 237, 285
Carturan, G. 493
Chaubey, D. 335
Chen, D.T.Y. 133
Chiu, J. 245
Comel, C. 51
Cordovilla, C.G. 421
Criado, J.M. 145
Crighton, J.S. 341

Datta, P.K. 467
De Bruijn, T.J.W. 393
De Jong, W.A. 393
Dobovišec, B. 99
Dohnálek, J. 375
Dollimore, D. 505

Emmerich, W.-D. 375

Fodor, L. 189
Fong, P.H. 133
Friman, M.-L. 567
Frosini, V. 491
Fujieda, S. 183
Fyans, R.L. 265

Gál, S. 189, 535
Gallagher, P.K. 13, 113
Garcia-Clavel, E. 577
Garn, P.D. 201, 593
Gauler, K.-D. 453
Geli, M. 403
Geoffroy, A. 305, 313
Glasson, D.R. 517
Goñi-Elizalde, S. 577
Gottardi, V. 493
Guggisberg, U. 461
Gygax, R. 541, 549
Gyorgy, E.M. 113

Habersberger, K. 387
Hädrich, W. 511
Hälldahl, L. 499
Haglund, B.O. 207
Hajduk, N. 41
Halle, R. 409
Halonbrenner, R. 445
Hara, Z. 219
Heidemann, G. 345
Henig, A. 453
Hörkkö, A. 567

Ichihashi, M. 195, 219

Jäth, M. 453
Jain, P.C. 335
Janik, G. 453
Jarmontowicz, A. 363
Jerman, Z. 139
Joannou, J. 237

Kaisersberger, E. 251, 511
Kambe, H. 587

Kamp, A.C.F. 237, 285
Keil, W. 571
Kishi, A. 195, 219
Kiss, G. 87
Koch, E. 75
Kołaczkowski, A. 399
Kolenda, Z.S. 41
Kraus, Z. 483, 489
Krstic, B. 299
Krug, D. 57
Krzywobłocka-L., R. 363
Kunimatsu, Y. 259
Kurpjuweit, U. 273, 571

Lallemant, M. 81
Le Parlouër, P. 169
Leskelä, M. 567
Li, K.M. 341
Liberti, L. 433
Ligethy, L. 427, 477
Liptay, G. 427, 477
Lombardi, G. 583
Louis, E. 421
Ludwig, W. 293
Luks, T. 207

Madhusudanan, P.M. 127
Maesono, A. 195, 219
Manley, T.R. 467
Marchetti, A. 493
Marik-Korda, P. 529
Marti, E. 305, 313
Maruta, M. 259
Maruyama, T. 415
Mayer, J.S. 265
McNeill, I.C. 319
Meisel, T. 87
Menis, O. 201
Mentlík, V. 483, 489
Meyer, M.W. 541, 549
Möhler, H. 453
Mokhlisse, A. 81
Moll, J. 57
Morgan, S.R. 381
Moriguchi, H. 163

Nagy, J. 477
Nair, C.G.R. 127
Nakanishi, M. 183
Niinistö, L. 567
Norwisz, J. 41

Oehme, F. 523
O'Neill, P. 517
Oswald, H.-R. 1, 63
Ottaway, M.R. 237

Parton, C. 505
Paulik, F. 69
Paulik, J. 69
Perron, W. 279
Petrick, H.-J. 149
Prasad, T.P. 119
Poukari, J. 567
Pysiak, J.J. 35

Ray, H.S. 555
Redfern, J.P. 237, 595
Regenass, W. 561
Reller, A. 63
Robens, E. 213
Rordorf, B.F. 305, 313

Rosina, A. 99
Roudergues, N. 81
Rumsey, J.A. 285

Saito, Y. 415
Sasaki, S. 415
Scarinci, G. 493
Schnabel, K.H. 387
Schulz, J.P. 225
Šesták, J. 29
Seybold, K. 87
Shimizu, S. 163
Simon, J. 535
Sørensen, O.T. 231, 499
Stäudel, L. 351
Stebnicka-Kalicka, I. 369
Stilkerieg, B. 75
Surinãch, S. 155
Swami, M.S.R. 119
Szelagiewicz, M. 305, 313
Sztatisz, J. 189

Takaoka, K. 195, 219
Taniguchi, M. 163
Tejerina, F. 155
Thiel, G. 351

Tou, J.C. 177
Townsend, D.I. 177

Van den Berg, P.J. 105, 393
Van Dooren, A.A. 93
Veress, G.E. 69
Verhoeff, J. 105
Vermande, P. 51
Veron, J. 51
Vobořil, M. 403

Walkov, W. 387
Wappler, E. 225, 273, 571
Warrington, S.B. 327
Weber, H. 461
Wendlandt, W.W. 175
Whiting, L.F. 177
Wiedemann, H.G. 201, 279
Wöhrmann, H. 351
Wong, S.P. 133

Yamada, K. 259

Zepf, D. 57
Živković, Ž.D. 99

## Subject Index

Additives 571
Adiabatic measuring cell 531
Adipic acid 93
Adsorption effects 213
   isotherms on gold 215
   on glass 214
   on polytetrafluorethylene 216
   quartz 214
Aging of oil 571
Alkyd resins 454
Alkyd/melamine resin varnishes 455
Aluminium reaction with additives 571
Al-Zn-Mg alloy 421
Ammonium nitrate 399
Analyzing explosives 449
Anthracene, enthalpy of evaporation and melting 313
   enthalpy of sublimation 310
   entropy of sublimation 310
   heat capacity of 305
   vapor pressure of 309
   vapor pressure of, by DSC and TG measurements 315, 316
Apparent activation entropy 120
Approximated and real melting curves 89
Approximation method 126
Area-thermocouples 294
Arrhenius equation 58
   equation, rearranged form 36
   law 28
   model in thermal analysis 69
$As_2Ge_3$, DTA of 297
Assessment of thermal hazards 564
Autoclaving of cement pastes and mortars 369
Automatic introducer of samples 169
Auto zero function 267

$BaCO_3$, thermal decomposition of 145
$Ba(OH)_2$, melting of 115
$Ba(OH)_2 \cdot H_2O$, kinetics of the dehydration 113
Bismuth molybdate catalysts 387
Boiling point method, vapor pressure by DSC and TG 314
Borchardt-Daniels method 108

Bulk-surface rubbers 477
Burning zone 439
Buoyancy of sample containes in air 207

Calcia stabilized zirconia (CSZ) 415
Calcium dichromate pentahydrate, TG-DTA curves 239
   hydroxide in cement pastes 381
   oxalate 60
Calculation of the Ge-Sb-Bi phase diagram 156
Calibration check 95
Calorimeter, accelerating rate 177
Calorimetric analysis of polymer fabrication 461
   purity test 87
   sensitivity 93
Catalyst-reactant interaction 387
Cement paste 375
   pastes and mortars 369
$CeO_2$-$Gd_2O_3$ powder mixture 233
Ceramic powder compacts 231
Certified reference materials 202, 205
Chains of crystallites 579
3-Channel DTA apparatus 263
Characteristic kinetic parameters 56
Characteristics of endothermic interface reactions 81
Characterization of bismuth molybdate catalysts 387
Chemical erosion of solids 8
Chlor-paraffines 571
Choice of kinetic equation 29
Chromindur 16
Clathrate complexes 9
Coats-Redfern method 108
Comparative thermal stability tests 546
Comparison of EGA with DTG data 117
Composition of hardened concrete 363
Compound fertilizers 570
Concentration of nitrating mixtures 523
Concrete 363, 375
Condensation in capillaries 216
Conductive radiation cell 304
Continuous monitoring method 246
   water reading detector 327

Controlled polymerisation 462
Convection currents in the furnace atmosphere 207
  effects in thermogravimetry 207
Conversion of $UF_6$ to $UO_2$ powder 499
Coordination compounds, thermal analysis of 6
"Correct" $E_a$ value 121
Correlation factors 41
Critical Rayleigh number 301
Crucible for thermomicroscopy 280
Crystallization of polypropylene 170
Curie temperature 15
Curing behavior of clear varnishes 453
Cutting oils 571
Cyclic and constant rate thermal analysis technique (CRTA) 145
Cyclic step heating 219

Decomposition of azodicarbon amid 172
  of magnesium hydroxide 141
  of ammonium uranly carbonate 499
  of aqueous manganese nitrate solutions 393
  of ammonium nitrate 399
  of solids 139
  of the solid solution 421
Degradation products of polyurethane 322
Dehydration of calcium oxalate monohydrate, kinetic of 120
  of $CaC_2O_4 \cdot H_2O$ by DTA 166
  of $CaSO_4 \cdot \frac{1}{2} H_2O$ by DTA 166
  of crystallization water in salts by DTA 163
  of $CuSO_4 \cdot H_2O$ by DTA 166
  of $ZnC_2O_4 \cdot 2H_2O$ by DTA 166
Desulphurisation of industrial flue gases 517
Determination of nitric acid 523
Deviation from zero line 102
Diamino-anthraquinone 551
Diazotization of aromatic nitrocompounds 552
Dicalcium silicate doped with $Ca_5(PO_4)_3OH$ 409
1,2-Dichlorethane, transmitted light intensity 291
  TG-DTA curves 244
"DIE" apparatus 530
Differential method 126
  determination of the state of insulations 489
  scanning calorimetry of hazardous chemical reactions 540
  scanning calorimetry of industrial chemical reactions 549
  thermal analysis of $Ca_2 SiO_4$ 409
Diffusion controlled evaporation, TG 313
Dilatometer, microcomputer-controlled 219
2,4-Dinitrotoluene 2,4,6,-trinitrotoluene 450
Divided sample holder for solid-vapor reactions 190
DSC of polycarbonate 17
DSC of polyethelene 18
  curves of reactant mixtures 542
  curves of thermal treated samples 542
  study of glasses 493
DTA and thermomicroscopy 279
  apparatus, high sensitivity multichannel 259
  curve corrections 341
  method for the study of surfactants at liquid surfaces 149
  microcomputer based 265
  of resin glassfibre 473
Dynamic derivatographic method 428
  elasticity 225
  Jahn-Teller effect 10

Earth sciences, thermal analysis in 587
Effect of experimental conditions 29
EGA, an oxygen sensor (OS) 415
  of semiconductors 18
Elasticity of plastics 273
Electrical properties of insulation 16
Electronic microbalance 238
Electronics industry, application of TA in the 13
Emanation thermal analysis and DTA 403
Emanation thermal analysis of cement 375
Endpoint of reactions 351
Energy of activation 58, 128
Enthalpy and melting temperature for Ge,Sb,Bi 158
Enthalpy of melting of the eutectic and of the pure substance 88
Entropy of activation 128
Epoxide resins 490
Equilibrium boiling point 151
  boiling temperature 150
  shrinkage-force-curves 345
Error analysis in thermal purity determinations 87
Errors in thermogravimetric experiments 213
Error sensitivity 89
Evaluation of thermal and kinetic data 562

Evaporation of water 84
Experimental data of acid catalyzed iodination of acetone 133
Experimental reaction rate 52

Fertilizer components and their mixtures 569
   a thermoanalytical study 567
Flexural strength 474
Flow method, partial pressure by 305, 313
Fluorination of corundum 357
Force on a cylinder surface of a crucible 209
Fractur stress 475

Gas inlet system 254
Ge-Sb-Bi ternary phase diagram 155
$Ge_{.25}Te_{.60}Se_{.15}$ 405
Glasses obtained from organometallic gels 493
Gold, melting of 271
Graphical approach of non-isothermal kinetic 558

Hardness of lacquer and rubber 225
Heat capacities and specific heats 563
   exchange type of titration calorimetry 184
   flow calorimetry, industrial experience 561
Heating rate 93
   the reactant mixture 542
Heat of reaction 563
   of vaporization of Cd in AgCd-solid 256
   transfer 299
Heterogeneous, endothermic reactions 87
   systems 51
High sensitivity detector 260
Hot stage microscopy 280, 285
   unit 289
Hydrated and carbonated limes 521
Hydrate decomposition of plant acids 327
Hydration of cement 375, 381
Hygrometer 327

Impurity histograms 91
Influence of experimental variables on DSC curves 93
   of mass and grain size, of the DTA curve 99
Information from thermoanalytic experiments 341
Infrared image furnace 195
   rapid heating system 219
Initial baseline deflection 95

Injection enthalpiemetry 524
Integral method 126
Interdependence between apparent and actual activation energies 30
Interface temperature analysis 83
   thermodynamic model 85
Investigation of secondary reactions 544
   of the intended reaction 542
Iodination of acetone 137
Isobaric TG and EGA 418
Isokinetic temperature 30, 36
Isothermal decorboxylation of calcium carbonate 199
   thermogravimetric investigations 429
   thermogravimetry 197

Jahn-Teller face transitions 10

Kinetic analysis from thermogravimetric traces 119
   analysis of thermodesorption of phenol 433
   aspects of decomposition reactions 536
   characterization of the oscillatory Belousov-Zhabotinsky reaction 75
   compensation effect 28
   of dehydration of hydrates 167
   of thermal dissocation of solids 35
   parameter of copper oxidation 41
   parameters from TG data 51, 126, 133
   parameters, thermal decomposition of aqueous manganese nitrate solutions 393
Kinetics for the thermal decomposition $Ba(OH)_2 \cdot H_2O$ 113
   from isothermal and non-isothermal TG 126
   of bubble contraction 505
$KNO_3$-$NaNO_3$, phase diagram 282
Kundsen effect 207

Linear elastic fracture mechanics (LEFM) 468
Long term drift 210
Low pressure plasma, thermal analysis in 8
   temperature regeneration of activated carbon 433

MacCallum-Tanner's method 44
Magnesite and dolomite 521
Magnetic reference materials 201, 593
Manganese nitrate solutions 393
Mass spectrometric analysis 448
Mathematical study of the rate equation 52

Maximum surface temperature  442
Mechanical propertics of polyester
   glass  467
Melting point of indium as a function of
   pressure  335
Memories of Kyoto  587
Method of quasi-isothermal dilatrometry
   (QID)  231
Mg-ferrite formation under the different
   partial pressures of $O_2$  29
MgO, low and high temperature form  139
Microcomputer controlled furnace  195
$Mn(NO_3)_2$, TG of  14
Modulus of elasticity  474
Morphology of polyethylene  427
Multistep analysis  273

$NaHCO_3$, decomposition of  142
Naturally aged condition  423
Ni-complexes, thermal analysis of  7, 10
Nitrating mixtures, determination of
   $HNO_3$  523
Non-isothermal conditions  69
   kinetic  555

"Open system" DSC  169
"Oregonator" mechanism  75
Organic synthesis  563
Organometallic gels  493
Oxygen in a molten soda lime glass  506

Parabolic equation constants  43
Particle size  95
Peak characteristics  96
   symmetry  100
   temperature for different heating
      rates  425
PET-fibres  345
Phase composition of autoclaved
   cement  369
Phase diagram of Ge-Sb-Bi  155
   of $KNO_3$-$NaNO_3$  282
   of $KNO_3$-$NH_4NO_3$  335
Phase study of the praseodymium-oxygen
   system  415
Phenacetine-benzamide system  87
Planc radiation law  141
Polycondesation of phenolic resin  170
Polyethylene, DTA of  264
Polyimide, TG-MS curves of  246, 249
Polymerization, major problems  461
   plant  463
   process control  466
Polystyrene pyrolysis, kinetic of  56
Polyurethane/polyester resin varnishes  453

Portland cement  363
Potassium nitrate  93
Praseodymium oxygen system  416
Preexponential factor  58
Presentation of data  341
Principle of direct enthalpimetry (DIE)  530
Publications Committee Report  599
Purity determination  87

Quasi-isothermal dilatometry  231
Quantitative DTA at elevated pressures
   335
Quantitative phase analysis of the system
   $CaSO_4 \cdot xH_2O$  511
Quicklimes  518

Rate of heat evolution  563
Ratio of rates  59
Reaction between fly-ash and calcium
   hydroxide  381
   index  59
   of sodium carbonate and acetic acid  191
   of sodium carbonate and formic acid  192
   zone of a burning cigarette  439
Reactivity of lime with sulphur dioxide  517
   of solids  1
Relationship between $\Delta T$ and average
   grain size of magnesite  102
Rubber vulcanized  225

Safety criteria  550
Sample loading position for temperature
   calibration  203
Saturation pressure  213
Scaling up and plant disign  565
Se, DTA of  296
Self calibration function  270
   igniting  446
Semiconductor glass $Ge_{.25}Te_{.60}Se_{.15}$  403
Shrinkage force  345
Silicon, deposition of  9
   rubber semiconductive  479
Simultaneous DTA and thermo-
   microscopy  279
   measurements of DSC, TG and
      TMA  273
   TG-DTA system  237
Single crystals, thermal analysis of  6
   phases and binary mixtures  568
Sintering of corundum with ammonium
   fluoride  357
Sodium acetate trihydrate, DSC and TG
   of  191
   citrate dihydrate  93
   formate, DSC and TG of  192

Solid-gas decomposition reaction 81
  solution of Al-Zn-Mg alloys 421
  state chemistry 1
  vapor reactions 189
S-shaped isotherms 213
Standardization, Committee Report 593
Standard or student's distribution 47
State of order of PET-fibres 345
Steam-hydrogen atmosphere 502
  different mixtures of 502
Stearic acetic coated calcium carbonate filler 239
Stress relaxation 225
Strontium oxalate 60
Strucutral and morphological study of $\beta$-FeOOH 579
  reaction mechanism 4, 5, 6
Structure of high-voltage insulations 483
Study of subsystems 78
Subambient thermal volatilization analysis 319
Sublimation of cylindrical samples 82
Sulfonation of aromatic nitrocompounds 553
Surface temperature of the coal 441

Taylor's expansion 133
Temperature calibration in TG 201, 597
  dependence of activation enthalpy 141
  distribution 439
  reference material for themogravimery 201
  Temperatures of the TG-standards 206
Tensile strength 473
Tertiary butylperpivalate, decomposition of 105
Test Program, Sixth International 202
TG curve of a $Mg(OH)_2$ 133
TG-DTA of natural plaster 515
TG-DTA system, simultaneous 237
TG investigation of oils 535
TG-mass spectrometry 245, 251
The amount polymerisation 464
Theoretical thermogravimetric analysis 51
Theory of analytical geometry 35
  of kinetic evaluation 500
Thermal analysis in earth sciences 583
  of silicone rubbers 477
Thermal and oxidative behaviour of edible oils 535
Thermal behaviour of a sample at different cooling rates 406
  conductivity detector (TCD) 415
  decomposition of fresh sunflower and rapeseed oils 537

decomposition of N-methyl-N-nitroso-p-toluene sulfonamide 177
Thermal decomposition of $CoSO_4 \cdot 7H_2O$ 175
  of $CuSO_4 \cdot 5H_2O$ 175
  of Ni $Pt(CN)_4 \, (NH_3)_2$ 67
  of $Ni(SCN)_2(NH_3)_2$ 64
  of $Ni(SCN)_2(NH_3)_4$ 65
  of $NiSO_4 \cdot 6H_2O$ 175
Thermal dehydration in the system $CaSO_4 \cdot xH_2O$ 511
Thermal expansion 219
Thermal investigation of polyethylene 427
Thermal profiles 83
Thermal properties of semiconductive rubbers 478
Thermal studies of contracting gas bubbles in molten glass 506
Thermal volatilization analysis (TVA) 320
Thermoanalytical characterization of varnishes 453
Thermoanalytical curves of stored and aerated sunflower and rapeseed oils 539
  investigations of polyester-glass 467
  study of $\beta$-FeOOH 577
  study on fertilizer components 567
Thermobalance temperature sensor 207
Thermodilatometry of chromindur 20
Thermogravimetric desorption of phenol 434
  kinetic analysis 434
Thermomagnetometry 15
Thermometric titrimetry 351
Thermomicroscopy and DTA 279
Thermooxidational ageing 489
Thermovoltaic detection 175
Thin film investigations by DTA 93
Titan powder, transformation on sintering 220, 221
Titration calorimetry 184
Topochemical mechanism 68
Topochemistry 2, 4
Topotaxy 2, 4
Total condensation method 246
Treatment of thermoanalytical data 341
Transmitted light thermomicroscopy 279
Transpiration method, partial pressure by 305, 313
Transport equations, flow method 305
«Trial and error» method 73

Values of $\Delta H$, $\Delta S$ and E for dehydration reactions 160
Vapour density 153

Varnishes formulation 453
Vulcanized silicone rubbers 477

Water concentrations in a gas stream 327
 determination by "DIE" method 529

X-Ray diffraction patterns of akaganéite 579
 hematite 579

Young module 225

Zawadzki-Bretsznajder relationship 36

BOOKS ARE SUBJECT TO
RECALL AFTER

DEC 2 3 1983
AUG 17 1984
DES 18 1985
MAY 5 1988
MAY 1 5 1989